Handbook of Chemical Looping Technology

Handbook of Chemical Looping Technology

Edited by Ronald W. Breault

WILEY-VCH

Editor

Dr. Ronald W. Breault
National Energy Technology Laboratory
3610 Collins Ferry Road
Morgantown, WV 26507
USA

Cover: © John Wiley & Sons, Inc.

Library of Congress Card No.: applied for

British Library Cataloguing-in-Publication Data
A catalogue record for this book is available from the British Library.

Bibliographic information published by the Deutsche Nationalbibliothek
The Deutsche Nationalbibliothek lists this publication in the Deutsche Nationalbibliografie; detailed bibliographic data are available on the Internet at <http://dnb.d-nb.de>.

Print ISBN: 978-3-527-34202-0
ePDF ISBN: 978-3-527-80932-5
ePub ISBN: 978-3-527-80934-9
oBook ISBN: 978-3-527-80933-2

Typesetting SPi Global, Chennai, India
Printing and Binding C.O.S. Printers Pte Ltd, Singapore

Printed on acid-free paper

10 9 8 7 6 5 4 3 2 1

Contents

Preface

Chemical looping as reported in this book and in others is an old technology with a new name. The earliest technologies using a chemical looping concept date back to the early parts of the twentieth century. Fan and Tong in Chapter 1 state that there is evidence that some of the aspects of chemical looping were even present in a calcium carbide process as early as 1897 with the generally accepted earliest involving a process to produce hydrogen by looping iron oxide in 1910.

Although there are other books on chemical looping, they tend to promote a specific technology and/or a specific application. To the contrary, this book presents an overview of chemical looping technologies. As such, the book is divided into four topical areas. The first three of these look at various aspects of chemical looping combustion, while the last section provides chapters on other chemical looping technologies, namely those applied to carbon dioxide capture.

In the first section, Chemical Looping Process Concepts, there are five chapters, each dedicated to a different technology/concept to promote the commercialization of chemical looping combustion. In the first chapter, LS Fan and Andrew Tong discuss the moving bed fuel reactor concept developed at Ohio State University. In the second chapter, Justin Weber of the National Energy Technology Laboratory presents a summary of the two fluidized bed configurations – single and double loop systems. In Chapter 3, Vincenzo Spallina and his coauthors Fausto Gallucci and Martin van Sint Annaland discuss the cyclic fixed bed process being developed at Eindhoven University of Technology. The fourth chapter presents the CLOU (Chemical Looping with Oxygen Uncoupling) process by Kevin Whitty of the University of Utah, JoAnn Lighty of Boise State University, and Tobias Mattisson of Chalmers University of Technology, being developed at the University of Utah. In the final chapter of the first section, Kunlie Liu, Liangyong Chen, and Zhen Fan present a pressurized chemical looping combustion process being developed at the University of Kentucky.

The second section of the book examines oxygen carrier performance through four chapters. Hanjing Tian of West Virginia University and coauthors Ranjani Siriwardane, of the National Energy Technology Laboratory (NETL), Esmail Monazam of REM Engineering Services, and Roald Breault of NETL present a summary of the performance of iron-based carriers for chemical looping combustion. Tobias Mattisson of Chalmers University of Technology and Kevin Whitty of the University of Utah present the second chapter in the section on oxygen carriers for chemical looping with oxygen uncoupling, a process that utilizes the thermal decomposition of the carrier to give up gaseous oxygen. In the third chapter of this section, Fanxing Li of North Carolina State University and Nathan Galinsky of NETL and the Oak Ridge Institute of Science and Engineering (ORISE) present a summary of mixed metal oxide carriers. In the last chapter of this section, Sam Bayham of NETL, Nathan Galinsky of NETL and ORISE, Esmail Monazam of REM Engineering Services, and Roald Breault of NETL present a summary of oxygen carrier structure and attrition. It should be pointed out that carrier attrition, if not overcome, will be the undoing of chemical looping technologies. This particular chapter identifies some properties that can be improved to overcome these shortcomings.

The third section of the book presents three chapters on commercial designs for chemical looping technologies. In the first chapter of this section, CFD simulations for a commercial unit are presented by Subhodeep Banerjee of NETL and ORISE, and Hongming Sun and Ramesh Agarwal, both of Washington University. The second chapter presents a cost comparison of a couple of technologies and is written by Robert Stevens of NETL, Dale Keairns of Deloitte Consulting, and Richard Newby and Mark Woods, both of KeyLogic Systems. In the last chapter of the section, Joanne Lighty of Boise State University and Zachary Reinking and Matthew Hamilton, both of University of Utah, present a summary on modeling and system simulations for a CLOU process.

The final section of the book presents two chapters on alternate chemical looping technologies. In these chapters, the focus is on CO_2 capture. In the first of these chapters, Yiang-Chen Chou, Wan-Hsia Liu, and Heng-Wen Hsu of Taiwan's Industrial Technology Research Institute present a summary of the calcium looping carbon capture process. Finally, in the last chapter, Hamid Arastoopour and Javid Abbasian of the Illinois Institute of Technology present a summary of the magnesium oxide process for CO_2 capture that they are developing.

Section 1

Chemical Looping Process Concepts

1

The Moving Bed Fuel Reactor Process

Andrew Tong, Mandar V. Kathe, Dawei Wang, and Liang-Shih Fan

The Ohio State University, Department of Chemical Engineering, 151 W. Woodruff Ave, Columbus, OH 43210, USA

1.1 Introduction

Chemical looping refers to the use of a chemical intermediate in a reaction-regeneration cycle to decompose one target reaction into two or more sub-reactions. The decomposition of the target reaction with a reactive chemical intermediate can decrease the process irreversibility, and, thus, increase the recoverable work from the system yielding a higher exergy efficiency. Further, when one or more of the reactant feedstocks consist of an inert substrate, the chemical looping reaction pathway is designed to prevent the direct contact of the inert with the desired product, minimizing the product purification steps required [1–3]. In 1987, Ishida et al. was the first publication to use the term, "chemical looping," referring to the use of a metal oxide as the chemical intermediate to perform oxidation–reduction reaction cycles for power generation applications [4]. However, Bergmann's invention of a calcium carbide production process using manganese oxide redox reaction cycles with carbonaceous fuels suggests that the chemical looping concept was in development as early as 1897 [5]. Table 1.1 summarizes the early developments of chemical looping processes in the twentieth century [6–9, 12–21]. Though several achieved pilot scale demonstration, no early chemical looping processes were able to achieve widespread commercial realization due to limitations in the oxygen carrier reactivity, recyclability, and attrition resistance and the reactor design for maintaining, continuous high product yield.

With growing concerns of greenhouse gas emissions, a renewed effort in developing chemical looping processes occurred at the start of the twenty-first century as reflected in the exponential growth of research publications [1]. As of 2012, over 6000 cumulative hours of operation of chemical looping processes for power generation with CO_2 capture have been demonstrated over fuel processing capacities ranging from 300 W_{th} to 3 MW_{th} [22]. Nearly all chemical looping processes at the pilot scale demonstration have adopted a fluidized bed reactor design for the conversion of the fuel source to CO_2/H_2O, or the fuel reactor [23]. Recent developers are investigating fixed bed reactors to perform

Handbook of Chemical Looping Technology, First Edition. Edited by Ronald W. Breault.

Table 1.1 Summary of early chemical looping process development.

Process/ developer	Lane [6–11]	Lewis and Gilliland	HYGAS	CO_2 acceptor	HyPr-Ring
Year developed	1910s	1950s	1970s	1960s–1970s	1990s
Feedstock	Syngas	Solid fuel	Syngas	Solid fuel	Solid fuel
Products	H_2	CO_2	H_2	H_2 rich syngas	H_2
Chemical intermediate	Fe_3O_4—Fe	CuO—Cu_2O or Fe_2O_3—Fe_3O_4	Fe_3O_4—Fe	$CaCO_3$—CaO	$CaCO_3$—CaO/ $Ca(OH)_2$
	ARCO GTG	**DuPont**	**Otsuka**	**Solar water splitting**	**Steinfeld**
Year	1980s	1990s	1990s	1980s	1990s
Feedstock	CH_4	C_4H_{10}	CH_4	H_2O	CH_4, iron ore
Products	C_2H_4	$C_4H_2O_3$	Syngas	H_2, O_2	Syngas, iron
Chemical intermediate	Supported Mn	VPO	Supported CeO_2	ZnO—Zn or Fe_3O_4—FeO/Fe	Fe_3O_4—Fe

the cyclic oxidation–reduction reactions with chemical looping oxygen carriers for power generation and chemical production applications [24–27]. Alternatively, chemical looping processes utilizing a moving bed fuel reactor are under development for full and partial fuel conversion for CO_2 capture/power generation and syngas production, respectively [23, 28, 29]. This chapter describes the use of moving reactors for chemical looping processes with specific application to syngas and power production with CO_2 capture using metal oxide materials as oxygen carrier chemical intermediates. Two modes of moving bed operation are discussed and their application for full and partial fuel oxidation. Reactor thermodynamic modeling combined with experimental results are provided.

1.2 Modes of Moving Bed Fuel Reactor Operation

As illustrated in Figure 1.1, the moving bed fuel reactor can be operated in the counter-current or co-current mode based on the gas–solid flow contact pattern with Fe-based oxygen carrier as the exemplary chemical intermediate [1]. The counter-current moving bed fuel reactor in Figure 1.1a achieves a high oxygen carrier conversion while maintaining high CO_2 product purity. The oxygen carrier conversion, as defined in Eq. (1.1), is the mass ratio of the amount of oxygen used from the oxygen carrier exiting the fuel reactor relative its maximum available oxygen.

$$X_O = \frac{m_{ox} - m_{red}}{m_{ox} - m_{red}^{full}} \times 100\% \tag{1.1}$$

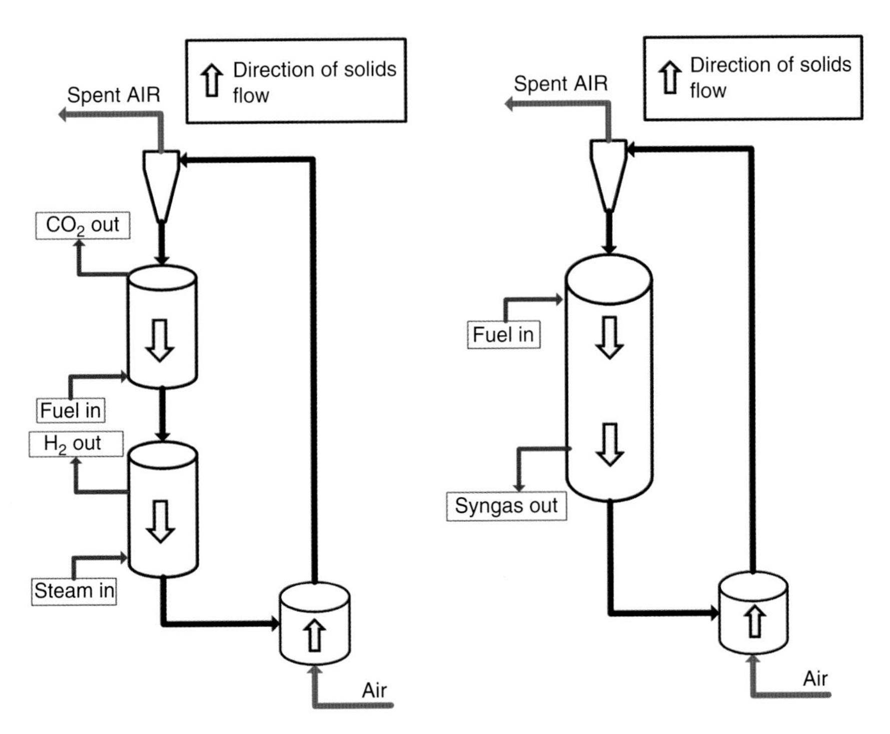

Figure 1.1 Conceptual design of a moving bed chemical looping processes with a counter-current (a) and co-current (b) fuel reactor for full fuel conversion to CO_2/H_2O and for fuel gasification/reforming to syngas, respectively.

where m_{ox} and m_{red} refer to the mass of the fully oxidized and the reduced sample at the outlet of the fuel reactor, respectively, and m_{red}^{full} refers to the mass of the sample at the fully reduced state (e.g. metallic iron for Fe-based oxygen carriers).

Figure 1.1b shows the co-current moving bed fuel reactor for partial oxidation of the solid or gaseous fuel source to syngas. The co-current process allows for accurate control of the oxygen carrier and fuel residence times, ratios, and distribution to maintain continuous high purity syngas. The present section discusses the advantages of each mode of the moving bed reactor operation and considers several applications for solid and gaseous fuel conversion for each.

1.2.1 Counter-Current Moving Bed Fuel Reactor:

In a counter-current moving bed operation of chemical looping process, the gas species in the fuel reactor travel the opposite direction relative to the solids flow. Further, the gas species operate below the minimum superficial gas velocity and, thus, travel only through the interstitial spaces of the packed moving bed of oxygen carrier solids. For full fuel conversion, the counter-current moving bed design is capable of maintaining high CO_2 purities and reducing the oxygen carrier to a low oxidation state, ideal for metal oxides with multiple oxidation states such as iron [30, 31]. Figure 1.2 is an example of operation lines for moving bed chemical looping fuel reactor and steam reactor. The figure illustrates

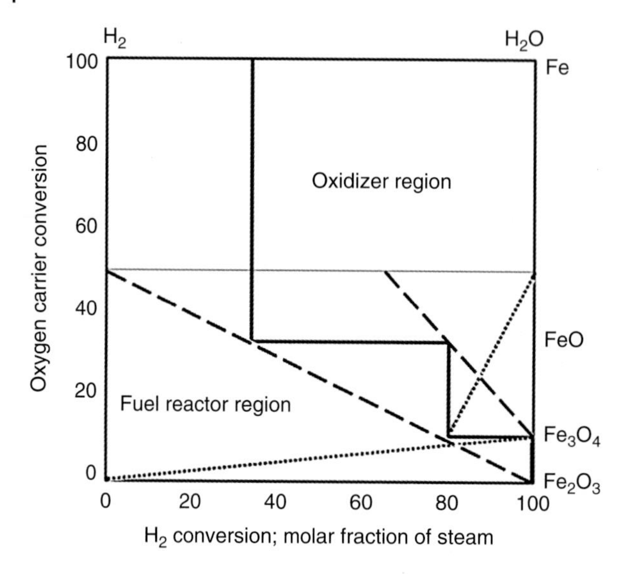

Figure 1.2 Operation lines for moving bed chemical looping fuel reactor and steam reactor.

the phase equilibrium of a Fe-based oxygen carrier particle at varying partial pressures (i.e. conversions) of the reducing gas at 850 °C. In the figure, the solid line represents the phase equilibrium of iron. The dashed line in the fuel reactor region represents the counter-current reactor operation while the dotted line represents the fluidized bed/co-current operation. The slope of the moving bed and fluidized bed operating lines are determined based on the oxygen balance between the oxygen carrier and the gas species. In the case of fluidized bed operation with iron-based oxygen carrier, the maximum oxygen carrier conversion achievable is 11% (i.e. reduction from Fe_2O_3 to Fe_3O_4), as a higher oxygen carrier conversion will result in a decrease in product purity from the fuel reactor. Further, the high extent of reduction of the iron oxide oxygen carrier achieved in the counter-current fuel reactor allows for thermodynamically favorable reaction of Fe/FeO with H_2O to produce H_2 via the steam–iron reaction. High purity H_2 production from a third reactor, i.e. the steam reactor, increases the product flexibility of the processes and can serve as an advanced approach for H_2 production with minimal process operations for product separation compared to traditional steam–methane reforming (SMR).

Figure 1.3 illustrates the design of the counter-current fuel reactor for solid fuel conversion to CO_2. Here, the fuel reactor is divided into two sections [32–34]. Once the solid fuel is introduced to the high temperature fuel reactor, it devolatilizes and the solid char species travel downward co-currently with the flow of oxygen carrier solids into the char gasification section. The volatiles travel upward counter-currently with the flow of the oxygen carrier. In the lower bed, the solid char is gasified using an enhancer gas consisting of CO_2 and/or H_2O recycled from the flue gas produced from the fuel reactor. The gasified char and volatile matter are polished to CO_2 and H_2O in the upper reactor bed. The

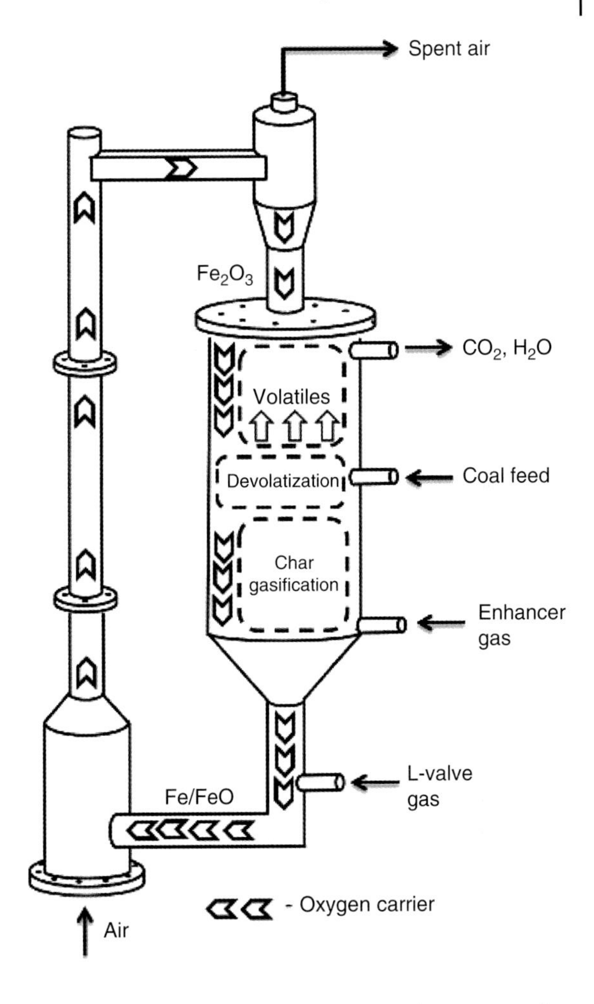

Figure 1.3 Conceptual design of the counter-current moving bed fuel reactor for solid fuel conversion to CO_2/H_2O.

packed moving bed reactor is designed to provide sufficient residence time for the solid fuel gasification and the fully oxidized Fe_2O_3 oxygen carrier entering the top section 1 reactor ensures CO_2 purity exiting the process is nearly 100% after H_2O is condensed out. No additional downstream conditioning equipment and/or molecular oxygen is required to fully oxide the solid fuel to CO_2, which translates to high process efficiency and reduced process capital costs. At Ohio State University (OSU), two chemical looping processes using a counter-current moving bed fuel reactor have been scaled to pilot plant demonstration for the conversion of gaseous fuels (the syngas chemical looping (SCL) process) and solid fuels (the coal direct chemical looping (CDCL), process) to H_2 and heat, respectively, with CO_2 capture. The SCL process is promising technology capable of reducing the H_2 separation costs compared to traditional coal gasification and the SMR process with natural gas. The CDCL process is an advance oxy-combustion technology for CO_2 capture from coal. Further details on the pilot plant developments are discussed in Section 1.4.

1.2.2 Co-current Moving Bed Fuel Reactor

In the co-current moving bed fuel reactor operation, the gas species travel in the same directions as the solid flow [23]. The gas flow rate is operated below the minimum fluidization velocity to ensure uniform gas velocity profile across the fuel reactor is achieved for precise and independent control of the gas and solid residence times. Co-current moving bed fuel reactors are generally used for the partial oxidation of carbonaceous fuels to a gaseous stream of high purity syngas. The optimal operating conditions derived from a phase diagram analysis correspond to a certain molar flow ratio of fuel to the oxygen carrier. The co-current contact pattern ensures a high syngas purity is achieved as the gaseous species are in direct contact with the reduced state of the oxygen carrier as it exits the system. The thermodynamic phase diagram of iron–titanium oxide, as shown in Figure 1.4a, indicates the necessary ratio of oxygen carrier to fuel input flow necessary to produce >90% purity syngas. Further, Figure 1.4b indicates the addition

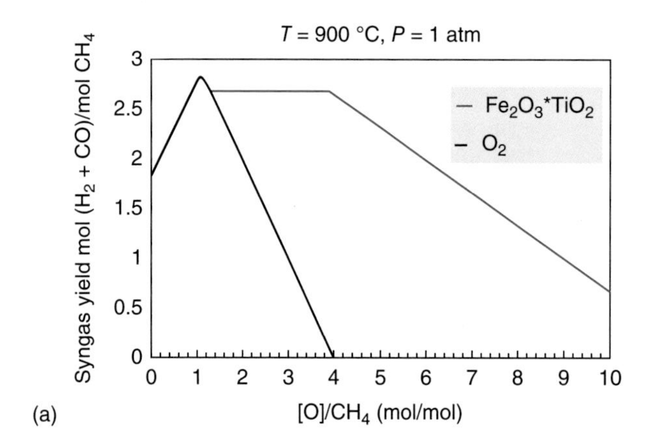

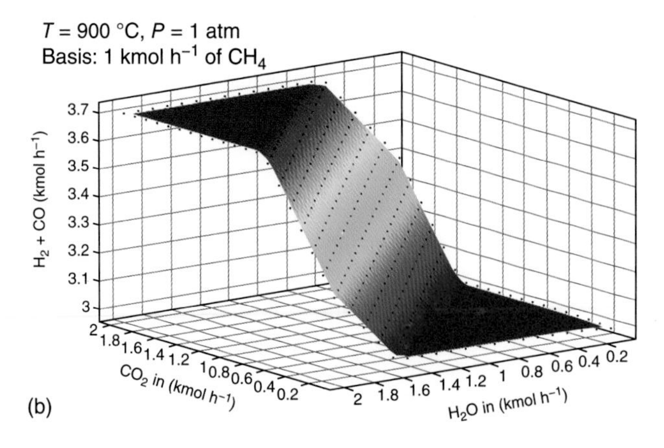

Figure 1.4 Syngas production purity at varying ratios of CH_4 and oxygen carrier flow in a co-current moving bed fuel reactor (a) and a 3-dimensional plot of the CO_2 and H_2O input and its impact on the syngas purity (b). Both results were simulated under isothermal operation conditions at 900 °C.

of steam and CO_2 can be used to adjust the ratio of CO/H_2 in the product stream while still maintaining high syngas purities.

Precise control of the oxygen carrier and gas residence time in the fuel reactor within a narrow distribution is necessary to maximize the product yield for chemical looping partial oxidation applications [23, 28, 35]. As illustrated in Figure 1.5a, fluidized bed operations are challenged with a wide residence time distribution for the solids due to the well-mixed nature of a bubbling fluidized bed. This result corresponds to wide distribution of metal oxide oxidation states in the reactor. The available lattice oxygen in the higher oxidation metal states can result in over conversion of the fuel to CO_2/H_2O, reducing the syngas selectivity. Further, the gas species exist both in the interstitial spaces emulsified with the solid media and in bubble phase generated when the superficial gas velocity exceeds the minimum fluidization velocity. Mass diffusion limitations between the bubble phase and emulsion phase can result in unconverted gaseous fuel species which can further reduce the syngas product yield. The packed moving bed is a possible design that can address these challenges. As illustrated in Figure 1.5b, the moving bed solids distribution exiting the fuel reactor is precisely controlled to a single oxygen carrier conversion value. The solids travel as a mass flow downwards and the superficial gas velocity is maintained below the minimum fluidization velocity preventing the formation of a bubble phase.

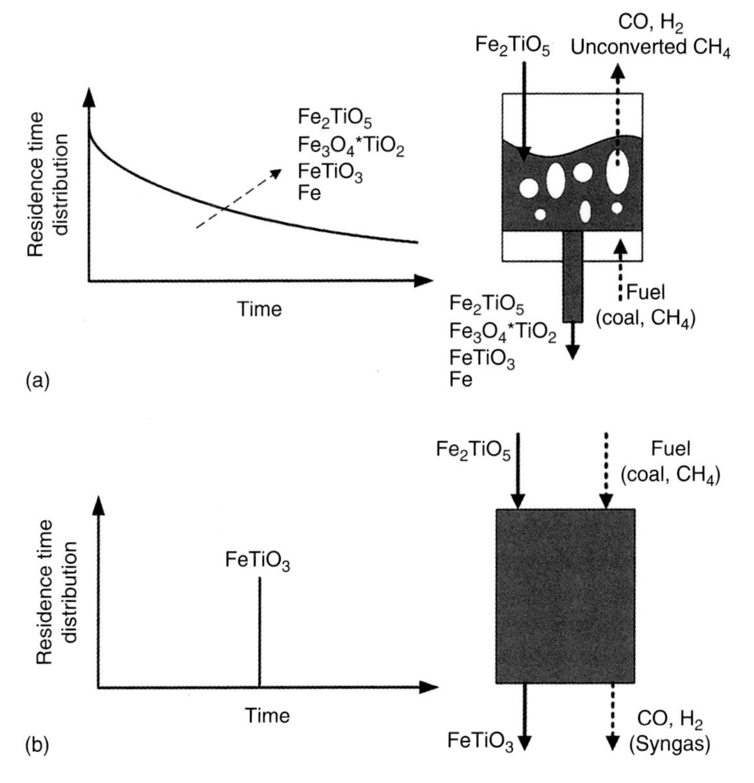

Figure 1.5 Conceptual Fe—Ti based oxygen carrier oxidation state distribution in a fluidized bed (a) and moving bed (b) fuel reactor.

As the moving bed design ensures a narrow distribution of oxygen carrier conversion along the height of the reactor, the chemical looping process design is simplified as the syngas yield from the fuel reactor is entirely thermodynamically driven. Here, reaction kinetics considerations are only necessary to supply sufficient residence time for the gas and solid phases to achieve the thermodynamically expected limits. Excessively high residence time of gaseous species will not impact the syngas yield as the gas composition at the outlet of the fuel reactor will be in equilibrium with the oxidation state of the oxygen carrier material used.

The co-current moving bed fuel reactor design is generally directed towards chemical looping processes for fuel gasification or reforming to syngas. OSU is developing the shale gas to syngas (STS) chemical looping reforming process and coal to syngas (CTS) chemical looping gasification processes. Both processes are completing sub-pilot demonstrations at the 15 and 10 kW_{th} capacities. Each process will be presented in Section 1.5.

1.3 Chemical Looping Reactor System Design Considerations for Moving Bed Fuel Reactors

As shown in Figure 1.1, the moving bed fuel reactor system comprises two or more reactors operated in moving bed and fluidized bed modes, respectively, for fuel conversion (fuel reactor) and oxygen carrier particle regeneration (air reactor). The reactors are connected using nonmechanical gas sealing devices and a gas–solid separator. The fuel reactor can be operated in counter-current or co-current moving bed mode, where the solid oxygen carrier particles travel downwards by gravity while the process gases flow upwards or downwards. The air reactor is a fluidized bed reactor which uses air for fluidization and regeneration. The air reactor is connected to a pneumatic riser to transport the oxygen carriers back to the fuel reactor with oxygen-depleted air from air reactor [36–38].

Design and sizing of the reactors and interconnecting gas-sealing devices are based on hydrodynamic calculations. The range of operating conditions, including temperature, fuel capacity, and residence times of gas and solid in each of the reactors, must also be identified for purpose of system design calculations. Based on the operating conditions, a performance model of the system can estimate the expected chemical reactions and process gas composition, and the gas flow rate in each of the reactors can be determined.

1.3.1 Mass Balance and Solids Circulation Rate

For a continuous steady-state operation of the chemical looping system, the amount of oxygen consumed by the fuel in the fuel reactor is the amount of the oxygen supplied by the oxygen carrier particles circulating in the system. The oxygen carried by the oxygen carrier particles is obtained from the air in the air reactor. It is essential for the chemical looping system to achieve a certain solids circulation rate so as to maintain the mass balance of the system. If the solids circulation rate is insufficient, the available lattice oxygen provided in the oxygen carrier particles will be incapable of performing the desired chemical reactions in the fuel reactor – i.e. the fuel will not be fully converted and/or utilized. If the solids circulation rate is too large, the sizes of the reactors and other auxiliary devices need to be designed unnecessarily larger, which increases the operational cost. For the gasification system, which is very sensitive to the oxygen carrier-to-fuel ratio, a higher than demand oxygen carrier particle circulation rate will reduce the fuel selectivity to syngas in the fuel reactor, and in turn lower the quality of syngas yield.

The amount of solids circulation rate of a chemical looping system is determined by the fuel processing capacity of the system, the desired composition of the product gas, and the material of the oxygen carrier particles and their properties such as the degree of reduction and amount of support.

1.3.2 Heat Management

The chemical reactions occurring in the moving bed fuel reactor between fuel and metal oxide are normally net endothermic, while, the regeneration reaction of reduced metal oxide by O_2 in the air flow in the air is exothermic, generating heat. According to the energy balance of the system for chemical looping combustion applications, the sum of the heat for the complete set of chemical reactions of chemical looping combustion system is equal to the heat of direct combustion of fuel with air. Thus, the heat releases from the air reactor is greater than the heat consumed in the fuel reactor. The exothermic heat resulting from the metal reaction with oxygen in chemical looping combustion system is used to provide the endothermic heat requirement for the fuel reactor and to produce electricity and/or steam.

In the air reactor, due to the large amount of heat generated by the re-oxidation reaction of oxygen carrier, the adiabatic temperature rise can occur sharply and result in the melting or softening of the oxygen carriers if the excess heat produced is not properly removed. Excess air input or cold shots, inert material loading in the oxygen carrier, and/or continuous heat removal via in-bed heat exchangers are generally accepted methods to mitigate and control the air reactor temperature. The softening temperature of the oxygen carrier is the maximum possible temperature at which the air reactor in the chemical looping combustion system is operable as the softening of the oxygen carrier can result in particle agglomeration and defluidization.

Similarly, in the fuel reactor, as the net reactions are endothermic, the temperature of the particles decrease, causing the chemical reaction to slow down. The temperature at which the desired chemical reactions are kinetically unfavorable is the minimum possible temperature for the operation of the fuel reactor in the chemical looping combustion system. Further, the operating temperature of the fuel reactor affects the phase equilibrium of the metal oxide. For example, a reasonable fuel reactor temperature for chemical looping combustion system shall provide a minimum amount of CO and H_2 exiting from the fuel reactor gas outlet and a maximum oxygen release from metal oxide. Depending on the reactions under consideration, the temperature in the chemical looping reactors may vary from 400 °C to more than 1000 °C.

For efficient heat integration of a chemical looping system, it is desirable that enough heat is transported from the air reactor to the fuel reactor by oxygen carrier particles. When the temperatures of solids, fuel, and product are determined, the required solids flow rate to carry enough heat to the fuel reactor is obtained. Compared to the mass flow rate of oxygen carrier required by mass balance of the chemical reactions, the difference is the amount of inert material needed so that both the heat in the fuel reactor and the particle temperature are successfully managed. Another favorable effect of loading inert materials such as Al_2O_3, SiC, and TiO_2 is to improve the particle recyclability. The mass ratio of the inert material and the metal oxide is called the support-to-oxygen carrier mass ratio.

1.3.3 Sizing of Reactors

The sizing process of the fuel reactor normally starts with the determination of the gas velocity, which is an important parameter for the smooth operation of the fuel reactor and is bounded by is operational mode, i.e. fluidized bed and moving bed, and the hydrodynamics of the particles, i.e. the minimum fluidization velocity. For the fuel reactor operated under fluidized bed mode, the gas velocity shall be higher than the minimum fluidization velocity. It is usually several times of the minimum fluidization velocity to utilize the advantages of the fluidized bed reactors including good gas–solids contact mode, uniformity, and intensive heat transfer between gas and solids. For a moving bed reactor, however, the gas velocity shall be less than the minimum fluidized velocity to avoid the fluidization of the bed material. The relationship between the gas velocity and the minimum fluidization velocity can be expressed as,

$$u_{g_{\text{reducer}}} = k \cdot u_{\text{mf}} \tag{1.2}$$

where $k > 1$ for fluidized bed reactor and $k < 1$ for moving bed reactor.

With the determination of the gas velocity, the volume fraction of the solids in the fuel reactor can be obtained from the operational mode and the hydrodynamic characteristics of the oxygen carrier particles. The cross-sectional area of the fuel reactor can also be determined based on the amount or flow of gas process through it. The volume of the reactor can then be determined. A requirement for the volume of the fuel reactor is that it should provide sufficient residence time to achieve the thermodynamic equilibrium of the conversion of oxygen carrier particles and fuel, which gives the following criteria:

$$\frac{\alpha_s \rho_s V_r}{\dot{m}_s} \geq T_s \tag{1.3a}$$

$$\frac{(1-\alpha_s)\rho_g V_r}{\dot{m}_g} \geq T_g \tag{1.3b}$$

where V_r is the volume of the reactor; α_s is the volume fraction of the solids, which is determined by the operational state of the fuel reactor; ρ_s and ρ_g are the densities of the oxygen carrier particles and process gas; $\dot{m}_s$and $\dot{m}_g$are the solids circulation rate of the system and the mass flow rate of the process gas; T_s and T_g are the required residence time for the oxygen carrier particles and gases, which are determined by the properties of the oxygen carrier particles and fuel, chemical reaction types, and the reactor operational conditions. The fuel reactor height is obtained from its required cross-sectional area to prevent particle fluidization and required volume to achieve the minimum gas and solid residence times.

1.3.4 Sizing of the Air Reactor

Because the regeneration reaction of reduced iron particles with air is a highly exothermic reaction in the air reactor, the generated heat has to be efficiently removed to maintain a constant reactor operating temperature and to avoid sintering and agglomeration of the oxygen carrier. Therefore, the air reactor is generally designed and operated as a fluidized bed because of its excellent gas–solids contact mode and effective heat transfer characteristics. The regenerated particles are then transported back to the fuel reactor via a lean phase pneumatic conveying riser.

The design basis for the air reactor and riser are closely dependent on each other, as the air introduced into the air reactor performs three functions: provides oxygen for oxygen carrier regeneration, fluidizes the oxygen carrier in the air reactor, and provides gas flow through the riser for the pneumatic transport of the oxygen carrier particles back to the fuel reactor. Under some cases, the amount of the air in the air reactor is adjusted to control the reactor temperature for the purpose of heat management. The required amount of oxygen consumption in the air reactor is determined by the fuel capacity and expected composition of the product gas. As the air reactor is operated under a fluidized bed, the gas velocity in it shall be higher than the minimum fluidization velocity, but less than the terminal velocity of the oxygen carrier particles. It is normally desired that the gas velocity is several times of the minimum fluidization velocity so as to operate the air reactor in a dense phase fluidized bed mode with a relatively compact reactor size. With the determination of the gas velocity, the volume fraction of the oxygen carrier particles in the air reactor and its cross-sectional area can be obtained. The volume and the height of the reactor can then be determined according to the requirement of providing sufficient residence time for the full regeneration of the oxygen carrier particles.

1.3.5 Gas Sealing

Gas sealing devices are required between the fuel and air reactors. Their role is to allow the oxygen carriers to flow through each reactor while keeping the gases in each reactor segregated. A reliable gas seal between each reactor is required for process safety and performance. A gas leakage from the air reactor to the fuel reactor or vice versa may result in the formation of an explosive mixture in the system, which poses a safety hazard. Further, an inefficient seal would cause fuel to leak into the air reactor and increase CO_2 emissions from the system thereby reducing the carbon capture efficiency or syngas yield. Also, leakage of air into the fuel reactor would result in formation of undesired products and contaminate the desired gaseous product.

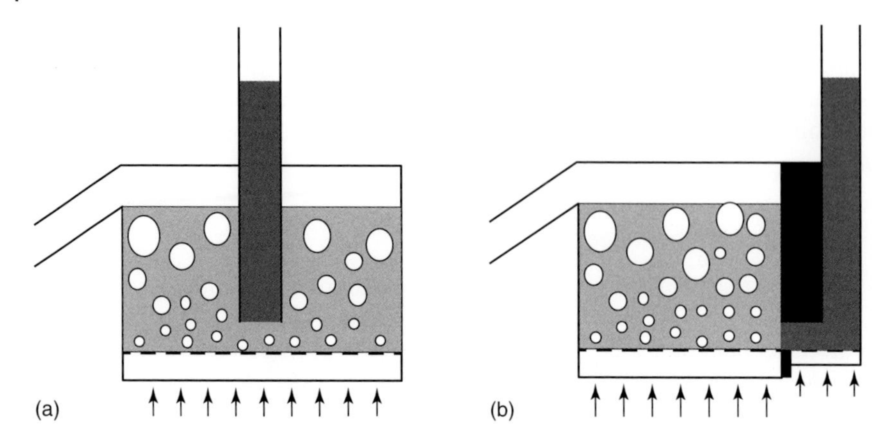

Figure 1.6 Automatic solid flow devices: (a) seal pot and (b) loop seal.

Automatic nonmechanical solid flow devices such as seal pots and loop seals, shown in Figure 1.6, are commonly used in chemical looping systems with fluidized bed fuel reactors for gas sealing purposes [38]. These nonmechanical devices are considered passive devices capable of maintaining the global solid circulation rate. However, they are generally not used for adjusting or modulating solid circulation rate in the process.

A seal pot is essentially an external fluidized bed into which the straight dip leg discharges solid particles. The solid particles and the fluidizing gas for the seal pot are discharged to the desired downstream vessel through an overflow transport line either designed as a downwardly angled pipe at the side or an overflow dip leg in the middle of the fluidized bed. With a seal pot, the solids in the dip leg rise to a height necessary to handle the pressure difference between the solids inlet and the outlet.

A loop seal is a variation of the seal pot that places the solids inlet dip leg at the side of the fluidized bed in a separate solids supply chamber. This allows for the solids return chamber to operate independently of the solids supply chamber, which results in a smaller device size, lower fluidization gas requirement, and higher efficiency. The height and the diameter of the solids supply chamber as well as its distance from the solids return chamber can be adjusted based on the process requirements necessary for balancing the pressure and handling the solids flow. Independent lubricating gas can also be added to the different locations of the solids supply chamber to assist in the operation of the loop seal.

1.3.6 Solids Circulation Control

The flow of oxygen carrier in chemical looping systems can be controlled either mechanically or non-mechanically. The use of mechanical valves to control the flow of solids was common during the early development of chemical looping processes since they allowed for maximum flexibility over the control of solids. The mechanical valves could also provide effective gas sealing between the reactors despite the pressure difference at the two ends of the valves. Although mechanical valves have been part of a number of successful tests of continuous chemical looping processes, they have serious drawbacks. Since chemical looping processes circulate a large amount of oxygen carrier particles at high temperatures, the material of construction of the valves and their internals would have to be able to withstand the temperature. In addition, the mechanical valves must also have a high abrasion resistance due to the large flow rate of solids. In addition to the large solids flow and harsh operating conditions experienced by the mechanical valve, its repeated opening and closing during operation would accelerate the rate of wear and tear with the possibility of mechanical failure occurring during operations. The mechanical valves become cost prohibitive due to the expensive material of construction required for withstanding the operating conditions, replacement frequency that would be associated with the purchase or maintenance of large mechanical valves. Thus, it would cost intensive to scale-up the chemical looping system with mechanical valves for long-term continuous operations.

An attractive alternative to mechanical valves for solids flow control is the use of nonmechanical solids flow control devices. These devices refer to valves with no internal mechanical moving parts and that only use aeration gases in conjunction with their geometric patterns to manipulate the solid particles flow through them. The nonmechanical solids flow control devices have no moving parts and thus have no issues of wear and tear, especially under extreme operating conditions such as elevated temperatures and pressures. Also, these devices are normally inexpensive as they are constructed from ordinary pipes and fittings. Due to their simplicity, the nonmechanical solids flow control devices can be quickly fabricated avoiding the long delivery times associated with mechanical valves. They are widely used in industries due to their advantages over mechanical solids flow control devices.

The most common types of nonmechanical solids flow control devices, also called nonmechanical valves, include the L-valve and J-valve, shown in Figure 1.7a. The principles of operation for these two types of valves are the same, with the only major difference being their shape and the direction of solids discharge. Solids flow through a nonmechanical valve is driven by the drag force on the particles caused by the slip velocity between them and the aeration gas. The aeration gas is added to the bottom portion of the standpipe section

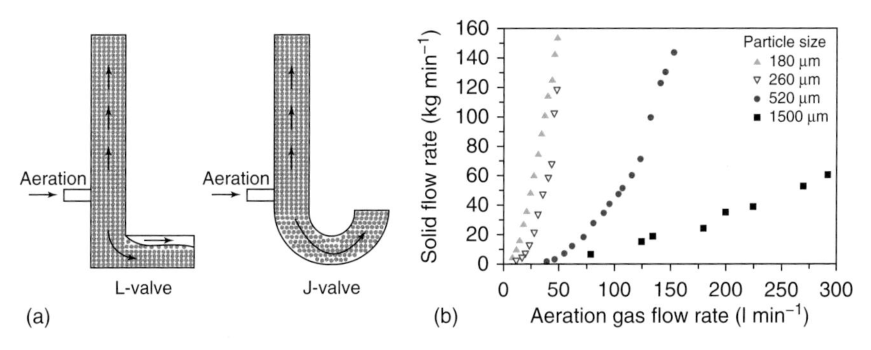

Figure 1.7 Nonmechanical L-valve and J-valve conceptual design (a) and solids flow rate as a function of aeration gas flow rate through an L-valve [36] (b).

of a nonmechanical valve and it flows downwards through the bend. The slip velocity between the gas and the solid particles produces a frictional drag force on the particles in the direction of gas flow. When the drag force exceeds the force required to overcome the resistance to solids flow around the bend, the solids begin to flow through the valve. A certain minimum amount of gas flow is required before the start of solids flow through the nonmechanical valve. Above this threshold amount of gas that is required to initiate solids flow, the solids flow rate varies proportionately to the aeration gas flow rate. A relationship between aeration gas flow rate and solids flow rate through an L-valve is shown in Figure 1.7b [36].

The actual aeration gas flow through the nonmechanical valve may be different from the amount of gas added externally through the aeration gas inlet port. It may be higher or lower than the amount of aeration gas externally injected, depending on the operating conditions of the system. In case the gas from the reactor leaks into the standpipe of a nonmechanical valve, the actual amount of aeration gas, Q_{ae}, would be the sum of the leaked gas flow into the standpipe, Q_{sp}, and the external aeration gas added from the aeration gas inlet port, Q_{ext}, as given in Eq. (1.4).

$$Q_{ae} = Q_{sp} + Q_{ext} \tag{1.4}$$

If the gas from the aeration gas inlet port leaked upward through the standpipe into the reactor, then the actual amount of aeration gas would be obtained from Eq. (1.5).

$$Q_{ae} = -Q_{sp} + Q_{ext} \tag{1.5}$$

where the negative sign in front of Q_{sp} denotes a change in the direction of aeration gas leakage compared to the previous case.

Nonmechanical valves have limitations in their operating capability based on the physical properties of the solid. Nonmechanical valves function smoothly for particles of Geldart Groups B and D but not as smoothly for particles of Geldart Groups A and C. Geldart Group A particles generally retain gas in their interstices and remain fluidized for a substantial period of time even after fluidizing gas is released from their fluidized state. Therefore, they can pass through the nonmechanical valves even after the aeration gas flow is stopped. The solids flow rate thus is not easily controlled for Geldart Group A particles. Geldart Group C particles are cohesive due to their relatively large inter-particle forces and thus are difficult to flow using aeration gas in nonmechanical valves. For a given solids flow rate, the required amount of aeration gas increases with the average particle size for solid particles of Geldart Groups B and D. This is due to the greater drag force to overcome to render the solid particles with larger diameter flow through the nonmechanical valve.

The solids flow rate pattern after the bend in the nonmechanical valve is generally in the form of pulses of relatively high frequency and short wavelength. This pulsating flow creates pressure fluctuations of a relatively steady pattern in the nonmechanical valve. The pressure drop across the valve is high when the particles stop and low when the particles surge. Increasing the length after the bend increases the solids flow rate pulses, which increases the chaotic pattern of pressure fluctuations. The total pressure drop across the nonmechanical valve also increases with the length after the bend. In some cases, additional gas is added to the section after the bend to prevent slug formation and induce solids flow. However, this increases the total amount of external gas, and hence, the operating costs. Based on the above issues with a longer pipe length after the bend, it can be noted that the horizontal section of the nonmechanical valve should be as short as possible to minimize the pressure fluctuations and the amount of aeration gas.

1.3.7 Process Pressure Balance

A good indication and reference on the good gas sealing and proper operation of the chemical looping system is the pressure balance of the system. The pressure drops through different sections/devices of the chemical looping system reflect the gas–solids contacting modes and solids fluidization conditions at different locations of the system.

An example of pressure profile in a chemical looping system is shown in Figure 1.8. The pressure drop through the riser, ΔP_{riser}, represented by the line P_H–P_A in the figure indicating a smooth and slight drop of the pressure gradient from the bottom to the top of the riser, where the solids volume fraction, α_s, is commonly less than 1%. Line P_G–P_H illustrates the pressure drop in the air reactor. The air reactor is normally a dense phase fluidized bed where the pressure drop equals to the solids holdup in the bed and is much larger than that in the riser. Line P_A–P_B–P_C describes the pressure profile across the gas seal between the cyclone and the fuel reactor. There is a point with a pressure

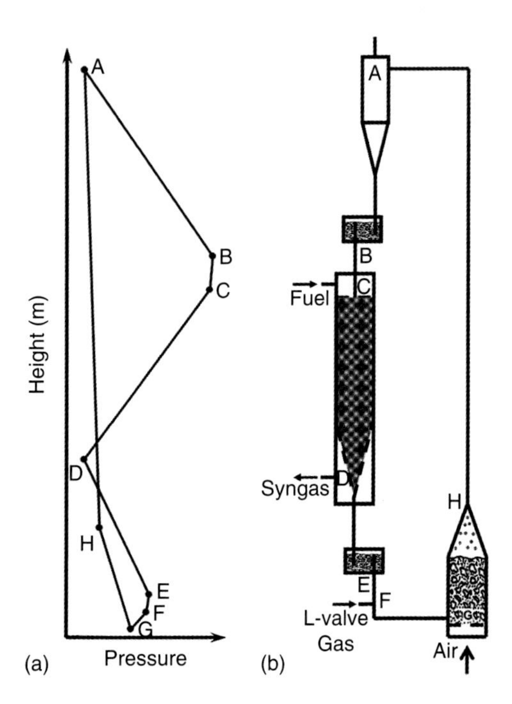

Figure 1.8 Two reactor chemical looping system with moving bed fuel reactor (b) and its pressure profile (a).

(P_B) relatively higher than the both ends of the gas seal so that the sealing gas can flow from the gas seal to both the upstream and downstream of the process and the gas mixing from either reactor to the other can be avoided. Line P_C–P_D illustrates the pressure drop across the fuel reactor which can be designed as a fluidized bed, a gas–solids counter-current moving bed or a gas–solids co-current moving bed. Line P_D–P_E–P_F represents the pressure profile across another gas seal which is located between the cyclone and the fuel reactor. The pressure distributions across these components form a closed loop and can be mathematically expressed as Eq. (1.6),

$$\Delta P_{\text{cyclone}} + \Delta P_{\text{seal 1}} + \Delta P_{\text{fuel reactor}} + \Delta P_{\text{seal 2}} = \Delta P_{\text{air reactor}} + \Delta P_{\text{riser}} \quad (1.6)$$

1.4 Counter-Current Moving Bed Fuel Reactor Applications in Chemical Looping Processes

Over the past 23 years, OSU has developed five chemical looping technologies to sub-pilot and pilot scale operations. The present section discusses the development of the counter-current moving bed fuel reactor systems of the SCL and CDCL processes for H_2 production and power generation with CO_2 capture, respectively [23].

1.4.1 Counter-Current Moving Bed Fuel Reactor Modeling

The counter-current moving bed reactor is designed to fully convert the fuel source to CO_2 and steam. Previous research was performed to simulate the reaction kinetics in combination with the moving bed reactor hydrodynamics to support the reactor size selection for optimum fuel conversion [39]. The oxygen carrier kinetics were simulated using an unreacting shrinking core model (USCM) with iron-oxide based oxygen carrier. The USCM constants were empirically quantified using experimental reduction rate data gathered from literature and OSU thermo-gravimetric analyzer studies. A one-dimensional moving bed fuel reactor model was then constructed and verified using 2.5 kW_{th} bench scale moving bed studies with H_2/CO as the reducing gases at steady state conditions. The reactor model matched well with experimental results and provided insight design of the reactor operating conditions and sizing. The results are shown in Figure 1.9. In the case of CO and H_2 as the reducing gas in the fuel reactor, a critical molar flow ratio of 1.4647 CO/H_2:Fe_2O_3 was observed where a plateau in solids conversion at 33% is observed in the middle section of the fuel reactor. This occurrence indicates that when the fuel to oxygen carrier approaches the critical ratio and the reactor is excessively long, the reduction of FeO to Fe is constrained. This is due to the equilibrium partial pressures of reducing gases in the middle of the fuel reactor inhibited the further reduction of the oxygen carrier. Once the oxygen carrier travels down close to the inlet of the reducing gas, the partial pressures favor the further reduction of FeO to Fe. To eliminate the formation of the plateau and to maintain high

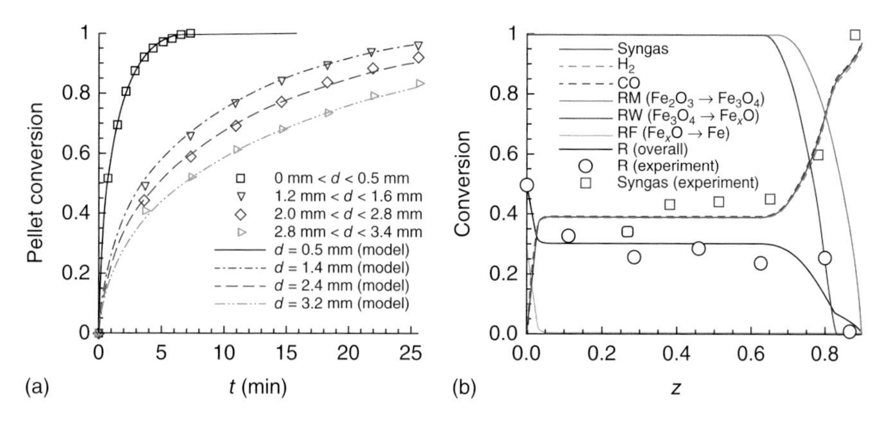

Figure 1.9 (a) Comparison between the kinetic unreacted shrinking core model and the experimental data for different Fe_2O_3 particle sizes; (b) moving bed fuel reactor model comparison with 2.5 kW_{th} bench unit test run at 1.46 CO/H_2:Fe_2O_3 molar flow ratio.

fuel conversion to CO_2, one should operate the counter-current moving bed reactor at a fuel:oxygen carrier flow ratio at a slightly lower value than the critical ratio. Parametric studies with the 25 kW_{th} sub-pilot counter-current moving bed reactor were performed to determine if the critical ratio also exists when using methane as the reducing gas in the fuel reactor. Figure 1.10 summarizes the solids profile for three test conditions with varying CH_4 to Fe_2O_3 flow ratio [40]. Here, a plateau is observed at a solid conversion of approximately 33% when the when the CH_4:Fe_2O_3 molar flow ratio is 0.466, but not observed when the ratio is decreased to 0.366. Therefore, a critical flow ratio exists for methane conversion to CO_2 between 0.366 and 0.466 similar to the model's prediction for CO conversion to CO_2. Note, the stoichiometric lattice oxygen requirement of the oxygen carrier for converting CH_4 to CO_2 and H_2O is four times greater than when CO and H_2 is used as the reducing gas.

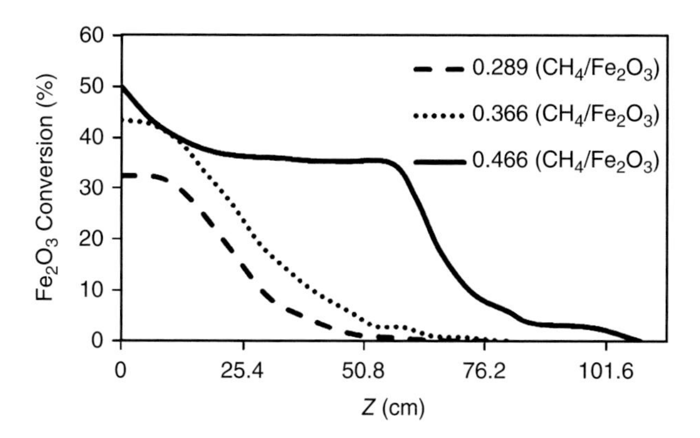

Figure 1.10 Oxygen carrier profile at steady state operation in the fuel react at varying CH_4:Fe_2O_3 molar flow ratios using a 25 kW_{th} sub-pilot chemical looping reactor system operating isothermally at 975 °C.

1.4.2 Syngas Chemical Looping Process

The SCL process, conceptually illustrated in Figure 1.1a, consists of a counter-current moving bed fuel reactor for CO_2 capture, a counter-current moving bed oxidizer for H_2 production, and a dense fluidized bed air reactor for oxygen carrier regeneration and heat production [1, 30, 40–43]. The counter-current moving bed design reduces the Fe-based oxygen carrier to Fe/FeO from the fuel reactor. The low oxidation state of the oxygen carrier entering the oxidizer is thermodynamically favored for H_2 production via the steam–iron reaction. The partially oxidized, Fe_3O_4, oxygen carriers are transported to the air reactor to be fully regenerated to Fe_2O_3 where the heat generated is used to compensate for the endothermic reactions in the fuel reactor and parasitic energy requirements of the overall processing plant.

In the case of integrated gasification combined cycle (IGCC), the SCL is considered a process intensification approach to replace the water gas shift and acid gas removal units with a single chemical looping reactor. A process flow diagram of IGCC plant incorporating the SCL reactor is illustrated in Figure 1.11. A techno-economic analysis (TEA) for IGCC–SCL power plant was performed in comparison to a conventional IGCC process with 90% CO_2 capture [44–48]. The results of TEA are summarized in Table 1.2. The SCL–IGCC system

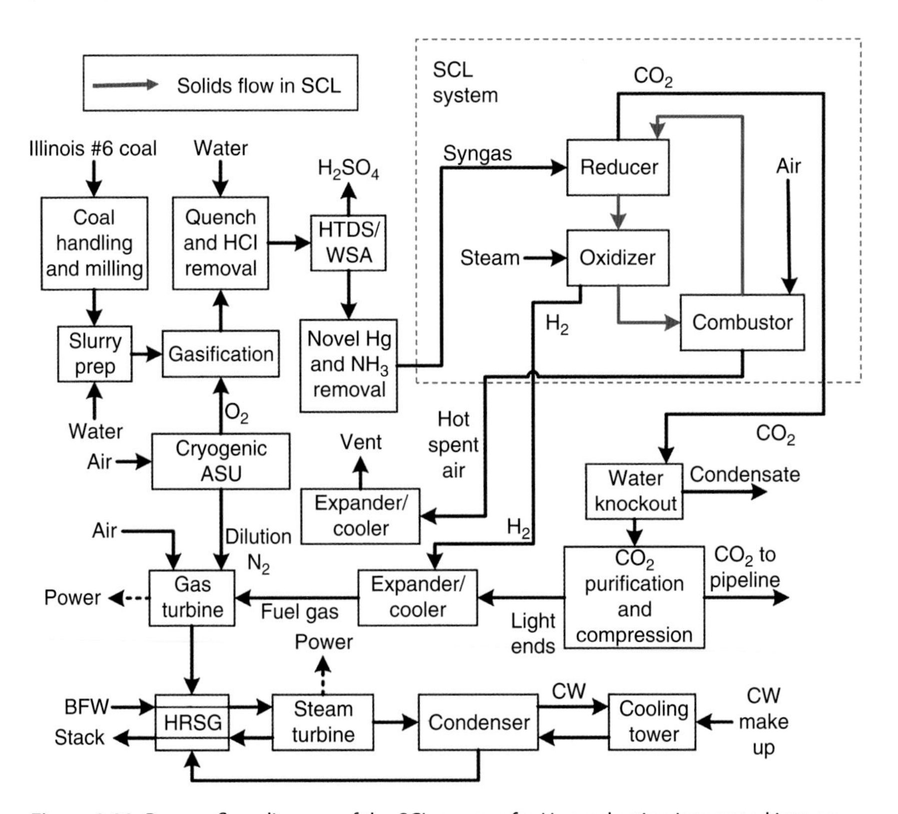

Figure 1.11 Process flow diagram of the SCL process for H_2 production integrated into an IGCC process.

Table 1.2 Economic comparison of the SCL plant for H_2 production with >90% carbon capture in an IGCC configuration.

Parameter (2011 $)	Baseline IGCC	SCL–IGCC
Total plant cost ($/kW)	3324	2934
Cost of electricity ($/MWh)	143.1	133.8

reduces the total plant costs by ~10%, resulting in a reduction in the first year cost of electricity (COE) from $143/MWh (2011 $) for the baseline case to $134/MWh (2011 $) when adopting the SCL reactor system. The case of natural gas as the feedstock, the SCL process is capable of replacing the conventional steam–methane reformer to produce high purity H_2 as an industrial gas product. The SCL's oxidizer reactor is capable of producing high purity H_2 without the need for additional gas–gas separation units, which represent nearly 75% of the capital cost in a conventional SMR process for industrial H_2 production.

Over 450 h of operation of the 25 kW_{th} SCL sub-pilot unit at the OSU clean energy research facility have been completed with smooth solids circulation and efficient reactor operation. Sample results from a continuous three-day demonstration, illustrated in Figure 1.12, indicate that nearly 100% of the simulated syngas was converted to CO_2 and H_2O in the fuel reactor with 99.99% purity H_2 produced from the oxidizer. Figure 1.13 summarizes three test conditions where the steady state fuel conversion results were compared to the theoretical expected fuel reactor performances for a counter-current moving bed and fluidized bed reactor.

A pressurized 250 kW_{th}–3 MW_{th} SCL pilot plant was constructed at the successful completion of the 25 kW_{th} sub-pilot demonstration. This pilot plant, constructed at the National Carbon Capture Center in Wilsonville, AL, represents the first large scale demonstration of a high pressure chemical looping process for high purity H_2 production. Figure 1.14a shows a picture of the constructed SCL pilot plant and Figure 1.14b shows sample results of the fuel reactor conversion

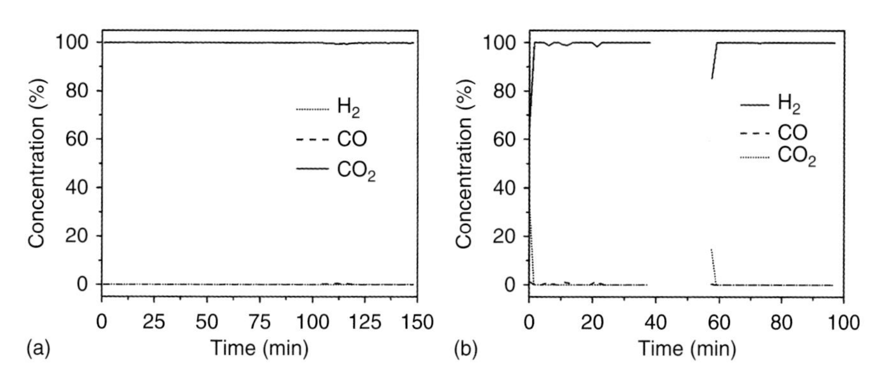

Figure 1.12 (a) Fuel reactor gas outlet composition obtained during sub-pilot scale demonstration of the SCL process; z-2. (b) Oxidizer outlet gas composition obtained during sub-pilot scale demonstration of the SCL process.

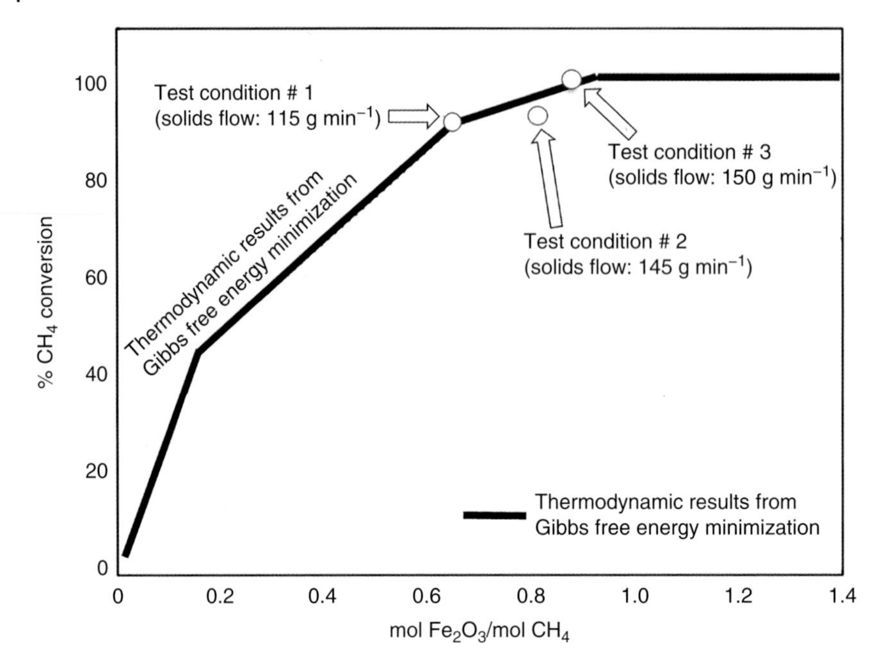

Figure 1.13 Analysis of syngas-conversion-sensitivity with respect to Fe_2O_3/syngas molar flow ratio for a fluidized bed reactor and a moving bed reactor.

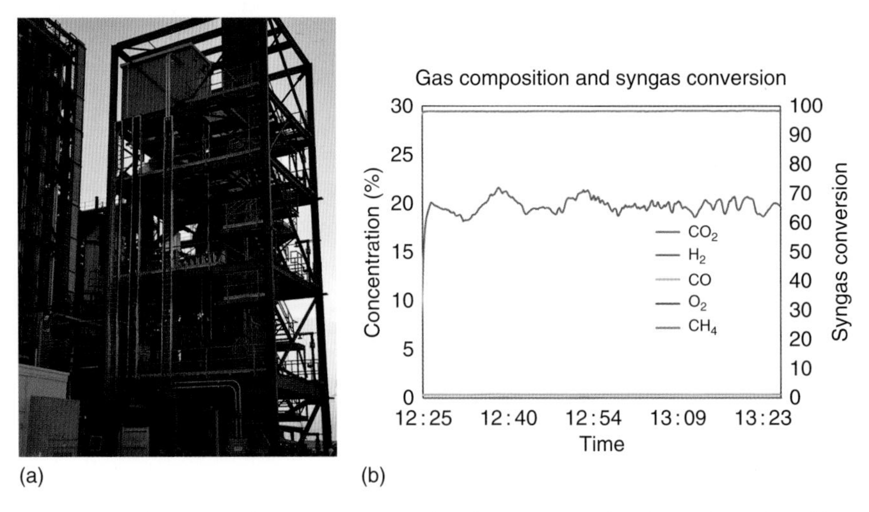

Figure 1.14 Photo of 250 kW_{th}–3 MW_{th} SCL pilot plant at NCCC (a) and sample results of the fuel reactor performance (b).

profile. The demonstration results, using the syngas as feedstock from Kellogg Brown & Root (KBR's) transport gasifier, are consistent with the thermodynamic predictions from ASPEN and sub-pilot scale experiments. Further operation of the SCL pilot plant are ongoing for continuous high purity H_2 production for gasified coal.

1.4.3 Coal Direct Chemical Looping Process Development

CDCL process represents an advanced oxy-combustion technology for CO_2 capture that does not require molecular oxygen produced from air separation unit (ASU). Figure 1.3 illustrates the fuel reactor design for the CDCL process for the counter-current operation to produce high purity CO_2 while maintaining high oxygen carrier conversion. Figure 1.15a is a simplified process flow diagram for the integration of the CDCL process with a supercritical steam cycle for power generation with >90% CO_2 capture. Babcock & Wilcox Power Generation Group (B&W) in collaboration with OSU performed a TEA of the CDCL process at the 550 MW_e capacity. The reactor sizing and conceptual design of the power plant were completed (shown in Figure 1.5b) and incorporated into the fixed capital cost model. Table 1.3 summarizes the economic assessment of the CDCL process in comparison to a pulverized coal power plant without CO_2 control and a plant with an amine scrubber for post combustion CO_2 capture [49–51]. The CDCL process achieves 96.5% carbon capture efficiency with a 26.8% increase in COE compared to a conventional pulverized coal supercritical steam power plant with no CO_2 control. When compared to the 63.7% increase in COE required for post combustion CO_2 capture with amine scrubbers, the CDCL process is considered a promising approach to mitigate CO_2 emissions in power generation from fossil fuels. The major cost savings achieved in the CDCL plant is due to the simplicity of the chemical looping reactor design to achieve full fuel conversion in a single loop. Further, Figure 1.15a and Table 1.3 indicate that the capital costs of the CDCL reactor system is offset by the replacement of the pulverized coal boiler. Additional equipment requirements for the CDCL process are limited to the CO_2 conditioning stream for sulfur removal and CO_2 compression for transportation.

The CDCL process has been demonstrated at the 25 kW_{th} sub-pilot scale for over 1000 h of operation with solid fuels ranging from woody biomass to anthracite coals [32–34, 52–55]. A continuous 200-h demonstration of sub-pilot unit was completed in 2012 showing nearly 100% coal conversion to CO_2 with no carbon carryover to the air reactor. Figure 1.16 shows sample data collected on the gas composition collected from the fuel and air reactor during the 200-h demonstration.

Recent efforts have been directed to characterize the fate of sulfur species in the coal to verify the necessity of a flue gas desulfurization (FGD) device on each of the gas outlets. The tests were conducted using sub-pilot CDCL reactor system shown in Figure 1.17a. The gas sampling conditioning system used for the fuel and air reactor was designed to ensure the sulfur species was maintained in the gas phase during condensate removal. Figure 1.18 summarizes the sulfur balance between the fuel and air reactor gas outlets and the residual amount in the ash when Powder River Basin (PRB) sub-bituminous coal is used as the fuel input. The results show that <5% of the sulfur is emitted from the air reactor gas outlet which corresponds to a lower emissions rate than the Environmental Protection Agency regulation requirement of <1.4 lb sulfur/MW_{gross}. Thus, the sub-pilot CDCL operations indicate the use of a FGD unit on the flue gas stream of the air reactor is not required. The success of the 25 kW_{th} sub-pilot CDCL testing unit led to the construction of a 250 kW_{th} pilot unit. Figure 1.17b shows

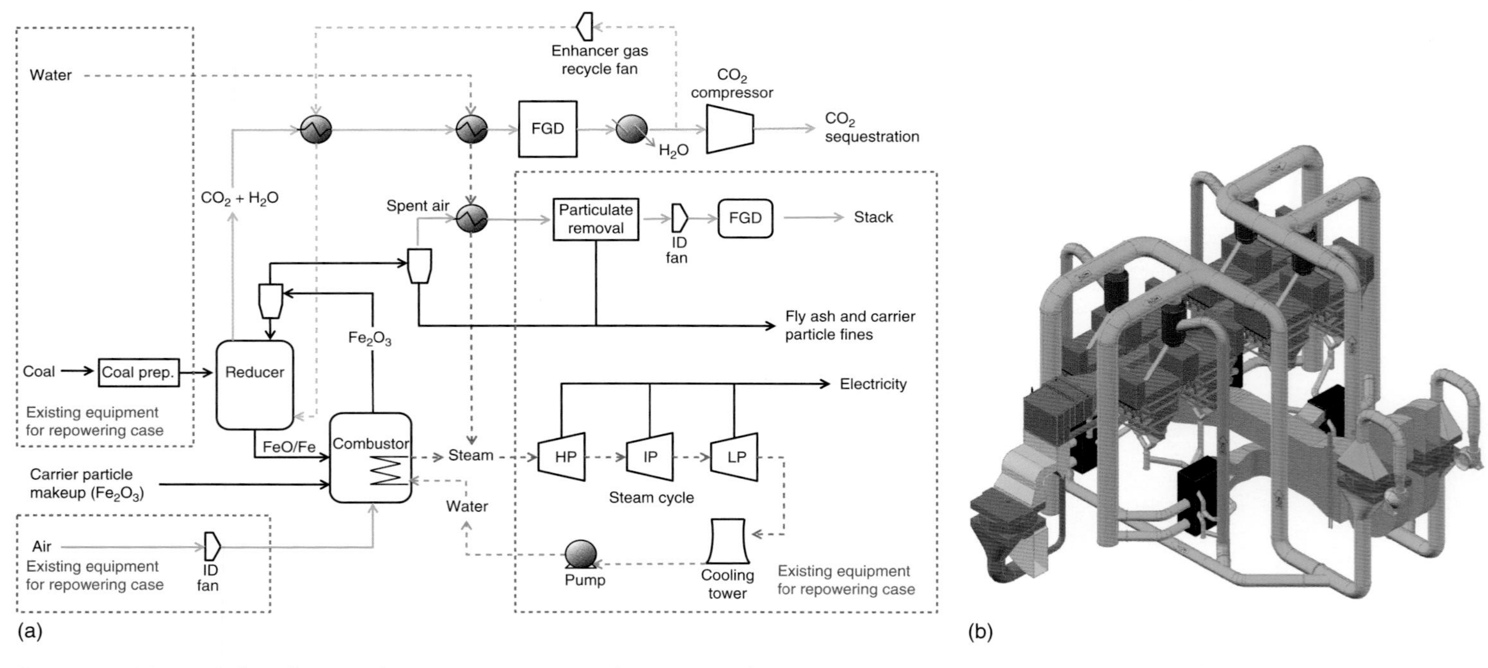

Figure 1.15 (a) Process flow diagram of CDCL process integrated a supercritical steam cycle. Items located with the blue dotted boxes indicate existing equipment in a conventional pulverized coal power plant. (b) Image of the conceptual design of a 550 MW_e CDCL plant.

Table 1.3 Comparison of the CDCL chemical looping technology for CO_2 capture with a base amine based CO_2 capture plant.

	Base plant	MEA plant	CDCL plant
Coal feed ($kg\,h^{-1}$)	185 759	256 652	205 358
CO_2 capture efficiency (%)	0	90	96.5
Net power output (MW_e)	550	550	550
Net plant HHV efficiency (%)	39.3	28.5	35.6
Cost of electricity ($/MWh)	80.96	132.56	102.67
Increase in cost of electricity (%)	—	63.7	26.8

MEA, mono-ethanol-amine and HHV, high heating value.

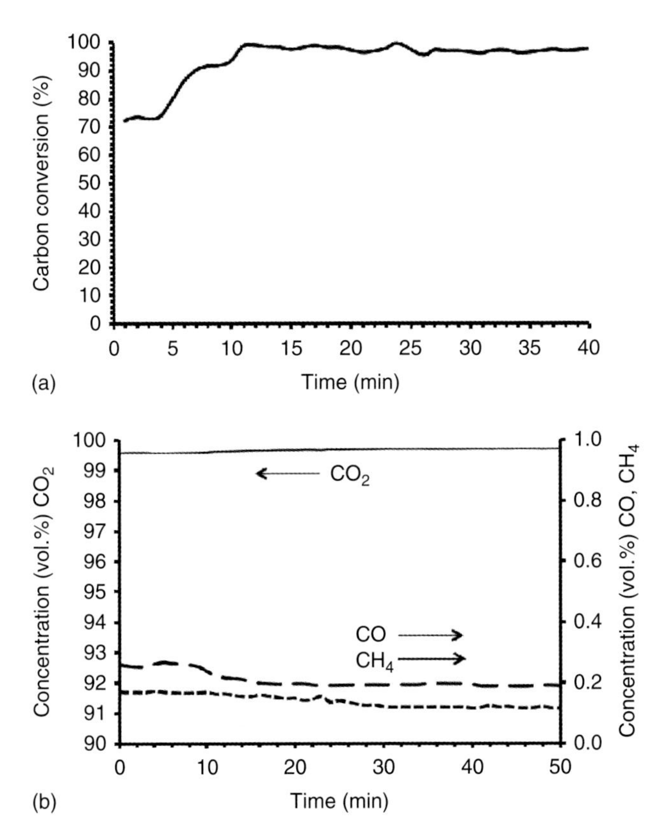

Figure 1.16 (a) Carbon conversion profile for CDCL operation, s-2; (b) fuel reactor gas outlet composition for long-term CDCL operation.

the constructed pilot unit at B&W's Research Center in Barberton, OH. The pilot unit operations will analyze the scale up factors for the coal feed distribution in the moving bed fuel reactor and the performance of the fuel reactor under adiabatic operating conditions. Construction and assembly of the unit was completed and the component and reactor commissioning activities have commenced. Unit testing with coal feed is anticipated to be completed in early 2017.

Figure 1.17 Photo of the 25 kW_{th} sub-pilot CDCL unit (a) and the 250 kW_{th} CDCL pilot unit (b).

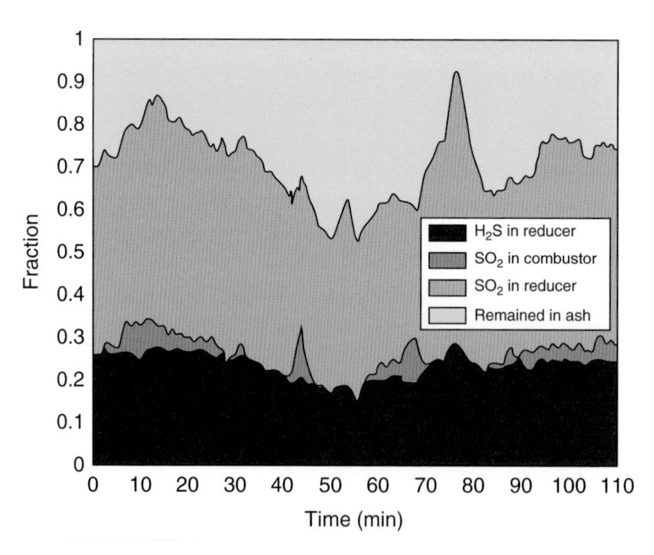

Figure 1.18 Sulfur balance in CDCL sub-pilot operation with PRB coal at 950 operating temperature.

1.5 Co-current Moving Bed Fuel Reactor Applications in Chemical Looping Processes

Co-current moving bed reactor system is directed for partial oxidation of solid and gaseous fuels to syngas. The present section describes two applications of the co-current moving bed fuel reactor for coal and natural gas conversion.

1.5.1 Coal to Syngas Chemical Looping Process

The CTS chemical looping process was developed for a high efficiency conversion of solid fuels to syngas [56]. The CTS process uses a co-current moving bed fuel reactor for producing syngas and a fluidized bed air reactor for regenerating the reduced oxygen carrier material. The CTS process produces a flexible ratio of H_2:CO syngas at >90 vol% purity, eliminating the need for molecular oxygen from an ASU and water–gas sift reactor for H_2 upgrading. A case study of this chemical looping process was performed where the CTS process was integrated with a 10 000 tonne d^{-1} methanol production plant to quantify the efficiency advantages associated with the CTS system compared to a conventional coal gasification process. The overall process flow for a coal to methanol plant, using the CTS coal gasification technology is shown in Figure 1.19. Table 1.4 summarizes the economic comparison of the CTS technology to a conventional gasification technology when integrated into a methanol production plant. As compared to the methanol synthesis plant using a conventional coal gasifier, a 28% reduction in the total plant capital cost is obtained when using the CTS technology. Further, the total coal input is reduced by ~14% due to higher carbon efficiency of the CTS process for coal gasification. The combined capital and operating cost reduction results in a 21% lower methanol required selling price than the baseline technology with 90% CO_2 capture. The CTS is also under

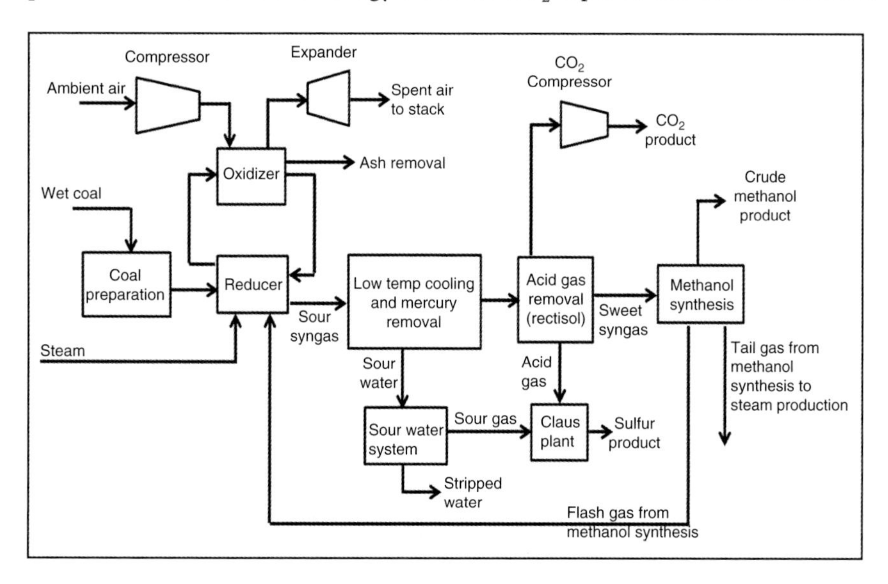

Figure 1.19 Coal to syngas process integrated in a methanol synthesis plant.

Table 1.4 Comparative summary of capital costs and cost of methanol production for the CTS case and the conventional baseline case.

Case (2011 MM$)	Baseline with 90% CO_2 capture	Baseline without CO_2 capture	CTS
CO_2 capture (%)	90	0	90
Total plant costs	4775	4568	3497
Total as spent capital	6852	6580	5003
Capital costs ($/gal)	1.23	1.18	0.89
Required selling price ($/gal)	1.78	1.64	1.41

development for power generation applications. Preliminary results have indicated that a CTS-IGCC plant can achieve the targeted regulation requirement for new coal-fired power plants of 1400 lbs CO_2/MWh_{gross} with CO_2 control equipment due to the high cold gas efficiency of the CTS process for syngas generation. This is obtained without the addition of molecular oxygen and is expected to provide significant cost savings in the electricity generation plant.

The CTS co-current moving bed concept was tested in a 2.5 kW_{th} bench-scale unit with sub-bituminous and bituminous coal over a range of operating conditions. Coal co-fed with steam and CH_4 were investigated to analyze their impact on the syngas yield and H_2:CO ratio. Representative data collected from two operating conditions for coal feed and coal and CH_4 co-feed are shown in Figure 1.20. From this figure, when only sub-bituminous coal was used as the reducing fuel in the fuel reactor, a syngas purity of ~90% was achieved – consistent with thermodynamically expected performance. A H_2:CO molar ratio of 0.65 was recorded at steady-state conditions, which was expected based on the H:C atomic ratio inherent in the coal feed. The amount of coal volatiles emitted from the fuel reactor was considered minimal due to the

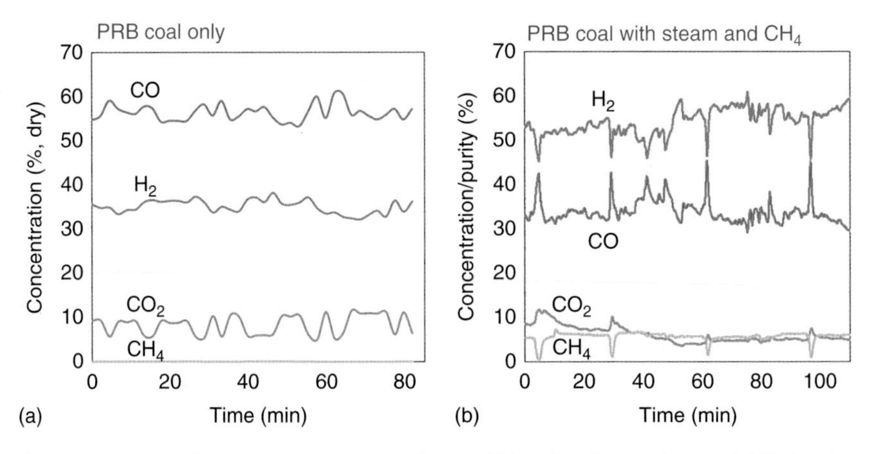

Figure 1.20 Exemplary syngas composition from CTS bench unit experiments. (a) PRB coal only. (b) PRB coal with steam and CH_4.

negligible concentrations of CH_4 observed. A carbon conversion of 93% was estimated for this test condition. Further experiments were performed for co-feeding CH_4 and/or steam to increase the ratio of H_2:CO in the syngas produced from the fuel reactor for demonstrating the versatility of the CTS technology for multiple product synthesis applications. Figure 1.20b, confirms the ratio can be adjusted by co-feeding these H_2 containing reactants. The specific test case presented in Figure 1.20b produced a syngas stream with a H_2:CO molar ratio of ~1.7, three times greater than the H_2:CO ratio produced when only coal feed is used, and with CO_2 concentrations of <10%. The results show the CTS process can adjust the H_2:CO ratio produced from the fuel reactor while maintaining a syngas purity of >90%. The versatility of the CTS process allows the process to be applicable to a range of high value chemical and fuels while eliminating the need for H_2 upgrading units such as the water–gas shift reactor and the use of molecular oxygen supplied from an ASU.

1.5.2 Methane to Syngas Chemical Looping Process

The chemical looping methane to syngas (MTS) process uses iron–titanium composite (ITCMO) materials to perform redox reactions that partially oxidize natural gas to syngas. The unique combination of a co-current downward moving bed and ITCMO particles enables the MTS process to eliminate the need for molecular oxygen, lower the temperature of operation, and significantly reduce the steam and natural gas consumption for an equivalent amount of liquid fuels production. The MTS process consists of a co-current moving bed fuel reactor integrated with a dense-phase fluidized bed air reactor to regenerate the reduced oxygen carriers. The co-current moving bed reactor operation ensures the syngas product produced achieves the thermodynamically expected performance based on the ITCMO oxygen carrier properties. The heat released from the regeneration of the oxygen carriers in the air reactor is used to compensate for the endothermic methane reforming reaction in the fuel reactor and any additional parasitic energy requirements in the overall chemical processing plant. The conditioned syngas produced from the MTS process can be used for industrial H_2 gas supply, liquid fuel synthesis, chemical production, or many other applications due the flexibility of the H_2:CO ratio produced.

Figure 1.21 illustrates the process flow diagram for the integration of the MTS process into a conventional gas to liquid (GTL) plant for the synthesis of liquid fuels at a 50 000 barrel d^{-1} production capacity. The baseline GTL plant contains an natural gas auto-thermal reformer (ATR) for syngas generation and recycles a fuel gas stream consisting mainly of light hydrocarbons (C_1–C_4). A chemical looping model was developed to compare the auto-thermal operation of the MTS process that is scaled to produce an identical amount of CO and H_2 to match the downstream liquid fuel synthesis units of the baseline comparison case. The performance results for the MTS plant using the co-current moving bed fuel reactor is summarized in Table 1.5. The metal oxide composites coupled with the co-current gas–solid contact pattern for the fuel reactor allows the MTS process to achieve a high syngas yield with less than 3% (v/v) CO_2 produced in the syngas product and a negligible carbon deposition on the oxygen carriers. Compared to

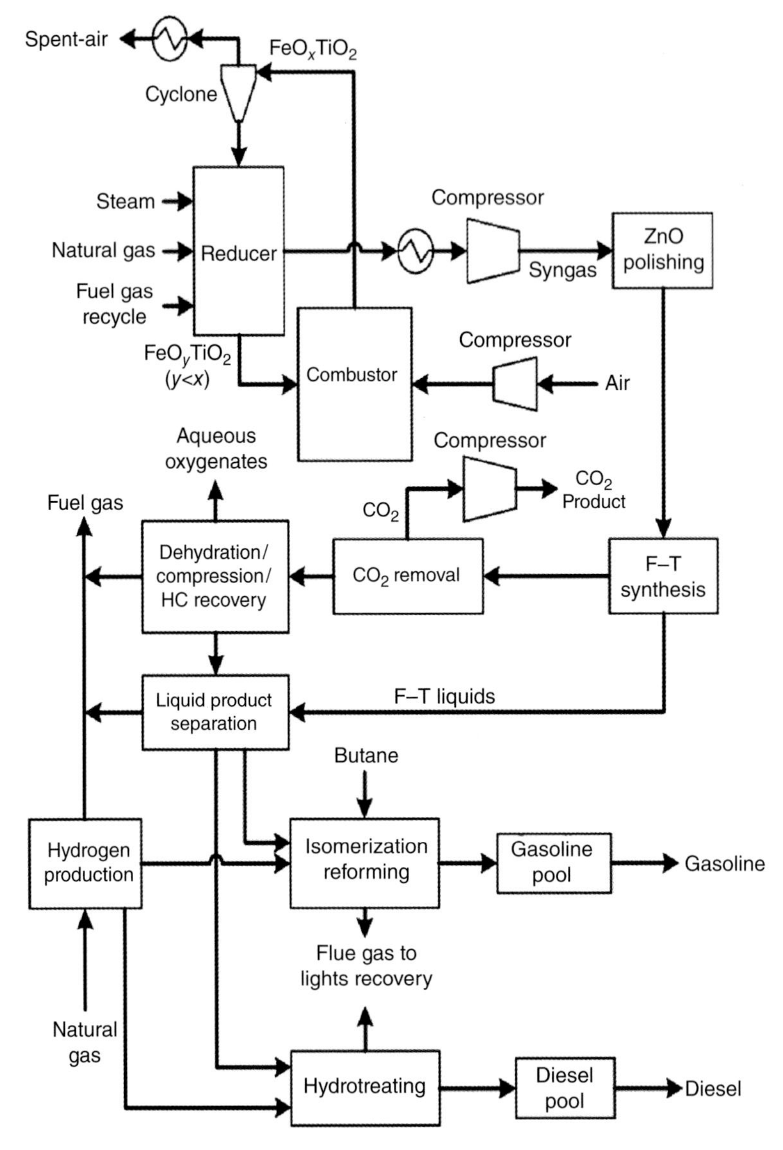

Figure 1.21 MTS process for syngas generation coupled with F–T complex for producing 50 000 bpd of liquid fuel.

conventional GTL process with ATR (utilizing 19 849 kmol h^{-1} of natural gas), the STS process requires 11% (v/v) less natural gas feed to produce an equivalent amounts of liquid fuel. Further, the MTS process eliminates the need for an energy intensive ASU. These combined benefits decrease the parasitic energy requirements of the syngas generation unit by 60% (kW_e basis). This results in twice the net power output generated from a GTL–MTS plant.

Table 1.5 Overall integrated performance of the STS process in a full-scale gas to liquids plant.

Component	Base case	MTS (10 atm)
Natural gas flow(kg h^{-1})	354 365	317 094
Natural gas flow (kmol h^{-1})	20 451	18 300
H_2O/C_{input}	0.68	0.249 9
H_2/CO	2.19	2.18
Stoichiometric number (S)	1.59	1.96
Total liquid fuel (gasoline + diesel) (bbl d^{-1})	50 003	50 003
Net plant power (kW_e)	40 800	85 000

The improved process efficiency of chemical looping system translates into an economic advantage in terms of total plant capital and operating costs. The capital cost reduction is driven by the process intensification where a single MTS reactor system can replace multiple unit operations in a conventional GTL plant, such as the ASU, the ATR for syngas generation, and the pre-reformer, into a single chemical looping unit operation. The elimination of these multiple unit operations translates to a higher capital cost savings, process thermal efficiency, and product yield. The economic analysis shows that the total plant cost for a 50 000 barrel d^{-1}y GTL plant can be reduced from ~$86 000 barrel d^{-1} for a conventional GTL plant with ATR (2011 $) to ~$70 000 barrel d^{-1} (2011 $) when incorporating the MTS chemical looping reactor system. The reduction in operating costs for the chemical looping system are mainly due to a reduction in natural gas flow. The higher methane conversion to syngas achieved by the MTS chemical looping process is due to the combination of thermodynamics of the ITCMO oxygen carrier and the co-current moving bed fuel reactor. At natural gas price of $2/MMBtu, a MTS–GTL process will remain economically competitive even when West Texas Intermediate crude oil prices are as low as $40/bbl.

1.5.3 CO_2 Utilization Potential

The performance estimates for the MTS shown in Table 1.5 assumed only natural gas and steam are used as the reacting gases in the fuel reactor [35, 57]. However, the quantity of syngas produced relative to the natural gas feed can be further increased by using CO_2 as a feedstock for the fuel reactor. The CO_2 reactant in combination with natural gas and steam co-feeding ensures the desired H_2:CO ratio syngas is generated for the required downstream chemical synthesis. Thus, the increase in the syngas production with CO_2 co-feed for a fuel reactor operation leads to further savings in natural gas flow beyond the 11% value shown in Table 1.5. CO_2 reaction parameter (CRP) is defined as CO_2 input to the fuel reactor divided by the CO_2 output from the fuel reactor. A CRP value of greater than 1 implies that the fuel reactor is consuming more than it is producing, acting

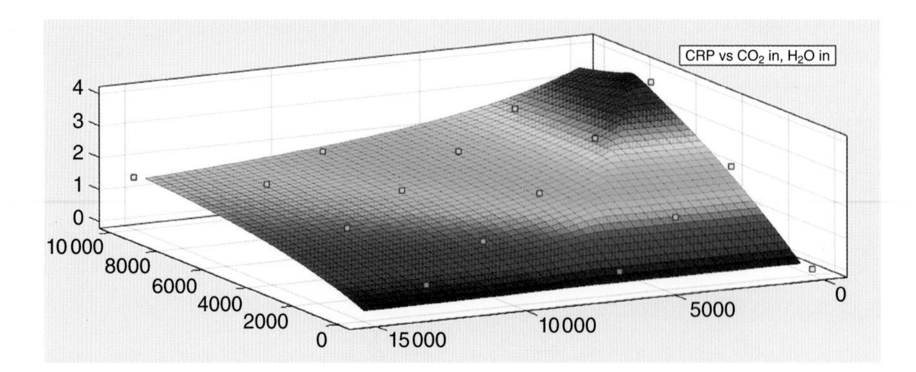

Figure 1.22 CRP variation as a function of steam molar input and CO_2 molar input at a natural gas flow of 15 300 kmol h^{-1}, Fe_2O_3:C molar ratio of 0.85, $P = 1$ atm and $T = 900\,°C$.

Table 1.6 MTS system performance with CO_2 co-injection for a natural gas flow of 15 300 kmol h^{-1}, Fe_2O_3:CH_4 ratio of 0.33 and a CRP = 2.

CO_2 in (kmol h^{-1})	CO_2 out (kmol h^{-1})	CRP	H_2O (kmol h^{-1})	H_2:CO
3 500	1 748	2.00	4 800	1.89
5 000	2 488	2.01	5 600	1.80
6 500	3 246	2.0	6 300	1.72
8 000	3 990	2.05	6 800	1.64
9 500	4 742	2.0	7 200	1.57
10 000	4 992	2.0	7 300	1.54

as a net CO_2 sink. The set of conditions where CRP is greater than 1 are typically quantified by plotting CRP values as a function of CO_2 and H_2O injection flows as shown in Figure 1.22. An example of the variation of syngas production performance with varying CRP values is shown in Table 1.6.

The MTS chemical looping system with CO_2 requires only 16 000 kmol h^{-1} of natural gas for producing 50 000 bpd of liquid fuel, which translates to a reduction in the natural gas consumption of 22% over the baseline system. A 22% reduction in natural gas flow results in an annual cost saving of \$60 million as compared to the baseline plant, when assuming a natural gas price of \$2/MMBtu and 90% production capacity. The process performance suggests that the MTS process is a potential approach to CO_2 utilization for chemical synthesis.

1.5.4 MTS Modularization Strategy

Chemical looping processes are inherently low capital cost intensive systems due to their ability to reduce the number of unit operations to generate the desired product. Thus, researchers have engaged in designing modular chemical looping

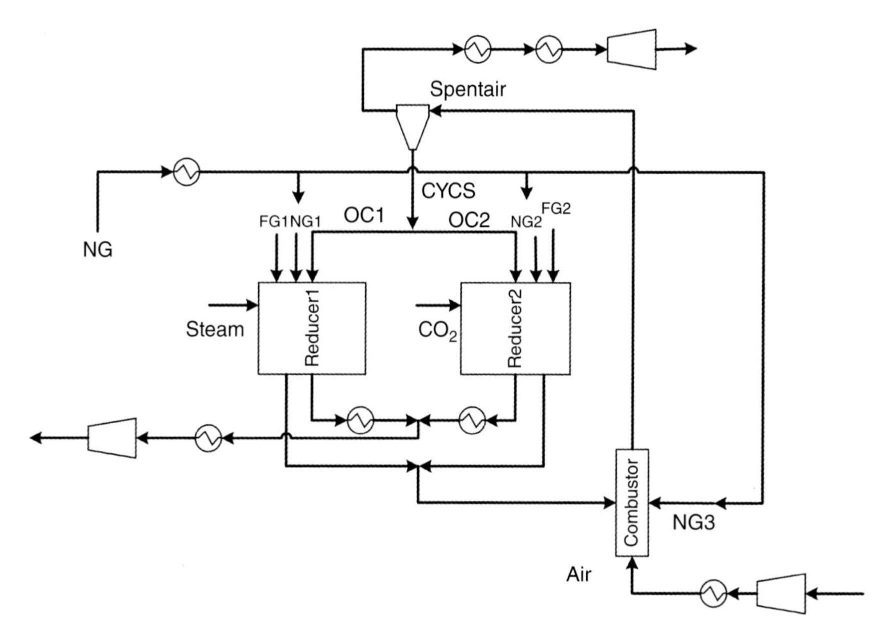

Figure 1.23 Chemical looping modular system for integration into a 50 000 bpd cobalt-based F–T process.

systems that can be cost-competitive processes even at smaller fuel processing capacities. Modular reactor designs are advantageous in increasing the operation flexibility of a chemical plant and can open opportunities for utilizing remote or stranded resources currently uneconomical to process. A modular reactor design for the MTS process suggests an increased syngas production efficiency is achieved by adjusting the fuel and reactant feed conditions in each module. Referring to Figure 1.23, the addition of CO_2 as a reactant results in a nonlinear syngas production trend from the fuel reactor, and, thus, can be exploited to maximize syngas production while maintaining the desired H_2/CO ratio [35]. The detailed material balance for the chemical looping module is shown in Table 1.7. The modularization strategy application reduces the natural gas consumption to 15 200 kmol h^{-1} for production of 50 000 barrel d^{-1} of liquid fuels.

A schematic design of the OSU 10 kW_{th} MTS fuel reactor test apparatus is given in Figure 1.24a. Multiple tests with the sub-pilot apparatus at varying CH_4: oxygen carrier flow ratios and reactant (i.e. CO_2 and H_2O) co-feed rates to verify that the experimental performance of the co-current moving bed reactor to match the thermodynamically expected syngas yields. Figure 1.24b is a sample gas profile from the co-current moving bed fuel reactor showing nearly full fuel conversion to syngas. Table 1.8 summarizes the operating conditions and results in comparison with the thermodynamic values. The results show the sub-pilot co-current moving bed fuel reactor can achieve >99% CH_4 feed conversion with 91.3% syngas purity and a $CO:H_2$ ratio of 1.89. Table 1.8 shows the experimental results match well with the thermodynamically expected syngas yield.

Table 1.7 Material balance for the two-reactor modular chemical looping system shown in Figure 1.23.

	Air	CO_2	CYCS	FG1	FG2	NG1	NG2	NG3	OC1	OC2	Spent air	Steam	Syngas
Temperature (°C)	900	900	1150	900	900	900	900	900	1150	1150	1150	900	755
Pressure (atm)	1	1	1	1	1	1	1	1	1	1	1	1	1
All flows ($kmol\,h^{-1}$)													
Total flow	30549	7800	89500	16293	6983	10123	4408	669	62650	26850	26263	16800	94835
CH4	0	0	0	520	223	9425	4104	623	0	0	0	0	359
H2O	0	0	0	21	9	0	0	0	0	0	1342	16800	11786
CO	0	0	0	2248	964	0	0	0	0	0	0	0	20720
CO2	0	7800	0	161	69	101	44	7	0	0	697	0	6918
H2	0	0	0	6602	2829	0	0	0	0	0	0	0	45491
Fe2O3	0	0	9948	0	0	0	0	0	6964	2984	0	0	0
N2	24134	0	0	6528	2798	162	71	11	0	0	24143	0	9556
O2	6415	0	0	0	0	0	0	0	0	0	79	0	0
n-C4H10	0	0	0	11	5	40	18	3	0	0	0	0	0
TiO2	0	0	79552	0	0	0	0	0	55686	23866	0	0	0
FeTiO3	0	0	0	0	0	0	0	0	0	0	0	0	0

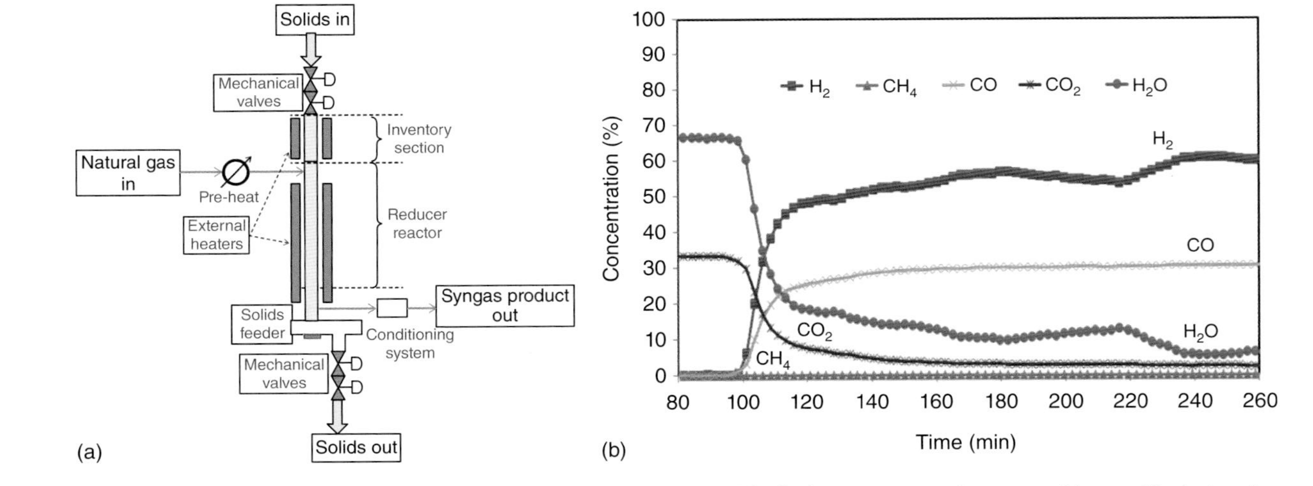

Figure 1.24 (a) Schematic of 10 kW_{th} co-current moving bed fuel reactor test unit and (b) fuel reactor gas outlet composition profile during demonstration of the 10 kW_{th} fuel reactor.

Table 1.8 Key experiment and simulated performance data of 10 kW_{th} sub-pilot STS reactor.

	Experimental results	Theoretical limit
[O]:CH_4	2.2	
Temperature (°C)	975	
CH_4 conversion (%)	>99.9	99.96
H_2:CO	1.89	1.91
CO:CO_2	11.8	12.2

1.6 Concluding Remarks

Moving bed fuel reactors are used in chemical looping processes as advanced energy conversion systems. The packed moving bed fuel reactor can be operated in the co-current and counter-current modes with respect to the gas–solid contact pattern. A counter-current moving bed fuel reactor is advantageous for full fuel conversion to CO_2 and H_2O while maximizing oxygen utilization of the oxygen carrier. The co-current moving bed fuel reactor is used for the partial oxidation of carbonaceous fuels to syngas where the precise control of the solid and gas residence times ensures the thermodynamic limits of the syngas purity and ratio produced can be achieved. Design considerations for sizing the moving bed fuel reactor and dense fluidized bed air reactor with respect to fuel process capacity, operating temperature, and oxygen carrier properties were discussed. Consideration was also provided for types of gas-sealing and solid flow control valves for integrated the operation of fuel and air reactors. The SCL and CDCL processes are exemplary chemical looping processes that use count-current moving bed fuel reactors. The CDCL process is an advanced oxy-combustion process for power generation capable of exceeding 90% CO_2 capture with minimal increase in COE (~26.8% compared to base pulverized coal power plant without CO_2 capture). Sub-pilot demonstrations confirm that nearly 100% CO_2 capture can be achieved and sulfur emissions from the air reactor are below the Environmental Protection Agency (EPA) emissions limit. The SCL has been scaled to a high pressure 250 kW_{th}–3 MW_{th} pilot plant and demonstrated for high purity H_2 production with CO_2 capture from gaseous fuels such as gasified coal syngas. The CTS and MTS process represent exemplary co-current moving bed fuel reactor chemical looping processes for the partial oxidation of solid and gaseous fuels, respectively, to syngas. Experimental studies at bench and 10 kW_{th} sub-pilot scale show the co-current moving bed fuel reactor design is capable of achieving product syngas purity and H_2/CO ratios at the thermodynamically predicted limits. When CO_2 is used as the feedstock to the co-current moving bed fuel reactor, the amount syngas produced can be increased substantially. A nonlinear relationship between the syngas produced relative to the CO_2 input can be exploited in a modular chemical looping reactor system design to substantially reduce the amount of natural gas consumed to produce an equivalent amount of syngas – over 25%

reduction in natural gas consumption compared to conventional ATR of natural gas. Thus, the simplicity of the moving bed fuel reactor design combined with its versatility and efficiency show this design is a viable approach chemical looping processes as proven by the experimental and modeling results.

References

1 Fan, L.-S. (2011). *Chemical Looping Systems for Fossil Energy Conversions.* Wiley.

2 Knoche, K. and Richter, H. (1968). Improvement of reversibility of combustion processes. *Brennstoff-Warme-Kraft* 20 (5): 205.

3 Richter, H.J. and Knoche, K.F. (1983). Reversibility of combustion processes. In: *Efficiency and Costing*, vol. 235, 71–85. American Chemical Society. doi: 10.1021/bk-1983-0235.ch003.

4 Ishida, M., Zheng, D., and Akehata, T. (1987). Evaluation of a chemical-looping-combustion power-generation system by graphic exergy analysis. *Energy* 12 (2): 147–154.

5 Bergmann, F.J. (1897). Process for the production of calcium carbide in blast furnaces. German Patent 29,384.

6 Hurst, S. (1939). Production of hydrogen by the steam–iron method. *J. Am. Oil Chem. Soc.* 16 (2): 29–35.

7 Gasior, S., Forney, A., Field, J. et al. (1961). *Production of Synthesis Gas and H_2 by the Steam–Iron Process.* Washington, DC: U.S. Department of the Interior, Bureau of Mines.

8 Lewis, W.K. and Gilliland, E.R. (1954). Production of pure carbon dioxide. US Patent 2,665,972.

9 Dobbyn, R., Ondik, H., Willard, W. et al. (1978). Evaluation of the Performance of Materials and Components Used in the CO_2 Acceptor Process Gasification Pilot Plant. Final Report. *DOE-ET-10253-T1*. Washington, DC: U.S. Department of Energy.

10 Teed, P.L. (1919). *The Chemistry and Manufacture of H_2*. Longmans, Green and Company.

11 Institute of Gas Technology (1979). Coal Gasification. *EF-77-C-01-2435*. Washington, DC: U.S. Department of Energy.

12 Lane, H. (1913). Process for the production of hydrogen. US Patent 1,078,686.

13 Institute of Gas Technology (1979). Development of the Steam–Iron Process for Hydrogen Production. *EF-77-C-01-2435*. Washington, DC: U.S. Department of Energy.

14 Otsuka, K., Wang, Y., Sunada, E., and Yamanaka, I. (1998). Direct partial oxidation of methane to synthesis gas by cerium oxide. *J. Catal.* 175 (2): 152–160.

15 Dudukovic, M.P. (2009). Frontiers in reactor engineering. *Science* 325 (5941): 698–701.

16 Nakamura, T. (1977). Hydrogen production from water utilizing solar heat at high temperatures. *Sol. Energy* 19 (5): 467–475.

17 Steinfeld, A. and Palumbo, R. (2001). Solar thermochemical process technology. In: *Encyclopedia of Physical Science and Technology*, vol. 15 (ed. R.A. Meyers), 237–256. Academic Press.

18 Jones, C.A., Leonard, J.J., and Sofranko, J.A. (1987). Fuels for the future: remote gas conversion. *Energy Fuels* 1 (1): 12–16.

19 Jones, C.A., Leonard, J.J., and Sofranko, J.A. (1987). The oxidative conversion of methane to higher hydrocarbons over alkali-promoted $MnSiO_2$. *J. Catal.* 103 (2): 311–319.

20 Steinfeld, A., Kuhn, P., and Karni, J. (1993). High-temperature solar thermochemistry: production of iron and synthesis gas by Fe_3O_4-reduction with methane. *Energy* 18 (3): 239–249.

21 Steinfeld, A. (2005). Solar thermochemical production of hydrogen – a review. *Sol. Energy* 78 (5): 603–615.

22 Adanez, J., Abad, A., Garcia-Labiano, F. et al. (2012). Progress in chemical-looping combustion and reforming technologies. *Prog. Energy Combust. Sci.* 38 (2): 215–282.

23 Fan, L.-S., Zeng, L., and Luo, S. (2015). Chemical-looping technology platform. *AIChE J.* 61 (1): 2–22.

24 Hamers, H., Gallucci, F., Cobden, P. et al. (2013). A novel reactor configuration for packed bed chemical-looping combustion of syngas. *Int. J. Greenhouse Gas Control* 16: 1–12.

25 Ortiz, M., Gallucci, F., Snijkers, F. et al. (2014). Development and testing of ilmenite granules for packed bed chemical-looping combustion. *Chem. Eng. J.* 245: 228–240.

26 Hamers, H.P., Romano, M.C., Spallina, V. et al. (2015). Energy analysis of two stage packed-bed chemical looping combustion configurations for integrated gasification combined cycles. *Energy* 85: 489–502.

27 Noorman, S. and van Kuipers, S.A. (2007). Packed bed reactor technology for chemical-looping combustion. *Ind. Eng. Chem. Res.* 46 (12): 4212–4220.

28 Luo, S., Zeng, L., Xu, D. et al. (2014). Shale gas-to-syngas chemical looping process for stable shale gas conversion to high purity syngas with a H_2:CO ratio of 2:1. *Energy Environ. Sci.* 7 (12): 4104–4117.

29 Tong, A., Bayham, S., Kathe, M.V. et al. (2014). Iron-based syngas chemical looping process and coal-direct chemical looping process development at Ohio State University. *Appl. Energy* 113: 1836–1845.

30 Li, F., Zeng, L., Velazquez-Vargas, L.G. et al. (2010). Syngas chemical looping gasification process: bench-scale studies and reactor simulations. *AIChE J.* 56 (8): 2186–2199.

31 Kathe, M.V., Empfield, A., Na, J. et al. (2016). Hydrogen production from natural gas using an iron-based chemical looping technology: thermodynamic simulations and process system analysis. *Appl. Energy* 165: 183–201.

32 Kim, H., Wang, D., Zeng, L. et al. (2013). Coal direct chemical looping combustion process: design and operation of a 25-kW_{th} sub-pilot unit. *Fuel* 108 (0): 370–384.

33 Bayham, S., McGiveron, O., Tong, A. et al. (2015). Parametric and dynamic studies of an iron-based 25-kW_{th} coal direct chemical looping unit using sub-bituminous coal. *Appl. Energy* 145: 354–363.

34 Bayham, S.C., Kim, H.R., Wang, D. et al. (2013). Iron-based coal direct chemical looping combustion process: 200-h continuous operation of a 25-kW_{th} subpilot unit. *Energy Fuels* 27 (3): 1347–1356.

35 Kathe, M., Empfield, A.M., Sandvik, P.O. et al. (2017). Utilization of CO_2 as a partial substitute for methane feedstock in chemical looping methane–steam redox processes for syngas production. *Energy Environ. Sci.* 10 (6): 1345–1349. doi: 10.1039/C6EE03701A.

36 Wang, D. and Fan, L.-S. (2015). L-valve behavior in circulating fluidized beds at high temperatures for group D particles. *Ind. Eng. Chem. Res.* 54 (16): 4468–4473.

37 Wang, D. and Fan, L.-S. (2014). Bulk coarse particle arching phenomena in a moving bed with fine particle presence. *AIChE J.* 60 (3): 881–892.

38 Knowlton, T.M. (1997). Standpipes and return systems. In: *Circulating Fluidized Beds* (ed. J.R. Grace, A.A. Avidan and T.M. Knowlton), 214–260. Dordrecht: Springer.

39 Zhou, Q., Zeng, L., and Fan, L.-S. (2013). Syngas chemical looping process: dynamic modeling of a moving-bed reducer. *AIChE J.* 59 (9): 3432–3443.

40 Tong, A., Zeng, L., Kathe, M.V. et al. (2013). Application of the moving-bed chemical looping process for high methane conversion. *Energy Fuels* 27 (8): 4119–4128.

41 Sridhar, D., Tong, A., Kim, H. et al. (2012). Syngas chemical looping process: design and construction of a 25 kW_{th} subpilot unit. *Energy Fuels* 26 (4): 2292–2302.

42 Tong, A., Sridhar, D., Sun, Z. et al. (2013). Continuous high purity hydrogen generation from a syngas chemical looping 25 kWth sub-pilot unit with 100% carbon capture. *Fuel* 103: 495–505.

43 Zeng, L., Tong, A., Kathe, M. et al. (2015). Iron oxide looping for natural gas conversion in a countercurrent moving bed reactor. *Appl. Energy* 157: 338–347.

44 Gerdes, K., Summers, W., and Wimer, J. (2011). Cost Estimation Methodology for NETL Assessments of Power Plant Performance. *Final Rep. DOE/NETL-2011/1455*. Washington, DC: U.S. Department of Energy.

45 Zoelle, A., Turner, M.J., and Chou, V. (2015). Quality Guidelines for Energy System Studies: Performing a Techno-Economic Analysis for Power Generation Plants. *DOE/NETL-2015/1726*. Washington, DC: U.S. Department of Energy.

46 Woods, M. and Matuszewski, M. (2012). Quality Guideline for Energy System Studies: Specifications for Selected Feedstocks. *DOE/NETL-341/011812*. Washington, DC: U.S. Department of Energy.

47 Turner, M.J. and Pinkerton, L. (2013). Quality Guidelines for Energy System Studies: Capital Cost Scaling Methodology. *DOE/NETL-341/0131113*. Washington, DC: U.S. Department of Energy.

48 Grant, T., Morgan, D., and Gerdes, K. (2013). Quality Guidelines for Energy System Studies: Carbon Dioxide Transport and Storage Costs in NETL Studies. *DOE/NETL-2013/1614*. Washington, DC: U.S. Department of Energy.

49 Li, F. and Fan, L.-S. (2008). Clean coal conversion processes – progress and challenges. *Energy Environ. Sci.* 1 (2): 248–267.

50 Fan, L.-S. and Li, F. (2010). Chemical looping technology and its fossil energy conversion applications. *Ind. Eng. Chem. Res.* 49 (21): 10200–10211.

51 Zeng, L., He, F., Li, F., and Fan, L.-S. (2012). Coal-direct chemical looping gasification for hydrogen production: reactor modeling and process simulation. *Energy Fuels* 26 (6): 3680–3690.

52 Luo, S., Li, J., Zhang, R. et al. (2015). Status and perspective of solid-fueled chemical looping technology. *Shiyou Xuebao, Shiyou Jiagong/Acta Petrolei Sinica (Petroleum Processing Section)* 31 (2): 426–435.

53 Zeng, L., Kathe, M.V., Chung, E.Y., and Fan, L.-S. (2012). Some remarks on direct solid fuel combustion using chemical looping processes. *Curr. Opin. Chem. Eng.* 1 (3): 290–295.

54 Li, F., Zeng, L., and Fan, L.-S. (2010). Biomass direct chemical looping process: process simulation. *Fuel* 89 (12): 3773–3784.

55 Kobayashi, N. and Fan, L.-S. (2011). Biomass direct chemical looping process: a perspective. *Biomass Bioenergy* 35 (3): 1252–1262.

56 Kathe, M.V., Xu, D., Hsieh, T.-L. et al. (2014). Chemical Looping Gasification for Hydrogen Enhanced Syngas Production with In-Situ CO_2 Capture. *Tech. Rep. OSTI: 1185194*. Washington, DC: U.S. Department of Energy.

57 Kathe, M., Fryer, C., Sandvik, P. et al. (2017). Modularization strategy for syngas generation in chemical looping methane reforming systems with CO_2 as feedstock. *AIChE J.* 63 (8): 3343–3360. doi: 10.1002/aic.15692.

2

Single and Double Loop Reacting Systems

Justin Weber

National Energy Technology Laboratory, U. S. Department of Energy, 3610 Collins Ferry Road, Morgantown, WV, 26507, USA

2.1 Introduction

Bubbling fluidized beds and circulating fluidized beds have now been developed for over a century, being used successfully at industrial scale with applications such as fluid catalytic crackers (FCCs) and coal combustion. Fluidized-type reactors excel at gas–solid reactions, providing good gas–solids contact, excellent solids mixing, fuel flexibility, and high heat transfer coefficients. Typical particle sizes for fluidized systems (100–500 μm) allow for large specific surface areas, allowing higher reaction rates. Solids can be added and removed continuously with non-mechanical valves controlling the flow of solids, facilitating continuous processes. Thus, it is not surprising that this is the most popular approach to chemical looping reactors, with both air and fuel reactors utilizing a fluidized or circulating fluidized bed. The use of a fluidized bed in a chemical looping-like process even dates to one of the original works, US Patent 2665971, filed May 12, 1946 [1]. One of the first continuous looping reactors, Chalmers University of Technology's 10 kW_{th} chemical looping reactor, consisted of two interconnected fluidized beds [2].

In many respects, a single loop chemical looping combustion process with bubbling fluidized bed reactors is similar to commercial catalytic crackers, Figure 2.1. Both have two bubbling fluidized bed reactors, connected with solids flow control devices, move significant quantalities of granular solids, and have a riser to lift solids. One reactor regenerates the solids while the other reactor cracks or oxidizes the fuel. The chemical looping process is also very similar to Dupont's process for the production of maleic anhydride via partial oxidation of butane, which used a vanadium phosphorus oxide catalyst circulated between a reactor that consisted of a riser and a fluid bed regenerator [4].

Several experimental units have been constructed around the world, ranging in size from 0.5 kW_{th} to a pilot-scale unit at 3 MW_{th} [5]. The units combine many different multiphase flow reactors, such as fluid beds, spouted beds, and risers. They also use a variety of gas sealing and flow control devices known as non-mechanical valves such as L-valves, loop seals, and seal pots. There is a plethora

Handbook of Chemical Looping Technology, First Edition. Edited by Ronald W. Breault.

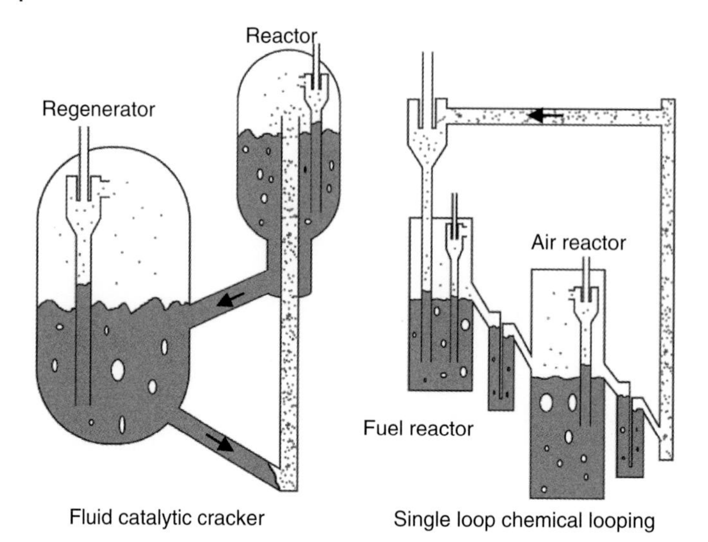

Figure 2.1 Simple sketches of a fluid catalytic reactor and a single loop chemical looping reactor. The chemical looping reactor sketch was inspired by the 10 kW_{th} unit at Chalmers University of Technology, Sweden, and the 10 kW_{th} unit at the Department of Energy and Environment, Instituto de Carboquímica, Spain. Source: Lyngfelt and Thunman 2005 [2] & Adanez et al. 2006 [3].

Figure 2.2 National Energy Technology Laboratory's 50 kW_{th} chemical looping reactor.

of potential designs that can be envisioned, with the goal of developing the highest performing system in terms of fuel conversion, carbon capture efficiency, reliability, and carrier life (attrition, agglomeration, etc.) (Figure 2.2).

Experiments have been conducted with various oxygen carriers, both commodity carriers such as natural ores and waste products, and completely manufactured carriers. These carriers include various metal oxides, such as iron, copper, nickel, and manganese, as well as calcium carbonate (limestone). A variety of fuels have also been tested including gaseous fuels (syngas, methane), liquid fuels (ethanol, kerosene, oil, sewage sludge), and solid fuels (biomass, petcoke, coal). These experimental units have amassed significant operating experience, demonstrating combustion efficiencies above 98% and carbon capture efficiencies above 97% (Table 2.1).

2.2 Reactor Types

Typical reactor types used in single and double loop chemical looping systems include fluidized beds, spouted beds, and risers, Figure 2.3. Reactors need to be appropriately designed for the oxygen carrier that will be used in the unit. Typically, fluidized beds are used with Geldart Group A and B particles, while spouted beds are used for group D particles. Risers are best used with smaller particles, group C, A, and B, because as the particles get larger, high gas velocities are required to lift the solids, increasing blower power requirements. Almost all chemical looping configurations have some type of riser because solids need to be lifted in order to take advantage of gravity to flow the solids between the reactors and around the loop.

Most of the constructed chemical looping systems consist of fluidized beds and risers (Alstom, Chalmers University of Technology, Technische Universitat Wien, Instituto de Carboquímica, Korea Institute of Energy Research, National Energy Technology Laboratory). There are a couple units that use a spouted bed for the fuel reactor (Southeast University, China, University of Kentucky).

One of the challenges with using these reactors is that they were developed and applied to reaction systems where the solid is the material that needs to be reacted with the gas. In coal combustion, the solid fuel is burned with gaseous oxygen and stays in the system until it has been converted to ash. In coal gasification, the coal is reacted with steam and carbon dioxide. In the fuel reactor in chemical looping combustion, the reaction process is the other way – the gaseous fuel (natural gas, volatiles, gasification products) need to react with solid. In other words, in chemical looping combustion, we care about converting all the *gaseous* fuel as opposed to coal combustion where we care about converting all the *solid* fuel.

This same issue was realized when the fluidized bed Fischer–Tropsch synthesis plant in Brownsville, TX, did not achieve the performance predicted based on the pilot plant [6]. Till that point, fluidized bed processes either did not require a high degree of conversion (FCC) or were easily achievable (roasting and drying) [6]. The failure of the process was attributed to bubble dynamics and their role in scale-up, which was eventually corrected, allowing the process to achieve the target performance [6].

This difference presents a significant challenge in the application of these reactors to chemical looping combustion. Any gaseous fuel that bypasses or does not encounter the solids will not be converted, decreasing the combustion efficiency of the process. If complete combustion does not occur, then downstream catalytic or oxygen-polishing steps might be required to meet or

Table 2.1 Unit Summary: Tabulated here are selected units from around the world that consist of a single loop or double loop design utilizing fluid beds, spouted beds, and/or risers as part of the loop with fuel inputs of 50 kW_{th} and greater. The entries are sorted by thermal size.

Group	Location	Size	Fuel(s)	Carrier(s)	Fuel conversion	Capture efficiency	Reactors
Alstom, USA [14]	Bloomfield, CT, USA	3 MW_{th}	Coal	Limestone			Two interconnected circulating fluidized beds
Chalmers University of Technology [31]	Göteborg, Sweden	1.4 MW_{th}	Biomass	Ilmenite, Mn ore	45–60%		Bubbling bed (fuel reactor (FR)), circulating fluidized bed (air reactor (AR))
University of Darmstadt [21, 32]	Darmstadt, Germany	1 MW_{th}	coal, torrefied biomass	ilmenite, iron ore	56–73%	46–66%	Two interconnected circulating fluidized beds
University of Utah [33, 34]	Salt Lake City, UT, USA	200 kW_{th}	CH_4, Coal	CuO			Two interconnected circulating fluidized beds
Korea Institute of Energy Research (KIER) [35]	Daejeon, Korea	200 kWth	CH_4	NiO	99.2%		Two interconnected circulating fluidized beds
SINTEF [9]	Trondheim, Norway	150 kW_{th}	CH_4	CuO	90–98%		Two interconnected circulating fluidized beds
Technische Universitat Wien [13, 36]	Vienna, Austria	140 kW_{th}	H_2, CO, CH_4	Ilmenite, NiO, CuO	92+%		Two interconnected circulating fluidized beds
Chalmers University of Technology [22]	Göteborg, Sweden	100 kW_{th}	Coal, petcoke, wood char	Ilmenite, iron ore, manganese ore [15]	75–84%	95.5–99%	Two interconnected circulating fluidized beds
Huazhong University of Science & Technology (HUST) [37]	Wuhan, China	50 kW_{th}	Coal	Iron ore			Bubbling bed (FR), Turbulent bed (AR)
Instituto de Carboquímica (ICB-CSIC) [19]	Zaragoza, Spain	50 kW_{th}	Coal	Ilmenite, iron ore, $Cu/Fe/MgAl_2O_4$	85–92%	63–90%	Turbulent bed (FR), Bubbling bed (AR)
Korea Institute of Energy Research (KIER)	Daejeon, Korea	50 kW_{th}	CH_4, syngas	NiO, CoO	99.4+%		Bubbling bed (FR), Bubbling bed (AR)
National Energy Technology Laboratory [26]	Morgantown, WV, USA	50 kW_{th}	CH_4	Hematite, promoted hematite, Cu/Fe	40–90%		Bubbling bed (FR), Turbulent bed (AR)
Southeast University [24]	Nanjing, China	50 kW_{th}	Coal	Iron ore	81–85%	83–87%	Turbulent bed (FR), Turbulent bed (AR)

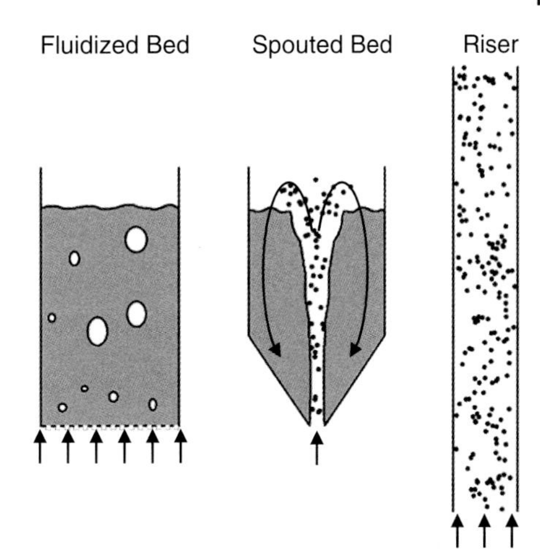

Figure 2.3 Typical reactor types.

exceed carbon dioxide pipeline standards. This could add significant costs to the overall system.

2.2.1 Fluid Beds

Fluid beds are simply cylindrical or rectangular containers that hold a "bed" of granular material. At the bottom of the bed is a gas distributor that uniformly introduces gas across the bed area. This gas moves through the bed at a given gas velocity, establishing a pressure gradient. Once the pressure drop across the bed is sufficient to support the weight of the particles, the bed is considered incipiently fluidized. The gas velocity at this point is known as the minimum fluidization velocity, U_{mf}. Fluidized beds are predominantly used with Geldart type B (typically 100–1000 μm in diameter), or sand-like particles. Typical oxygen carriers fall into this category of solids.

Fluid beds are attributed with good gas–solid contact as well excellent solids mixing. The mixing of solids allows fluid beds to have a near uniform temperature as well as excellent heat transfer. However, gas can bypass contact with the solids if it travels through the bubble phase. This gas will not have a chance to react with the solids, causing gaseous fuel to be emitted from the reactor.

As fluid beds scale and the diameters and bed heights increase, this issue will get worse because bubble diameters are related to the bed height and reactor diameter. As the bed height increases, the bubbles have more opportunity to coalesce into larger bubbles, providing more volume for gas bypassing, similarly for bed diameter. Baffles, variable diameters, or staged reactors may be needed to ensure that the gaseous fuel contacts the solids, maximizing the conversion of the fuel.

2.2.2 Spouted Beds

In many respects, spouted beds are similar to fluid beds except that they typically operate with larger Geldart group D materials (generally, 1 mm particle diameters and above). A high gas velocity spout is established in the center of the bed, carrying and ejecting solids into the freeboard. The solids that are carried into the freeboard develop a fountain, where the solids rain back down in the annulus. This flow pattern establishes a strong recirculation of solids in the bed that is proportional to the spout gas velocity.

It is uncertain whether a spouted bed could perform well as a fuel reactor. Generally, the spout is a high velocity region with very low solids concentrations. Any gaseous fuels that enter the spout will have little chance to interact with the solid carrier, leaving the bed unreacted [7].

Regardless, there are groups investigating the use of a spouted bed as a fuel reactor for solid fuels. Shen et al. at Southeast University demonstrated coal combustion efficiencies upward of 92% and carbon capture efficiencies of 80% in a 1 kW_{th} spouted bed reactor [8].

2.2.3 Risers

Risers operate at much higher gas velocities as compared to fluid beds, above the transport velocity of the granular material. At these gas velocities, the solids are dispersed across the diameter, with average solids concentrations typically less than 10%, although due to interactions between the wall and the solids, solids concentrations at the wall can be much higher, even packed. Depending on the gas velocity (U_g) and solids flux (G_s), the riser can be operated in many different flow regimes including dilute transport, transport, core–annular, and dense suspension.

Risers are found in almost all multiphase flow processes such as fluid catalytic cracking, circulating fluidize bed combustors, and transport integrated gasification (TRIG™) gasifiers. Similarly, almost all proposed and constructed chemical looping reactors have a riser because the solids need to be elevated or raised at some point in the process so that gravity can be utilized to move the solids through the rest of the loop. Utilizing gravity to move the solids allows gas to move in the opposite direction of the solids velocity. This opposing flow is critical to establishing gas sealing between the reactors. Some designs utilize the component as a reactor (fuel and/or air reactor) as well.

Risers that are typically used as reactors in industry are focused on converting the solids as opposed to the gas, leading to some concern about the effectiveness of the riser to convert all the gas in a chemical looping system. However, there are several chemical looping reactors that use a riser as the fuel reactor, successfully converting the gaseous fuel, such as CH_4 or syngas, to CO_2 and H_2O. The 150 kW_{th} reactor at SINTEF has demonstrated CH_4 fuel conversions of 98% [9].

In an attempt to increase the gas–solid contact and increase the particle residence time in the riser, the addition of internals has been proposed [10]. These internals cause disturbances in the riser flow, increasing the solids concentration, Figure 2.4.

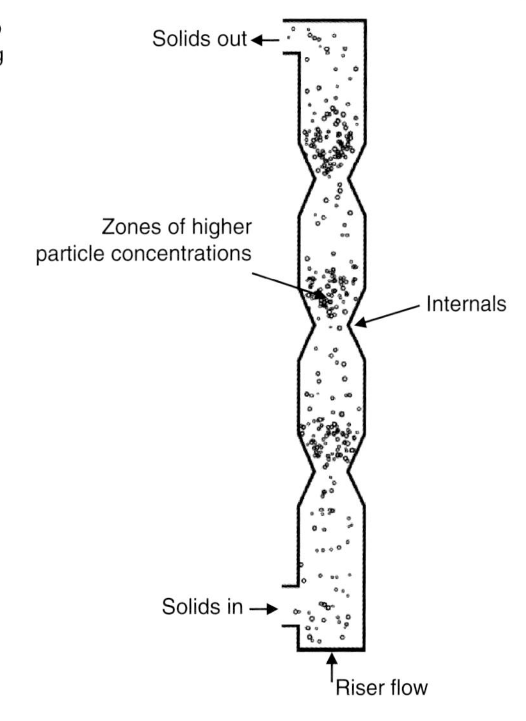

Figure 2.4 Riser internals proposed to increase solids concentrations, leading to increased gas/solids contact. Source: Guío-Pérez et al. 2014 [11].

2.3 Gas Sealing and Solids Control

Gas sealing between the reactors is important to achieve and maintain high carbon capture efficiencies. If any carbon-containing gases leak to the air reactor, those gases will be readily combusted with oxygen and be mixed in a predominantly nitrogen-based gas stream, decreasing the carbon capture efficiency of the process. Similarly, nitrogen needs to be prevented from leaking into the fuel reactor from the air reactor. If significant nitrogen does leak into the fuel reactor, the carbon dioxide purity will decrease, preventing the ability to utilize the stream for other applications or sequester the carbon dioxide, defeating the advantage that chemical looping has over other capture technologies.

Various devices can be used to help prevent gas leakage, such as loop seals, seal pots, and non-mechanical valves. These devices establish pressure gradients that discourage the flow of gas from one reactor to another. In many configurations, the highest pressure in the unit will be located in one of these devices. Gas injected on either side of this highest pressure will stay on that side, as long as these units are operated in a controlled manner so that solids columns are not lost. If solids columns are lost and the highest pressure is no longer located in the non-mechanical valve, then gas can backflow through the valve. These non-mechanical valves have been demonstrated to seal the gases between the two reactors with experimental carbon capture efficiencies reported in the literature

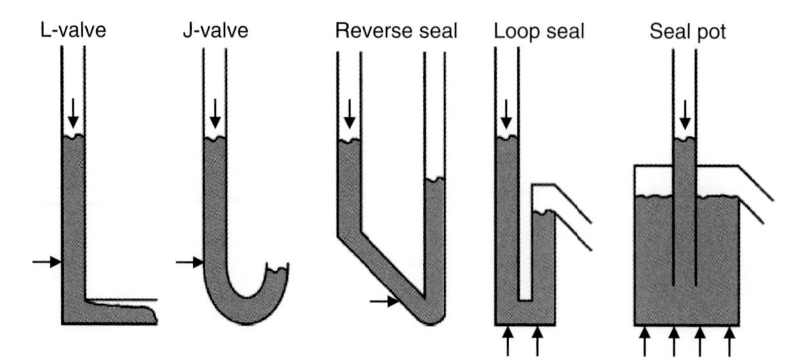

Figure 2.5 Non-mechanical valves.

typically greater than 90%, even with solids fuels. Some units have demonstrated reaching carbon capture efficiencies higher than 98% [12].

Non-mechanical valves also provide a restriction where the flow of solids through the component can be controlled by varying the amount of gas or aeration. These are particularly useful components when the flow rate of solids needs to be controlled. With small benchtop reactors, these valves typically consist of a small diameter (around 2.5 cm) pipe with long horizontal sections. At these sizes, the propensity of the solids to bridge the pipe and plug are high, leading to operational issues. These issues will diminish as units are scaled from benchtop to pilot-scale reacting systems.

Typical non-mechanical valves include L-valves, J-valves, and reverse seal. Loop seals and seal pots can also be operated as a valve. The shape of the valve flow path is typically described by the name, Figure 2.5.

2.4 Single Loop Reactors

By combining at least two reactors with a couple of gas sealing devices, a simple single loop reacting system can be configured, Figure 2.6. Single loop reactors are a class of reacting systems where the solids follow one flow path between the two (or more) reactors. In these configurations, recirculation of oxygen carriers in a particular reactor, either the fuel reactor or the air reactor, does not occur. Systems that utilize a fluidized bed or spouted bed as the fuel reactor typically follow this design.

2.5 Double (or More) Loop Reactors

Double (or more) loop reactors are a class of reacting systems where the solids flow path splits, allowing for re-circulation of oxygen carrier in a particular reactor, typically the fuel reactor. This allows for the residence time of the oxygen carrier to be significantly increased without increasing the size (diameter and/or height) of the reactor. A good example of this reactor configuration

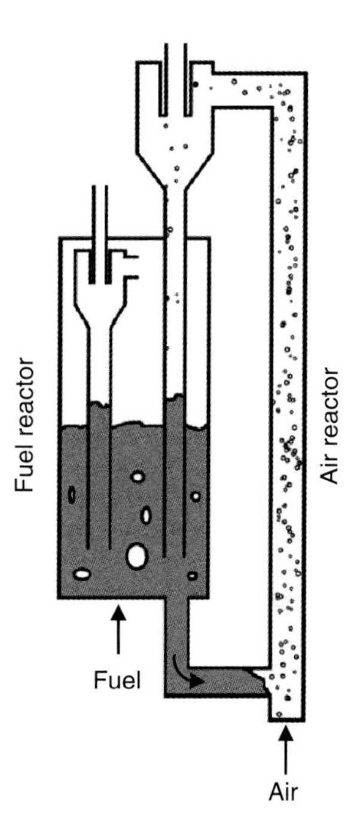

Figure 2.6 Single loop reactor.

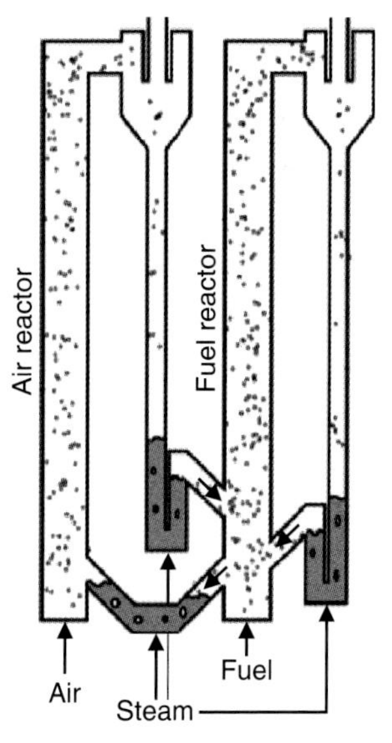

Figure 2.7 Double loop reactor, 140 kW_{th} unit at Technische Universitat Wien, Austria. Source: Kolbitsch et al. 2009 [13]. Reproduced with permission of American Chemical Society.

is the 140 kW_{th} unit at Technische Universitat Wien, Austria, Figure 2.7 [13]. This unit was designed to operate in both combustion and reforming regimes with gaseous fuels and nickel-based oxygen carriers. Alstom also has a double loop unit, very similar to the Technische Universitat Wien unit, operating with limestone as the oxygen carrier at a 3 MW_{th} scale [14].

A number of groups, comprising almost all chemical looping reactor units larger than 100 kW_{th}, have converged to the two interconnected circulating fluid bed design including Alstom, Darmstadt, SINTEF, Chalmers, and Technische Universitat Wien [9, 12, 13, 15]. This design has demonstrated good performance on both gaseous fuels and solid fuels with a variety of oxygen carriers including Fe, Cu, Ni, Mn, and Ca-based materials, both manufactured and natural. This design typically operates at higher solids circulation rates and lower oxygen carrier inventories as compared to units that utilized fluid beds and spouted beds. This design also allows re-purposing of existing circulating fluidized beds by adding another circulating fluid bed to the original unit.

2.6 Solid Fuel Reactors

It is generally assumed that in bubbling and circulating fluidized beds, solid–solid reactions are irrelevant [12]. Thus, devolatilization of the solid fuel followed by gasification must happen to produce gaseous products that can then react with the oxygen carrier. As such, reactor concepts usually take two different approaches. The indirect process has a dedicated gasifier where the solid fuel is gasified. This syngas is then fed into the fuel reactor and reacted with the oxygen carrier compared to the direct process where solid fuel is directly fed into the fuel reactor. It is generally accepted that coal direct chemical looping is advantageous because the products of gasification can immediately react with the oxygen carrier, helping promote the forward gasification reactions. Coal direct chemical looping also avoids the costs associated with an air separation unit (ASU) and gasifier.

Solid fuels present several more challenges that need to be overcome when used in chemical looping reactor systems. Solid fuels create solid waste (ash) that needs to be removed from the system. Solid fuels also introduce more elements that need to be dealt with including sulfur, fuel-bound nitrogen, and alkali metals. Devolatilization can occur rapidly when coal is introduced to the bed, causing gas bypassing. Finally, unreacted solid carbon can easily be leaked to the air reactor. If carbon does leak to the air reactor, it will be burned and the resulting carbon dioxide will not be captured.

2.6.1 Volatiles

As "cold" coal is introduced into a reactor operating between 800 and 1000 C, rapid devolatilization can occur, producing significant quantities of gas including H_2O, H_2, CO, CO_2, CH_4, and other larger hydrocarbons. If enough gas is produced at the coal feed site, larger bubbles or channels could develop, causing the unconverted species to bypass the bed. Several experimental units have experienced this issue including the 100 kW_{th} unit at Chalmers, specifically concluding that "contact between oxygen carrier and volatiles is a major issue for the gas conversion" [12]. If unconverted fuel (H_2, CO, CH_4, etc.) does leave the fuel reactor, the plant efficiency decreases. The outlet of the fuel reactor might also not meet CO_2 pipeline requirements, requiring additional flue gas cleanup.

To guarantee that the fuel reactor outlet stream does not have significant levels of unconverted volatiles, some proposed systems have included

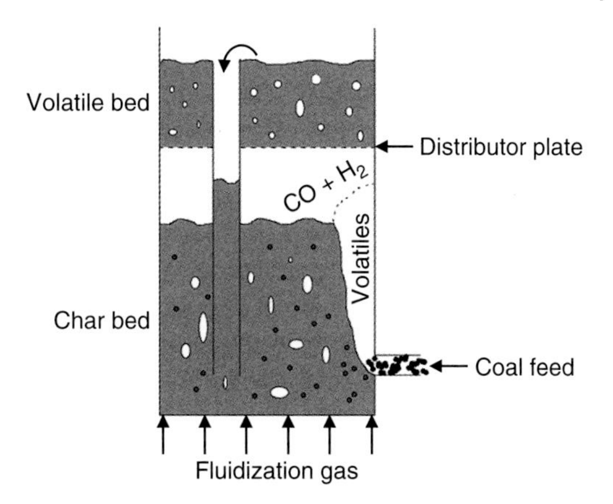

Figure 2.8 Two stage fuel reactor.

"oxygen-polishing" units where pure oxygen is introduced into the fuel reactor exhaust to oxidize any unconverted gas. An ASU is required to produce the oxygen, reducing the efficiency of the process.

Some researchers have proposed a two-stage approach to the fuel reactor design, Figure 2.8 [16, 17]. This two-stage approach also has the advantage of the fresh oxidized oxygen carrier encountering the volatiles first, providing the best thermodynamic environment for converting the volatiles, before overflowing into the char bed.

Gu et al. constructed a 2 kW_{th} two-stage fuel reactor, operating on syngas and CH_4 [18]. By utilizing two beds, CO conversion from syngas was improved by 13–16% points. With the dual-stage configuration, syngas conversion reached 99.2% while CH_4 conversion reached 92.9%.

2.6.2 Carbon Leakage

As the oxygen carrier leaves the fuel reactor, char and ash will be dragged along with the carrier. If the char makes it to the air reactor, the char will be readily combusted in an oxygen-rich environment. The carbon dioxide formed will be mixed with nitrogen and emitted from the air reactor into the atmosphere. Any carbon leaked to the air reactor will not be captured, lowering the carbon capture efficiency of the process.

There are two ways to prevent significant amounts of carbon from leaking to the fuel reactor: (i) increase the reaction performance to react with the carbon before it has the chance to leave the fuel reactor and (ii) insert a carbon separation device between the fuel and air reactors.

Pérez-Vega et al. have shown that as the temperature in the fuel reactor increases, resulting in faster kinetics, the carbon capture efficiency of the unit increases [19]. They theorized that by further increasing the temperature in the fuel reactor to 1010 °C, a carbon capture efficiency of 95% could be achieved [19].

Many researchers have proposed and proved that separation of a lighter and smaller char particle from the larger and heavier oxygen carrier using an aerodynamic separation device is possible [20]. The stream of char, and most likely ash, could then be returned to the fuel reactor, selectively increasing the residence

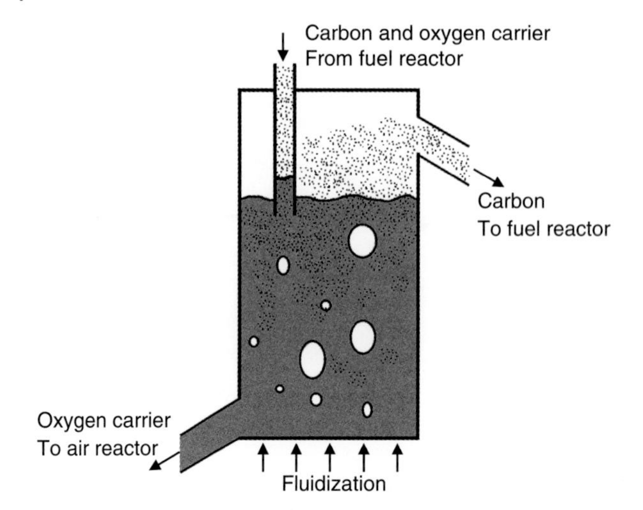

Figure 2.9 Carbon separation in a fluidized bed.

time of the char in the fuel reactor. This allows the char to have more time to gasify in the fuel reactor, increasing the carbon capture efficiency and combustion efficiency.

Most researchers have proposed utilizing a fluidized bed or fluidized bed-like device to cause segregation and separation of the char from the oxygen carrier, Figure 2.9 [20]. There are only a handful of reacting units that have implemented such a device [19, 21, 22]. These devices use hydrodynamics to separate the char from the oxygen carrier.

2.6.3 Ash Separation

Similar to the char separation, ash needs to be removed from the carrier stream. Ash that does not have any carbon left on it can be allowed to enter the air reactor without impacting the carbon capture efficiency. However, ash cannot be allowed to accumulate in the system. Ash can build up and could affect fluidization quality, or cause agglomerations to form, impacting the flow of solids around the unit. The composition of the ash could also affect the performance of the oxygen carrier, deactivating sites.

Ash could be separated aerodynamically from the oxygen carrier, similar to the carbon stripper, since ash is significantly lighter and smaller than the oxygen carrier. This process could be performed as another unit operation, or the primary carrier recirculation cyclone could be specifically designed to allow ash to leave the system but retain the oxygen carrier.

2.7 Pressurized Reactors

A number of chemical looping system studies have indicated that by operating at pressure, cycle and combustion efficiencies could be increased above atmospheric processes. By combining the chemical looping process with a modern combined cycle, cycle efficiencies could be drastically improved as compared to conventional steam cycles. Zerobin and Pröll suggest that for gaseous chemical looping combustion systems to be competitive with gas turbine combined cycles (GTCCs), chemical looping reactors need to be combined with gas turbines and heat recovery steam generators (HRSGs), requiring the process to operate upward of 2 MPa and 1200 °C [23].

Reaction performance, especially with solid fuels, could improve with pressure. Tests with a 50 kW_{th} system operating on coal showed that by increasing the operating pressure from 0.1 to 0.5 MPa, the carbon conversion of the coal was increased from 81% to 85% [24].

The challenges with operating a multiphase flow process at pressure include high-pressure solid fuel feeding and high-pressure carrier make-up feeding. If the exhaust is expanded through a gas turbine, removal of all particulates before the turbine as well as minimization of the pressure drop between the compressor and turbine become critical. However, there already exists industrial-scale experience with several demonstrations of pressurized fluid bed combustors (PFBCs), such as the Tidd demonstration project, which was a 70 MW unit that operated at 1.2 MPa [25].

2.8 Solid Circulation Rate

The circulation of solids between the two reactors moves the oxygen that is required from the air reactor to the fuel reactor to oxidize the fuel. The required circulation rate can be calculated based on the amount of oxygen required to oxidize the fuel (CH_4, coal, etc.), the fuel conversion, and the amount of oxygen available. Consider the following generalized fuel ($C_xH_yO_z$) reacting with a generic oxygen carrier (MeO_i):

$$C_xH_yO_z + a\mathrm{MeO}_i \leftrightarrow xCO_2 + \frac{y}{2}H_2O + a\mathrm{Me}_{i-1}$$

where a is the stoichiometric oxygen carrier to fuel ratio given by

$$a = 2x + \frac{1}{2}y - z$$

For H_2 a would be 1, for CO a would 1, and for CH_4 a would be 4. Coal could also be expressed such as a Pittsburgh #8 bituminous coal, $C_{0.55}H_{0.41}O_{0.04}$, ignoring other elements such as nitrogen and sulfur. The generic oxygen carrier, MeO_i, can represent various oxygen carriers by providing the correct i, which is the number of moles of oxygen per mole of metal. For example, CuO can be expressed by setting $i = 1$, whereas Fe_2O_3 can be expressed by setting $i = \frac{3}{2}$. The mass flow rate of required solids from the air reactor to the fuel reactor, assuming complete conversion of the fuel and the oxygen carrier leaving the air reactor is oxidized to MeO_i, can be calculated by

$$\dot{m}_{\mathrm{MeO}} = a\mathrm{MW}_{\mathrm{MeO}_i}\dot{N}_{C_xH_yO_z}$$

where $\mathrm{MW}_{\mathrm{MeO}_i}$ is the molecular weight of MeO_i and $\dot{N}_{C_xH_yO_z}$ is the molar flow rate of the fuel, $C_xH_yO_z$. Owing to inefficiencies of the reactions in both the fuel and air reactors, this calculated value would be the minimum circulation rate. Typical circulation rates will need to be higher than this. If inert materials are part of the oxygen carrier, such as an Al_2O_3 supported CuO, these inert materials need to be properly accounted for.

Oxidation reactions of the oxygen carrier are always exothermic; however, reactions in the fuel reactor could be endothermic. Copper oxygen carriers exhibit exothermic reactions with CH_4 in the fuel reactor whereas iron-based oxygen carriers have a strong endothermic reaction with CH_4 in the fuel reactor, Table 2.2. If the reactions in the fuel reactor are endothermic, then the circulation of solids also needs to move enough energy from the air reactor to the fuel reactor to sustain the reactions. This adds additional requirements that need to be satisfied by the circulation rate.

The constant pressure, adiabatic temperature change can be calculated by equating the enthalpy of the reactants to the enthalpy of the products and solving for the temperature of the products.

$$h_{\text{reac}}(T_i, P) = h_{\text{prod}}(T_f, P)$$

As an example, using the CH_4 and Fe_2O_3 reaction as shown in Table 2.2 where Fe_2O_3 has a temperature of 1000 °C leaving the air reactor and CH_4 has a temperature of 25 °C, the resulting constant pressure adiabatic temperature, T_f, would be 870 °C. This results in a temperature change of 130 °C just due to reactions. By doubling the amount of Fe_2O_3, which does not react and only serves the purpose of moving energy from the air reactor to the fuel reactor, this temperature drop can be reduced to 65 °C. The adiabatic constant pressure temperature change as a function of equivalence ratio is depicted in Figure 2.10.

In the case of Fe_2O_3 reduction with CH_4, the circulation of solids to satisfy the heat balance becomes more important than the transfer of oxygen. If the temperature in the fuel reactor cannot be maintained, then the reaction rates will slow down, possible leading to poor fuel conversions. If the oxygen carrier is not being reduced, then the heat released in the air reactor will decrease, leading to lower air reactor temperatures. The process could continue to lose temperature due to this feedback loop, extinguishing the reactions.

Measuring the circulation rate in real time during the operation of the unit presents a significant challenge. Knowing what the circulation rate is helps control the process and verify that enough solids are being moved to satisfy oxygen

Table 2.2 Global metal oxide reactions and combustion heat at standard conditions (298.15 K, 0.1 MPa).

		ΔH_c^0 (kJ mol^{-1})
CuO/Cu	$CH_4 + 4CuO \leftrightarrow 4Cu + CO_2 + 2H_2O$	−178.0
	$O_2 + 2Cu \leftrightarrow 2CuO$	−312.1
Fe_2O_3/Fe_3O_4	$CH_4 + 12Fe_2O_3 \leftrightarrow 8Fe_3O_4 + CO_2 + 2H_2O$	141.6
	$O_2 + 4Fe_3O_4 \leftrightarrow 6Fe_2O_3$	−471.6

Source: Abad et al. 2007 [38]. Reproduced with permission of Elsevier.

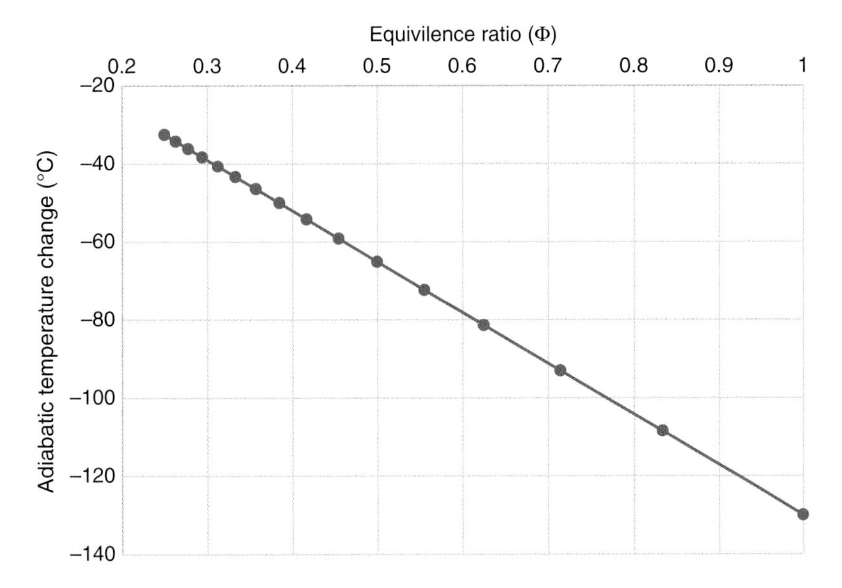

Figure 2.10 Constant pressure adiabatic temperature change from a stream of Fe_2O_3 at 1000 °C due to reduction reactions with CH_4, forming Fe_3O_4.

requirements in the fuel reactor and energy balance requirements if the oxygen carrier is endothermic. The easiest way to estimate the solids circulation rate is to look at the pressure drop across various portions of the loop that have high velocity and uniform flow of the particles such as the cross-over entering a cyclone [26]. Advanced instrumentation that can perform this measurement at chemical looping temperatures and conditions is desired. One promising concept is the use of microwaves to measure the velocity of solids and estimate the solids concentration, leading to a mass flow rate of solids [27].

2.9 Lessons Learned

As new units become operational and experience continues to grow, many lessons have been learned. These lessons are specific to these "small" bench-scale to pilot units and are not necessarily applicable to industrial-scale units. The key issues with the operation of these single and double loop chemical looping reactors are general gas/solid multiphase flow issues, not necessarily specific to chemical looping. Unless otherwise noted, these lessons learned have come from the author's experience operating the 50 kW_{th} unit at the National Energy Technology Laboratory in Morgantown, WV.

2.9.1 Solids and Pressure Balance Control

The biggest issue regarding the sustained operation of these units is the simultaneous control of solids flow around the loop and the pressure balance. The flow of solids between the air and fuel reactors controls the transfer of oxygen and in the case of endothermic carriers, such as iron, the transfer of energy to sustain the reactions in the fuel reactor. The flow of solids is also critical to maintaining appropriate solids heights in non-mechanical valves, which is required to establish pressure gradients to prevent gases from flowing in the wrong direction. The flow of solids becomes even more critical with double or more loop systems.

All chemical looping systems have at least two reactor outlets, one from the air reactor and one from the fuel reactor. This presents a significant challenge to maintaining and controlling the pressure balance. This pressure balance is critical, especially with small units. This control is primarily handled by adjusting the back pressure on the gas outlets and the heights of the stand pipes. Minor changes in the backpressure can have significant consequences on the flow of solids and gas around the loop. This is one of the reasons why it is imperative to remove all solids from the reactor exhausts as soon as possible. Any buildup of material causing restrictions will affect the ability to control the pressure balance.

One of the major operational challenges of Alstom's 3 MW_{th} calcium unit is the control of solids in the dip legs [14]. If sufficient inventory is not maintained in the dip legs (stand pipe), then the gas can flow in the wrong direction, up the dip leg, spoiling the vortex in the cyclone, resulting in significant solids loss. An analogous situation has been observed at the 1 MW_{th} Darmstadt unit. Significant particles losses through a high-efficiency cyclone were observed during shake down tests [28]. This issue was attributed to the backflow of gases through the loop seal, spoiling the cyclone vortex.

Having robust controls that can handle the simultaneous control of the solids flow rate, while maintaining solids heights, the pressure balance during steady-state operation, transitions between conditions, and upset events, is critical to successful long duration operation.

2.9.2 Solids in Reactor Exhaust

As soon as an oxygen carrier is introduced to the system, it will deposit everywhere in the system. The solids will also start coming out of all the reactor exhausts both during normal operation (attrition) and during any upset event. Exhaust lines need to be able to handle these solids.

To overcome this issue, secondary cyclones were installed on all three exhaust lines as close to the unit as possible. These cyclones are effective at removing most of the solids. Next, candle filters were installed downstream of the secondary cyclones for removing any further solids from the exhaust stream. Filters were installed in parallel so that filter banks could be taken off line and changed. These changes have significantly improved the operability of the unit.

2.9.3 Condensation in Exhaust

When water is produced from reactions and/or steam is introduced as a fluidizing gas, water will be present in the exhaust lines. It is imperative that all solids be removed from the exhaust gas before the temperature of the exhaust and exhaust piping can fall below a temperature at which the water vapor can condense. If liquid water does collect in the exhaust when solids are present, a sludge is formed, which can cause significant restrictions in the exhaust piping. This sludge is particularly hard to remove with simple gas purges and may require disassembly of the piping to clean, which typically is not an option while the unit is operating. These restrictions can lead to backpressure control issues, eventually leading to pressure imbalances and process upsets.

2.9.4 Self-Fluidization

One issue that could happen that is specific to chemical looping is the possibility for self-fluidization, especially with solid fuels. When solid fuel and carrier are mixed, reactions leading to the production of gaseous species can occur. If these reactions happen vigorously, enough gas could be produced to aerate or even fluidize solids without the need for additional fluidization gas. This presents serious problems in non-mechanical valves where the control of solids depends on the ability to control the amount of aeration. Losing control of a valve could cause solids control issues, ultimately leading to process instabilities and upsets.

Alstom experienced this phenomenon in their 3 MW_{th} unit and attributes it to major control issues of their fuel reactor seal pot control valve [14]. The seal pot control valve became ineffective at controlling the flow of solids, leading to the inability to control the height of solids in the reducer dip leg [29].

2.9.5 Cyclones

A single well design cyclone is not efficient enough for preventing significant inventory losses during normal operation of the process. At least two cyclones in series are required. This is evident with the following example.

Suppose a single loop chemical looping unit contains 100 kg of oxygen carrier inventory. The unit circulates the solids at a rate of 1000 $kg\,h^{-1}$. There is a single cyclone that removes solids from the air reactor exhaust to return the solids to the fuel reactor. The cyclone has been designed to be 99.9% efficient across the particle size distribution. Ignoring attrition, the rate of solids loss from that cyclone can be calculated by

$$1000 \left[\frac{\mathrm{kg}}{\mathrm{h}}\right] \times (1 - 0.999) = 1 \left[\frac{\mathrm{kg}}{\mathrm{h}}\right]$$

At a loss rate of 1 $kg\,h^{-1}$, the entire inventory of solids will have been lost in 100 hours (~4.2 days). This is an extremely high loss rate, especially considering that the carrier needs to stay in the unit for as long as possible. By adding another 99.9% efficient cyclone downstream of the primary cyclone, this loss rate is dramatically reduced to 0.001 $kg\,h^{-1}$, allowing operation of the unit for 100 000 hours (4166 days).

Several units have experienced operational issues and high solids loss rates, attributed to low cyclone efficiencies. The 150 kW_{th} unit at SINTEF has attributed their high particles losses to a poorly designed fuel reactor cyclone [9]. The 200 kW_{th} unit at the University of Utah has also struggled with cyclone performance [30].

2.10 Summary

The single loop and double (or more) loop chemical looping reactors take advantage of over 100 years' worth of multiphase flow reactor development. These chemical looping units, operating over a range of reactor types, process configurations, oxygen carriers, and fuels, are accumulating significant operational experience. This experience proves that the processes can be controlled and operated successfully, and achieve the goals of capturing carbon dioxide from the combustion of fossil fuels without the need of expensive post combustion capture or ASUs. There still are significant issues and risk with appropriately designing the process and controlling the process. However, combustion efficiencies above 98% and carbon capture efficiencies above 97% have been demonstrated, showing significant promise in the technology.

Acknowledgements

The author would like to acknowledge Thomas O'Brian, Joseph Mei, Askar Fahr, and the late Adel Sarofim for their historical knowledge and support in the development of chemical looping processes.

References

1 Gilliland, E.R. and Lewis, W. K. (1954). Production of pure carbon dioxide. US Patent 2,665,971.
2 Lyngfelt, A. and Thunman, H. (2005). Construction and 100 h of operational experience of a 10-kW chemical looping combustor. In: *The CO_2 Capture and Storage Project (CCP) for Carbon Dioxide Storage in Deep Geologic Formations for Climate Change Mitigation*, Capture and Separation of Carbon Dioxide From Combustion Sources, vol. 1, 625–646. London: Elsevier Science.
3 Adanez, J., Gayan, P., Celaya, J. et al. (2006). Chemical looping combustion in a 10 kW prototype using a CuO/Al_2O_3 oxygen carrier: effect of operating conditions on methane combustion. *Ind. Eng. Chem. Res.* 45: 6075–6080.
4 Contractor, R.M. (1999). Dupont's CFB technology for maleic anhydride. *Chem. Eng. Sci.* 54 (22): 5627–5632.
5 Adanez, J., Abad, A., Garcia-Labiano, F. et al. (2012). Progress in chemical-looping combustion and reforming technologies. *Prog. Energy Combust. Sci.* 38: 215–282.
6 Geldart, D. (1986). Introduction. In: *Gas Fluidization Technology* (ed. D. Geldart), 1–10. Wiley.
7 Nagarkatti, A. and Chatterjee, A. (1974). Pressure and flow characteristics of a gas phase spout-fluid bed and the minimum spout-fluid condition. *Can. J. Chem. Eng.* 52: 185–195.
8 Shen, L., Wu, J., and Xiao, J. (2009). Experiments on chemical looping combustion of coal with a NiO based oxygen carrier. *Combust. Flame* 156 (3): 721–728.

9 Langørgen, Ø., Saanum, I., and Haugen, N.E.L. (2017). Chemical looping combustion of methane using a copper-based. *Energy Procedia* 114: 352–360.
10 Guío-Pérez, D.C., Marx, K., Pröll, T., and Hofbauer, H. (2011). Fluid dynamic effects of ring-type internals in a dual circulating fluidized bed system. *Proceedings of the 10th International Conference on Circulating Fluidized Bed Technology (CFB10)*, Sunriver, OR, USA.
11 Guío-Pérez, D.C., Pröll, T., and Hofbauer, H. (2014). Influence of ring-type internals on the solids residence time distribution in the fuel reactor of a dual circulating fluidized bed system for chemical looping combustion. *Chem. Eng. Res. Des.* 92 (6): 1107–1118.
12 Linderholm, C. and Schmitz, M. (2016). Chemical-looping combustion of solid fuels in a 100 kW dual circulating fluidized bed system using iron ore as oxygen carrier. *Environ. Chem. Eng.* 4: 1029–1039.
13 Kolbitsch, P., Bolhàr-Nordenkampf, J., Pröll, T., and Hofbauer, H. (2009). Comparison of two Ni-based oxygen carriers for chemical looping combustion of natural gas in 140 kW continuous looping operation. *Ind. Eng. Chem. Res.* 48: 5542–5547.
14 Chamberland, R., Andrus, H., and Edberg, C. (2015). Alstom's chemical looping combustion technology with CO_2 capture for new and existing coal-fired power plants. U.S. DOE/NETL CO_2 Capture Technology Meeting, Pittsburgh, PA.
15 Linderholm, C., Schmitz, M., Diermann, M. et al. (2017). Chemical-looping combustion of solid fuel in a 100 kW unit using sintered manganese ore as oxygen carrier. *Int. J. Greenhouse Gas Control* 65: 170–181.
16 Thon, A., Kramp, M., Hartge, E.-U. et al. (2014). Operational experience with a system of coupled fluidized beds for chemical looping combustion of solid fuels using ilmenite as oxygen carrier. *Appl. Energy* 118: 309–317.
17 Coppola, A., Solimene, R., Bareschino, P., and Salatino, P. (2015). Mathematical modeling of a two-stage fuel reactor for chemical looping combustion with oxygen uncoupling of solid fuels. *Appl. Energy* 157: 449–461.
18 Gu, H., Shen, L., Zhang, S. et al. (2018). Enhanced fuel conversion by staging oxidization in a continuous chemical looping reactor based on iron ore oxygen carrier. *Chem. Eng. J.* 334: 829–836.
19 Pérez-Vega, R., Abad, A., García-Labiano, F. et al. (2016). Coal combustion in a 50 kW_{th} chemical looping combustion unit: seeking operating conditions to maximize CO_2 capture and combustion efficiency. *Int. J. Greenhouse Gas Control* 50: 80–92.
20 Monazam, E., Breault, R., Weber, J., and Layfield, K. (2017). Elutriation of fines from binary particle mixtures in bubbling fluidized bed cold model. *Powder Technol.* 305: 340–346.
21 Ströhle, J., Orth, M., and Epple, B. (2014). Design and operation of a 1 MW_{th} chemical looping plant. *Appl. Energy* 113: 1490–1495.
22 Markström, P., Linderholm, C., and Lyngfelt, A. (2013). Chemical-looping combustion of solid fuels – design and operation of a 100 kW unit with bituminous coal. *Int. J. Greenhouse Gas Control* 15: 150–162.
23 Zerobin, F. and Pröll, T. (2017). Potential and limitations of power generation via chemical looping combustion of gaseous fuels. *Int. J. Greenhouse Gas Control* 64: 174–182.

24 Xiao, R., Chen, L., Saha, C. et al. (2012). Pressurized chemical-looping combustion of coal using an iron ore as oxygen carrier in a pilot-scale unit. *Int. J. Greenhouse Gas Control* 10: 363–373.

25 U.S. Department of Energy National Energy Technology Laboratory (2001). *Tidd PFBC Demonstration Project*. Morgantown, WV: USDOE.

26 Bayham, S., Straub, D., and Weber, J. (2017). *Operation of the NETL Chemical Looping Reactor with Natural Gas and a Novel Copper–Iron Material*. Morgantown, WV: National Energy Technology Laboratory.

27 Chorpening, B.T., Spencer, M., Stehle, R.C. et al. (2016). Doppler sensing of unsteady dense particulate flows. 2016 IEEE Sensors, Orlando, FL, USA.

28 Abdulally, I., Beal, C., Andrus, H. et al. (2012). Alstom's chemical looping prototypes, program update. 37th International Technical Conference on Clean Coal & Fuel Systems, Clearwater, FL, USA.

29 Sloan, D.G., Andrus, H.E., and Chapman, P.J. (2014). Alstom limestone chemical looping system: experiments and isothermal simulation. 2014 Multiphase Flow Science Conference, Morgantown, WV.

30 Whitty, K. and Lighty, J. (2017). Integrated oxygen production and CO_2 separation through chemical looping combustion with oxygen uncoupling. 2017 NETL CO_2 Capture Technology Project Review Meeting, Pittsburgh, PA.

31 Berdugo Vilches, T., Lind, F., Ryden, M., and Thunman, H. (2016). Experience of more than 1000 h of operation with oxygen carriers and solid biomass at large scale. 4th International Conference on Chemical Looping, Nanjing, China.

32 Ohlemüller, P., Ströhle, J., and Epple, B. (2017). Chemical looping combustion of hard coal and torrefied biomass in a 1 MW_{th} pilot plant. *Int. J. Greenhouse Gas Control* 65: 149–159.

33 Whitty, K.J., Lighty, J.S., and Fry, A. (2016). Development and scale-up of copper-based chemical looping with oxygen uncoupling. 4th International Conference on Chemical Looping, Nanjing China.

34 Whitty, K.J., Hamilton, M.A., Merrett, K.M. et al. (2016). Copper-based chemical looping with oxygen uncoupling: process development, reactor scale-up, pilot-scale studies and system modeling. 33rd Annual International Pittsburgh Coal Conference, Cape Town, South Africa.

35 Baek, J.-I., Kim, U., Jo, H. et al. (2016). Chemical looping combustion development in Korea. 4th International Conference on Chemical Looping, Nanjing, China.

36 Kolbitsch, P., Bolhàr-Nordenkampf, J., Pröll, T., and Hofbauer, H. (2010). Operating experience with chemical looping combustion in a 120 kW dual circulating fluidized bed (DCFB) unit. *Int. J. Greenhouse Gas Control* 4: 180–185.

37 Ma, J., Zhao, H., Niu, P. et al. (2016). Design and operation of a 50 kW_{th} chemical looping combustion (CLC) reactor using coal as fuel. 4th International Conference on Chemical Looping, Nanjing, China.

38 Abad, A., Adánez, J., García-Labiano, F. et al. (2007). Mapping of the range of operational conditions for Cu-, Fe-, and Ni-based oxygen carriers in chemical-looping combustion. *Chem. Eng. Sci.* 62 (1–2): 533–549.

3

Chemical Looping Processes Using Packed Bed Reactors

Vincenzo Spallina[1], Fausto Gallucci[2], and Martin van Sint Annaland[2]

[1] *University of Manchester, Group of Catalysis and Porous Materials, School of Chemical Engineering and Analytical Science, Manchester, England, M1 3AL, United Kingdom*
[2] *Eindhoven University of Technology, Chemical Process Intensification Group, Department of Chemical Engineering and Chemistry, Eindhoven, 5612 AZ, The Netherlands*

3.1 Introduction

Dynamically operated packed bed reactors (PBRs) are being increasingly studied for chemical looping applications [1]. Compared to the most conventional interconnected fluidized bed reactors (FBRs) where the solids are circulating between the fuel and air reactors, in PBRs the solids remain stationary in the bed and the gases are alternatively switched between reduction and oxidation stages. Therefore, no solids circulation is required and operation at pressurized conditions can be more easily accomplished. On the other hand, depending on the final application, when using packed beds some auxiliary components, such as valves, have to withstand high pressure and temperature (>20 bar; >1200 °C), which will increase the costs of the system. In PBRs relatively low gas velocities should be used to avoid excessive pressure drops; however, PBRs can be operated with large particle sizes. PBRs for chemical looping combustion (CLC) consist of at least two (or three) steps in which the oxygen carrier (OC) is first oxidized with air (oxidation), after which it is reduced with a fuel (reduction). A heat removal stage can follow either the oxidation stage or the reduction stage (depending on the OC used); see Figure 3.1.

In a PBR for CLC, two different front velocities are generated along the bed: (i) the reaction front, which proceeds very fast along the reactor determining the gas–solid conversion and the heat generation stored in the bed and (ii) a heat front, which is significantly slower, where only heat transfer occurs and at which the solid material is cooled down to the inlet temperature of the gas.

A schematic representation of the evolution of the reaction/heat fronts in the axial profiles is shown in Figure 3.2 for the oxidation reaction assuming an infinite reaction rate, neglecting the gas phase heat capacity, the radial dispersion as well as the mass and heat axial dispersion (referred to as the "sharp front approach"). At t_0 the gas starts to react with the particle (exothermic reaction) and the heat produced is stored in the material (t_1); once the reaction has reached the end

Handbook of Chemical Looping Technology, First Edition. Edited by Ronald W. Breault.

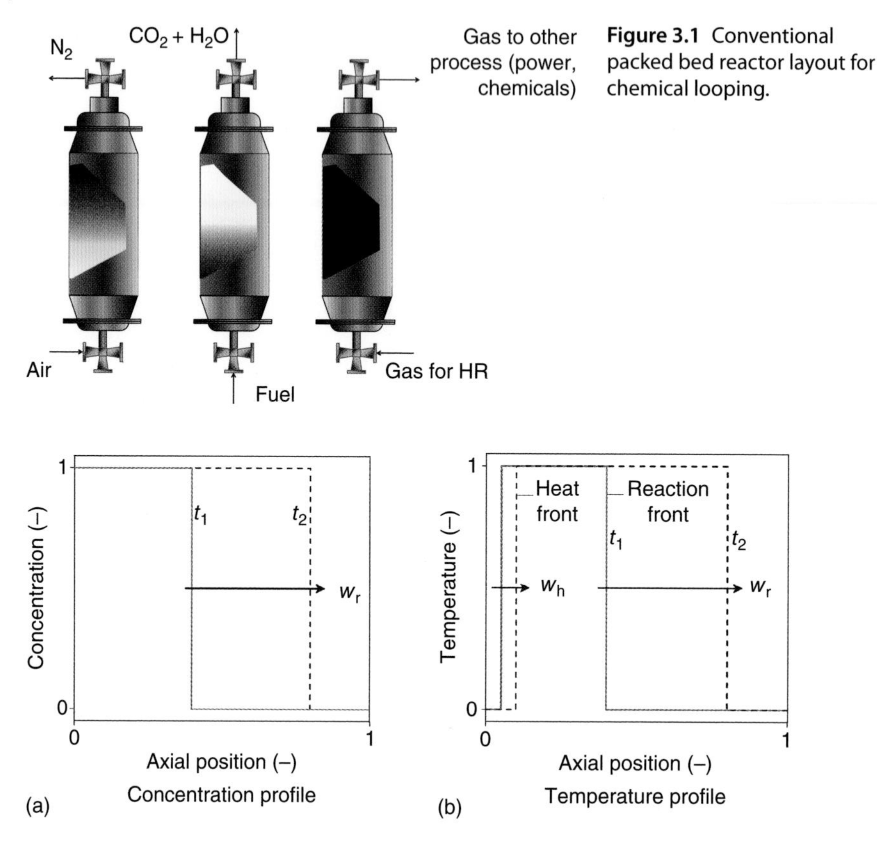

Figure 3.1 Conventional packed bed reactor layout for chemical looping.

Figure 3.2 Schematic representation of the evolution of the (dimensionless) axial profile of (a) the gaseous reactant concentration and (b) the temperature. Source: Noorman et al. 2007 [1]. Copyright 2007. Reprinted with permission from American Chemical Society.

of the reactor, the heat front has moved over only a short distance from the beginning of the bed (t_2). Because the reaction front moves much faster than the heat front, the PBR can be used as heat storage, from which it is possible to produce a high-temperature gas stream (for power production), or it can be used to drive endothermic reactions. After the solid material is completely oxidized (and the heat removed), the reactor is switched to the reduction phase, where the solid is reduced by converting a fuel, particularly natural gas or syngas, into H_2O and CO_2 (for the case of CLC). A purge cycle is required to prevent the fuel coming in direct contact with the air.

Considering the heat produced by the oxidation reaction (using air), the total amount of heat accumulated can be derived from Equation (3.1), where w_h and w_r are the heat and reaction front velocities respectively, as calculated from Equations (3.2) and (3.3). With these equations it is possible to derive an explicit equation for the maximum temperature rise in the bed, reported in Equation (3.4). Remarkably, the maximum solid temperature rise (ΔT_{MAX}) does not depend on gas flow rate and it is only influenced by the properties of the gas phase (c_{p,O_2}, MW_{O_2}, y_{O_2}) and the solid material properties ($c_{p,s}$, $y_{OC,act}$, $MW_{OC,act}$). This relation is valid as long as the reaction front velocity is faster than the heat front velocity and the maximum temperature depends on the initial solid temperature. Fernández et al. [2] have demonstrated that in the presence of a large gas flow rate (highly diluted), the heat front can become faster than the reaction front and in this case the initial solid temperature (T_0) does not affect the maximum temperature of the reactor.

$$\frac{\rho_g v_g y_{O_2}}{MW_{O_2}}(-\Delta H_{R,ox}) = \varepsilon_s \rho_s c_{p,s}(w_h - w_r)(T_{MAX} - T_0) \tag{3.1}$$

$$w_h = \frac{\rho_g v_g c_{p,g}}{\varepsilon_s \rho_s c_{p,s}} \tag{3.2}$$

$$w_r = \frac{\rho_g v_g y_{O_2} MW_{OC,act}}{\varepsilon_s \rho_s y_{OC,act} \xi MW_{O_2}} \tag{3.3}$$

$$T_{MAX} - T_0 = \Delta T_{MAX} = \frac{(-\Delta H_{R,ox})}{\frac{c_{p,s} MW_{OC,act}}{y_{OC,act}\xi} - \frac{c_{p,g} MW_{O_2}}{y_{O_2}}} \tag{3.4}$$

Hamers et al. [3] have quantified the maximum temperature rise for different OCs as a function of the active weight contents in order to determine the appropriate operating conditions for the reactor design and operation. As can be discerned from Figure 3.3a for the case of oxidation with air, Ni- and Cu-based OC can reach very high temperature rises (>700 °C) with less than 25% of active weight content in the particles. In the case of reduction (Figure 3.3b), the temperature rise depends on the fuel composition; however, only the Cu-based OC shows a remarkable temperature rise.

PBRs have been proposed for different processes and applications. In this work, the latest progress on PBR for chemical looping processes will be discussed. In the first part, the different OCs that have been proposed, including the main properties of the solid materials and the kinetic modeling, will be discussed; in the second part, the processes proposed will be discussed and the most up-to-date results will be presented: the main processes are related to CLC, chemical looping reforming (CLR), chemical looping for H_2 production (CL-H_2), and the sorption-enhanced reforming assisted with chemical looping combustion (SER-CLC).

3.2 Oxygen Carriers for Packed Bed Reactor

For a PBR, the ideal oxygen carrier material has a high activity at low temperatures and a high selectivity, and is resistant to poisoning. In addition, the structural integrity of the OC is important for multi-cycle operation to maintain the performance with sufficient lifetime. In general, PBRs are operated with bigger particles (to minimize the pressure drop), which could lead to intra-particle diffusion limitations [4].

Although most of the research in material development has been focused on OCs to withstand fluidization conditions, operated with either gas or solid fuels [5], in the last years different OC formulations have been studied and modeled for specific application in PBRs. Results from natural ilmenite ($FeTiO_3$) was studied in Schwebel et al. [6] showing a large solid conversion (40%) with high CO_2 yields

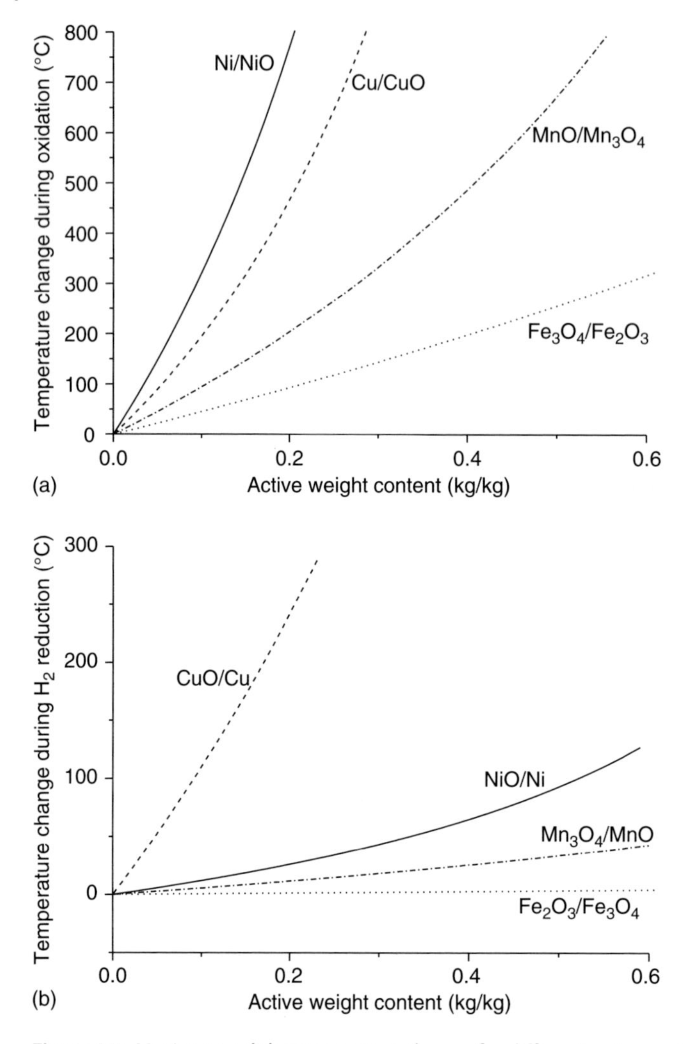

Figure 3.3 Maximum solids temperature change for different active weight contents for different oxygen carrier pairs during oxidation with air (a) and reduction with hydrogen (b). Source: Hamers et al. 2013 [3]. Copyright 2013. Reproduced with permission from Elsevier.

for coarse particles (particle diameter d_p, 1–1.4 mm) in a PBR, and no effects on the OC conversion in the presence of different gases (CO, H_2, CH_4) were found. However, after 52 cycles, 1/3 of the material in the bed was sintered, mostly because of the high-temperature operation. Jacobs et al. [7] have performed an extensive thermal and mechanical analysis of synthetic ilmenite extrudates. The authors have tested five different compositions based on ilmenite and an additive (titanium oxide, manganese oxide, nano-sized titanium oxide). It was found that the presence of MnO as dopant (between 10 and 15 wt% basis) resulted in the formation of an iron–manganese mixed oxide upon sintering, and in terms of mechanical and thermal strengths the performance was well above the critical value (2 daN mm^{-1}). The kinetics of the same material has been studied by

Ortiz et al. [8] using syngas during the reduction with thermogravimetric analysis (TGA) and a differential reactor to study the effect of catalytic reactions, particularly water gas shift (WGS). The reactive experiments were performed in the temperature range of 600–1100 °C using 5–50 vol.% of reactants. The activated ilmenite pellets showed high reactivity with H_2 and O_2, but slow reactivity with CO. The shrinking core model, with mixed control of chemical reaction and diffusion through the product layer, was used to describe the ilmenite redox kinetics. Ilmenite did exhibit a significant catalytic activity for the WGS reaction, but lower than that for conventional iron-based high-temperature WGS catalysts, as also discussed in Schwebel et al. [9].

A detailed study on the reactivity of Ni-based OCs (Ni/Al_2O_3 and $Ni/Al_2O_3{-}SiO_2$) measured by TGA coupled with detailed heterogeneous particle modeling to evaluate the effects of physical and chemical properties on particle reactivity was presented by Han et al. [10]. In their study, the internal mass transfer limitations were found to be significant for $d_p > 1$ mm. The experiments and modeling proposed have shown also reactivity changes due to differences in spinel fraction and sintering effects, proving that the model can be used to track the performance of OCs throughout their use. Recently, the authors have carried out a kinetics model review using a structural identifiability analysis to determine whether it is theoretically possible to uniquely estimate the kinetic parameters of the candidate reaction models [11]. Ni supported with bentonite for packed bed applications has been considered in the study of Ryu et al. [12]. In the presence of CH_4, carbon deposition was found and the OC particle tested showed no agglomeration or breakage up to 900 °C, but at 1000 °C sintering took place and particle lumps were formed. Ni/Al_2O_3 and CuO/Al_2O_3 were tested and modeled in Hamers et al. [13] for 1.1 mm particle to be used in CLC with PBRs at different operating pressures (up to 20 bar in high pressure thermogravimetric analysis (HP-TGA)) showing a decrease in the kinetics at higher pressures (but at constant partial pressure), which the authors tentatively attributed to a reduced number of oxygen vacancies in the material. The influence of pressure has been investigated for CuO/NiO-based OCs in Nordness et al. [14] using a fixed bed reactor (2.2 g of solid inventory), in which a large carbon deposition was found when working at higher pressures. An experimental investigation of CuO supported on Al_2O_3 for PBRs has been discussed in San Pio et al. [15]: in their work, very large morphological changes were observed, which became more and more pronounced for particles having experienced a larger number of redox cycles, such as nonhomogeneous Cu distribution and no Cu migration across the particle. Moreover, it was found that the formation of $CuAl_2O_4$ spinel increases after repeated cycles, decreasing the reaction rate without altering the final conversion.

3.3 Chemical Looping Combustion

Both the CLC process and its integration in power plants have been presented in the last decade by several authors from an experimental and numerical point of view. Most of these studies have focused on different options for the heat management of the reactors in order to improve the performance at the process and plant levels.

At the process level, the research is mostly focused on determining optimal operating conditions to reach a high CO_2 yield with limited fuel slip, a high OC conversion, and stable material performance after repeated cycles, while using different gases as feedstock. At the plant level, the main analysis is related to the optimal integration of the chemical looping process within the power production unit. This includes the assessment of the overall energy efficiency, CO_2 capture rate, and dynamic operation, and their effect on the final efficiency penalty as well as a comparison with other CO_2 capture technologies already proposed in the literature, including chemical looping with FBRs.

In the case of CLC, two main reactor heat management strategies have been identified. In the first case, also called HR-MeO (Figure 3.4a), after the oxidation with air, the heat stored inside the reactor is removed by feeding additional air, which is later fed to the power cycle and therefore the heat removal phase occurs when the OC is oxidized; after that, the bed is reduced again by feeding the fuel. This strategy is suitable when the OC shows a high reactivity also at low temperatures, e.g. 600–800 °C, as in the case of Ni- and Cu-based OCs. In the second case (Figure 3.4b, HR-Me), the heat removal is carried out after the reduction by feeding an inert gas that does not react with the reduced OC such as N_2. This heat management strategy has been proposed for less reactive OCs, as in the case of $FeTiO_3$ (ilmenite).

For each of the heat management strategies, different variations have been proposed with the aim of reducing the fuel slip, to produce a large stream of high-temperature gas for the gas turbine at a temperature of about 1200 °C, to minimize the cost of the OCs, and to reduce the variability of the gas conditions for proper integration with other plant components.

The HR-MeO strategy has been originally presented by Noorman et al. [4]. Hamers et al. [16] have discussed heat management using Cu-based material (12.5 wt% on Al_2O_3) in the presence of syngas (H_2/CO). CuO reacts with syngas and a complete fuel conversion is achieved. Owing to the relative low melting temperature of Cu (around 1035 °C) the maximum temperature of the bed has to be kept lower to avoid excessive overheating and damaging of the reactor/OC;

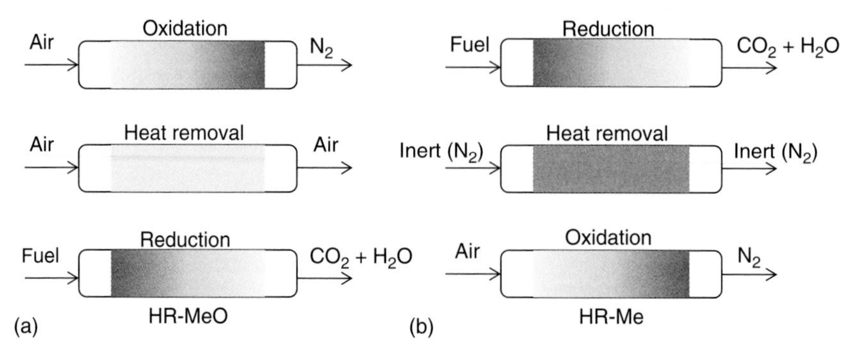

Figure 3.4 Reactor heat management strategies.

therefore, the maximum air temperature is taken equal to 850 °C. The reactor modeling and heat management have been discussed also in presence of NiO/Al_2O_3 OC in Hamers et al. [17]. For the latter case, the number of reactors required for the complete process assessment have been quantified as a function of the reactor design variables; about 20 rectors ($L/D = 2$ and 8 m length) are required, resulting in about 1200 tons of OC (in which 180 ton of NiO is needed) for an 854 MW_{LHV} integrated gasification combined cycle (IGCC) power plant. The same heat management strategy has been discussed also in Fernández et al. [18] using CH_4 as fuel. In their study, the authors have designed the complete process for 500 MW_{LHV} where at least five reactors operated in parallel with 6.7 m internal diameter and 10 m length are required. This configuration has been also studied with counter-current feeding [19] to reduce the risk of carbon deposition; however, this option seems infeasible if considered in the complete process, because there is an accumulation of heat in the central part of the reactor, which would overheat the materials as presented in Spallina et al. [20]. Carbon deposition on Ni-based OCs has been investigated by Diglio et al. [21]. In order to improve the performance of the entire process, a two-stage CLC heat management strategy has been proposed by Hamers et al. [3], in which the cycle is carried out using two reactors in series using two different OCs in these reactors. The first reactor is operated with CuO (12.5 wt% on Al_2O_3) in a temperature range between 450 and 850 °C, while the second reactor is operated with Mn_3O_4 (30 wt% on Al_2O_3) in the range of 850–1200 °C. The minimum temperature in the second reactor is obtained by using hot air as feed gas at 850 °C produced during the heat removal stage of the first bed (Figure 3.5). Owing to the different heat management strategies, the high-temperature

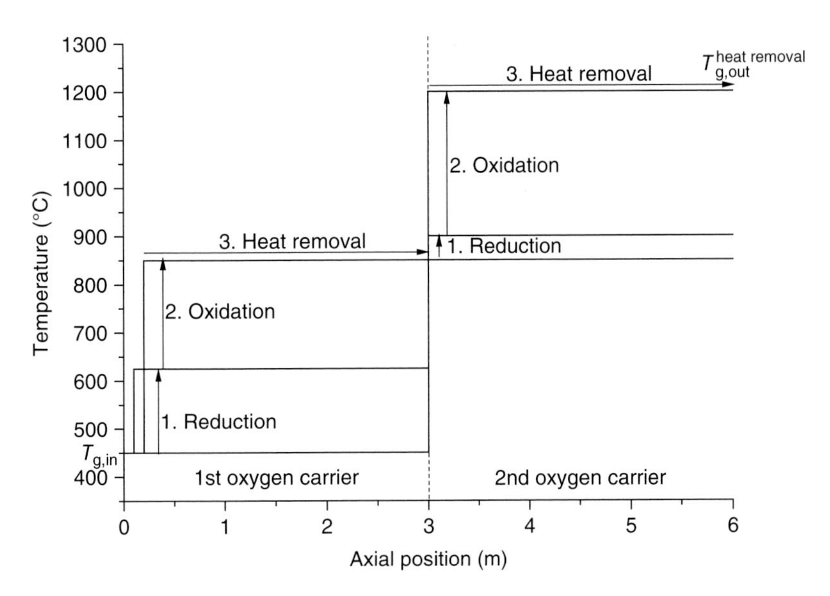

Figure 3.5 Schematic axial temperature profile along the PBR in the two-stage packed-bed CLC concept. Source: Hamers et al. 2015 [22]. Copyright 2015. Reproduced with permission from Elsevier.

pressurized air is produced at 1200 °C during the heat removal, which is suitable for an efficient combined cycle. This strategy has been also experimentally tested and validated in a 20 kW_{th} PBR operated up to 7 bar and 1100 °C [23] using a combination of Cu-based and Ni-based OCs.

Spallina et al. [20] have carried out a comparison of different heat management strategies to operate CLC with ilmenite at 20 bar using coal-derived syngas based on the HR-Me configuration. As previously anticipated, the heat removal phase is carried out after the reduction and the resulting gases at the outlet of the reduction and heat removal are above 1200 °C, while it decreases from 900 to 450 °C during the oxidation (see Figure 3.6a). In this way, the reduction reaction occurs when the bed is at its maximum temperature after the oxidation. This strategy is possible for power production only using CO-rich syngas because the maximum solid temperature does not change significantly during the reduction (about 10 °C or lower), as shown in Figure 3.6b. The heat removal stage cannot be carried

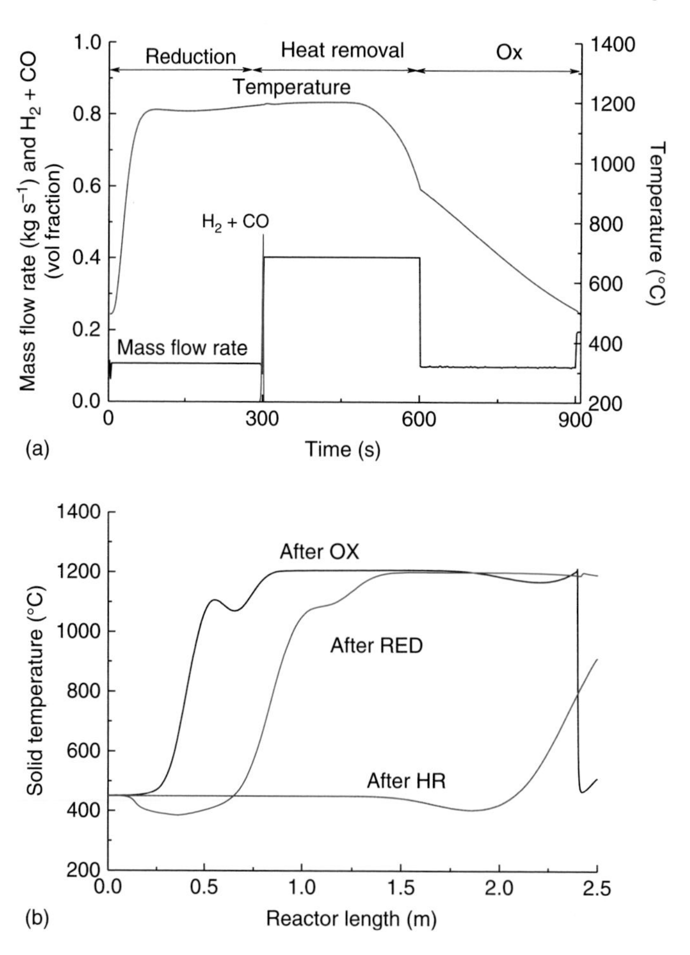

Figure 3.6 (a) Gas temperature and flow rate at the reactor outlet and (b) axial solid temperature profiles after each step. Source: Spallina et al. 2013 [20]. Copyright 2013. Reproduced with permission from Elsevier.

out with air because the solid is completely reduced, so an inert gas (i.e. N_2) has to be used. The same strategy has also been considered for the case where the oxidation is carried out with counter-current feeding, which has the advantage of reducing the temperature of the CO_2–H_2O stream at the reactor outlet and increases the amount of N_2 for the heat removal with about 30%. Gallucci et al. [24] have demonstrated the feasibility of using ilmenite for packed bed reactor CLC and validated the reactor model, which has subsequently been used for the design and simulation of the complete process.

This heat management strategy has also been studied to determine the maximum number of reactors to be used for an 850 MW_{LHV} plant [25]. One important assumption for this system was based on a maximum pressure drop of 8% (with $d_p = 5$ mm) to limit the cost of compression. A low effectiveness factor has been calculated for the large particle diameter, as also discussed in Han et al. [26]. It was found that 14–16 reactor units (specifically three in oxidation, three in reduction, the remaining in heat removal) are required with an internal diameter of 5.5 m and a reactor length of 11 m, as shown in Figure 3.7.

Since the outlet conditions of the system are varying due to the dynamic operation (Figure 3.8a), an operation strategy has been proposed to reduce the fluctuations: the reactors that are in a certain phase are operated with a phase displacement (as shown in Figure 3.7), so that the outlet stream is not the same in each reactor at the same time, with which it is possible to average the temperature and composition (Figure 3.8b).

The complete energy analysis of chemical looping combustion processes with packed bed reactors has been carried out mostly for IGCC [17, 22, 27, 28], where different heat management strategies were considered. In these works, an optimization of the operating conditions of the complete plant has been carried out with respect to gasification conditions, operating temperature and pressure of the gas turbine, maximum temperature and pressure levels of the steam cycle, and desulfurization processes. The results and a comparison of the different plant integration options have been listed in Table 3.1. The comparison is carried out with respect to the benchmark technologies based on a state-of-the-art IGCC

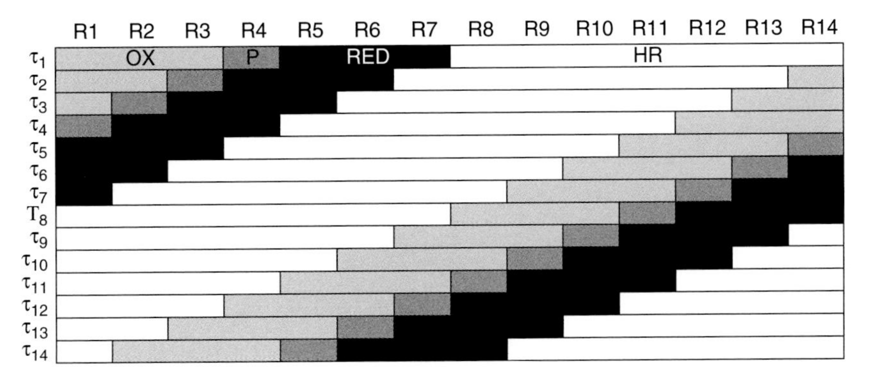

Figure 3.7 Sequence of operation of the 14 reactors (R1–R14) during the cycle time in co-current configuration. OX, oxidation; P, purge; RED, reduction; and HR, heat removal. Source: Spallina et al. 2015 [25]. Copyright 2015. Reproduced with permission from Elsevier.

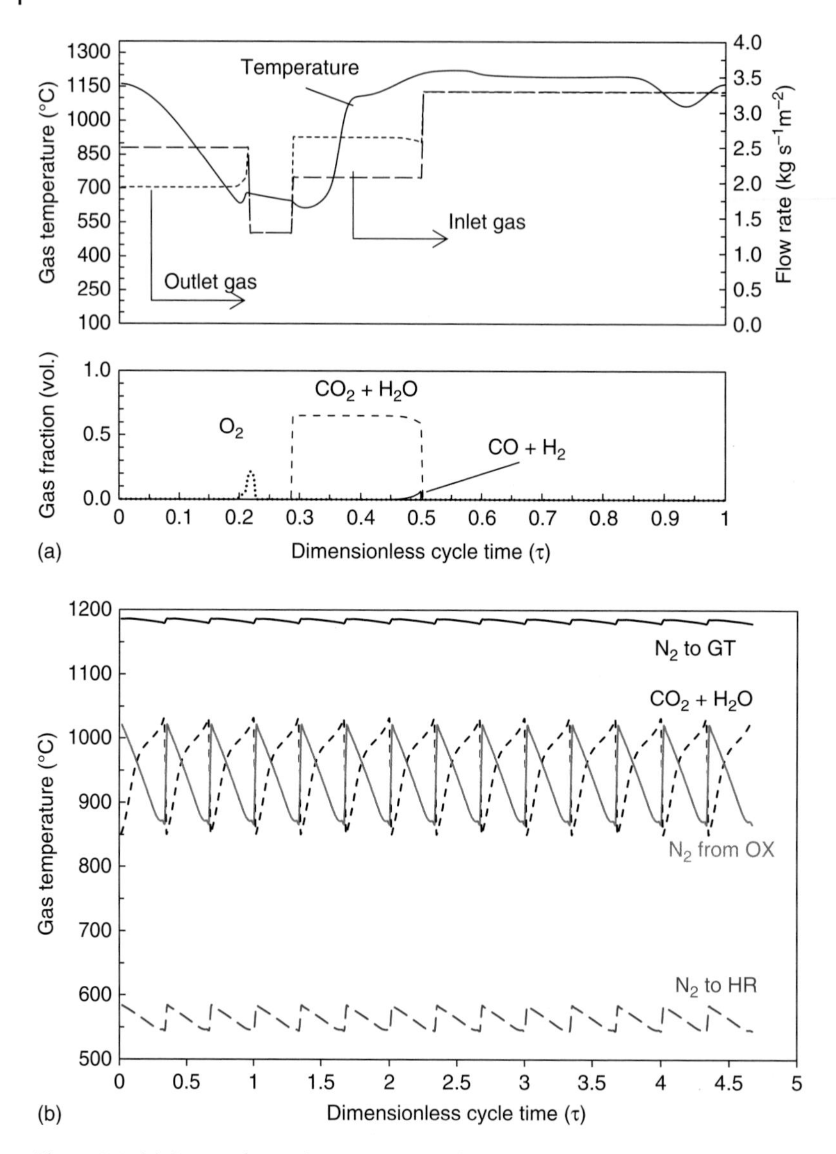

Figure 3.8 (a) Gas conditions (temperature and composition) at the reactor outlet and (b) resulting gas condition after the overall mixing with all the streams from other reactors operated in the same step. Source: Spallina et al. 2015 [25]. Copyright 2015. Reproduced with permission from Elsevier.

with and without CO_2 capture, as calculated in Spallina and coworkers [27]. The coal gasification is based on an entrained flow, oxygen blown, and a dry feed Shell-type gasifier, where coal feeding is carried out using CO_2 lock-hoppers to avoid excessive syngas dilution with N_2 (as in the conventional IGCC). Syngas cooling is used to produce high-pressure steam fully integrated with the power island. Sulfur compounds (mostly H_2S) are separated through a Selexol® process. Power production is based on a combined cycle: the gas turbine is calculated to reproduce the performance of an advanced state-of-the-art industrial heavy-duty machine, currently used in large-scale natural gas-fired combined cycle power plants [29]. The heat recovery steam generator (HRSG) is based on three pressure levels in which the maximum pressure of the steam is 160 bar and the maximum temperature at the turbine inlet is 565 °C. All these assumptions are based on the EBTF report [30]. In the case of IGCC with CO_2 capture, two desorption columns are used to obtain a selective separation of H_2S and CO_2 (up to 95%) and the regeneration is carried out also using low-pressure steam for stripping. The main performance parameters used for the comparison are electric efficiency with respect to the thermal input (lower heating value (LHV) basis), CO_2 capture rate with respect to the carbon at the inlet of the plant, and CO_2 avoidance with respect to the benchmark technology in which CO_2 capture is not considered. Another important parameter is the specific primary energy consumption for CO_2 avoidance (SPECCA), which identifies the energy cost (in terms of primary energy) of the complete plant to capture 1 kg of CO_2 and indicates the effectiveness of the CO_2 capture technology with respect to other technologies. It is calculated as follows:

$$\text{SPECCA}\left[\frac{\text{MJ}_{\text{LHV}}}{\text{kg}_{\text{CO}_2}}\right] = 3600 \cdot \frac{\left(\frac{1}{\eta} - \frac{1}{\eta_{\text{ref}}}\right)}{(e_{\text{CO}_2,\text{ref}} - e_{\text{CO}_2})} \tag{3.5}$$

where the η and η_{ref} represent the electrical efficiencies of the plant with CO_2 capture and the reference plant without CO_2 capture (IGCC), while the $e_{CO_2,ref}$ and e_{CO_2} are the specific CO_2 emissions from the reference plant without CO_2 capture and the plant considered with CO_2 capture ($kg_{CO_2}/MWh_e - 1$). Apart from IGCCs, the comparison includes also the integration of pressurized dual FBRs for CLC, which is also calculated with the same set of assumptions.

As can be seen from Table 3.1, CO_2 capture using CO_2 absorption (pre-combustion capture) implies a CO_2 avoidance >80% with an efficiency penalty of around 10% points with respect to IGCC without CO_2 capture (first column in Table 3.1). This results in a SPECCA of 3.34 MJ $kg_{CO_2}^{-1}$. The integration of CLC increases the CO_2 avoidance >95% with a minimum reduction in efficiency of 4–5% points, and therefore the SPECCA ranges from 0.8–1.3 MJ $kg_{CO_2}^{-1}$. This is made possible by lower costs for separation and CO_2 compression. On the other hand, the electricity production from the gas turbine decreases because of the lower turbine inlet temperature and lower flow rate during gas expansion. CLC using PBRs shows a similar performance compared to pressurized FBRs in which the main difference is in the gross power production share (gas turbine versus steam turbine). This is explained by the different heat management of the two systems. Some of the heat management strategies allow reducing the temperature of the CO_2, which results in an increased amount of gas to be used for the heat removal and consequently for the gas turbine. In the case of the FBR the temperature difference between air and fuel reactors (assumed always <200 °C) is dictated by the solids circulation rate. Depending on the selected OC, the HR-MeO heat management strategy can be used, where both options, viz the single-stage and two-stage heat management configurations, have been

Table 3.1 Summary of the energy balance and plant performance for different chemical looping combustion plants.

	IGCC [28]	IGCC [28]	FBR [17]	CLC1 [17]	CLC2 [22]	CLC3 [28]	CLC4 [28]	CLC5 [27]	CLC6 [27]
Oxygen carriers	N/A	Selexol	NiO	NiO	CuO Mn_3O_4	$FeTiO_3$	$FeTiO_3$	Different	Different
Heat management	N/A	N/A	Fluid-bed	HR-MeO	HR-MeO	HR-Me	HR-Me	HGD	FGD
Thermal input LHV (MW_{LHV})	812.5	898.8	853.9	853.9	853.9	853.9	853.9	853.9	853.9
Gas turbine (MW_e)	261.6	263.9	192.1	225.1	194.0	175.1	215.3	225.2	226.8
Steam turbine (MW_e)	179.5	161.2	224	186.8	212.3	239.8	197.6	192.8	192.7
Gross power output (MW_e)	441.1	425.1	416.1	411.9	406.3	414.9	412.9	418	419.5
Syngas blower (MW_e)	−1	−1.1	−0.8	−0.8	−0.8	−1.0	−1.0	−1.5	−1.5
Steam cycle pumps (MW_e)	−2.9	−3.49	−4.0	−3.8	−3.8	−4.9	−3.7	−3.5	−3.4
N_2 compressor (MW_e)	−34.2	−29.8	—	—	—	—	—	—	—
ASU (MW_e)	−29.6	−32.7	−35.1	−33.9	−33.9	−33.8	−33.8	−33.9	−33.9
LH CO_2 compressor (MW_e)	—	—	−3.1	−3.1	−3.1	−3.1	−2.8	−3.1	−3.1
Acid gas removal (MW_e)	−0.4	−14.7	−0.4	−0.4	−0.4	−0.4	−0.4	−1.9	−1.9
CO_2 compressor (MW_e)	—	−19.7	−11	−11	−11	−11.0	−14.3	−11.3	−11
N_2 compressor gasifier (MW_e)	—	—	−1.3	−1.3	−1.3	−1.4	−1.4	−1.5	−1.5
Heat rejection (MW_e)	−5.5	−6.3	−3.7	−3.6	−4.1	−4.2	−4.2	−3.7	−3.7
Other auxiliaries, BOP, (MW_e)	−3.2	−3.6	−3.4	−3.4	−3.4	−4.6	−4.6	−3.4	−3.4
Net power generated (MW_e)	367.3	317.3	353.3	350.6	344.5	350.5	346.7	354.3	356.1
Electrical efficiency (%)	45.20	35.30	41.37	41.05	40.34	41.05	40.60	41.50	41.71
CO_2 capture efficiency (%)	—	93	97.1	97.1	97.1	96.1	96.1	97	97
CO_2 purity (%)	—	98.2	96.6	96.7	96.7	96.5	96.5	96.7	96.7
CO_2 emission (kg_{CO_2}/MWh_e)	769.8	101.4	24.5	24.7	25.1	33.4	33.8	25.5	25.2
CO_2 avoided (%)	—	84.7	96.8	96.8	96.7	95.1	95.0	96.7	96.7
SPECCA ($MJ_{LHV}\ kg_{CO_2}^{-1}$)	—	3.34	0.89	0.98	1.29	1.1	1.2	0.86	0.8

evaluated: the use of CuO-based OC requires a second stage to increase the temperature of the bed up to 1200 °C without damaging the solid material. In the case of the NiO-based reactor, the overall cost of the OC becomes dominant in the economics of the plant. In the case of an OC with a low reactivity (e.g. $FeTiO_3$), the HR-Me heat management needs to be implemented without relevant changes in the final performance, but with increased power cycle complexity due to the presence of a semi-closed gas combined cycle layout. The performance of the system has been investigated with two alternative sulfur removal treatments: in the first case, hot gas desulfurization (600–800 °C) has been used upstream the CLC reactors, while in the second case the analysis has been carried out assuming sulfur-tolerant OCs and using a SO_2 removal downstream from the CO_2-rich gases of the CLC reactors. In terms of net electrical efficiency, the performance improves by 1% point, regardless of whether the high-pressure FBR or PBR is used.

The integration of PBRs for CLC using natural gas fueled power plants has been discussed in Fernández et al. [18] using ilmenite and in Chen et al. [31] using the 30% (wt% basis) Ni-based OC. Plant performances are in the range of 48.2% with 100% of CO_2 capture. The authors have presented dynamic simulations of the integrated plant showing that the natural gas combined cycle (NGCC) performance is only slightly affected by the batch nature of the operation with fixed bed CLC reactors.

3.4 Chemical Looping Reforming

One of the main disadvantages of CLC is represented by the need to produce high-temperature gas for the gas turbine and the large number of reactors required to carry out heat removal without excessive pressure drops. The large number of reactors is also dictated by the overall heat balance: from an energy point of view, CLC releases the overall heat of reaction, which needs to be removed by heating the feed gas from about 450 up to 1200 °C.

An interesting alternative is represented by the CLR system. In this case, the plant is designed to produce reformed syngas. In the CLR process, the heat removal occurs by combining exothermic reactions (CLC) with endothermic reactions (hydrocarbons reforming).

To achieve the reforming, the OC has also catalytic activity for steam reforming reactions. Diglio et al. [32] developed a 1D reactor model to simulate multiple cycles of a 22 cm long reactor with 6 cm diameter and presented the results of the cyclic process, where the reduction gas is first completely oxidized (gas–solid reactions) and thereafter the reformed syngas is produced. Recently, Spallina et al. [33] have presented an experimental and numerical analysis of a possible route for the CLR process integrated in chemical plants. In our work, the CLR process with PBRs has been investigated using a commercial Ni-based catalyst at atmospheric pressure. A schematic diagram of the CLR process with PBRs is presented in Figure 3.9.

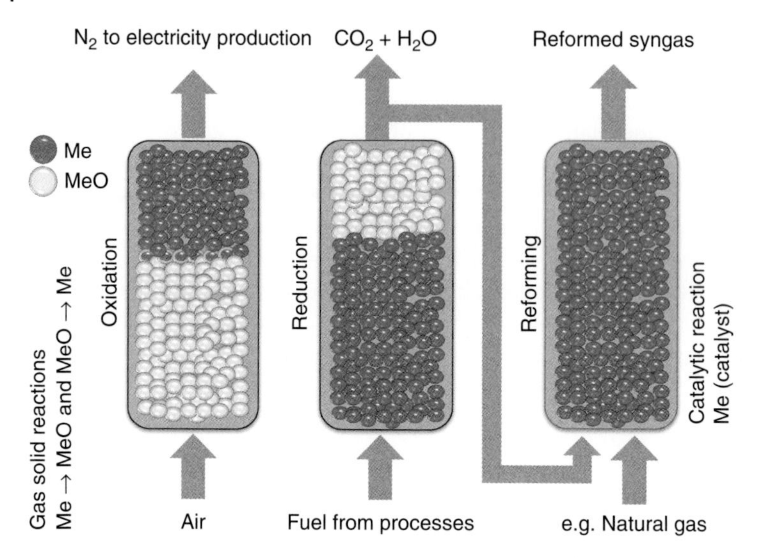

Figure 3.9 Schematic process of chemical looping reforming with PBRs. Source: Spallina et al. 2017 [33]. Copyright from Creative Commons Attribution-NonCommercial-No Derivatives License (CC BY NC ND). Reproduced with permission from Elsevier.

In the case of Ni-based OC, the reactor operated in *oxidation* converts the Ni into NiO by reaction with air while releasing N_2. The reactor operated in *reduction* converts the OC back to Ni using low-grade fuel available in the plant from downstream processes producing CO_2 and H_2O. Finally, a third reactor operated in *reforming*, where the OC is in Ni form, is fed with natural gas together with recirculated CO_2/H_2O from the reduction phase and additional steam to convert CH_4 into reformed syngas. Some additional H_2O is required to increase the CH_4 conversion; however, part of the reforming reactants are taken from the reduction outlet. In the case where only H_2O/CO_2 is used, the reformate syngas is richer in CO.

With respect to the possible integration of CLR with PBRs in chemical plants, a simplified plant flowsheet is depicted in Figure 3.10, illustrating the proposed process integration. Four different applications are considered. In the case of pure H_2 production, after the reforming the resulting syngas is cooled down to the desired temperature and fed to downstream plant units. One or two conventional WGS stages are required, depending on the CO content after the reforming. Complete CH_4 and CO conversion to H_2 is not strictly required because the pressure swing adsorption (PSA) off-gas (and some additional fuel) is required for the reduction (a similar concept has been proposed by Ryden and Lyngfelt [34] using FBRs). After the WGS reactors, a conventional state-of-the-art PSA process is used to meet the required H_2 purity. In the case of CH_3OH production, the reformate is compressed to the operating pressure for the methanol synthesis (up to 50–100 bar) and it is converted into a CH_3OH-rich stream. Part of the incondensable gases (mostly CH_4 and unconverted CO_2, H_2, and CO) from the methanol separation unit are recirculated back to the CH_3OH reactor to increase the CH_3OH yield and the remaining part is used

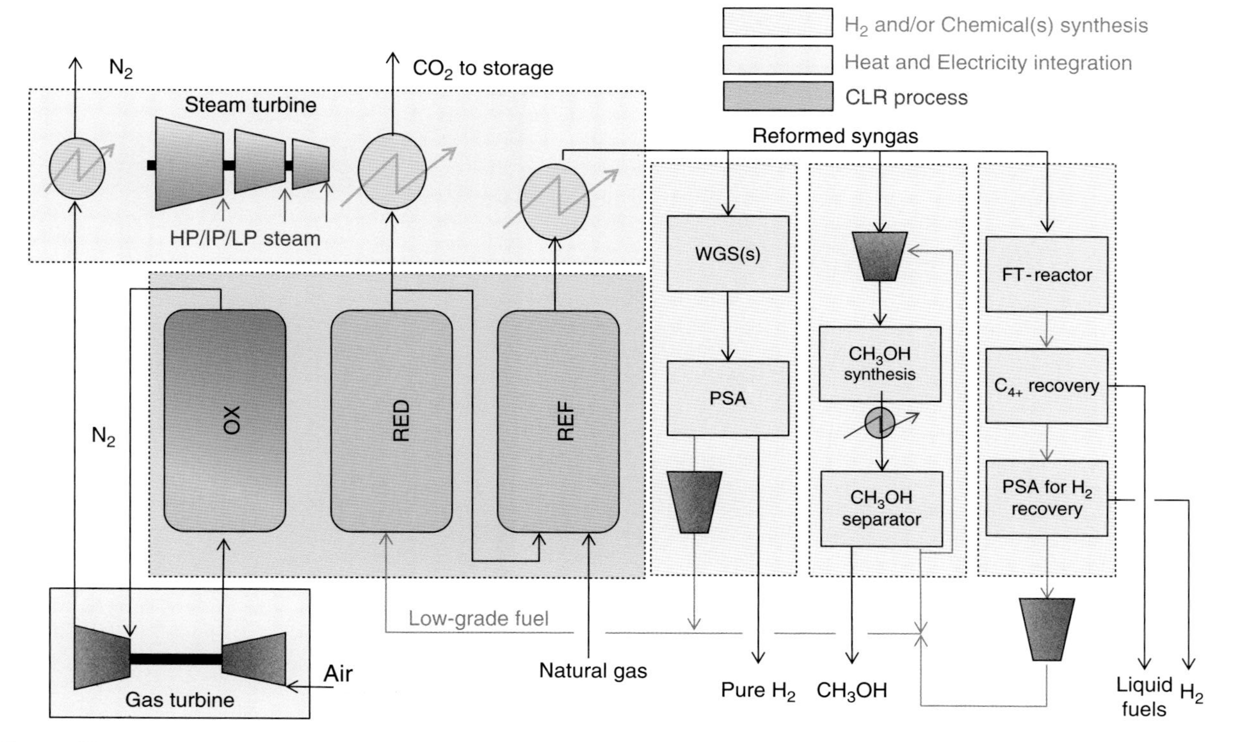

Figure 3.10 Chemical looping reforming integrated with H_2, CH_3OH, and Fischer–Tropsch processes.

for the reduction of the OC. In the case of integration with a Fischer–Tropsch (FT) process, the reformate syngas is sent to the reactor where the syngas is converted into naphtha and other C^{4+}. Depending on the FT conversion, a large amount of H_2 is also produced, which can be recovered in a dedicated PSA and used for other hydro-treating processes downstream while the remaining light fuels can be sent for reduction. The air flow rate to the reactor operated in oxidation is compressed to the CLR operating pressure and the remaining N_2 is then expanded in a gas turbine, reducing the electricity demand of the plant. Another option for chemicals production is the combination of pure H_2 production (from the PSA) with high-pressure N_2 after the oxidation, which could be used in an NH_3 synthesis reactor. The low-grade fuel resulting from the conversion/separation processes abovementioned are used for the reduction and converted into H_2O and CO_2. Part of the stream that is not recirculated to the reforming unit is then cooled down to ambient temperature providing some heat recovery, and the CO_2 is separated, compressed, and sent to the final storage.

As for the conventional reforming, the CO/H_2 ratio depends on the reactants and operating conditions. In the case of CLR, the CO/H_2 ratio depends on the amount of H_2O/C and CO_2/C at the reforming inlet. This increases the flexibility of the process with respect to the different processes listed above. Moreover, the integration of CLR is suitable for revamping some existing plants with the advantage of keeping the same units for the chemical synthesis. Several advantages can be considered:

- With respect to conventional fired tubular reforming (FTR) plants, the CLR process does not require an external furnace and high-temperature heat transfer surface because the process is autothermal.
- Compared to autothermal reforming (ATR), no air separation unit (ASU) is needed to produce high-purity oxygen and, in the case of an air-blown reformer, the syngas is not diluted with N_2.
- With respect to the dual fluidized bed CLR, the CLR with PBRs maintains high hydrogen production efficiency and can be operated at higher pressure. Moreover, intrinsic CO_2 separation is achieved in the CLR with PBRs, which delivers a high-pressure CO_2 stream without the need for a dedicated process based on ammines or other noncommercial technologies as proposed in another work [35].
- With respect to the integrated reformer with CLC [34, 36] and all the other alternative technologies mentioned, the steam consumption can be reduced owing to the CO_2/H_2O recirculation. A higher H_2 production efficiency is also expected because the oxidation and reduction phases take place at lower temperatures, reducing the overall heat duty required for the process.

On the other hand, high temperature packed bed processes are characterized by intrinsic dynamic behavior and thus require a large number of reactors and a proper heat management strategy.

The validation of the process has been carried out using a Ni-based OC (19 wt% supported on $CaAl_2O_4$) at atmospheric pressure in a small PBR. The reactor design and heat management strategies have been presented for H_2 and CH_3OH production plant in Spallina et al. [33] respectively for a medium-sized

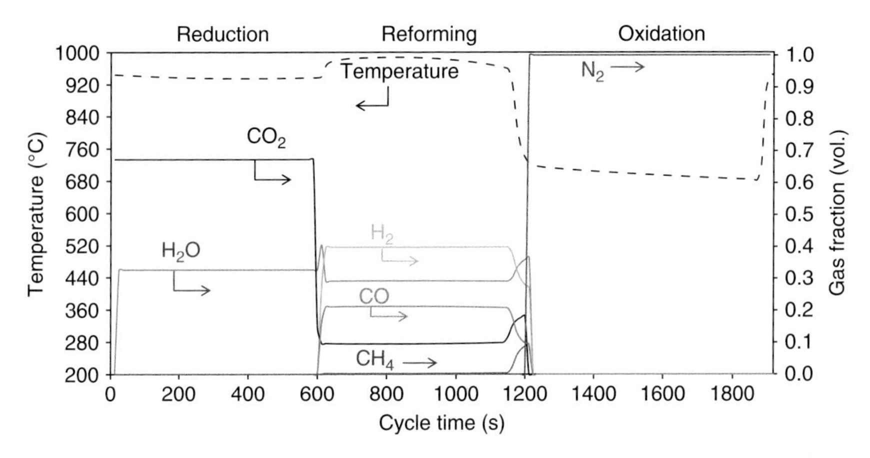

Figure 3.11 Gas conditions (temperature and composition) at the reactor outlet during the complete cycle. Spallina et al. 2017 [33]. Copyright from Creative Commons Attribution-NonCommercial-No Derivatives License (CC BY NC ND). Reproduced with permission from Elsevier.

H_2 plant (30 000 $Nm^3\ h^{-1}$) and large-scale CH_3OH plant (5000 tons d^{-1} usually referred as metric tons per day (MTPD)). On the basis of a simplified plant analysis, proper inlet stream flow rates and compositions of the PBRs have been determined and used for the modeling of the reactor.

In the case of H_2 production, the streams at the reactor outlet are shown in Figure 3.11: during oxidation, only CO_2 and H_2O are produced (temperature of about 920 °C) starting from a PSA-off gas at the inlet (CH_4 2.5%, CO 14.3%, CO_2 54%, H_2 29.2%); during the reforming, the CH_4 conversion is almost complete except in the last part of the cycle due to a decrease in the temperature in the bed; during oxidation, the O_2 in the air oxidizes the OCs and the N_2 is released at a temperature close to 700 °C. In this process, the heat removal occurs by combining it with endothermic reactions. In Figure 3.12, the axial solid temperature profiles are shown at different moments in the cycle. After the reduction, the bed is at high temperature and the heat front moves from the inlet (left side) to about 30% of the entire bed with a partial effect in cooling the bed just at the end of the reduction. During the reforming the bed is cooled by two main mechanisms: (i) convective cooling by the gas flow, which moves the heat front to the reactor outlet blowing the heat out of the bed (similarly to the CLC heat removal phase) and (ii) cooling by the endothermic reforming reactions, which consumes heat by converting CH_4 into H_2 and CO. This second mechanism generates a second heat front inside the reactor.

On the basis of a preliminary plant assessment, the integration of CLR already results in advantages with respect to conventional technologies used for H_2 or CH_3OH production. In the case of H_2 production, the reforming efficiency (LHV basis) is >75% with respect to about 74% of the state-of-the-art FTR and 69% of FTR with pre-combustion CO_2 capture. Additionally, the presence of a more efficient thermodynamic cycle (gas turbine + steam cycle) results in the production of 30 $GWh_{el}\ kg_{H_2}^{-1}$ (CLR + PBR) with respect to about 0.3 $GWh_{el}\ kg_{H_2}^{-1}$ (FTR)

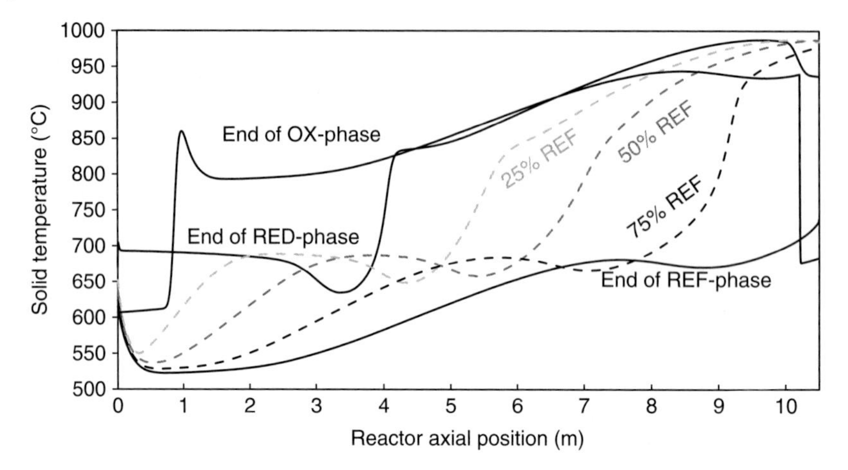

Figure 3.12 Axial solid temperature profiles at the end of the reduction (RED), oxidation (OX) and during the reforming (REF) phases. Spallina et al. 2017 [33]. Copyright from Creative Commons Attribution-NonCommercial-No Derivatives License (CC BY NC ND). Reproduced with permission from Elsevier.

and the consumption of 10 $GWh_{el}\ kg_{H_2}{}^{-1}$ (FTR + methyl di-ethanol-amine (MDEA)). In the case of CH_3OH, the preliminary assessment has shown a similar gas to liquid (GTL) efficiency (carbon conversion of 83% on mol basis). However, the specific electricity production is 0.5–1 $GWh_{el}\ kg_{CH_3OH}{}^{-1}$ in the case of CLR, while it drops to −0.33 $GWh_{el}\ kg_{CH_3OH}{}^{-1}$ for conventional GTLs plants where reformed syngas is produced in a two-stage reforming integrated with an ASU, as proposed by Haldor Topsøe [37]. Overall, in terms of energy conversion and recovery, the CLR shows a first law efficiency of 82.1% with respect to 77.9% of conventional technologies. The carbon that has not been converted into CH_3OH is completely captured (for CLR), while in the case of a conventional plant, the CO_2 is released to the atmosphere. These figures confirm the promising results for the application of chemical looping in relevant industrial processes.

3.5 Other Chemical Looping Processes

Apart from CLC and reforming, other processes partly involving chemical looping with PBRs have been recently proposed and studied. In this paragraph, a brief description and literature overview of these new processes will be presented. In particular, two other processes will be considered for the application of chemical looping integrated with other processes, viz the steam-iron process and sorption-enhanced reforming (SER).

3.5.1 Chemical Looping for H_2 Production

The combination of chemical looping and the steam-iron process (CL-H_2) is based on the use of Fe-based OCs, where the metal oxide is first fully reduced in a fuel reactor and subsequently oxidized in two separate reactors: in the first reactor the Fe/FeO is converted into Fe_3O_4 by reaction with steam to produce pure H_2 and, consecutively, in the second reactor the OCs reacts with air to complete the oxidation to Fe_2O_3 [38–41]. The integration of the steam-iron process has been discussed also for power and H_2 co-production using natural gas [42, 43] or solid fuels [44–46]. The reactions occurring during the chemical looping for H_2 production are listed in Table 3.2. Typically, the process is based on two counter-current moving bed reactors allowing for higher iron-oxide conversion compared to FBRs, while achieving complete fuel conversion. This system has been proposed and extensively studied mostly at the Ohio State University [47, 48]. Bohn et al. [49] and Lorente et al. [50] have proposed a steam-iron process using packed bed reactors as in Figure 3.13a. Bohn et al. have successfully demonstrated the three-stage chemical looping process for combined pure hydrogen production and CO_2 capture in a reactor of 10.2 mm diameter and 20 cm length (using 20 g of Fe_2O_3) while Lorente et al. have discussed the steam-iron process in a PBR as a means for H_2 storage.

Most of the research on OC design for H_2 production has been devoted to Fe-based OCs. Iron oxides have been successfully deposited on high surface area alumina supports in an attempt to prevent excessive sintering during reduction; however, the formation of more complex structures makes the chemistry of the system more complex [51]. A different OC is represented by Fe-containing perovskites, which are also defined as non-stoichiometric oxides. In this case, the amount of H_2 recovered during the oxidation with H_2O is not strictly determined by the oxygen chemical potential of the phase change [52]. De Vos et al. [53] have also studied and compared the mechanical resistance of Al_2O_3 and $MgAl_2O_4$ as support for Fe-based OC.

Voitic et al. [54] have studied the feasibility of compressed hydrogen production without additional gas compression based on the steam-iron process. Experiments were performed in a laboratory-scale test rig using a fixed bed to evaluate

Table 3.2 Main reactions prevailing during chemical looping for H_2 production.

Gas–solid reactions	Heat of reaction	
$H_2 + 3Fe_2O_3 \rightarrow H_2O + 2Fe_3O_4$	$\Delta H^\circ_{298} = -5.8\ kJ\ mol^{-1}_{H_2}$	R1
$CO + 3Fe_2O_3 \rightarrow CO_2 + 2Fe_3O_4$	$\Delta H^\circ_{298} = -46.98\ kJ\ mol^{-1}_{CO}$	R2
$CH_4 + 12Fe_2O_3 \rightarrow CO_2 + 2H_2O + 8Fe_3O_4$	$\Delta H^\circ_{298} = 141.63\ kJ\ mol^{-1}_{CH_4}$	R3
$H_2 + Fe_2O_3 \rightarrow H_2O + 2FeO$	$\Delta H^\circ_{298} = 38.42\ kJ\ mol^{-1}_{H_2}$	R4
$CO + Fe_2O_3 \rightarrow CO_2 + 2FeO$	$\Delta H^\circ_{298} = -2.72\ kJ\ mol^{-1}_{CO}$	R5
$H_2 + \frac{1}{3}Fe_2O_3 \rightarrow H_2O + \frac{2}{3}Fe$	$\Delta H^\circ_{298} = 32.92\ kJ\ mol^{-1}_{H_2}$	R6
$CO + \frac{1}{3}Fe_2O_3 \rightarrow CO_2 + \frac{2}{3}Fe$	$\Delta H^\circ_{298} = -8.21\ kJ\ mol^{-1}_{CO}$	R7
$H_2O + 3FeO \leftrightarrow H_2 + Fe_3O_4$	$\Delta H^\circ_{298} = -60.56\ kJ\ mol^{-1}_{H_2O}$	R8
$H_2O + \frac{3}{4}Fe \leftrightarrow H_2 + \frac{1}{4}Fe_3O_4$	$\Delta H^\circ_{298} = -37.77\ kJ\ mol^{-1}_{H_2O}$	R9
$O_2 + 4FeO \rightarrow 2Fe_2O_3$	$\Delta H^\circ_{298} = -560.5\ kJ\ mol^{-1}_{O_2}$	R10
$O_2 + 4Fe_3O_4 \rightarrow 6Fe_2O_3$	$\Delta H^\circ_{298} = -471.96\ kJ\ mol^{-1}_{O_2}$	R11

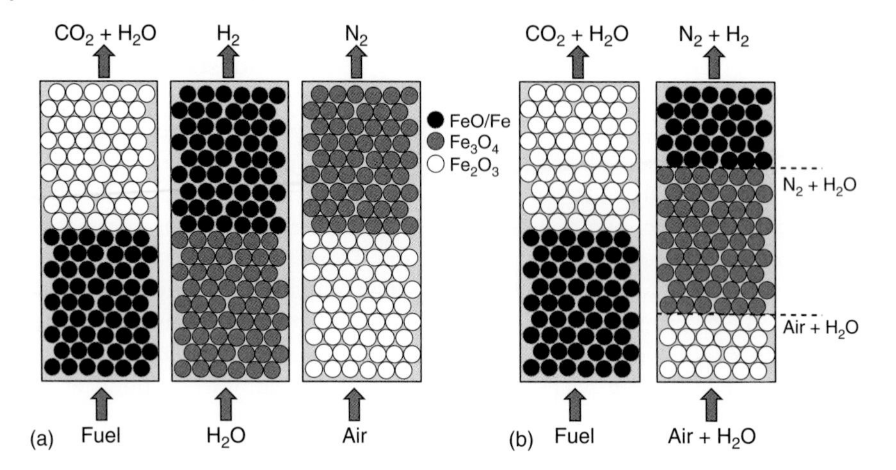

Figure 3.13 (a) Chemical looping for pure H_2 production and (b) chemical looping for H_2/N_2 production.

the influence of high-pressure H_2 production (from 8 to 23 bar) at intermediate temperature (750 °C) on the cycle stability, on the conversion efficiency, and on the structural integrity of an Fe_2O_3–Al_2O_3 (90 + 10 wt%) OC. The OC conversion was initially stable for 15 redox cycles (84%) with only small losses (0.8%) per cycle due to thermal sintering, which was independent from the different pressure levels of the preceding oxidation. The evaluation of the pressurized oxidation did not reveal any performance decrease as well.

A different combination of chemical looping and the steam-iron process has been discussed and presented in our previous work [55] and is shown in Figure 3.13b. In this case, the process is designed for the production of a H_2/N_2-rich stream. With respect to the conventional CL-H_2 process, the steam and air oxidation occurs in the same reactor by feeding a mixture of air and H_2O. This process was referred to as PCCL (pre-combustion chemical looping). The proof of principle of the two reactors concept has been carried out in a μPBR filled with 7 g of Fe_2O_3/Al_2O_3 placed in a furnace. The gases at the outlet of the reactor have been cooled down and after water separation the remaining gases have been analyzed with a mass spectrometer coupled with a CO analyzer. A sequence of reduction and combined steam–air oxidation cycles has been carried out using different inlet fuel compositions based on $CO/H_2/CH_4$ gases diluted with CO_2. The results of the selected experiments are shown in Figure 3.14. They show the complete conversion of CH_4 while feeding a mixture of CH_4 and CO_2 (CH_4/CO_2 molar ratio equal to 1) at 850 °C. During the reduction time, only CO_2 is detected and the CH_4 concentration decreases. After the OC is completely reduced the composition at the breakthrough is mostly based on H_2 and CO and no CH_4 appears: this behavior can be explained by the catalytic activity of the reduced OC for methane reforming. Carbon deposition does not occur at these conditions and this is confirmed also by the oxidation where no CO_2 is detected except for a small amount already present in the air as well as based on a small carbon balance error, which is below 3%.

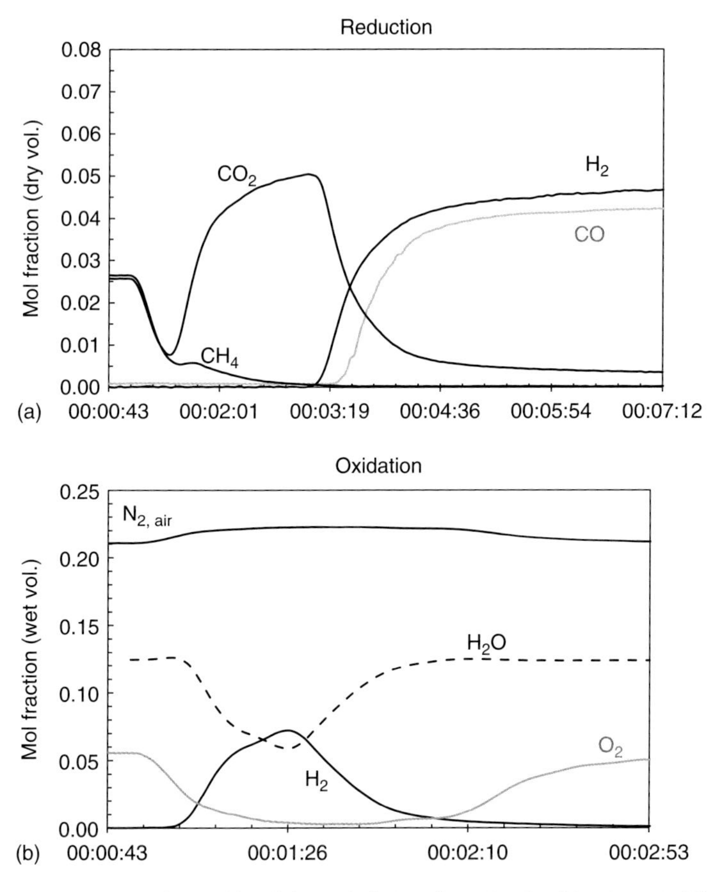

Figure 3.14 Isothermal breakthrough during the reduction (a) using 0.32 Nl min^{-1} of syngas at 850 °C and oxidation (b). Source: Spallina et al. 2016 [55]. Copyright 2016. Reproduced with permission from Elsevier.

The same trend has been noticed while feeding a mixture of $CO/H_2/CO_2$ for the reduction in which the reverse WGS reaction occurs. At the oxidation stage, the resulting composition shows that the longer the reduction time, the higher the amount of H_2 that can be obtained during the oxidation. When the oxidation is carried out with a small amount of air and large amount of H_2O (60 ml min^{-1}), the reaction front of the steam oxidation is faster than the reaction front of the O_2. However, in case a similar number of moles of atomic oxygen from air and H_2O are fed to the system (as in Figure 3.14), the amount of H_2 produced does not change significantly and the O_2 breakthrough is faster.

The heat management strategy of the complete process has been studied and presented in Spallina et al. [56], where different active weight contents and flow rates have been considered. The complete process has been designed for two different feed gases: in the first case, natural gas (diluted with H_2O and CO_2) is used for the reduction and the axial solid temperature profiles after reduction and oxidation are shown in Figure 3.15a; in the second case, syngas from coal gasification

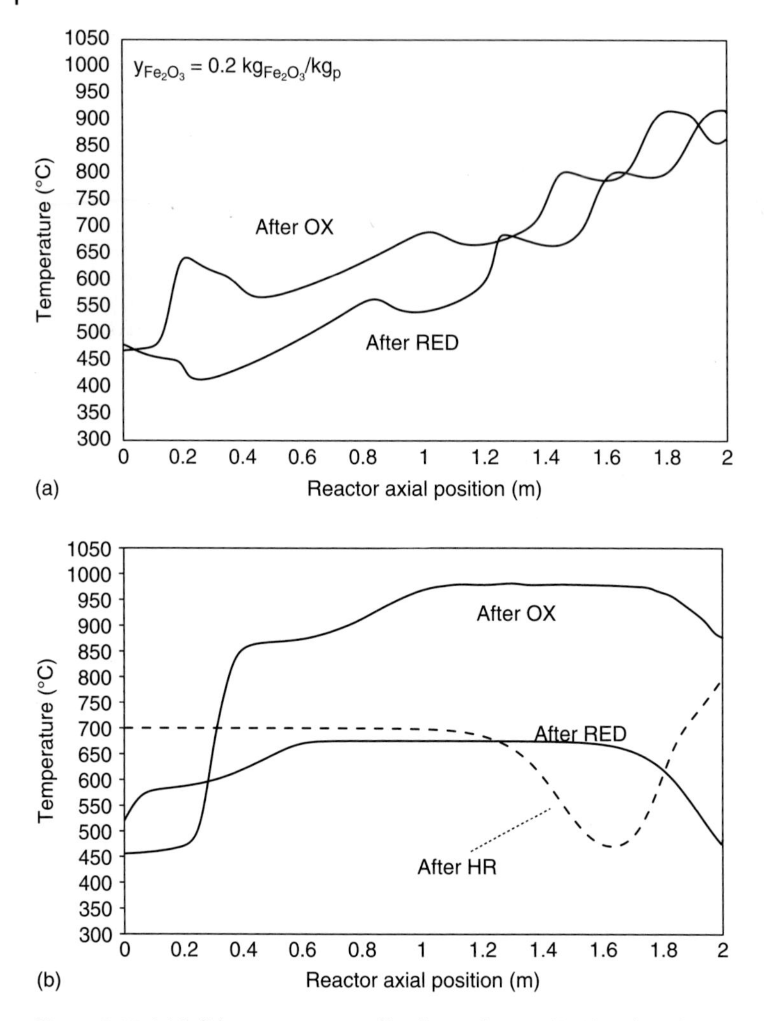

Figure 3.15 (a) Solid temperature profile after reduction (RED) and oxidation (OX) using natural gas and (b) syngas from coal gasification.

after sulfur removal has been considered (Figure 3.15b). As for CLC, the overall reaction may also require a heat removal stage to reach cyclic behavior. In the case of natural gas combustion, the heat of reaction is lower and therefore only oxidation and reduction are enough to thermally balance the complete process. In the case of syngas from coal gasification, the higher heat of reaction requires an additional heat removal stage in which air is used and therefore the heat removal occurs after the oxidation.

The complete energy balance of the two plants has been presented and compared with a state-of-the-art NGCC and IGCC power plant with and without CO_2 capture in [55, 57], as reported in Table 3.3.

In case of syngas from coal gasification, two configurations have been considered with and without heat removal (which depend on the amount of H_2 produced during the oxidation). The comparison was carried out with reference

Table 3.3 Overall energy balance for the PCCL concept and comparison with benchmark technologies for different plant layouts based on IGCC (left side) and natural gas (right side).

Configuration name	Units	IGCC	IGCC	IG-CLC	IG-PCCL	IG-PCCL	NGCC	NGCC	PCCL-HP	PCCL-LP
CO_2 capture technology		N/A	Selexol	CLC	CL	CL	N/A	MEA	CL	CL
				Medium	High	Medium		Low	High	Low
Gas turbine										
GT net power production	MW_e	301.8	270	176.4	208.2	246.2	355.63	355.63	319.54	319.54
Other PCCL compressors/expanders	MW_e	—	—	—	−6.7	−6.8	—	—	−10.38	19.62
Steam cycle										
ST net power production	MW_e	186.3	150	236.6	204	171.1	189.95	138.74	174.45	134.09
Gasification + AGR										
ASU + other consumption	MW_e	−34.6	−34.5	−37.2	−37.2	−37	—	—	—	—
Acid gas removal	MW_e	−0.4	−0.4	−0.4	−0.4	−0.4	—	—	—	—
N_2 compressors	MW_e	−42.6	−26.2	−1.4	−1.4	−1.4	—	—	—	—
CO_2 treating system										
CO_2 compression	MW_e	—	−18.8	−14.2	−9.9	−14	—	−14.77	−5.46	−10.68
CO_2 separation	MW_e		−13.7	—	—	—	—	−3.04	—	—
Balance of plant	MW_e	−3.8	−2.9	−6.8	−5.7	0	−2.44	−12.67	−5.7	−5.77
Overall net power	MW_e	406.7	323.5	353	351.1	357.7	543.15	463.9	472.42	456.77
Thermal input, MW_{LHV}	MW_{th}	860	860	860	860	860	929.65	929.65	929.65	929.65
Net electric efficiency	%	47.3	37.6	41	40.8	41.6	58.4	49.9	50.8	49.1
Carbon capture rate	%	—	89.7	96.1	95.8	95.8	—	91.2	100	100
CO_2 avoided	%	—	85.6	95.1	—	—	—	89.7	100	100
CO_2 specific emissions	kg_{CO_2}/MWh_e	736	96	33.4	35.7	35	350.6	36.12	0	0
SPECCA	$MJ_{LHV}\ kg_{CO_2}^{-1}$	—	3.07	1.66	1.73	1.5	—	3.35	2.63	3.32

to an IGCC with and without CO_2 capture and an IGCC integrated gasification chemical looping combustion (IG-CLC), as described in the Section 3.3. With respect to the IG-CLC, the two integrated gasification pre-combustion chemical looping (IG-PCCL) plants present the same thermal efficiency ranging from 40.8% to 41.6%; however, the power share differs significantly: in the case of IG-CLC the electric power from the gas turbine was limited by the maximum temperature of the CLC reactors (1200 °C), while in this case the production of a H_2/N_2 fuel allows considering the maximum temperature of the gas turbine, up to 1300 °C; on the other hand, the steam cycle shows a lower power production, because a lower amount of heat is recovered by producing HP-steam from the reactor operated in reduction and a higher steam consumption is required at the inlet of the reactor operated in oxidation to form H_2. In the case of natural gas, two configurations have been compared with a state-of-the-art NGCC and an NGCC with CO_2 post-combustion capture through chemical absorption using mono-ethanol-amine (MEA). The two NG-fuelled PCCL plants differ in the possible recovery from the CO_2-rich stream at the reactor outlet: in the first case the stream from the reactor is cooled down producing high-pressure steam and after H_2O condensation the CO_2 is compressed starting from 20 bar; in the second case, the CO_2-rich stream is first expanded up to 5 bar to produce additional electricity (about 30 MW_e) and after the complete cooling and H_2O condensation the CO_2 is compressed up to 110 bar. The different heat integration results in a different steam cycle power production with an effect on the electrical efficiency of about 1% point higher in case the gas leaving the reactor operated in reduction is cooled at high pressure.

In terms of SPECCA, for the PCCL with IGCC, the same conclusions can be drawn as for the case of IG-CLC, in which the lower energy penalty and the higher CO_2 avoidance can reduce the SPECCA of 50%. In the case of natural gas, the better SPECCA results mostly from the complete CO_2 separation, in comparison to the MEA process, but the energy penalty is substantially the same as for the NGCC + MEA, which is however a mature technology.

Another important advantage of this technology is the possibility to use it for NH_3 production in which the H_2/N_2 streams are produced at the outlet of the oxidation. With respect to a conventional natural gas plant for NH_3, the chemical looping plant does not require WGS reactors, a CO_2 absorption unit as well as a methanator reactor for the abatement of the CO.

3.5.2 Cu–Ca Process for Sorption Enhanced Reforming

Chemical looping with packed bed reactors has been recently proposed also in combination with SER through the FP7 European project ASCENT[1]. The system has been proposed in 2010 by Abanades et al. [58] and the main schematics are represented in Figure 3.16.

The SER process is a combination of steam methane reforming (endothermic reaction) and CO_2 sorption in CaO to form $CaCO_3$ (exothermic reaction), resulting in an overall autothermal system. In order to separate pure CO_2, the

1 http://www.ascentproject.eu

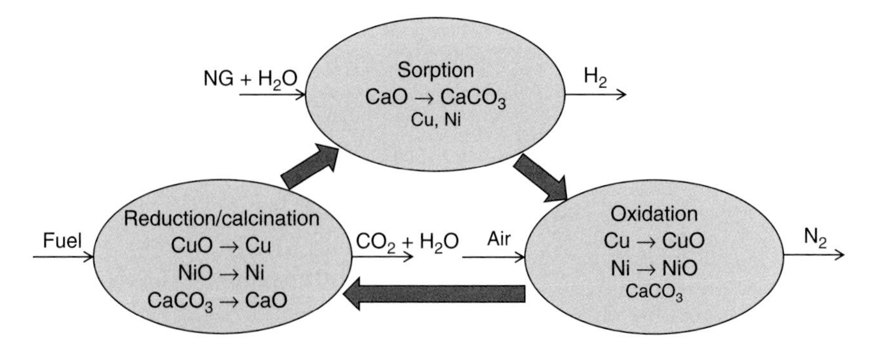

Figure 3.16 Schematic of the Cu–Ca process. Source: Abanades et al. 2010 [58]. Copyright 2010. Adapted with permission from American Chemical Society.

reverse reaction (called calcination) is required where $CaCO_3$ decomposes into $CaO + CO_2$, which is an endothermic reaction favored at high temperatures (800–900 °C) and it is typically carried out in combination with a combustion process. To avoid CO_2 dilution with N_2 from air, the SER process is integrated with an ASU to carry out the calcination by burning some fuel with pure O_2 [59]. However, the ASU results in a high efficiency penalty of around 7–10% points and it increases the capital expenditure (CAPEX) of the plant.

The integration of SER with CLC would represent a possible alternative to the standard SER so that the ASU can be replaced by the use of CuO.

The complete process is carried out in three different steps as illustrated in Figure 3.16:

(a) *Sorption.* natural gas and H_2O react to produce syngas and the CO_2 formed reacts with CaO to form $CaCO_3$ as in the SER process. This reaction will require a suitable reforming catalyst (e.g. noble transition metals or Ni) as Cu may not provide sufficient activity.
(b) *Oxidation.* in this step, the oxidation of Cu to CuO with air takes place, at conditions such that no major decomposition of $CaCO_3$ occurs (this will be favored by high-pressure operation and high temperatures).
(c) *Reduction/calcination.* the calcination of $CaCO_3$ to form CaO and gaseous CO_2 and the reduction of CuO with a fuel gas (using a syngas) occur simultaneously. The reduction of CuO is exothermic and thus the heat of reaction for the calcination is provided in situ while forming a CO_2-rich stream not diluted with N_2.

The combination of SER with CLC has been discussed in Fernández et al. [2] to define a preliminary process layout and the best operating conditions of the complete process. In their work, the authors have determined the required CaO/Cu ratio to make the entire process feasible and have defined the operating windows for the process, so that the reactors can operate close to thermally neutral conditions. In the case of oxidation, the authors have found that a large N_2 recycle is required to avoid an excessive increase in the temperature, which would favor the calcination reaction, thereby reducing the CO_2 capture rate. Alarcón and Fernández [60] have investigated the reactor modeling of the combined reduction and calcination process at different temperatures, different syngas compositions, and $CuO/CaCO_3$ ratios. The results presented by the authors show that the optimal operating temperature is around 900 °C and the use of a CO-rich gas reduces $CuO/CaCO_3$ ratio. However, in the presence of a higher CuO content, the temperature increases with a possible damage of the Cu-based material, while in the case of a relatively small amount of CuO, the heat produced is insufficient to calcine the entire bed, reducing the sorption capacity of the bed. In Fernández et al. [61], the validation of the model has been carried out with experiments in a laboratory-scale fixed bed reactor (i.d. × L, 38 mm × 920 mm). Martini et al. [62] have presented the complete process modeling and reactor behavior using a more detailed 1D pseudo-homogeneous model and a simplified sharp-front approach method and have investigated the effect of different operating conditions on the heat/reaction front velocities for the different steps.

The complete SER/CLC process design and thermodynamic assessment for H_2 production have been carried out in Martínez et al. [63] reaching overall H_2 efficiencies of 77% with a 94% carbon capture rate demonstrating the promising nature of this technology.

3.6 Conclusions

The present chapter has discussed the application of PBRs for chemical looping processes for carbon-free power, hydrogen, and chemicals production from fossil fuels. With respect to conventional FBRs that have been amply studied especially for CLC, the use of PBRs for chemical looping allows easier operation at higher pressures, while the materials are not being fluidized reducing the risks of particle damaging and losses avoiding solids circulation. However, because of the dynamic operation, a relatively large number of reactors need to be operated in parallel. Moreover, several costly components (e.g. high-temperature valves) pose several technical questions, which are currently under investigation in order to make the technology more mature for further commercialization.

Much research has been carried out over the last years to investigate the development, the feasibility, and the integration of packed bed chemical looping technologies. Recent advances in terms of OC development and manufacturing, as well as modeling and experimental validation have been mostly focused on CLC processes, focusing in particular on the design and heat management of the reactors. The reactor concept is now at the early stage of demonstration and new research is required to study the material properties and the effect of dynamic operation on the system performance and techno-economic feasibility of the technology. To achieve high efficiencies in fuel conversion and a high-temperature gas production for a gas turbine, PBRs have been considered with different heat removal layouts depending on the type of OCs used. In the case of CLR, the heat removal occurs in combination with endothermic reactions and the results are very interesting not only for H_2 production but also for other GTLs processes, because no ASU is required and the thermal integration shows some improvement in terms of electricity consumption and steam-to-export.

In recent years also other processes with chemical looping in packed bed reactors have been proposed and studied, in particular, the steam-iron process combined with chemical looping for the production of high-purity H_2 for different applications including proton exchange membrane (PEM) fuel cells with integrated CO_2 capture. In addition, the combination of air and steam oxidation in the same unit has also been proposed and successfully tested at the laboratory scale as a possible solution to produce a H_2/N_2 stream for CO_2-free electricity or NH_3 production. Finally, the integration of SER-CLC using Cu as OC represents another interesting option for H_2 production with integrated CO_2 capture with promising results in terms of process efficiency and carbon capture rates.

Nomenclature

Abbreviations

AGR	Acid gas removal
ASU	Air separation unit
ATR	Autothermal reforming
CLC	Chemical looping combustion
CLR	Chemical looping reforming
FTR	Fired tubular reforming
IGCC	Integrated gasification combined cycle
LP/IP/HP	Low/intermediate/high pressure
LHV	Lower heating value
LT/IT/HT	Low/intermediate/high temperature
MEA	Mono-ethanol-amine
MDEA	Methyl di-ethanol-amine
MTPD	Metric tons per day
NGCC	Natural gas combined cycle
PBR	Packed bed reactor
PSA	Pressure swing adsorption
SER	Sorption enhanced reforming
SPECCA	Specific primary energy consumption for CO_2 avoided
WGS	Water gas shift

Variables and units

A	Area (m^2)
b	Gas–solid stoichiometric factor ($mol_s\ mol_g^{-1}$)
Bi	Biot number (–)
C	Concentration ($mol\ m^{-3}$)
c_p	Heat capacity ($J\ kg^{-1}\ K^{-1}$)
d	Diameter (m)
d_p	Particle diameter (m)
E_A	Activation energy ($kJ\ mol^{-1}$)

MW	Molecular weight ($kg\,mol^{-1}$)
n	Reaction order (–)
n_i	Gas flux of ith-component ($mol\,m^{-2}\,s^{-1}$)
P	Pressure (bar)
r_i	Reaction rate ($mol\,m^{-3}\,s^{-1}$)
R	Gas constant ($J\,mol^{-1}\,K^{-1}$)
R_i	Thermal resistance (KW^{-1})
r_i	Reaction rate ($mol\,m\,r^{-3}\,s^{-1}$)
r_0	Particle radius (m)
t	Time (s)
T	Temperature (K or °C)
v_g	Gas velocity ($m\,s^{-1}$)
X	Solid conversion (–)
y_i	Mass fraction (–)
ΔH_R	Heat of reaction ($kJ\,mol^{-1}$)

Subscripts

η	efficiency (–)
ε_g	Void fraction (–)
$\varepsilon_{s,p}$	Solid porosity (–)
ζ	Stoichiometric factor ($mol_g\,mol_s^{-1}$)
ρ	Density ($kg\,m^{-3}$)

References

1 Noorman, S., Annaland, M.V.S., and Kuipers, H. (2007). Packed bed reactor technology for chemical-looping combustion. *Ind. Eng. Chem. Res.* 46: 4212–4220.

2 Fernández, J.R., Abanades, J.C., Murillo, R., and Grasa, G. (2012). Conceptual design of a hydrogen production process from natural gas with CO_2 capture using a Ca–Cu chemical loop. *Int. J. Greenhouse Gas Control* 6: 126–141.

3 Hamers, H.P., Gallucci, F., Cobden, P.D. et al. (2013). A novel reactor configuration for packed bed chemical-looping combustion of syngas. *Int. J. Greenhouse Gas Control* 16: 1–12.

4 Noorman, S., Gallucci, F., van Sint Annaland, M.M., and Kuipers, J.A.M. (2011). A theoretical investigation of CLC in packed beds. Part 2: reactor model. *Chem. Eng. J.* 167: 369–376.

5 Adanez, J., Abad, A., Garcia-Labiano, F. et al. (2012). Progress in chemical-looping combustion and reforming technologies. *Prog. Energy Combust. Sci.* 38 (2): 215–282.

6 Schwebel, G.L., Filippou, D., Hudon, G. et al. (2014). Experimental comparison of two different ilmenites in fluidized bed and fixed bed chemical-looping combustion. *Appl. Energy* 113: 1902–1908.

7 Jacobs, M., Van Noyen, J., Larring, Y. et al. (2015). Thermal and mechanical behaviour of oxygen carrier materials for chemical looping combustion in a packed bed reactor. *Appl. Energy* 157: 374–381.

8 Ortiz, M., Gallucci, F., Melchiori, T. et al. (2016). Kinetics of the reactions prevailing during packed-bed chemical looping combustion of syngas using ilmenite. *Energy Technol.* 4: 1137–1146.

9 Schwebel, G.L., Leion, H., and Krumm, W. (2012). Comparison of natural ilmenites as oxygen carriers in chemical-looping combustion and influence of water gas shift reaction on gas composition. *Chem. Eng. Res. Des.* 90: 1351–1360.

10 Han, L., Zhou, Z., and Bollas, G.M. (2014). Heterogeneous modeling of chemical-looping combustion. Part 2: particle model. *Chem. Eng. Sci.* 113: 116–128.

11 Han, L., Zhou, Z., and Bollas, G.M. (2016). Model-based analysis of chemical-looping combustion experiments. Part I: structural identifiability of kinetic model for NiO reduction. *AIChE J.* 62: 2419–2431.

12 Ryu, H.-J., Bae, D.-H., and Jin, G.-T. (2003). Effect of temperature on reduction reactivity of oxygen carrier particles in a fixed bed chemical-looping combustor. *Korean J. Chem. Eng.* 20: 960–966.

13 Hamers, H.P., Gallucci, F., Williams, G. et al. (2015). Reactivity of oxygen carriers for chemical-looping combustion in packed bed reactors under pressurized conditions. *Energy Fuel* 29: 2656–2663.

14 Nordness, O., Han, L., Zhou, Z., and Bollas, G.M. (2016). High-pressure chemical-looping of methane and synthesis gas with Ni and Cu oxygen carriers. *Energy Fuel* 30: 504–514.

15 San Pio, M.A., Roghair, I., Gallucci, F., and van Sint Annaland, M. (2016). Investigation on the decrease in the reduction rate of oxygen carriers for chemical looping combustion. *Powder Technol.* 301: 429–439.

16 Hamers, H.P., Gallucci, F., Cobden, P.D. et al. (2014). CLC in packed beds using syngas and CuO/Al_2O_3: model description and experimental validation. *Appl. Energy* 119: 163–172.

17 Hamers, H.P., Romano, M.C., Spallina, V. et al. (2014). Comparison on process efficiency for CLC of syngas operated in packed bed and fluidized bed reactors. *Int. J. Greenhouse Gas Control* 28: 65–78.

18 Fernández, J.R. and Alarcón, J.M. (2015). Chemical looping combustion process in fixed-bed reactors using ilmenite as oxygen carrier: conceptual design and operation strategy. *Chem. Eng. J.* 264: 797–806.

19 Han, L. and Bollas, G.M. (2016). Chemical-looping combustion in a reverse-flow fixed bed reactor. *Energy* 102: 669–681.

20 Spallina, V., Gallucci, F., Romano, M.C. et al. (2013). Investigation of heat management for CLC of syngas in packed bed reactors. *Chem. Eng. J.* 225: 174–191.

21 Diglio, G., Bareschino, P., Mancusi, E., and Pepe, F. (2017). Numerical assessment of the effects of carbon deposition and oxidation on chemical looping combustion in a packed-bed reactor. *Chem. Eng. Sci.* 160: 85–95.

22 Hamers, H.P., Romano, M.C., Spallina, V. et al. (2015). Energy analysis of two stage packed-bed chemical looping combustion configurations for integrated gasification combined cycles. *Energy* 85: 489–502.

23 Kooiman, R.F., Hamers, H.P., Gallucci, F., and van Sint Annaland, M. (2015). Experimental demonstration of two-stage packed bed chemical-looping combustion using syngas with CuO/Al_2O_3 and $NiO/CaAl_2O_4$ as oxygen carriers. *Ind. Eng. Chem. Res.* 54: 2001–2011.

24 Gallucci, F., Hamers, H.P., van Zanten, M., and van Sint Annaland, M. (2015). Experimental demonstration of chemical-looping combustion of syngas in packed bed reactors with ilmenite. *Chem. Eng. J.* 274: 156–168.

25 Spallina, V., Chiesa, P., Martelli, E. et al. (2015). Reactor design and operation strategies for a large-scale packed-bed CLC power plant with coal syngas. *Int. J. Greenhouse Gas Control* 36: 34–50.

26 Han, L., Zhou, Z., and Bollas, G.M. (2013). Heterogeneous modeling of chemical-looping combustion. Part 1: reactor model. *Chem. Eng. Sci.* 104: 233–249.

27 Hamers, H.P., Romano, M.C., Spallina, V. et al. (2015). Boosting the IGCLC process efficiency by optimizing the desulfurization step. *Appl. Energy* 157: 422–432.

28 Spallina, V., Romano, M.C., Chiesa, P. et al. (2014). Integration of coal gasification and packed bed CLC for high efficiency and near-zero emission power generation. *Int. J. Greenhouse Gas Control* 27: 28–41.

29 Chiesa, P., Lozza, G., and Mazzocchi, L. (2005). Using hydrogen as gas turbine fuel. *J. Eng. Gas Turbines Power* 127: 73.

30 EBTF (2011). European best practice guidelines for assessment of CO_2 capture technologies.

31 Chen, C., Han, L., and Bollas, G.M. (2016). Dynamic simulation of fixed-bed chemical-looping combustion reactors integrated in combined cycle power plants. *Energy Technol.* 4: 1209–1220.

32 Diglio, G., Bareschino, P., Mancusi, E., and Pepe, F. (2015). Simulation of hydrogen production through chemical looping reforming process in a packed-bed reactor. *Chem. Eng. Res. Des.* 105: 137–151.

33 Spallina, V., Marinello, B., Gallucci, F. et al. (2017). Chemical looping reforming in packed-bed reactors: modelling, experimental validation and large-scale reactor design. *Fuel Process. Technol.* 156: 156–170.

34 Ryden, M. and Lyngfelt, A. (2006). Using steam reforming to produce hydrogen with carbon dioxide capture by chemical-looping combustion. *Int. J. Hydrog. Energy* 31: 1271–1283.

35 Ortiz, M., Abad, A., de Diego, L.F. et al. (2011). Optimization of hydrogen production by chemical-looping auto-thermal reforming working with Ni-based oxygen-carriers. *Int. J. Hydrog. Energy* 36: 9663–9672.

36 Pans, M.A., Abad, A., de Diego, L.F. et al. (2013). Optimization of H_2 production with CO_2 capture by steam reforming of methane integrated with a chemical-looping combustion system. *Int. J. Hydrog. Energy* 38: 11878–11892.

37 Aasberg-Petersen, K., Nielsen, C.S., Dybkjær, I., and Perregaard, J. (2013). *Large Scale Methanol Production from Natural Gas*, 1–14. Haldor Topsøe.

38 Piotrowski, K., Mondal, K., Lorethova, H. et al. (2005). Effect of gas composition on the kinetics of iron oxide reduction in a hydrogen production process. *Int. J. Hydrog. Energy* 30: 1543–1554.
39 Singh, a., Al-Raqom, F., Klausner, J., and Petrasch, J. (2012). Production of hydrogen via an iron/iron oxide looping cycle: thermodynamic modeling and experimental validation. *Int. J. Hydrog. Energy* 37: 7442–7450.
40 Gupta, P., Velazquez-Vargas, L., and Fan, L. (2007). Syngas redox (SGR) process to produce hydrogen from coal derived syngas. *Energy Fuel* 2900–2908.
41 Plou, J., Duran, P., Herguido, J., and Peña, J.A. (2012). Steam-iron process kinetic model using integral data regression. *Int. J. Hydrog. Energy* 37: 6995–7004.
42 Lozza, G., Chiesa, P., Romano, M.C., and Savoldelli, P. (2006). Three reactors chemical looping combustion for high efficiency electricity generation with CO_2 capture from natural gas. Proceedings of ASME Turbo Expo 2006, GT2006-90345. pp. 1–11.
43 Chiesa, P., Lozza, G., Malandrino a, R.M., and Piccolo, V. (2008). Three-reactors chemical looping process for hydrogen production. *Int. J. Hydrog. Energy* 33: 2233–2245.
44 Li, F., Zeng, L., and Fan, L.-S. (2010). Biomass direct chemical looping process: process simulation. *Fuel* 89: 3773–3784.
45 Sorgenfrei, M. and Tsatsaronis, G. (2013). Design and evaluation of an IGCC power plant using iron-based syngas chemical-looping (SCL) combustion. *Appl. Energy* 113: 1958–1964.
46 Cormos, C. (2010). Evaluation of iron based chemical looping for hydrogen and electricity co-production by gasification process with carbon capture and storage. *Int. J. Hydrog. Energy* 35: 2278–2289.
47 Sridhar, D., Tong, A., Kim, H. et al. (2012). Syngas chemical looping process: design and construction of a 25 kWth subpilot unit. *Energy Fuel.*
48 Tong, A., Sridhar, D., Sun, Z. et al. (2013). Continuous high purity hydrogen generation from a syngas chemical looping 25 kWth sub-pilot unit with 100% carbon capture. *Fuel* 103: 495–505.
49 Bohn, C.D., Müller, C.R., Cleeton, J.P. et al. (2008). Production of very pure hydrogen with simultaneous capture of carbon dioxide using the redox reactions of iron oxides in packed beds. *Ind. Eng. Chem. Res.* 47: 7623–7630.
50 Lorente, E., Peña, J.A., and Herguido, J. (2011). Cycle behaviour of iron ores in the steam-iron process. *Int. J. Hydrog. Energy* 36: 7043–7050.
51 Bohn, C.D., Cleeton, J.P., Müller, C.R. et al. (2010). Stabilizing iron oxide used in cycles of reduction and oxidation for hydrogen production. *Energy Fuel* 24: 4025–4033.
52 Murugan, A., Thursfield, A., and Metcalfe, I.S. (2011). A chemical looping process for hydrogen production using iron-containing perovskites. *Energy Environ. Sci.* 4: 4639.
53 De Vos, Y., Jacobs, M., Van Der Voort, P. et al. (2017). Optimization of spray dried attrition-resistant iron based oxygen carriers for chemical looping reforming. *Chem. Eng. J.* 309: 824–839.
54 Voitic, G., Nestl, S., Lammer, M. et al. (2014). Pressurized hydrogen production by fixed-bed chemical looping. *Appl. Energy* 157: 399–407.

55 Spallina, V., Gallucci, F., Romano, M.C., and van Sint Annaland, M. (2016). Pre-combustion packed bed chemical looping (PCCL) technology for high efficient H_2-rich gas production processes. *Chem. Eng. J.* 294: 478–494.

56 Spallina, V., Van Etten, M.P.C., Gallucci, F., and Annaland, M.V.S. (2016). Heat management of pre-combustion chemical looping technology using packed bed reactors. 4th International conference on chemical looping, Nanjing, China, 26–28 September 2016, pp. 1–13.

57 Spallina, V., Gallucci, F., van Sint Annaland, M., et al. (2015). Thermodynamic analysis of a pre-combustion chemical looping process for power production integrated with coal gasification combined cycle. 40th International Technical Conference on Clean Coal Fuel Systems, Clearwater, Florida, USA, 31 May–4 June 2015.

58 Abanades, J.C., Murillo, R., Fernández, J.R. et al. (2010). New CO_2 capture process for hydrogen production combining Ca and Cu chemical loops. *Environ. Sci. Technol.* 44: 6901–6904.

59 Martínez, I., Romano, M.C., Chiesa, P. et al. (2013). Hydrogen production through sorption enhanced steam reforming of natural gas: thermodynamic plant assessment. *Int. J. Hydrog. Energy* 38: 15180–15199.

60 Alarcón, J. and Fernández, J.R. (2015). $CaCO_3$ calcination by the simultaneous reduction of CuO in a Ca/Cu chemical looping process. *Chem. Eng. Sci.* 137: 254–267.

61 Fernández, J.R., Alarcon, J.M., and Abanades, J.C. (2016). Investigation of a fixed-bed reactor for the calcination of $CaCO_3$ by the simultaneous reduction of CuO with a fuel gas. *Ind. Eng. Chem. Res.* 55: 5128–5132.

62 Martini, M., van den Berg, A., Gallucci, F., and van Sint Annaland, M. (2016). Investigation of the process operability windows for Ca–Cu looping for hydrogen production with CO_2 capture. *Chem. Eng. J.* 303: 73–88.

63 Martínez, I., Romano, M.C., Fernández, J.R. et al. (2014). Process design of a hydrogen production plant from natural gas with CO_2 capture based on a novel Ca/Cu chemical loop. *Appl. Energy* 114: 192–208.

4

Chemical Looping with Oxygen Uncoupling (CLOU) Processes

Kevin J. Whitty[1], JoAnn S. Lighty[2], and Tobias Mattisson[3]

[1] *University of Utah, Department of Chemical Engineering, 50 S. Central Campus Drive, Room 3290, Salt Lake City, UT, 84112, USA*
[2] *Boise State University, Department of Mechanical and Biomedical Engineering, College of Engineering, 1910 University Drive, Boise, ID, 83725, USA*
[3] *Chalmers University of Technology, Department of Energy and Environment, Division of Energy Technology, Hörsalsvägen 7B, Göteborg, 41296, Sweden*

4.1 Introduction

Chemical looping with oxygen uncoupling (CLOU) is a variant of chemical looping combustion in which oxygen is spontaneously liberated from the oxygen carrier as gaseous O_2 in the fuel reactor. This allows heterogeneous combustion of solid fuel by O_2, which is much more efficient than typical chemical looping systems that require the fuel to be in gaseous form to react with the solid oxygen carrier. Experimental studies have demonstrated that coal conversion in a CLOU system is as much as 50 times faster than conventional chemical looping combustion under similar conditions [1–4].

The key to CLOU is the active oxygen-carrying metal oxide system, in particular the thermodynamics with respect to oxygen concentration in the temperature range of solid fuel combustion. In the fuel reactor of a conventional non-CLOU chemical looping combustion system, oxygen remains bound to the solid oxygen carrier as a metal oxide. Because solid–solid reactions are inefficient, the fuel must be in gaseous form to react with the oxidized metal. That works well with gaseous fuels such as natural gas, but in order to process a solid fuel such as coal or petroleum coke, the char must be gasified to produce combustible syngas either in a separate reactor upstream of the fuel reactor or in situ within the fuel reactor itself (so-called iG-CLC, for in situ gasification chemical looping combustion). This is depicted schematically in Figure 4.1a. Gasification of solid char by steam or CO_2 is relatively slow at typical fuel reactor temperatures, so conversion rates are limited by the slow gasification reactions. In contrast, because CLOU oxygen carriers release gaseous O_2 in the fuel reactor, char rapidly combusts to form CO_2 as shown in Figure 4.1b. Combustion is much faster than gasification, so overall conversion is much faster in a CLOU system.

The ability of certain metal oxides to spontaneously release gaseous oxygen in the fuel reactor was first recognized by researchers at Chalmers University of

Handbook of Chemical Looping Technology, First Edition. Edited by Ronald W. Breault.

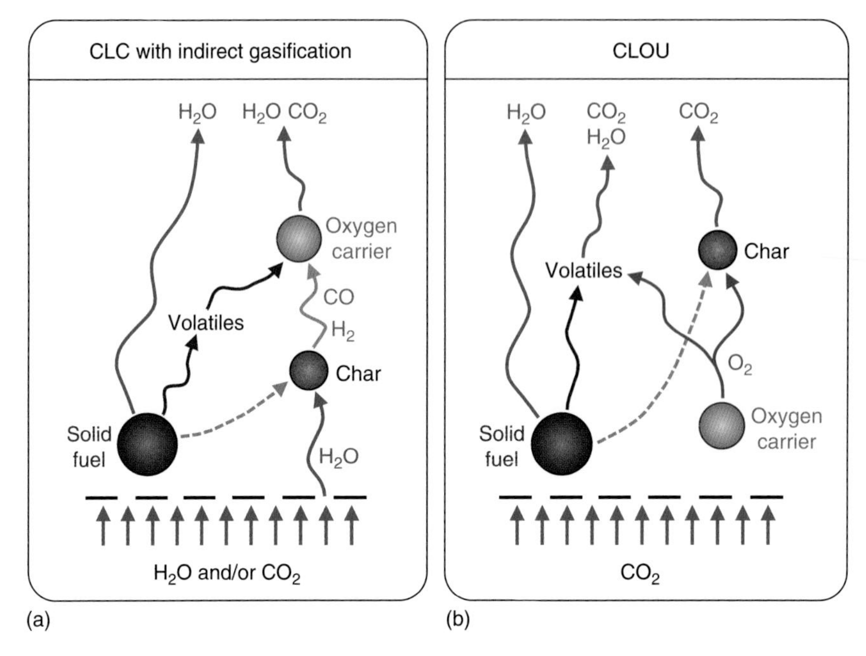

Figure 4.1 (a) Standard chemical looping combustion, which relies on gasification of char to form CO and H_2 to react with the oxygen carrier. (b) CLOU, which releases gaseous O_2 to combust the char. Source: Adánez et al. 2012 [5]. Adapted with permission of Elsevier.

Technology, who demonstrated CLOU of petroleum coke using a copper-based oxygen carrier in a laboratory-scale batch fluidized bed equipped with an analyzer to measure O_2, CO_2, CO, and CH_4 in the product gas [1, 6]. Gas concentration profiles from this experiment are shown in Figure 4.2. The system was initially at 885 °C and fluidized with air to fully oxidize the copper to cupric oxide (CuO). The steady-state oxygen concentration in the product gas was stable at 21%. At 100 seconds, the fluidizing gas was switched to nitrogen. Although no oxygen was being introduced into the system, the exit gas contained approximately 1% O_2, which was being generated by the oxygen carrier as the CuO naturally reduced to cuprous oxide (Cu_2O) in the oxygen-starved environment according to Reaction (4.1).

$$4CuO \rightarrow 2Cu_2O + O_2 \tag{4.1}$$

The product of the reaction is gaseous oxygen and at 885 °C its equilibrium concentration is approximately 1 vol%. At time 180 seconds, 0.1 g of petroleum coke was added to the fluidized bed. Over the next few seconds, as volatiles were released from the fuel particles, CO_2 and CO were observed in the product gas and the O_2 concentration dropped to zero. CO resulted because the rate of carbon monoxide production during devolatilization exceeded the rate of O_2 release by the oxygen carrier. After a few more seconds, CO was no longer observed and the oxygen concentration increased. Over the next 200 seconds, CO_2 continued to be produced and O_2 remained below its equilibrium concentration. It was during this time that the key CLOU reaction was taking place. Specifically, O_2 released by Reaction (4.1) was combusting the petcoke char. The rate of O_2 generation

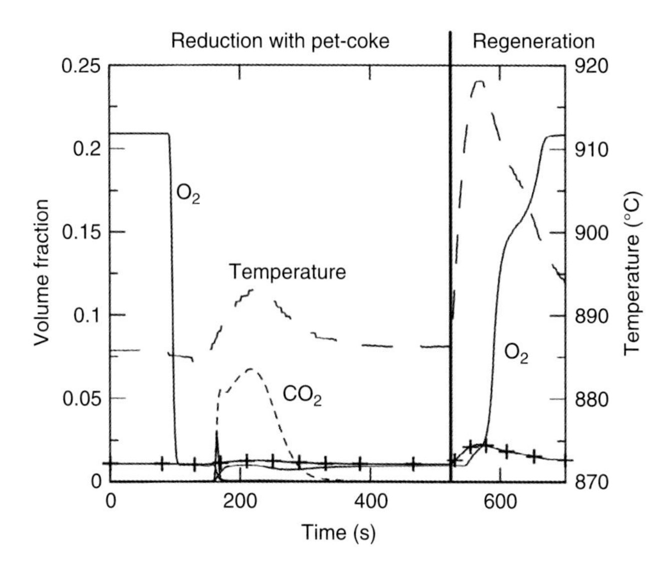

Figure 4.2 Gas concentration profile for conversion of 0.1 g petroleum coke by a copper-based CLOU carrier in a laboratory-scale batch fluidized bed initially at 885 °C. The fluidizing gas was switched from air to nitrogen at 100 seconds and fuel was dropped into the bed at 180 seconds. The fluidizing gas was switched back to air at 520 seconds. Source: Mattisson et al. 2009 [6]. Adapted with permission of Elsevier.

exceeded that of the char combustion, as evidenced by the continued presence of O_2 in the product gas. As the char burned out, the CO_2 concentrations decreased and the O_2 concentration increased until it reached its equilibrium value of 1%. Finally, at 520 seconds, the fluidizing gas was switched to air. The O_2 concentration remained below 21% for about two minutes as the Cu_2O reacted with oxygen to form CuO. A large temperature increase was also observed during this period due to the exothermic oxidation reaction. After the copper was all fully oxidized the exit O_2 concentration became 21%.

In the years since the proof-of-concept demonstration of CLOU, the process has been intensively studied, with the focus on the development of oxygen carriers [3, 7–26], operation of laboratory-scale chemical looping systems using CLOU carriers [2, 27–36], reaction analysis [37–44], and modeling of CLOU-based chemical looping systems [4, 45–47]. Several reviews of chemical looping combustion development also summarize advancements of the technology [5, 48–50]. This chapter provides an overview of CLOU, addresses considerations for design of reactors specific for CLOU, and presents the state of development of the CLOU process. Oxygen carriers for CLOU are addressed in detail in Chapter 7 and simulation of CLOU reactors is discussed in Chapter 12.

4.2 Fundamentals of the CLOU Process

The key to CLOU is the metal–metal oxide system chosen for the oxygen carrier. For a given metal element denoted Me, the general oxidation reaction is

$$Me_xO_{y-2} + O_2 \rightarrow Me_xO_y \tag{4.2}$$

As with any chemical looping combustion oxygen carrier, the forward reaction takes place in the air reactor and will always be exothermic. The distinguishing characteristic of a CLOU oxygen carrier is that reduction of the metal does not require reaction with a gaseous fuel such as natural gas or synthesis gas, but instead occurs naturally by the reverse of the reaction above, liberating gaseous O_2. The reverse reaction, or "oxygen uncoupling" reaction, is endothermic with the heat required matching that released during oxidation.

For an oxygen carrier to function in a CLOU process, the thermodynamics of the metal must be such that at temperatures typical for combustion of a fuel (approximately 800–1100 °C), the metal is in an oxidized form in the air reactor where the O_2 concentration is relatively high. In the fuel reactor, which is typically fluidized with steam and/or carbon dioxide and has a low concentration of O_2, the metal of a CLOU carrier should be in a more reduced form than in the air reactor. As a guide, the equilibrium O_2 concentration for the reversible oxidation/reduction reaction of a CLOU oxygen carrier should span the range from a few percent to 21% within the temperature range 800–1100 °C.

Three monometal oxide systems based on copper, manganese, and cobalt have been identified as being thermodynamically suitable for CLOU [1]. The oxygen uncoupling reactions that take place in the fuel reactor are shown below, along with the energy required to drive the reaction at 900 °C, per molecule of O_2 released.

$$\text{Copper: } 4\text{CuO} \leftrightarrow 2\text{Cu}_2\text{O} + \text{O}_2 \quad \Delta H_{900\,°\text{C}} = 262.5 \text{ kJ mol}^{-1}\text{ O}_2 \tag{4.3}$$

$$\text{Manganese: } 6\text{Mn}_2\text{O}_3 \leftrightarrow 4\text{Mn}_3\text{O}_4 + \text{O}_2 \quad \Delta H_{900\,°\text{C}} = 193.3 \text{ kJ mol}^{-1}\text{ O}_2 \tag{4.4}$$

$$\text{Cobalt: } 2\text{Co}_3\text{O}_4 \leftrightarrow 6\text{CoO} + \text{O}_2 \quad \Delta H_{900\,°\text{C}} = 407.1 \text{ kJ mol}^{-1}\text{ O}_2 \tag{4.5}$$

The equilibrium O_2 partial pressures for these reactions are displayed in Figure 4.3. On the left side of the curves, the more oxidized form of the metal is stable. Upon crossing to the right side of the curve either by increasing the temperature or by reducing O_2 partial pressure, the reaction reverses and O_2 is spontaneously released. As long as there is fuel available to rapidly consume the O_2, the oxygen partial pressure in the fuel reactor will remain low and the oxygen carrier will continue to release oxygen.

Figure 4.4 shows conditions for the specific case of a copper-based CLOU system and combustion of coal. Coal combustion in fluidized bed systems is efficient at temperatures around 900–950 °C, which coincides with the temperature at which the equilibrium O_2 partial pressure of the CuO/Cu_2O system ranges from 1.5% to 4.5% O_2. The O_2 concentration entering the air reactor is 21%. As the oxygen carrier is oxidized, the O_2 concentration of the gas decreases as it flows through the reactor. The exit concentration of oxygen may be as low as 5%, but cannot become any lower than the equilibrium concentration at the given temperature. There needs to be some finite difference between exit and equilibrium concentrations to ensure a reaction driving force throughout the reactor.

In the fuel reactor of the system represented in Figure 4.4, the O_2 concentration is very low, below the equilibrium line and the reduced form of the carrier is favored. The CLOU Reaction (4.1) takes place, producing gaseous oxygen. The O_2 reacts with the coal through combustion, keeping the O_2 partial pressure low and

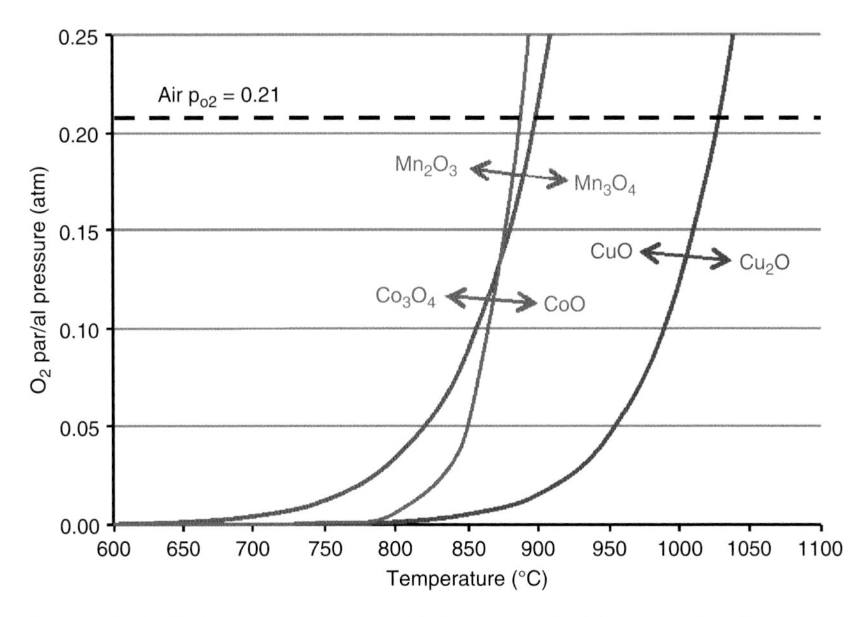

Figure 4.3 Equilibrium partial pressure of O_2 for the metal oxide systems CuO/Cu_2O, Mn_2O_3/Mn_3O_4, and Co_3O_4/CoO.

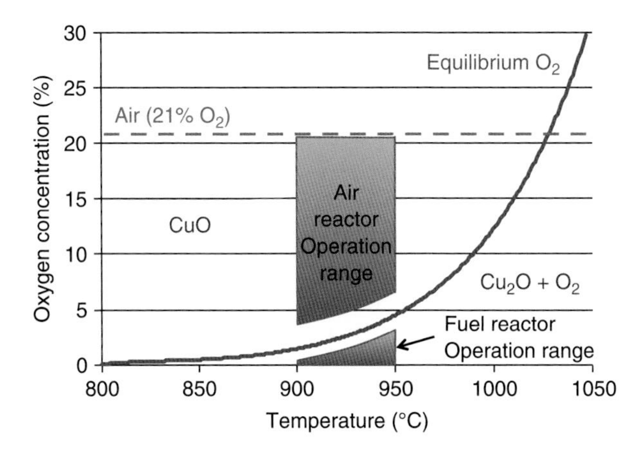

Figure 4.4 Typical operating regions for the air and fuel reactor of a copper-based CLOU system processing coal.

perpetuating oxygen release. It should be noted that further reduction of Cu_2O to Cu by heterogeneous reaction with coal volatiles may occur in the same manner as with non-CLOU chemical looping. But the expectation is that full reduction to Cu is limited and any copper will be re-oxidized to Cu_2O or CuO in the air reactor.

4.2.1 CLOU Oxygen Carriers

Oxygen carrier materials for CLOU share the same basic characteristics as conventional (non-CLOU) oxygen carriers and are distinguished only by the active metal or metals used. As noted earlier, the thermodynamics of CLOU carriers result in spontaneous reduction in a low O_2 environment, so that heterogeneous reaction with a reducing gas (fuel) is not required. Chapter 7 discusses oxygen carrier materials for CLOU systems in detail.

4.2.2 CLOU Oxygen Carrier Oxidation

The oxidation behavior of CLOU oxygen carriers in an air reactor is different from that of non-CLOU carriers. While the intrinsic rate of oxidation for most metals increases with temperature, the reversible reaction and comparatively near-equilibrium conditions for CLOU carriers result in two competing influences of temperature on the reaction rate. This can be shown by considering a simplified rate expression for the oxidation of a metal in a reversible reaction:

$$\begin{aligned} \text{rate}_{\text{oxidation}} &= (\text{rate constant}) \times (\text{driving force})^{\text{reaction order}} \\ &= \left(Ae^{-E_a/\text{RT}}\right)\left(p_{O_2,\text{actual}} - p_{O_2,\text{equil}}\right)^n \end{aligned} \tag{4.6}$$

where E_a is the activation energy, R is the ideal gas constant, T is the absolute temperature, $p_{O_2,\text{actual}}$ is the actual partial pressure of oxygen in the system, and $p_{O_2,\text{equil}}$ is the equilibrium partial pressure of oxygen at the given temperature. The quantity $Ae^{-E_a/RT}$ represents the rate constant at a given temperature while the driving force is $(p_{O_2,\text{actual}} - p_{O_2,\text{equil}})$.

Although not explicitly indicated in Eq. (4.6), there are two aspects of the rate expression that are affected by temperature. First, there is the classical exponential influence of temperature addressed by the $Ae^{-E_a/\text{RT}}$ term. Secondly, there is the influence of temperature on oxygen equilibrium partial pressure $p_{O_2,\text{equil}}$ that is depicted in Figure 4.3. For most systems, the equilibrium partial pressure of O_2 is zero over the temperature range of interest, so that the dependency of rate on oxygen concentration is not affected by changes in temperature. But because CLOU by design operates near the equilibrium conditions of the metal oxide system, the equilibrium concentration of oxygen must be taken into consideration. As seen in Figure 4.3, the equilibrium O_2 partial pressure for CLOU materials can be significant at high temperature. The result is that the concentration driving force $(p_{O_2,\text{actual}} - p_{O_2,\text{equil}})$ becomes less as temperature increases and the resulting reduction in oxidation rate (Eq. (4.6)) can exceed the positive influence of the increase in rate due to temperature resulting from the thermodynamics of the system and represented by the Arrhenius expression.

The reduction in oxidation rate with increasing temperature has been observed experimentally for copper-based oxygen carriers [4, 11–13, 42, 44, 51]. As seen in Figure 4.5a, the observed rate of oxidation achieves a maximum at approximately 890 °C. Above that temperature, the driving force difference between the supplied O_2 partial pressure of 0.21 atm and the equilibrium partial pressure becomes less (Figure 4.5b), so the rate decreases resulting in an apparent negative activation energy at higher temperatures. Various experimental methods can be used to isolate the influences of temperature on oxidative driving force and the intrinsic chemical reaction. For example, the supplied O_2 partial pressure can be varied for experiments at a given temperature to study the influence of oxidation driving force on the rate. In addition, the supplied O_2 partial pressure can be

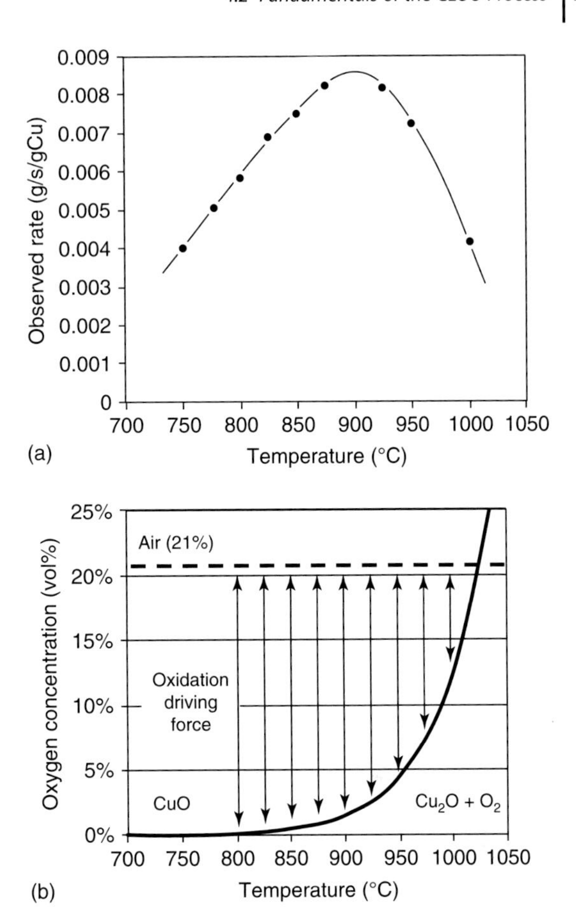

Figure 4.5 (a) Measured oxidation rate of Cu_2O oxidation of a copper-based CLOU oxygen carrier and (b) depiction of oxidation driving force for Cu_2O/CuO system.

varied with temperature to hold the oxidation driving force constant, thus allowing one to focus on the intrinsic chemical kinetics of the oxidation reaction [42].

Several studies have been performed to characterize the rate of oxidation of CLOU oxygen carriers, primarily focusing on copper-based materials [4, 13, 18, 38, 40, 42, 44, 48]. Empirical and mechanistic models have been proposed and fit to available experimental data.

4.2.3 CLOU Oxygen Carrier Reduction ("Uncoupling")

The most important characteristic of CLOU oxygen carriers is their ability to release gaseous O_2 in the fuel reactor. The rate of oxygen uncoupling is related to the equilibrium partial pressure of O_2 at the particle temperature. As observed in Figure 4.3, the equilibrium partial pressure increases with temperature and experimental results confirm that the rate of the metal reduction reaction responsible for O_2 release increases with temperature (Figure 4.6) [10, 40, 43, 44]. Similar to oxidation, the driving force for O_2 release is the difference between the equilibrium O_2 partial pressure and the actual O_2 partial pressure in the gas surrounding the oxygen carrier particle, $(p_{O_2,equil} - p_{O_2,surr})$. Studies in which the uncoupling reaction has been performed in gas environments with a finite oxygen

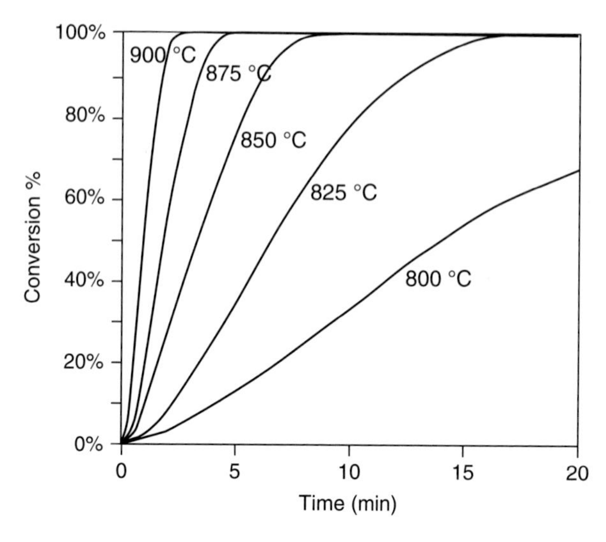

Figure 4.6 Reduction of CuO to Cu_2O in nitrogen as a function of temperature for a CLOU oxygen carrier of 45% CuO content on zirconia. Thermogravimetric analysis data.

concentration lower than the equilibrium concentration have proved that the rate decreases when oxygen is present in the surrounding atmosphere [43]. It is generally believed that the O_2 concentration in the fuel reactor will be quite low, below 0.5 vol%, since the oxygen present will readily react with volatile matter and char. However, at higher temperatures where oxygen release is rapid, and with comparatively unreactive fuels, the O_2 content in the gas surrounding the particles may exceed 1%, which would then slow the rate of oxygen release.

Several investigations have focused on characterization of CLOU carrier oxygen release, and models describing reduction and expressions to model the rate of oxygen release have been developed [40, 43, 44].

4.3 CLOU Reactor Design

Design of fluidized bed reactors for conventional (non-CLOU) chemical looping combustion of gaseous and solid fuels has been the subject of many investigations and a variety of different systems have been built and tested. Generally speaking, the designs involve either a bubbling or circulating fluidized bed air reactor coupled to a bubbling or circulating fuel reactor with overflow pipes (bubbling bed reactors) or cyclones (circulating bed reactors) to separate oxygen carrier particles from the exit gas. Such systems are described in Chapter 2. CLOU reactor systems share many design characteristics with non-CLOU chemical looping systems and in some regards the only difference between a conventional non-CLOU reactor and a CLOU reactor is the nature of the oxygen carrier. However, the unique chemical behavior of CLOU oxygen carriers described earlier creates special challenges – and opportunities – for chemical looping reactors. Some of the important considerations are described in the following sections, with emphasis on fluidized bed-based systems.

4.3.1 Fuel Flow and Overall Balances

Flow rates and material balances of a CLOU system are determined in the same manner as for any chemical looping combustion system. System design of fluidized bed chemical looping systems is addressed in Chapter 2 and the methodology for determining reactor flow rates and material balances is described in a number of references, e.g. Lyngfelt et al. [52]. Primary mass balance considerations are reviewed here.

The feed rate of solid fuel to a CLOU system (kg s^{-1}) depends on the overall desired thermal input, or fuel power P (MW), and the heating value of the fuel, H_{fuel} (MJ kg^{-1}).

$$\dot{m}_{\text{fuel}} = \frac{P}{H_{\text{fuel}}} \tag{4.7}$$

Assuming the fuel is completely converted in the chemical looping system, the air reactor must be fed enough oxygen to combust the fuel. Defining S_r as the stoichiometric ratio of air required to completely combust the fuel (kg air/kg fuel), the minimum flow rate of air to the air reactor would be $\dot{m}_{\text{fuel}}S_r$ if all incoming oxygen were to react. However, as described earlier, the gas exiting the air reactor of a CLOU system must have some concentration of O_2 to ensure that the air reactor is overall oxidizing with respect to the thermodynamic equilibrium of the oxygen carrier and that there is an oxidation driving force throughout the reactor. For an air reactor exit O_2 volume fraction $y_{O_2,\text{exit}}$, the required mass flow of air to the air reactor is

$$\dot{m}_{\text{air}} = \dot{m}_{\text{fuel}}S_r \frac{0.21(1 - y_{O_2,\text{exit}})}{0.21 - y_{O_2,\text{exit}}} \tag{4.8}$$

Oxygen is transported from the air reactor to the fuel reactor by the oxygen carrier particles. The required mass flow rate of oxygen carrier depends on the oxygen carrying capacity of the material, which in turn depends on the metal oxide system used and the concentration of the active metal on the carrier. The oxygen carrying capacity of the oxygen carrier, also known as the oxygen ratio, R_o, is the mass fraction of oxygen available for reaction when the carrier is in its oxidized state:

$$R_o = \frac{m_{\text{OC,ox}} - m_{\text{OC,red}}}{m_{\text{OC,ox}}} \tag{4.9}$$

where subscripts "ox" and "red" represent the oxidized and reduced states. The oxygen carrying capacities of the pure metal oxide systems described previously are shown in Table 4.1.

A useful convention is the degree of oxidation of the oxygen carrier, denoted X:

$$X = \frac{m_{\text{OC,actual}} - m_{\text{OC,red}}}{m_{\text{OC,ox}} - m_{\text{OC,red}}} \tag{4.10}$$

For a fully oxidized carrier $X = 1$ and for a fully reduced carrier $X = 0$. Owing to the mixed reactor nature of fluidized bed systems and kinetics of the oxygen carrier reactions, the oxygen carrier is generally not fully oxidized in the air reactor nor fully reduced in the fuel reactor. The respective degrees of oxidation and

Table 4.1 Oxygen carrying capacity of pure metal oxide systems based on oxidized state.

Metal	Oxidized state	Reduced state	Oxygen carrying capacity (R_o)
Copper	CuO	Cu_2O	0.101
Manganese	Mn_2O_3	Mn_3O_4	0.034
Cobalt	Co_3O_4	CoO	0.066

reduction in the air and fuel reactors, and in particular the difference between those, ΔX, affect the oxygen carrier circulation rate required to transport sufficient oxygen to the fuel reactor for combustion.

$$\Delta X = X_{AR,exit} - X_{FR,exit} \tag{4.11}$$

The mass flow rate of oxygen carrier to the fuel reactor depends on the mass flow of oxygen required to completely combust the fuel, the oxygen carrying capacity, and the conversion of the oxygen carrier:

$$\dot{m}_{OC\ to\ FR} = \frac{0.233\dot{m}_{fuel}S_r[1 - (1 - X_{AR,exit})R_o]}{\Delta X R_o} \tag{4.12}$$

The factor 0.233 is the mass fraction of oxygen in air. The mass flow rate of oxygen carrier from the fuel reactor back to the air reactor is slightly less since oxygen is consumed in the fuel reactor:

$$\dot{m}_{OC\ to\ AR} = \dot{m}_{OC\ to\ FR} - 0.233\dot{m}_{fuel}S_r \tag{4.13}$$

4.3.2 Energy Considerations

The function of a chemical looping combustion system is to produce heat and power through the production of high pressure steam, which is fed to a turbine/generator in a conventional Rankine steam cycle. Steam can be produced by heat extraction in the dense bed or freeboard region of the fluidized beds. Downstream of the reactors, flue gas heat recovery can also be used to generate steam.

The air reactor of a chemical looping system will always release heat since oxidation of metals is an exothermic process. Heat released per mole of O_2 consumed for the monometallic oxygen carriers discussed earlier is shown in Reactions (4.3)–(4.5), noting that the oxidation reaction is the reverse of that shown. Without any measures to remove heat within the air reactor, the energy released through oxidation will increase the temperature of the bed relative to that of the material entering from the fuel reactor. The magnitude of the temperature increase depends on material heat capacity and the oxygen carrying capacity R_o of the active oxygen carrier, with higher oxygen capacity resulting in larger temperature increase since there is less inert solid to absorb the heat.

The fuel reactor involves not only reduction of the oxygen carrier, which is an endothermic process with the same amount of heat required as given off during oxidation in the air reactor, but also combustion of the fuel, an exothermic process. The overall heat demand or release depends on the relative heats of reaction for the oxygen carrier and the fuel. For illustrative purposes, consider the typical bituminous coal with a heating value (heat of combustion) of 25 000 kJ kg^{-1} requiring 2 kg (62.5 mol) O_2 per kg coal for complete combustion. For a copper-based CLOU oxygen carrier, heat required for reduction (Reaction (4.1)) is $62.5 \times 262.5 = 16\,400$ kJ kg^{-1} of fuel, but combustion produces 25 000 kJ kg^{-1} of heat for a net of 8600 kJ heat release per kg of fuel. The heat for reduction of manganese is lower than that for copper, so even more heat is released per mole of O_2 generated. However, as seen in Table 4.1 the oxygen carrying capacity for the Mn_2O_3/Mn_3O_4 system is 3.4 wt% compared to 10.1% for the CuO/Cu_2O system, so more oxygen carrier is required to deliver the required oxygen and the increase in bed temperature would be less.

In terms of energy release in the fuel reactor, a CLOU system is no different than a conventional non-CLOU chemical looping system as long as the metal oxide system, the extent of reaction and fuel conversion are the same. The only difference is the reaction pathway responsible for converting the fuel. For example, if the CuO/Cu_2O system were used to convert coal (represented by carbon, C) in a non-CLOU configuration, the coal would have to first be gasified by steam or CO_2 in what is sometimes called in-situ gasification CLC (iG-CLC), and the resulting H_2 and CO would react heterogeneously with the oxygen carrier. As shown below for steam gasification, the overall reaction is the same for conversion of carbon through release of O_2 in a CLOU system, and the overall energy required is identical.

Steam gasification:

$$C + H_2O \rightarrow CO + H_2 \quad \Delta H_{r,900°C} = 141.8 \text{ kJ mol}^{-1}$$
$$CO + 2CuO \rightarrow Cu_2O + CO_2 \quad \Delta H_{r,900°C} = 150.7 \text{ kJ mol}^{-1}$$
$$H_2 + 2CuO \rightarrow Cu_2O + H_2O \quad \Delta H_{r,900°C} = 117.6 \text{ kJ mol}^{-1}$$
$$\overline{C + 4CuO \rightarrow 2Cu_2O + CO_2 \quad \Delta H_{r,900°C} = 126.5 \text{ kJ mol}^{-1}}$$

CLOU:

$$4CuO \rightarrow 2Cu_2O + O_2 \quad \Delta H_{r,900°C} = 262.5 \text{ kJ mol}^{-1}$$
$$C + O_2 \rightarrow CO_2 \quad \Delta H_{r,900°C} = 389.0 \text{ kJ mol}^{-1}$$
$$\overline{C + 4CuO \rightarrow 2Cu_2O + CO_2 \quad \Delta H_{r,900°C} = 126.5 \text{ kJ mol}^{-1}}$$

Mattisson et al. [1] analyzed the expected change in fuel reactor temperature in an adiabatic system for the monometallic oxygen carriers shown in Reactions (4.3)–(4.5). The results are shown in Figure 4.7. The heat required for reduction of CuO and Mn_2O_3, per mole of O_2 released, is lower than that released by combustion of carbon, resulting in an overall temperature rise. In extreme cases, the bed temperature will increase by several hundred degrees if there is no heat recovery within the reactor. On the other hand, reduction of Co_3O_4 requires more heat than provided by the carbon combusted, resulting in a temperature decrease.

4.3.3 Air Reactor Design

The air reactor of a chemical looping system is responsible for oxidizing the oxygen carrier and its size and design depend on the nature of the carrier and the

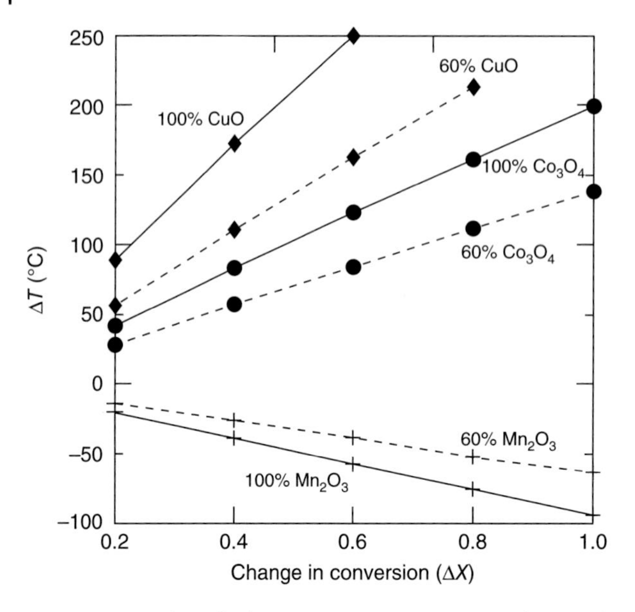

Figure 4.7 Resulting fuel reactor temperatures versus degree of conversion of monometallic CLOU oxygen carriers for combustion of carbon (C) and fluidization with CO_2 preheated to 400 °C. It is assumed that the support for the 60% carriers is ZrO_2 and that particles enter the fuel reactor at 913, 775, and 850 °C for the CuO, Mn_2O_3, and Co_3O_4 carriers, respectively. Source: Mattisson et al. 2009 [1]. Adapted with permission of Elsevier.

overall fuel processing rate of the system. The air fed to the air reactor should, at a minimum, contain enough oxygen to completely combust the fuel. As with any type of combustion system, feeding just that minimum amount would result in the effluent from the air reactor containing no oxygen. Regardless of whether the system is a conventional non-CLOU chemical looping combustor or a CLOU-based system, more air than the minimum is fed to ensure that the oxidation reactions are efficient throughout the volume of the air reactor. The design flow rate of air is determined by Eq. (4.8).

The thermodynamic equilibrium behavior of CLOU oxygen carriers creates additional design challenges for the air reactor and places a constraint on the oxygen concentration exiting the reactor. To ensure that conditions in the air reactor promote oxidation of the oxygen carrier, the O_2 concentration throughout the entire reactor must exceed the equilibrium concentration for the particular carrier metal and temperature. Referring to Figure 4.3, for example, one can see that it would not be possible to operate the air reactor of a Mn_2O_3/Mn_3O_4-based CLOU system to have an exiting O_2 concentration of 5% at 850 °C, since the equilibrium O_2 partial pressure at that temperature is 8.8%. Compared to a non-CLOU carrier that has an equilibrium O_2 partial pressure of essentially zero, the Mn-based CLOU system would require an air flow about 72% higher to provide enough oxygen to completely oxidize the carrier material at 850 °C.

Another practical consideration involves the driving force for the oxidation reaction. As noted previously, the driving force for the reaction is the difference between the local O_2 concentration and the equilibrium concentration. In order to ensure that there is appreciable driving force throughout the reactor, CLOU air reactors are generally designed such that the exit O_2 concentration is at least 2%-units above the equilibrium concentration. For the Mn-based system operating at 850 °C, the target O_2 exit concentration would thus be 10.8% and the air flow rate would need to be more than double that calculated based on complete oxidation of the oxygen carrier, assuming 0% O_2 in the exit.

It is also clear that the driving force for oxidation, and thus the rate of oxidation, is not uniform throughout the air reactor. Oxidation will be most efficient at the air inlet location where the O_2 concentration is near 21%. As the air flows through the reactor and O_2 is consumed by the oxygen carrier, the oxidation rate will decrease. The region of the reactor farthest from the air inlet will experience the lowest rates of oxidation and in extreme cases where the exit O_2 concentration is near the equilibrium concentration, a significant fraction of the reactor may contribute comparatively little to oxidation of the carrier. Thus, it is important to design the air reactor size and air flow to ensure that (i) the driving force for the oxidation reaction remains appreciable, (ii) the residence of the oxygen carrier in the air reactor is sufficient to achieve the necessary degree of oxidation, and (iii) sufficient oxygen is introduced to completely burn the fuel in the fuel reactor, taking into consideration the equilibrium O_2 partial pressure at the exit.

It is not necessary to completely oxidize the oxygen carrier in the air reactor. Studies have shown that the rate of carrier oxidation becomes less as the material is converted, i.e. that the reaction is of an order greater than zero in the active metal [3, 42]. For example, in the conversion of Cu_2O to CuO for copper-based CLOU, the reaction is approximately first order in Cu_2O and the rate slows significantly at high conversions (Figure 4.8). Achieving 99% or even 95% conversion to CuO, therefore, would require a significant amount of time and an unnecessarily large reactor. A more sensible approach is to balance the reactor size with solids residence time and circulation rate to operate within the range of conversions where the oxidation reaction is still reasonably fast. Although the optimum degree of carrier oxidation will depend on specific properties of the material including the reaction order in the active metal, a good rule of thumb is to aim for 80% of the carrier material being in the fully oxidized state. Targeting higher conversions results in diminishing returns in terms of capital investment and the inventory of oxygen carrier required.

A final consideration for design of the air reactor is the generally high degree of heat release associated with the oxidation reactions and the impact this has on reactor temperature. For example, approximately 825 kJ of heat is generated per kg CuO formed in a copper-based CLOU system. For an oxygen carrier composed of 60% CuO on a zirconia support, heat release would increase the bed temperature by about 250 °C [1]. Higher temperatures will increase the equilibrium O_2 partial pressures (Figure 4.3), which will in turn reduce the driving force and effectiveness of the oxidation reactions as described above. This can be overcome by controlling the temperature in the air reactor by reducing the degree of air preheating or by extracting heat from the bed. Operating with high solids circulation rates and lower loadings of active metal on the oxygen

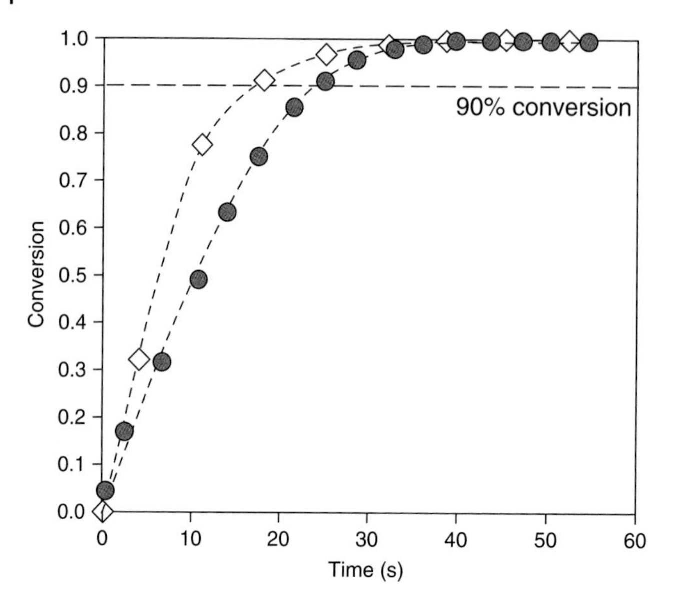

Figure 4.8 Oxidation of two copper-based oxygen carriers in air at 925°C based on TGA experiments. The time required to achieve the final 10% conversion is about as long as that required to achieve the first 90% conversion. Data from Clayton et al. 2014 [42].

carrier and/or lower extents of oxidation will also reduce the temperature rise associated with the reaction.

4.3.4 Fuel Reactor Design

The fuel reactor of a CLOU system behaves differently from the fuel reactor of a conventional chemical looping system. Owing to the release of gaseous O_2 by the oxygen carrier, it offers an overall more reactive environment, particularly for conversion of solid fuel char. In the same manner as observed in the experiment described in Figure 4.2, the local oxygen concentration can be at most as much as the equilibrium partial pressure at the temperature of the oxygen carrier material (Figure 4.3). If fuel is available to react with the oxygen, then the local O_2 concentration depends on the fuel-to-oxygen carrier ratio as well as the relative rates of oxygen consumption by combustion and oxygen release by the carrier. The rate of combustion, in turn, depends on the local O_2 concentration. Low fuel concentrations and comparatively unreactive fuels such as petcoke char result in O_2 concentrations near the equilibrium concentration. High fuel concentrations or highly reactive fuels such as biomass char rapidly consume the available O_2, so that the concentration will be near zero.

The O_2 concentration in a CLOU fuel reactor varies by location. Near the distributor at the bottom of a fluidized bed, the O_2 concentration is low even if little or no fuel is present, simply because oxygen release is not instantaneous and there will have been too little time for appreciable concentrations to build up in the incoming fluidizing gas (typically steam and/or recycled CO_2). The region around the fuel feeding point will also have low O_2 concentrations since volatiles released by the fuel readily combust homogeneously. Near the top of the bed, O_2 concentrations will be higher since volatiles concentrations will be low and only the fuel char will be available to consume the O_2.

Balancing the fuel char inventory with the available released oxygen is one of the biggest challenges of CLOU fuel reactor design and operation. Ideally, the fuel reactor will have a very low char concentration to minimize unwanted transport of char to the air reactor, where it will combust but the CO_2 will not be captured. However, low char concentrations do not effectively consume the O_2 released from CLOU carriers, so that a finite concentration of O_2 will exist throughout most of the bed. At the top of the bed near the gas exit, fuel volatiles will not be available and the gas will have had enough time since entering the system to develop an appreciable concentration of oxygen. It is therefore possible for the gas exiting the fuel reactor to contain unreacted oxygen, since O_2 generation is relatively uniform throughout the bed and the ability to exactly balance O_2 generation with char combustion at the top of the bed is limited. If the product gas from the fuel reactor contains an undesirably high concentration of O_2, it is possible to address that by the so-called natural gas polishing, whereby a stoichiometric amount of natural gas is fed into the hot product gas line to combust with the available O_2. This is similar to the strategy of oxygen polishing proposed for conventional chemical looping systems that can have undesirably high concentrations of combustible compounds such as CO and methane in the fuel reactor product gas.

The temperature of the fuel reactor is quite constant throughout the volume of the reactor, since the concentration of char as well as production of O_2 is relatively uniform throughout the bed. Individual particles may exhibit different temperatures. Char combustion is a highly exothermic reaction and the temperature of burning char particles will be higher than that of the oxygen carrier particles, which undergo endothermic reactions as O_2 is generated. But in a well-mixed fluidized bed differences in local particle temperatures will not be observed or affect the performance of the system.

4.3.5 Loop Seal Design

Non-mechanical valves such as loop seals are used in circulating fluidized beds or dual fluidized bed systems to allow solids to flow while minimizing pass-through of gas from the destination reactor to the source reactor. For example, a loop seal is typically installed on the solids transfer line from the fuel reactor to the air reactor to keep air from flowing through the line and entering the fuel reactor. Particular attention must be paid to the design of loop seals in CLOU systems, especially those that use cyclones for gas–particle separation. The nature of CLOU oxygen carriers to release O_2 when in an oxygen-free environment can complicate design and operation.

To illustrate, consider a dual fluidized bed CLOU system that uses a cyclone on the air reactor exit to separate oxygen carrier particles from the gas. The solids exit the bottom of the cyclone and pass through a loop seal to the fuel

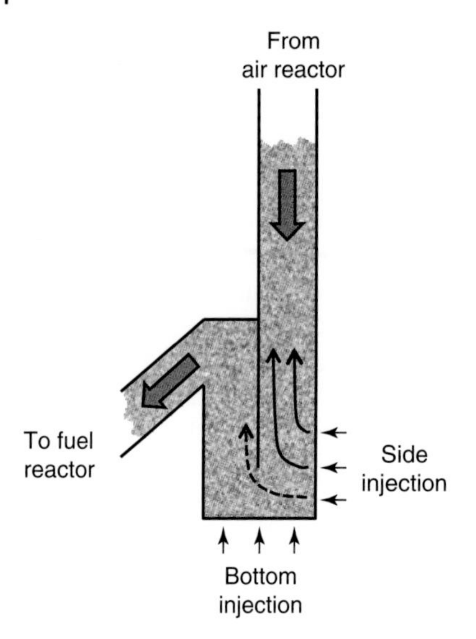

Figure 4.9 Schematic of a loop seal below the air reactor cyclone and feeding into the fuel reactor.

reactor (Figure 4.9). If the loop seal is fluidized with air, the air can flow with the solids to the fuel reactor. Oxygen in that air will assist combustion in the fuel reactor, but nitrogen will also enter the fuel reactor, diluting the CO_2-rich fuel reactor product gas. If the loop seal is instead fluidized with steam or CO_2, the freshly oxidized oxygen carrier will begin to release O_2 already in the loop seal, including in the column of solids in the dip leg below the cyclone. Gas flowing upward through that dip leg into the cyclone will carry with it oxygen that was meant to be released in the CLOU fuel reactor, reducing efficiency and necessitating a higher circulation rate to achieve the desired oxygen delivery to the fuel reactor.

One solution is to use two different gases in the loop seal. Air can be used as the side injection gas to keep the solids column below the cyclone flowing and fully oxidized. The primary loop seal fluidizing gas (bottom injection in Figure 4.9) can then be steam or CO_2. Although O_2 will start to be released once the particles exit the air-rich environment, it will flow with the steam or CO_2 and will ultimately end up in the fuel reactor. Optimizing injection locations for the different gases might require some trial and error. For example, if air were used for side injection at the bottom location in the loop seal of Figure 4.9, it might be swept with the solids toward the fuel reactor along the dashed line, resulting in nitrogen entering the fuel reactor and diluting the CO_2 product gas.

Similar issues must be considered for the loop seal connecting the outlet of the fuel reactor cyclone to the air reactor. If that loop seal is fluidized with air and the air flows with the solids into the air reactor, there is no concern. But if air on the inlet side of the loop seal flows through the column of solids below the cyclone, it will end up in the fuel reactor product gas where oxygen and nitrogen are not desired. Fluidizing with recycled CO_2 is a better solution in terms of fuel reactor product gas quality, but most of the loop seal CO_2 will flow into the air reactor where it will not be captured. A suitable solution is to fluidize the loop seal from the fuel reactor to the air reactor with steam. Although additional O_2 production may take place in the loop seal if the oxygen carrier is not entirely in its reduced state, the extent of oxidation will be limited and the material will be re-oxidized in the air reactor.

Another method to limit O_2 production in the loop seals is to operate them slightly cooler than the reactors themselves. As noted previously, the rate of O_2 production decreases sharply with temperature, so that a reduction of 100 or even 50 °C would help minimize O_2 generation. Of course, any active cooling would have to be balanced against the additional thermal load necessary to reheat the solids upon entering the reactor.

4.3.6 Sulfur

The attractiveness of CLOU is its superior performance for processing solid fuels. Many of the fuels that are candidates for chemical looping combustion, including coal and petroleum coke, can contain 2–6% sulfur by weight. It is important to understand the impact of sulfur on the performance of CLOU systems and on the quality of the CO_2-rich fuel reactor product gas.

Adánez-Rubio et al. [32] and Pérez-Vega et al. [34] studied the fate of sulfur in a small dual fluidized bed system processing lignite containing 5.2% sulfur and using a copper-based CLOU carrier. Under the conditions tested, nearly all sulfur exited the fuel reactor as SO_2. Consideration must therefore be given to how SO_2 in the CO_2 product stream will be handled in industrial-scale systems. A positive conclusion of these studies was that oxygen carrier performance was not affected during 15 hours of continuous operation.

Arjmand et al. [35] examined performance of perovskite-type CLOU oxygen carriers in sulfur-containing environments. The presence of calcium in the oxygen carrier reduced activity due to the formation of calcium sulfate. Similar behavior has been identified for mixed metal carriers, with nickel and iron being susceptible to reaction with sulfur [21]. Manganese carriers are expected to be inert toward sulfur under chemical looping combustion conditions, since manganese sulfates and sulfides are not stable at high temperature [21].

For the CuO/Cu_2O and Mn_2O_3/Mn_3O_4 systems, thermodynamic analysis and laboratory-scale experiments suggest that sulfur should not pose a serious threat to performance. But other metals in the oxygen carrier may react with sulfur. Systems including oxygen carrier, coal ash, and sulfur at high temperature are complex, and unreactive compounds can form that will reduce oxygen carrier capacity and reactivity. It is therefore recommended that oxygen carrier materials be experimentally tested to understand long-term performance in sulfur-containing environments.

4.4 Status of CLOU Technology Development

Since the concept of CLOU was first introduced in 2009, interest has grown and today there are many groups researching the technology. Much of the research focuses on development of robust, inexpensive CLOU oxygen carriers. Chapter 7 describes the production, properties, and performance of CLOU oxygen carriers in detail. This section focuses on CLOU reactor systems and highlights some of the important work that has been performed.

4.4.1 Laboratory-Scale CLOU Testing

To properly design a dual fluidized bed CLOU reactor system, it is important to understand the rates of chemical processes that occur within the reactors. A straightforward method for determining rates of oxygen carrier oxidation and reduction is through thermogravimetric analysis (TGA), in which the mass of a small sample of oxygen carrier is tracked as it is exposed to alternating gas atmospheres at a specified temperature. The oxidizing gas is typically air or a mixture of 5–15% O_2 in nitrogen to represent conditions within the air reactor after a portion of the incoming O_2 has been removed by reaction with the oxygen carrier. For CLOU, the reducing gas is generally pure nitrogen to represent conditions in the fuel reactor where the O_2 content is near zero due to its rapid consumption by char and volatiles. The oxygen carrying capacity of the material can be directly determined by measuring the weight difference between the oxidized and reduced states and the rates of those reactions can be evaluated from the slopes of the respective mass-versus-time curves (Figure 4.10). TGAs can also be programmed to continuously cycle the gas to evaluate long-term performance of oxygen carriers over multiple cycles.

The rates of oxidation and reduction have been measured by TGA for many different CLOU oxygen carriers [3, 4, 8, 11–14, 18, 22, 23, 30, 42–44]. Although reactivity depends on the nature and physical characteristics of the carrying material, in general copper-based oxygen carriers require 1–2 minutes for 90% oxidation of Cu_2O to CuO or for 90% reduction of CuO to Cu_2O in nitrogen at 900–950 °C [4, 13, 34, 40, 42, 43]. Less attention has been given to the study of CLOU reaction kinetics for manganese-based oxygen carriers, but it appears that rates of oxidation and reduction of these materials are similar.

Another common experimental technique for evaluating the performance of CLOU oxygen carriers, which more closely represents conditions in a chemical looping system than TGA, is to use a small batch fluidized bed such as that depicted in Figure 4.11 to test 10–50 g samples of oxygen carrier particles [2, 3, 6, 8, 11, 13, 14, 18, 20, 22, 25, 29, 39]. In a typical experiment, fluidizing gas is cycled between nitrogen and air or an air/nitrogen mixture to approximate conditions that materials are exposed to as they move between the fuel and air reactors. Through analysis of O_2 in the effluent gas, the time required for complete oxidation or complete reduction of the oxygen carrier can be determined and qualitative information regarding reaction rates can be discerned. The system can be programmed to alternate between oxygen-containing and inert gas to test the performance over multiple cycles. A thermocouple positioned within the bed will indicate a temperature rise during oxidation and

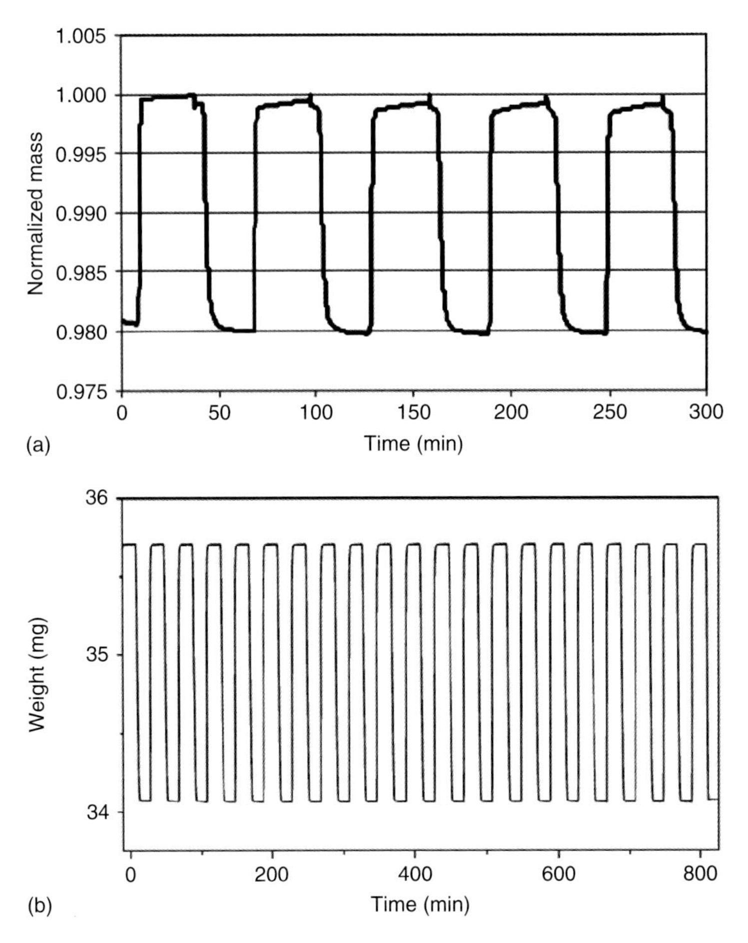

Figure 4.10 TGA curves of oxygen carrier oxidation and reduction. (a) Mass increase during oxidation in air and decrease during reduction in N_2 for an oxygen carrier of 20% CuO on SiO_2. (b) Multiple cycles of a 45% CuO on ZrO_2 oxygen carrier.

a temperature decrease during reduction. The magnitude of the temperature change can be tracked over multiple cycles to provide an indirect indication of whether oxygen carrying capacity becomes less over time. To gain an understanding of fuel conversion in a CLOU environment, small batches of solid fuel can be fed to the bed while fluidizing with an inert gas as was done in the proof-of-CLOU concept experiment described earlier and shown in Figure 4.2. The degree of conversion can be tracked by monitoring CO and CO_2 in the gas exiting the reactor. Comparative studies of oxygen carrier attrition can also be performed in these small-scale fluidized bed systems by measuring the mass of fine particulate captured on filters downstream of the reactor.

4.4.2 Development-Scale and Pilot-Scale Systems

The CLOU process has been tested at various scales with different dual fluidized bed reactor designs. Notable systems that have been used to experimentally

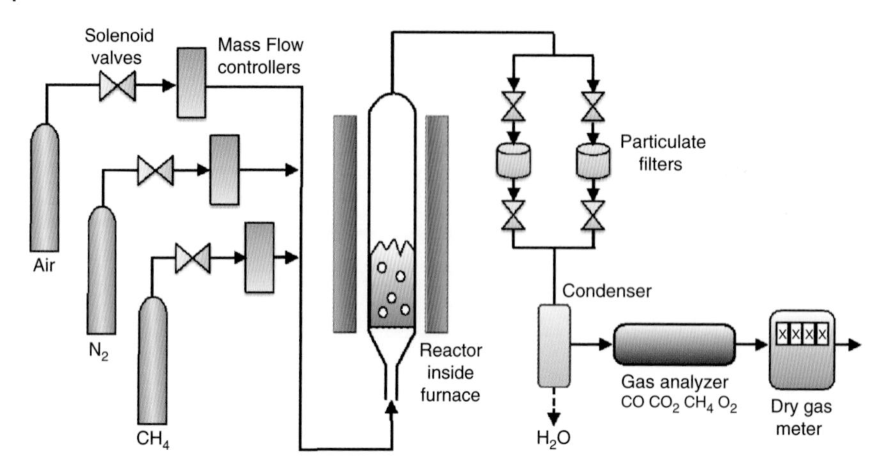

Figure 4.11 Batch fluidized bed experimental system.

research CLOU in two-reactor systems (air reactor and fuel reactor) are described below, in the order from small to large scale.

Researchers at Chalmers University of Technology have investigated the oxidation and uncoupling of various Cu- and Mn-based CLOU oxygen carriers in a small two-compartment fluidized bed reactor with a design thermal input of 300 W (Figure 4.12) [10, 26]. One section of the reactor represents the air reactor and is fluidized with air, and the other section, representing the fuel reactor, is fluidized with CO_2. The reactor is designed as one compact, integrated unit that is placed into a box furnace. Approximately 150–200 g of oxygen carrier can be

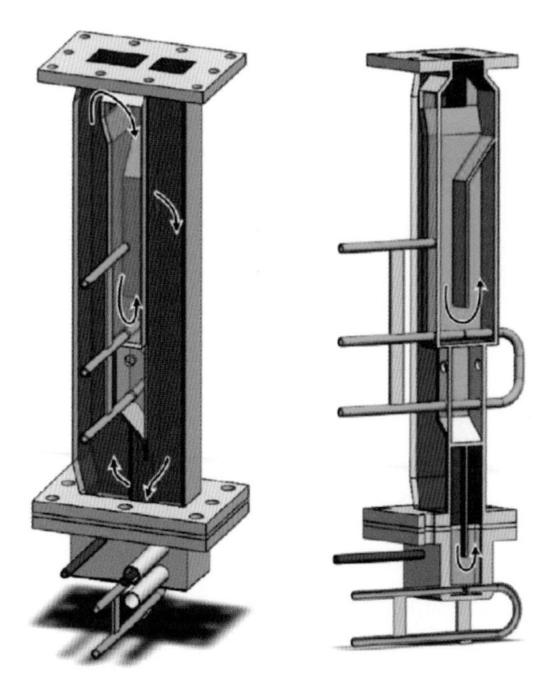

Figure 4.12 Chalmers 300 W chemical looping test reactor.

tested in the system and the rate of carrier oxidation and oxygen uncoupling can be determined as functions of temperature and O_2 concentration in the air reactor. The system does not have the capability to feed solid fuel, but is useful for evaluating uncoupling in an oxygen-free environment and to study the long-term performance of oxygen carrier materials.

The Spanish National Research Council Instituto de Carboquímica (CSIC-ICB) in Zaragoza, Spain, operates a 1.5 kW_{th} dual fluidized bed reactor for chemical looping studies (Figure 4.13). The air and fuel reactors of the system are housed in electric furnaces and a screw feeder system allows solid fuel to be fed into the bottom of the fuel reactor. The system is designed for continuous operation, and tests in excess of 15 hours duration have been performed. CSIC-ICB has investigated the performance of copper-based CLOU carriers during combustion of different coals [28, 30, 31] and biomass [33, 36], and has used the dual bed reactor system to study the fate of sulfur, nitrogen, and mercury emissions during CLOU processing of lignite coal [32, 34]. The 1.5 kW_{th} reactor has demonstrated excellent performance when operated with a $MgAl_2O_3$-supported copper-based oxygen carrier and is able to achieve greater than 95% CO_2 capture with nearly complete conversion of the fuel.

Since 2004, Chalmers University of Technology has operated a dual-bed reactor designed for 10 kW fuel input (Figure 4.14). The fuel reactor is a bubbling bed system with an overflow pipe to transfer solids to the air reactor via a loop seal. The lower part of the air reactor is of large enough diameter to operate in bubbling bed mode, but at the top of the reactor the diameter reduces to increase the velocity to transport particles up to a cyclone for particle–gas separation. Particles are returned from the cyclone to the fuel reactor via a loop seal. The 10 kW system has been used for evaluation of copper, manganese, and mixed metal oxide CLOU oxygen carriers, with particular focus on uncoupling reactions in the fuel reactor [26].

Recently, CSIC-ICB commissioned a new dual fluidized bed chemical looping system for solid fuels (Figure 4.15). The system is designed to operate either in conventional chemical looping combustion (CLC) or CLOU mode and has a design power rating of 50 kW_{th} when operating with a CLOU oxygen carrier [54]. Both the air and fuel reactors are relatively wide at the bottom and function as bubbling fluidized beds. The diameters of the reactors reduce above the bubbling bed section and the associated velocity increase causes the upper sections of the reactors to function as risers to carry particles to cyclones for gas–solid separation. The outlet of the fuel reactor incorporates a unique double loop seal design, which allows a portion of the solids to be returned to the fuel reactor rather than being sent to the air reactor, which can help improve residence time and conversion in the fuel reactor. The system also incorporates a carbon stripper to improve conversion of fuel char carried with the bed solids from the fuel reactor to the air reactor. The carbon stripper functions by separating the comparatively small and light char particles from the oxygen carrier particles by elutriation and then returns the char particles back to the fuel reactor.

Vienna University of Technology commissioned the first chemical looping combustion reactor rated at more than 100 kW. The dual circulating fluidized bed (DCFB) reactor (Figure 4.16) is rated at 120 kW and uses two circulating

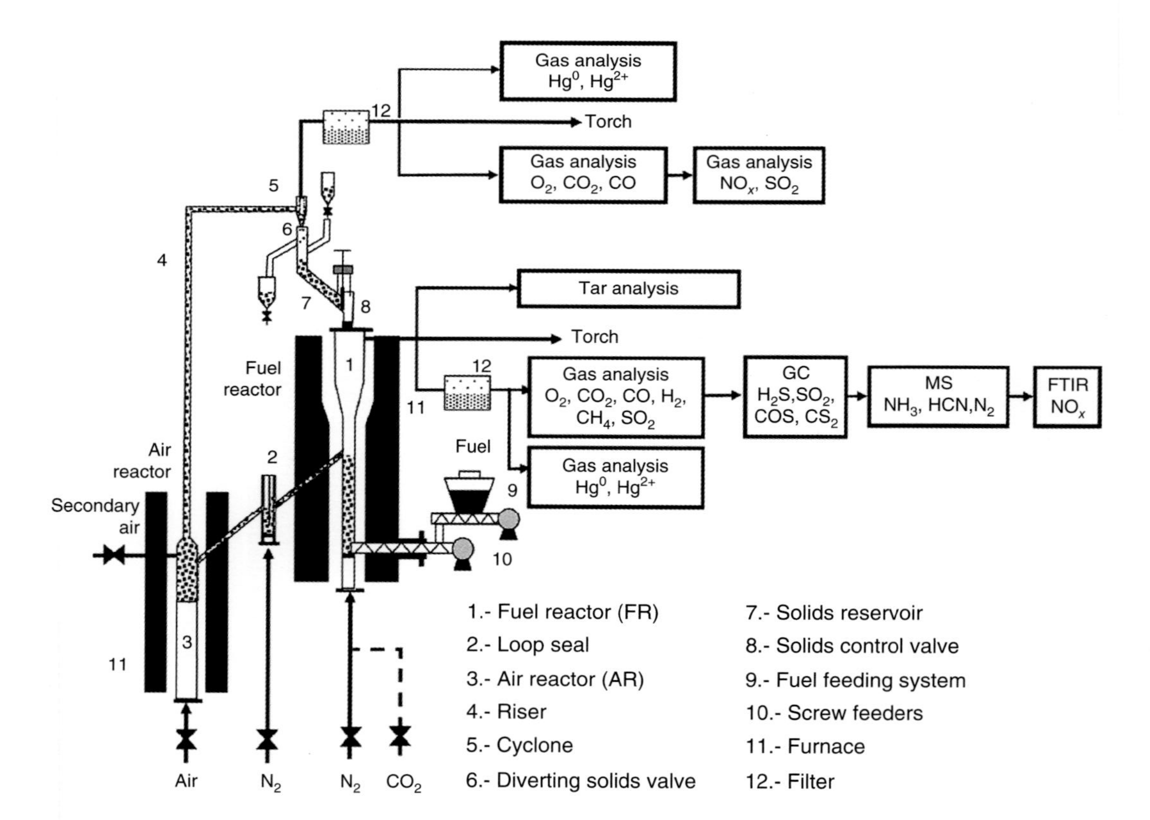

Figure 4.13 1.5 kW_{th} dual fluidized bed CLC reactor at CSIC-ICB in Zaragoza, Spain. Source: Pérez-Vega et al. 2016 [34]. Reproduced with permission of Elsevier.

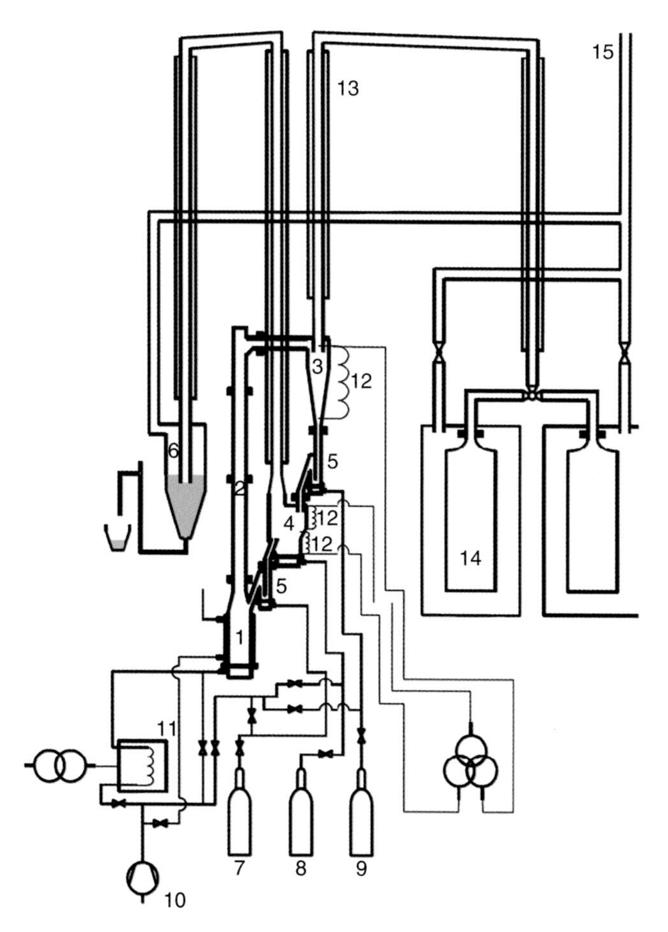

Figure 4.14 Chalmers 10 kW chemical looping test reactor. (1) Air reactor, (2) riser, (3) cyclone, (4) fuel reactor, (5) upper and lower particle locks, (6) water seal, (7) nitrogen, (8) natural gas, (9) nitrogen, (10) air, (11) preheater, (12) heating coils, (13) finned tubes for cooling of gas streams, (14) filters, and (15) connection to chimney. Source: Linderholm et al. 2008 [53]. Reproduced with permission of Elsevier.

fluidized bed systems connected by a lower loop seal to transfer solids from the fuel reactor to the air reactor. Air for the air reactor is fed through nozzles within the bed region and oxygen carrier particles are separated from the gas stream by a cyclone and then transferred to the fuel reactor via a loop seal. Particles separated from the fuel reactor cyclone are returned back to the fuel reactor. The Vienna University of Technology reactor has been used extensively for studies of natural gas chemical looping, but has also been used for evaluation of copper- and manganese-based CLOU carriers [26].

The University of Utah has designed and built a semi-pilot scale chemical looping process development unit (PDU) with a nominal power rating of 220 kW_{th}. A schematic diagram of the system is shown in Figure 4.17 and Figure 4.18 shows a photograph of the unit. The system was designed specifically for CLOU processing of coal but can also be used as a non-CLOU system for solid fuels or natural gas. The air and fuel reactors are both refractory-lined, circulating

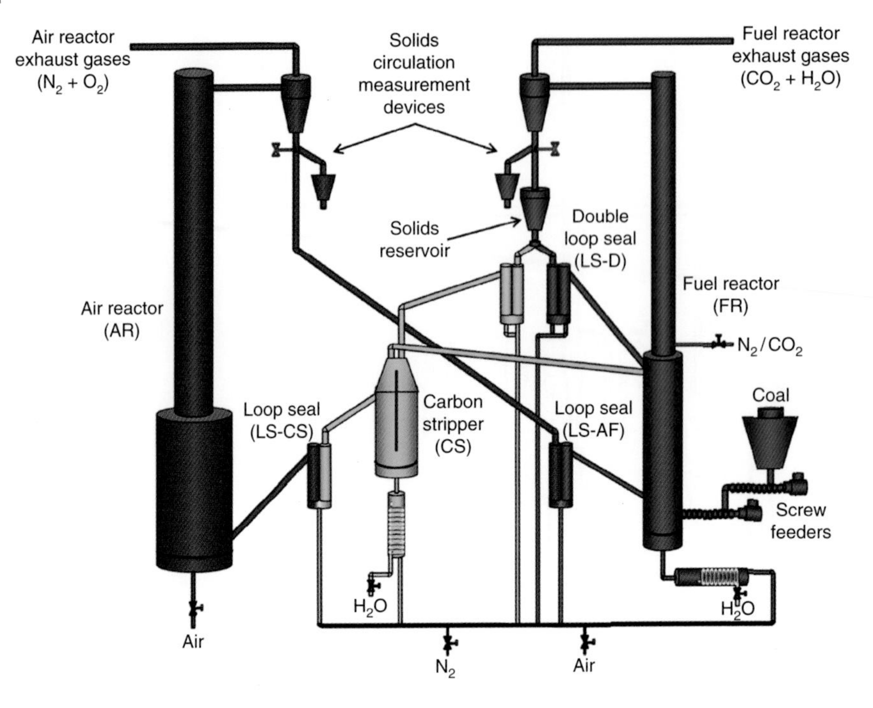

Figure 4.15 Design of CSIC-ICB's 50 kW_{th} dual fluidized bed CLC reactor. The system is designed for both conventional CLC and CLOU operation. Source: Abad et al. 2015 [54]. Reproduced with permission of Elsevier.

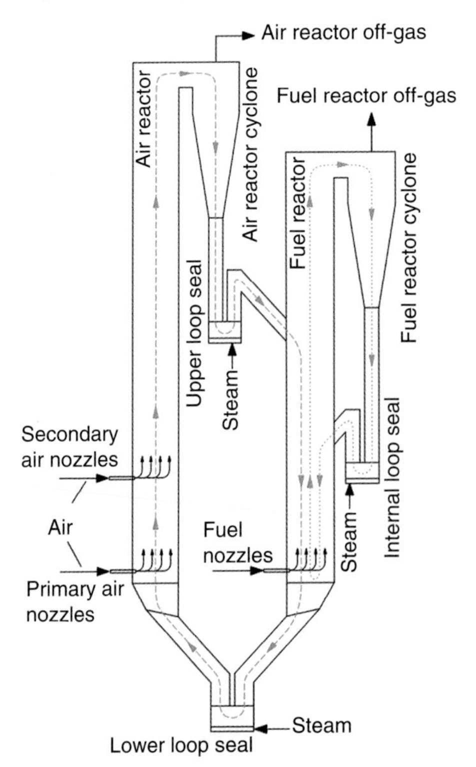

Figure 4.16 Schematic Vienna University of Technology 120 kW chemical looping combustion DCFB reactor. Source: Mayer et al. 2015 [55]. Reproduced with permission of Elsevier.

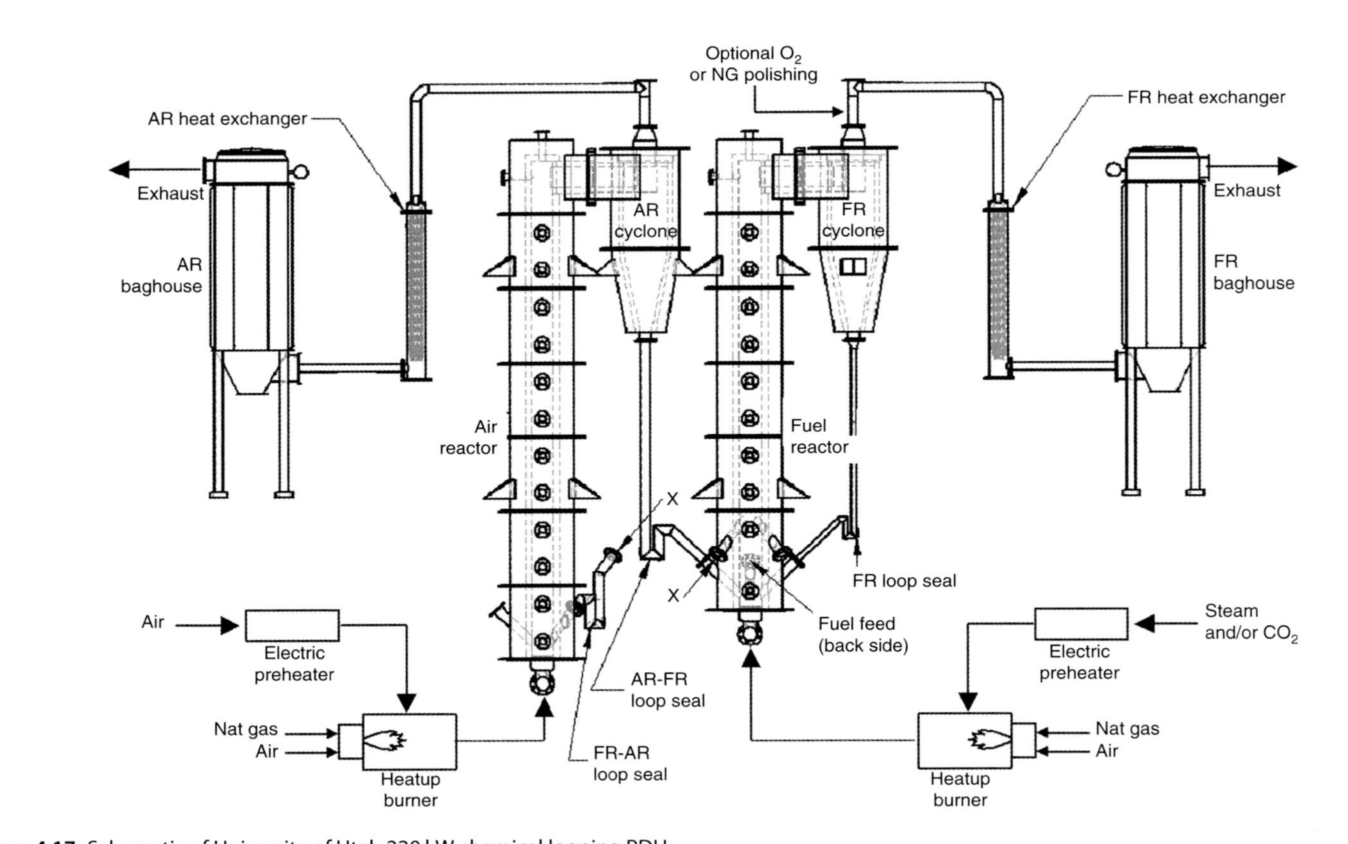

Figure 4.17 Schematic of University of Utah 220 kW chemical looping PDU.

Figure 4.18 Photograph of University of Utah chemical looping PDU.

fluidized bed systems with an internal diameter of 0.25 m and heights of 6.0 m (air reactor) and 5.3 m (fuel reactor). All particles from the air reactor cyclone flow through a loop seal into the fuel reactor, but the fuel reactor has an internal circulation system such that particles from its cyclone pass through a loop seal and are returned to the bottom of the fuel reactor. Transfer of solids from the fuel reactor to the air reactor is achieved through a connection in the lower part of the bed that is diametrically opposite to the fuel feed point. This configuration provides operational flexibility and helps maintain system stability by avoiding tight coupling between the reactors that would result if all solids from the fuel reactor cyclone were fed directly back to the air reactor. The University of Utah PDU was designed to incorporate many design aspects that larger demonstration or commercial systems would have. The DCFB design was chosen because it is more easily scaled and has a smaller footprint than bubbling bed systems. Rather than having metal reactors heated from the outside by electric heaters, the air and fuel reactors are formed from several layers of cast refractory cement in metal shells similar to industrial-scale systems. This allows comfortable operation at very high temperatures but the system requires approximately two days of preheating before chemical looping operation can begin. Product gas from the air and fuel reactors is cooled in variable load heat exchangers and bag houses on the effluent streams capture particles elutriated from the two reactors. Oxygen carrier circulation rates as high as 12 tons per hour can be achieved.

4.5 Future Development of CLOU Technology

In less than 10 years, chemical looping with oxygen technology has progressed from a conceptual idea to technology that is being demonstrated in pilot-scale systems. Most of the development effort has focused on CLOU oxygen carriers and today there are several candidates that offer high reactivity with little or no loss of oxygen carrying capacity over hundreds of oxidation/reduction cycles at laboratory scale. Physical robustness of CLOU carriers under high temperature fluidized bed conditions remains a challenge, but better formulations and methods of production are improving carrier lifetimes. The cost of CLOU oxygen carriers also needs to be reduced to make the technology competitive. New low-cost, mixed metal oxide oxygen carriers have shown promise and can help drive down the cost of CLOU technology.

Most importantly, as with all chemical looping combustion technologies, CLOU needs to be demonstrated under industrially relevant conditions. This involves long-term testing with low-cost, industrially scalable oxygen carriers processing coal or other relevant solid fuels. The overall objective of CLC technology is energy production with high CO_2 capture, so demonstration systems should incorporate heat recovery through steam generation within or immediately downstream of the reactors. Finally, it is important that the potential of CLC and CLOU to achieve carbon-free energy production be communicated to ensure strong governmental, industrial, and public support.

References

1 Mattisson, T., Lyngfelt, A., and Leion, H. (2009). Chemical-looping with oxygen uncoupling for combustion of solid fuels. *Int. J. Greenhouse Gas Control* 3: 11–19.

2 Leion, H., Mattisson, T., and Lyngfelt, A. (2009). Using chemical-looping with oxygen uncoupling (CLOU) for combustion of six different solid fuels. *Energy Procedia* 1 (1): 447–453.

3 Mei, D., Mendiara, T., Abad, A. et al. (2015). Evaluation of manganese minerals for chemical looping combustion. *Energy Fuels* 29: 6605–6615.

4 Eyring, E.M., Konya, G., Lighty, J.S. et al. (2011). Chemical looping with copper oxide as carrier and coal as fuel. *Oil Gas Sci. Technol.* 66: 209–221.

5 Adánez, J., Abad, A., Garcia-Labiano, F. et al. (2012). Progress in chemical-looping combustion and reforming technologies. *Prog. Energy Combust. Sci.* 38: 215–282.

6 Mattisson, T., Leion, H., and Lyngfelt, A. (2009). Chemical-looping with oxygen uncoupling using CuO/ZrO_2 with petroleum coke. *Fuel* 88: 683–690.

7 Shulman, A., Cleverstam, E., Mattisson, T., and Lyngfelt, A. (2009). Manganese/iron, manganese/nickel, and manganese/silicon oxides used in chemical-looping with oxygen uncoupling (CLOU) for combustion of methane. *Energy Fuels* 23: 5269–5275.

8 Adánez-Rubio, I.I., Gayán, P., García-Labiano, F. et al. (2011). Development of CuO-based oxygen-carrier materials suitable for chemical-looping with oxygen uncoupling (CLOU) process. *Energy Procedia* 4: 417–424.

9 Arjmand, M., Azad, A.-M., Leion, H. et al. (2011). Prospects of Al_2O_3 and $MgAl_2O_4$-supported CuO oxygen carriers in chemical-looping combustion (CLC) and chemical-looping with oxygen uncoupling (CLOU). *Energy Fuels* 25: 5493–5502.

10 Rydén, M., Lyngfelt, A., and Mattisson, T. (2011). $CaMn_{0.875}Ti_{0.125}O_3$ as oxygen carrier for chemical-looping combustion with oxygen uncoupling

(CLOU) – experiments in a continuously operating fluidized-bed reactor system. *Int. J. Greenhouse Gas Control* 5: 356–366.
11 Adánez-Rubio, I., Gayán, P., Abad, A. et al. (2012). Evaluation of a spray-dried $CuO/MgAl_2O_4$ oxygen carrier for the chemical looping with oxygen uncoupling process. *Energy Fuels* 26: 3069–3081.
12 Gayán, P., Adánez-Rubio, I., Abad, A. et al. (2012). Development of Cu-based oxygen carriers for chemical-looping with oxygen uncoupling (CLOU) process. *Fuel* 96: 226–238.
13 Wen, Y.Y., Li, Z.S., Xu, L., and Cai, N.S. (2012). Experimental study of natural Cu ore particles as oxygen carriers in chemical looping with oxygen uncoupling (CLOU). *Energy Fuels* 26: 3919–3927.
14 Peterson, S.B., Konya, G., Clayton, C.K. et al. (2013). Characteristics and CLOU performance of a novel SiO_2-supported oxygen carrier prepared from CuO and β-SiC. *Energy Fuels* 27: 6040–6047.
15 Azimi, G., Leion, H., Rydén, M. et al. (2013). Investigation of different Mn–Fe oxides as oxygen carrier for chemical-looping with oxygen uncoupling (CLOU). *Energy Fuels* 27: 367–377.
16 Azimi, G., Rydén, M., Leion, H. et al. (2013). $(Mn_zFe_{1-z})_yO_x$ combined oxides as oxygen carrier for chemical-looping with oxygen uncoupling. *AIChE J.* 59: 582–588.
17 Mohammad Pour, N., Leion, H., Rydén, M., and Mattisson, T. (2013). Combined Cu/Mn oxides as an oxygen carrier in chemical looping with oxygen uncoupling (CLOU). *Energy Fuels* 27: 6031–6039.
18 Xu, L., Wang, J., Li, Z., and Cai, N. (2013). Experimental study of cement-supported CuO oxygen carriers in chemical looping with oxygen uncoupling (CLOU). *Energy Fuels* 27: 1522–1530.
19 Arjmand, M., Leion, H., Mattisson, T., and Lyngfelt, A. (2014). Investigation of different manganese ores as oxygen carriers in chemical-looping combustion (CLC) for solid fuels. *Appl. Energy* 113: 1883–1894.
20 Zhao, H., Wang, K., Fang, Y. et al. (2014). Characterization of natural copper ore as oxygen carrier in chemical-looping with oxygen uncoupling of anthracite. *Int. J. Greenhouse Gas Control* 22: 154–164.
21 Rydén, M., Leion, H., Mattisson, T., and Lyngfelt, A. (2014). Combined oxides as oxygen-carrier material for chemical-looping with oxygen uncoupling. *Appl. Energy* 113: 1924–1932.
22 Galinsky, N., Mishra, A., Zhang, J., and Li, F. (2015). $Ca_{1-x}A_xMnO_3$ (A = Sr and Ba) perovskite based oxygen carriers for chemical looping with oxygen uncoupling (CLOU). *Appl. Energy* 157: 358–367.
23 Shafiefarhood, A., Stewart, A., and Li, F. (2015). Iron-containing mixed-oxide composites as oxygen carriers for chemical looping with oxygen uncoupling (CLOU). *Fuel* 139: 1–10.
24 Sundqvist, S., Arjmand, M., Mattisson, T. et al. (2015). Screening of different manganese ores for chemical-looping combustion (CLC) and chemical-looping with oxygen uncoupling (CLOU). *Int. J. Greenhouse Gas Control* 43: 179–188.
25 Tian, X., Zhao, H., Wang, K. et al. (2015). Performance of cement decorated copper ore as oxygen carrier in chemical-looping with oxygen uncoupling. *Int. J. Greenhouse Gas Control* 41: 210–218.

26 Mattisson, T., Adánez, J., Mayer, K. et al. (2014). Innovative oxygen carriers uplifting chemical-looping combustion. *Energy Procedia* 63: 113–130.
27 Adánez-Rubio, I., Abad, A., Gayán, P. et al. (2012). Identification of operational regions in the chemical-looping with oxygen uncoupling (CLOU) process with a Cu-based oxygen carrier. *Fuel* 102: 634–645.
28 Abad, A., Adánez-Rubio, I., Gayán, P. et al. (2012). Demonstration of chemical-looping with oxygen uncoupling (CLOU) process in a 1.5 kW_{th} continuously operating unit using a Cu-based oxygen-carrier. *Int. J. Greenhouse Gas Control* 6: 189–200.
29 Arjmand, M., Leion, H., Mattisson, T., and Lyngfelt, A. (2013). ZrO_2-supported CuO oxygen carriers for chemical-looping with oxygen uncoupling (CLOU). *Energy Procedia* 37: 550–559.
30 Adánez, J., Gayán, P., Adánez-Rubio, I. et al. (2013). Use of chemical-looping processes for coal combustion with CO_2 capture. *Energy Procedia* 37: 540–549.
31 Adánez-Rubio, I., Abad, A., Gayán, P. et al. (2013). Performance of CLOU process in the combustion of different types of coal with CO_2 capture. *Int. J. Greenhouse Gas Control* 12: 430–440.
32 Adánez-Rubio, I., Abad, A., Gayán, P. et al. (2014). The fate of sulphur in the Cu-based chemical looping with oxygen uncoupling (CLOU) process. *Appl. Energy* 113: 1855–1862.
33 Adánez-Rubio, I., Abad, A., Gayán, P. et al. (2014). Biomass combustion with CO_2 capture by chemical looping with oxygen uncoupling (CLOU). *Fuel Process. Technol.* 124: 104–114.
34 Pérez-Vega, R., Adánez-Rubio, I., Gayán, P. et al. (2016). Sulphur, nitrogen and mercury emissions from coal combustion with CO_2 capture in chemical looping with oxygen uncoupling (CLOU). *Int. J. Greenhouse Gas Control* 46: 28–38.
35 Arjmand, M., Kooiman, R.F., Rydén, M. et al. (2014). Sulfur tolerance of $Ca_{(x)}Mn_{(1-y)}M_{(y)}O_{(3-d)}$ (M = Mg, Ti) perovskite-type oxygen carriers in chemical-looping with oxygen uncoupling (CLOU). *Energy Fuels* 28: 1312–1324.
36 Mendiara, T., Adánez-Rubio, I., Gayán, P. et al. (2016). Process comparison for biomass combustion: in situ gasification-chemical looping combustion (iG-CLC) versus chemical looping with oxygen uncoupling (CLOU). *Energy Technol.* 4: 1130–1136.
37 Siriwardane, R., Tian, H., Miller, D. et al. (2010). Evaluation of reaction mechanism of coal–metal oxide interactions in chemical-looping combustion. *Combust. Flame* 157: 2198–2208.
38 Sahir, A.H., Lighty, J.S., and Sohn, H.Y. (2011). Kinetics of copper oxidation in the air reactor of a chemical looping combustion system using the law of additive reaction times. *Ind. Eng. Chem. Res.* 50: 13330–13339.
39 Shulman, A., Cleverstam, E., Mattisson, T., and Lyngfelt, A. (2011). Chemical-looping with oxygen uncoupling using Mn/Mg-based oxygen carriers – oxygen release and reactivity with methane. *Fuel* 90: 941–950.
40 Sahir, A.H., Sohn, H.Y., Leion, H., and Lighty, J.S. (2012). Rate analysis of chemical-looping with oxygen uncoupling (CLOU) for solid fuels. *Energy Fuels* 26: 4395–4404.

41 Arjmand, M., Keller, M., Leion, H. et al. (2012). Oxygen release and oxidation rates of $MgAl_2O_4$-supported CuO oxygen carrier for chemical-looping combustion with oxygen uncoupling (CLOU). *Energy Fuels* 26: 6528–6539.

42 Clayton, C.K., Sohn, H.Y., and Whitty, K.J. (2014). Oxidation kinetics of Cu_2O in oxygen carriers for chemical looping with oxygen uncoupling. *Ind. Eng. Chem. Res.* 53: 2976–2986.

43 Clayton, C.K. and Whitty, K.J. (2014). Measurement and modeling of decomposition kinetics for copper oxide-based chemical looping with oxygen uncoupling. *Appl. Energy* 116: 416–423.

44 Adánez-Rubio, I., Gayán, P., Abad, A. et al. (2014). Kinetic analysis of a Cu-based oxygen carrier: relevance of temperature and oxygen partial pressure on reduction and oxidation reactions rates in chemical looping with oxygen uncoupling (CLOU). *Chem. Eng. J.* 256: 69–84.

45 Peltola, P., Tynjälä, T., Ritvanen, J., and Hyppänen, T. (2014). Mass, energy, and exergy balance analysis of chemical looping with oxygen uncoupling (CLOU) process. *Energy Convers. Manag.* 87: 483–494.

46 Sahir, A.H., Dansie, J.K., Cadore, A.L., and Lighty, J.S. (2014). A comparative process study of chemical-looping combustion (CLC) and chemical-looping with oxygen uncoupling (CLOU) for solid fuels. *Int. J. Greenhouse Gas Control* 22: 237–243.

47 Peltola, P., Ritvanen, J., Tynjälä, T., and Hyppänen, T. (2015). Fuel reactor modelling in chemical looping with oxygen uncoupling process. *Fuel* 147: 184–194.

48 Moghtaderi, B. (2012). Review of the recent chemical looping process developments for novel energy and fuel applications. *Energy Fuels* 26: 15–40.

49 Mattisson, T. (2013). Materials for chemical-looping with oxygen uncoupling. *ISRN Chem. Eng.* 2013: 1–19.

50 Imtiaz, Q., Hosseini, D., and Müller, C.R. (2013). Review of oxygen carriers for chemical looping with oxygen uncoupling (CLOU): thermodynamics, material development, and synthesis. *Energy Technol.* 1: 633–647.

51 Chuang, S.Y., Dennis, J.S., Hayhurst, A.N., and Scott, S.A. (2010). Kinetics of the oxidation of a co-precipitated mixture of cu and Al_2O_3 by O_2 for chemical-looping combustion. *Energy Fuels* 24: 3917–3927.

52 Lyngfelt, A., Leckner, B., and Mattisson, T. (2001). A fluidized-bed combustion process with inherent CO_2 separation; application of chemical-looping combustion. *Chem. Eng. Sci.* 56: 3101–3113.

53 Linderholm, C., Abad, A., Mattisson, T., and Lyngfelt, A. (2008). 160 Hours of chemical-looping combustion in a 10 kW reactor system with a NiO-based oxygen carrier. *Int. J. Greenhouse Gas Control* 2 (4): 520–530.

54 Abad, A., Pérez-Vega, R., de Diego, L.F. et al. (2015). Design and operation of a 50 kW_{th} chemical looping combustion (CLC) unit for solid fuels. *Appl. Energy* 157: 295–303.

55 Mayer, K., Penthor, S., Pröll, T., and Hofbauer, H. (2015). The different demands of oxygen carriers on the reactor system of a CLC plant – results of oxygen carrier testing in a 120 kW_{th} pilot plant. *Appl. Energy* 157: 323–329.

5

Pressurized Chemical Looping Combustion for Solid Fuel

Liangyong Chen[1], Zhen Fan[1], Rui Xiao[2], and Kunlei Liu[1,3]

[1] *University of Kentucky, Center for Applied Energy Research, 2540 Research Park Drive, Lexington, KY, 40511-8410, USA*
[2] *Southeast University, Key Laboratory of Energy Thermal Conversion and Control, Ministry of Education, School of Energy and Environment, Sipailou 2#, Nanjing, Jiangsu Province, 210096, PR China*
[3] *University of Kentucky, Department of Mechanical Engineering, 351 Ralph G. Anderson Building, Lexington, KY, 40506-0503, USA*

5.1 Introduction

Coal-based pressurized chemical looping combustion combined cycle (PCLC-CC) is being considered attractive for large-scale power generation systems with CO_2 capture because of the superior plant efficiency and low costs of electricity and CO_2 capture as compared to the state-of-the-art generation – ambient-pressure coal-based chemical looping combustion (CLC) and pulverized combustion equipped solely with steam cycle. This concept was first reported by Southeast University (SEU), China, in which direct CLC combustion of the solid fuel under elevated pressure [1] is integrated with combined cycle for power generation from coal. The technology of direct solid-fueled PCLC using iron-based material as oxygen carrier (OC) has been tested there, in a laboratory-scale fixed bed reactor and a 100 kW_{th} pressurized facility, respectively.

The Center for Applied Energy Research at University of Kentucky (UK-CAER), USA, also developed a detailed direct coal-fueled CLC plant, which has a plant configuration similar to the pressurized fluidized bed combined cycle (PFBC), and is composed of a PCLC unit, flue gas turbine (FGT), heat recovery steam generator (HRSG), and steam cycle. In this plant, the direct coal-fueled PCLC unit serves as a combustor to generate high-temperature and -pressure flue gas. This flue gas is used to drive an FGT and HRSG coupled with a steam turbine (ST) for large-scale power generation. In the fuel reactor (FR), a CO_2 stream is produced, which is compressed for sequestration after water condensation and heat recovery for stream production steps. The cost-effective iron-based OCs from solid waste (red mud) or natural ore (ilmenite) are selected to increase reaction kinetics and to reduce operational costs. Thermodynamic analysis [2] predicts that the UK-CAER process at 550 MWe could provide more than 90% CO_2 capture, greater than 95% CO_2 purity, and a net plant efficiency of

Handbook of Chemical Looping Technology, First Edition. Edited by Ronald W. Breault.

more than 43% higher heating value (HHV) with CO_2 pressurized to 2215 psi. The direct coal-fueled PCLC had been examined experimentally at UK-CAER using various iron-based OCs and coal chars. Additionally, a 50 kW_{th} PCLC facility consisting of the interconnected spouted bed FR and fast fluidized bed air reactor (AR) has been constructed and under commissioning in April 2018. Reactor design, technical approaches for enhancing fuel utilization and reaction kinetics, emission mitigation, and avoiding agglomeration will be validated on 50 kW_{th} PCLC facility. Performance evaluation from this facility will be extended for the next level of scale-up and near-future commercialization.

5.2 Coal-Based Pressurized Chemical Looping Combustion Combined Cycle

5.2.1 Concept

There are two approaches for the use of PCLC-CC with coal as fuel. The first one, called syngas-fueled PCLC-CC, is first to convert the coal into syngas in a stand-alone gasifier and subsequently to introduce the syngas product to the PCLC system. The second approach has a similar deployment to that of the classic solid-fueled CLC as shown in Figure 5.1 [3], and the combustor system is operated at elevated pressure. The operation of both AR and FR at elevated pressure could allow several advantages, such as improved fuel utilization, reduced dimension, and incorporation of a combined cycle power generation scheme [1]. This process possesses higher plant efficiency than the syngas-fueled PCLC by eliminating the air separation unit (ASU) for the external gasifier. Because of this, the technology presented in this chapter pertains to the direct coal-fueled PCLC. For the two abovementioned processes, interconnected fluidized bed reactors or moving bed reactors are preferred for the AR or FR.

Figure 5.2 illustrates the chemical process mechanism in the FR of the direct coal-fueled PCLC. The solid fuels are directly fueled to the FR, where the major chemical reactions include coal gasification, hydrocarbon oxidation, OC reduction, and water–gas shift reaction (WGSR). The majority of coal char is first gasified by steam or CO_2 into syngas. Meanwhile, the reducing components of syngas are burnt by sufficient amount of metal-based OCs surrounding the coal char particles, generating CO_2 and H_2O. A certain amount of CO or H_2 remains in the CO_2/H_2O stream, mainly depending on the thermodynamic equilibrium and

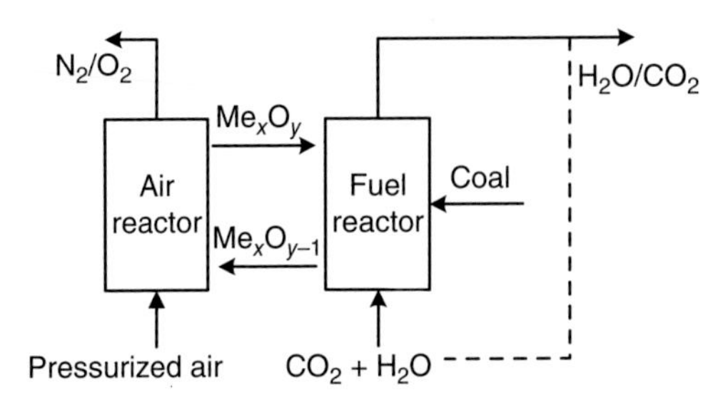

Figure 5.1 Schematic layout of the direct coal-fueled PCLC system. Source: Juan Adanez et al. 2012 [3]. Reproduced with permission from Elsevier.

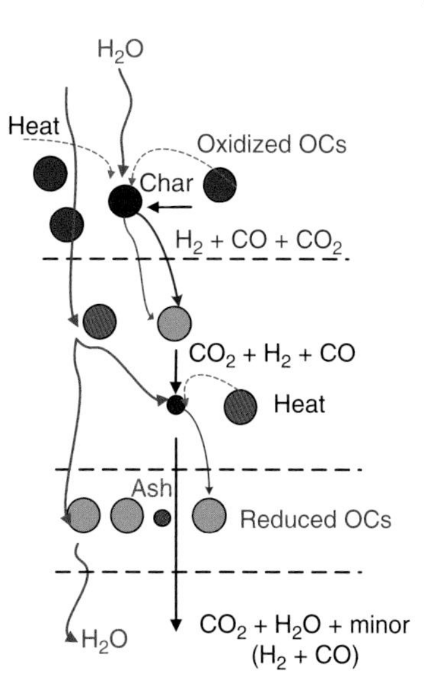

Figure 5.2 Chemical process involved in the fuel reactor.

OC reactivity and selectivity. Coal gasification rate is significantly improved due to the elevated pressure of the gasification agent and the lowered inhibitions from CO and H_2 products. Thus, the reaction between the two solids, coal/char and OCs, occurs via the phase transfer of intermediate gas. Steam and/or CO_2 are generally supplied as gasification or fluidization agent. In the above process, the homogeneous WGSR, catalyzed or uncatalyzed by OCs, influences the composition of the final gas products. For the iron-based OCs, OC reduction proceeds according to the following reverse reaction in the FR.

Fe_2O_3 OC:

$$3Fe_2O_3 + H_2 \leftrightarrow 2Fe_3O_4 + H_2O \tag{5.1}$$

$$3Fe_2O_3 + CO \leftrightarrow 2Fe_3O_4 + CO_2 \tag{5.2}$$

Ilmenite OC:

$$Fe_2TiO_5 + TiO_2 + H_2 \leftrightarrow 2FeTiO_3 + H_2O \tag{5.3}$$

$$Fe_2TiO_5 + TiO_2 + CO \leftrightarrow 2FeTiO_3 + CO_2 \tag{5.4}$$

Accordingly, the oxygen-depleted particles are regenerated to the oxidized form at higher temperature in the AR:

$$4Fe_3O_4 + O_2 \rightarrow 6Fe_2O_3 \tag{5.5}$$

$$4FeTiO_3 + O_2 \rightarrow 2Fe_2TiO_5 + 2TiO_2 \tag{5.6}$$

5.2.2 Process of the Direct Coal-Fueled PCLC Developed by UK-CAER

The UK-CAER process of the direct coal-fueled PCLC-CC is shown in Figure 5.3, which is originally targeted to improve the plant efficiency and CO_2 capture while eliminating the costly coal-pretreatment step to avoid OC agglomeration. It is comprised of a PCLC island, FGT, HRSG, ST, and CO_2 compressor.

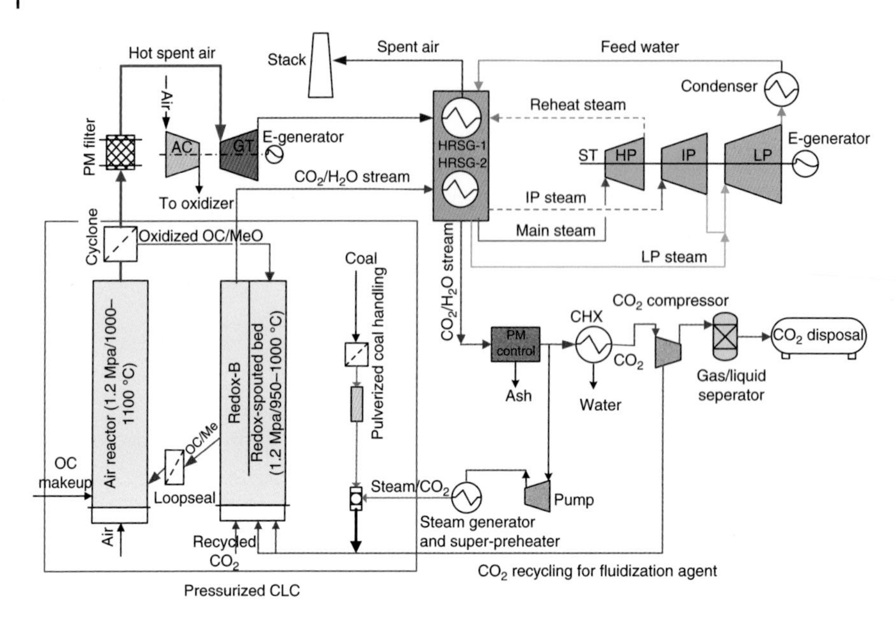

Figure 5.3 Coal-based pressurized chemical looping combustion combined cycle proposed by UK-CAER.

The PCLC-CC process begins by feeding coal particles via a draft tube with hot steam/CO_2 into the spouted bed FR (or fluidized bed FR), where it is first mixed with the hot, freshly oxidized iron-based OCs and begins pyrolysis, gasification, and combustion by OC materials. The contents of the FR, both gas and solid, will further mix with the recycle gas (recycled CO_2/H_2O stream) in a bubbling fluidization regime. In this manner, the residual carbon and remaining syngas are fully converted to CO_2/H_2O. Here, the recycled steam/CO_2 stream is used to maintain a state of bubbling fluidization and improve the gasification rate of the residual carbon. Furthermore, with application of density and particle size difference between OC and coal, in this bubbling regime, the majority of coal ash is separated from OCs, e.g. the bed materials, before the reduced OCs return to the AR via overflow. The flow of OCs returning to the AR is controlled by a loop seal. The "clean" OC then flows by gravity to the oxidation stage where it is "combusted" with air and recovered to the fully oxidized form before starting a new cycle. The AR is usually configured to be a fast fluidized bed (reaction zone) and a riser, with the latter providing the drive force for the global solid circulation.

The PCLC island generates two gas streams when it is operated at high temperature (1000–1050 °C for the AR and 950–1000 °C for the FR) and elevated pressure such as 1.2 MPa, but depending on the pressure limit of the available FGT. The first is a high-volume/high-pressure/hot spent air stream from the AR that is devoid of CO_2, which will pass through a cyclone for solid recirculation and a ceramic candle filter or a multi-stage, micro-channel cyclone unit for solids removal. Then the hot spent air is used to drive an FGT (Brayton cycle) followed by a HRSG for electricity generation (HRSG-Rankine cycle). The second is a lower volume stream from the FR, consisting primarily of CO_2, H_2O, and a very small quantity of CO and H_2. The CO_2/H_2O stream is expanded and also heat-recovered by the steam cycle (via the other HRSG), then particulates and water vapor are removed by a filter and a condenser prior to the CO_2 compression strain, leaving a CO_2-rich gas stream for sequestration (>95% CO_2 purity). Part of the CO_2/H_2O stream is recycled for fluidization of the FR bed materials. Obviously, the arrangements in Figure 5.3 provide the bases for further optimization of heat and mass integration, and the details should comply with application requirement.

The direct coal-fueled PCLC-CC for power generation has the following advantages and technical features:

1. Noticeable reduction to cost of electricity (COE) and CO_2 capture due to high plant efficiency and the avoidance of costly ASU, and coal pyrolyzer and tar cracker without comprising on concerns with OC agglomeration.
2. Simplicity with only one solid recirculation loop but possessing appropriate physical arrangements of the FR, which enable sufficient fuel utilization and high purity of CO_2 stream.
3. A relatively small volume of coal impurity-containing gas produced in the FR that is less costly to treat.
4. Combined cycle that eliminates the need to install heat-transfer surfaces inside pressure vessels for temperature control, thereby eliminating their associated corrosion/erosion problems.
5. Cost-effective iron-based OC (red mud OC or ilmenite OCs) with high reactivity to significantly improve coal char gasification, high resistance to water vapor without loss of reactivity, high mechanical and thermal stability, and high selectivity of converting syngas to CO_2/H_2O.
6. Reduced reactor size through high reaction kinetics and lower power requirements for compression of the enriched CO_2 stream due to the elevated pressure (>1.2 MPa).
7. Taking advantage of differential density and particle size between coal ash and iron-based OCs to realize a simple and effective in-line ash separation.

5.2.3 Process of the Direct Coal-Fueled PCLC at SEU, China

SEU, China, had tentatively developed a combined cycle using direct CLC of the solid fuel under elevated pressure. The overall process is similar to the natural gas CLC combined cycles. The PCLC island consists of two pressurized reactors, where the FR is a pressurized spout-fluidized bed composed of a mixer and a riser, and the AR is a moving bed reactor. The authors argued that the spout-fluid bed provides strong solids mixing and enough residence time both for coal gasification and OC reduction. The whole PCLC Island is contained in a big pressurized vessel. A gas turbine (GT), expanding depleted air from the AR, and a CO_2 turbine, expanding the FR outlet gas, are adopted. The exhaust gas heat from GT is recovered in HRSG to drive an ST. No calculated plant efficiency was provided for this system. The detailed plant configuration can be found in Ref. [1].

5.3 Fundamentals and Experiments of Pressurized Chemical Looping Combustion

Several reactions are involved in the direct coal-fueled PCLC: (i) solid fuel conversion to gaseous fuel via devolatilization and gasification, (ii) OC reduction by intermediate gaseous fuel, (iii) the solid–solid reaction between solid fuel and OC, which occurs via direct contact at high temperature, (iv) the homogeneous WGSR in the FR, and (v) OC regeneration in the AR. This section focuses on OC reduction and oxidation (ii and v), because they are the two major chemical reactions dominaing the overall CLC process. The WGSR takes place between the gas products and just influences the final gas composition [4]. The solid–solid reaction between solid fuel and OC could be negligible at the typical CLC operational temperature [5]. This section also describes two pilot-scale experiments that are being used to establish the baseline of direct coal-fueled PCLC.

5.3.1 Transient Oxidation of Magnetite to Hematite in PCLC

The transient oxidation of magnetite (Fe_3O_4) to hematite (Fe_2O_3) has been theoretically analyzed at UK-CAER with emphasis on the influence of environmental conditions with the view of using pure magnetite as OC to reduce the complexity of analysis. The parameters of interest are the oxidation rates and the burnout times of the particles at high temperature under high pressure environments. The results presented here just provide a fundamental understanding of the oxidation mechanisms of Fe_3O_4 particles and necessary information for the design and operation of PCLC facilities.

The reduced OC entering the AR is assumed to be a spherical, porous Fe_3O_4 particle of radius r_s and surface temperature T_s in a quiescent environment of temperature T_∞ and oxidizer mass fraction Y_∞. OC oxidation occurs both on external particle surface and within its pores. During oxidation, the particle size remains constant while its bulk density increases. Oxidation involves a two-step mechanism [6, 7] with the exothermic oxidation of Fe_3O_4 to metastable maghemite, γ-Fe_2O_3, immediately followed by an exothermic transition from γ-Fe_2O_3 to stable α-Fe_2O_3. In analyses, an Arrhenius-type oxidation mechanism has been employed, and the influence of the radiation and heating from environment is considered.

Figure 5.4 shows the analytical results: (a) Burnout time and final particle temperature at the end of oxidation for different particles at $T_\infty = 1273$ K and various initial particle temperatures $T_{s,0}$. (b) The mass oxidation rates at different pressure, $r_s = 100$ μm, $T_{s,0} = 300$ K, and $T_\infty = 1273$ K. (c) Burnout times and final particle temperatures of various Fe_3O_4 particles oxidized in air with $T_\infty = 1273$ K and $T_{s,0} = 873$ K under different pressures. (d) Environmental temperature influences on the burnout times and final particle temperature, $r_s = 100$ μm, $T_{s,0} = 300$ K.

High pressure enhances the oxidation rates. For example, at $T_s = 1000$ K, the oxidation rates under high pressure (10 atm) are almost doubled as compared to atmospheric pressure. The burnout time decreases monotonically as the

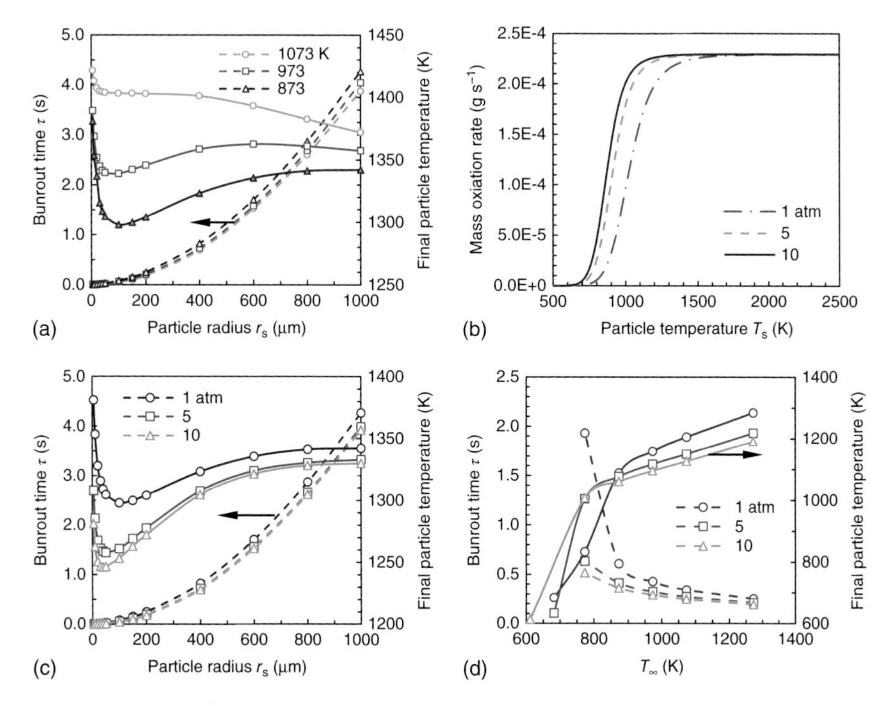

Figure 5.4 Oxidation rates, burnout time, and final particle temperature during Fe_3O_4 OC regeneration.

particle sizes decrease and is shortened under high pressure. Higher initial particle temperatures result in shorter burnout times, but this influence is not so significant. The final particle temperature gradually decreases as the particle size decreases from 1000 μm, then increases rapidly, and the lowest final temperature occurs at around $r_s = 100$ μm particles. Environmental temperature is expected to affect the burnout times significantly. As T_∞ increases, the burnout time decreases and final particle temperature increases. When the T_∞ is below 900 K, the final particle temperature is greater under high pressure than at 1 atm. But when the T_∞ is above 1000 K, this pressure influence is reversed, i.e. the final particle temperature under high pressure is lower than that at low pressure.

5.3.2 The Solid Behaviors in the Solid-Fueled PCLC (Fuel Reactor Side)

A comprehensive research investigation was performed experimentally at UK-CAER to understand the complex reactions in the FR of solid-fueled PCLC in the range of 1–6 bars. A series of tests were performed in thermogravimetric analysis (TGA) and fixed and fluidized bed reactors, respectively, to demonstrate the effects of operational pressure, coal char reactivity, and different iron-based OCs behaviors.

5.3.2.1 Materials

Three iron-based OCs, including activated ilmenite ore and two red mud OCs, were selected for experimentation. Before ilmenite ore was used, it was stabilized

Table 5.1 Chemical composition and physical property of iron-based OCs.

		RM-1	RM-2	Ilmenite
Composition	Fe_2O_3	48.35	45.47	52.46
	SiO_2	11.68	9.69	5.4
	Al_2O_3	17.07	21.76	4.32
	TiO_2	6.19	8.95	33.04
	CaO	12.77	5.08	0.99
	MgO	0.48	0.27	3.26
	Na_2O	2.31	8.07	—
	K_2O	0.22	0.09	—
	Balance	0.93	0.62	0.53
Particle density	$g\,cm^{-3}$	3.53	3.64	4.09
Bulk density	$g\,cm^{-3}$	1.30	1.15	2.20
d_{50}	μm	254	266	310
BET	$m^2\,g^{-1}$	0.246	0.311	0.284

in air at 950 °C for 24 h, and then activated by 10 redox cycles in a fluidized bed reactor. The two RM OCs, RM-1 and RM-2, were prepared from two different bauxite residuals of aluminum industry by freeze-granulation, and they were calcinated at 1150 °C for 6 h to obtain a desirable mechanical strength. The chemical and physical properties and scanning electron microscope (SEM) images of these OCs are presented in Table 5.1 and Figure 5.5. The active content of the two RM OCs is Fe_2O_3, while those in the activated ilmenite are Fe_2TiO_5 and a small amount of Fe_2O_3. The reaction during reduction and oxidation proceeds according to Equations (5.1)–(5.6). The pre-determined reactivity of these OCs with wet simulated syngas in the TGA apparatus, designated as OC conversion (X_{OC}) vs. time, is shown in Figure 5.6.

Three different coal chars, prepared from Powder River Basin (PRB), Eastern Kentucky, and Western Kentucky coals (EKY and WKY), were used as solid fuels. The proximate and ultimate analyses of coal chars are presented in Table 5.2. Coal char was used instead of coal just because char gasification is the major step occurring in the CLC FR and because the complexity of the experiments due to the flash volatilization when coal is fed into the reactor and the potential agglomeration caused by coal tar produced from initial coal combustion could be avoided.

5.3.2.2 Experiment Setup

A fluidized bed facility, consisting of a fluidized bed reactor, gas feeding setup, flue gas cleaning system, gas analysis system, and data acquisition system, was used to perform the experiments of the coal char-fueled PCLC, as shown in Figure 5.7. All CLC experiments were performed at 950 °C using 50 vol% steam balanced by N_2 as gasification agent. Pressures of 1, 2, 4, and 6 bars (absolute pressure) were adopted to investigate the effects of operational pressure. To avoid segregation

Figure 5.5 SEM images of iron-based OCs: (a) RM-1, (b) RM-2, and (c) activated ilmenite.

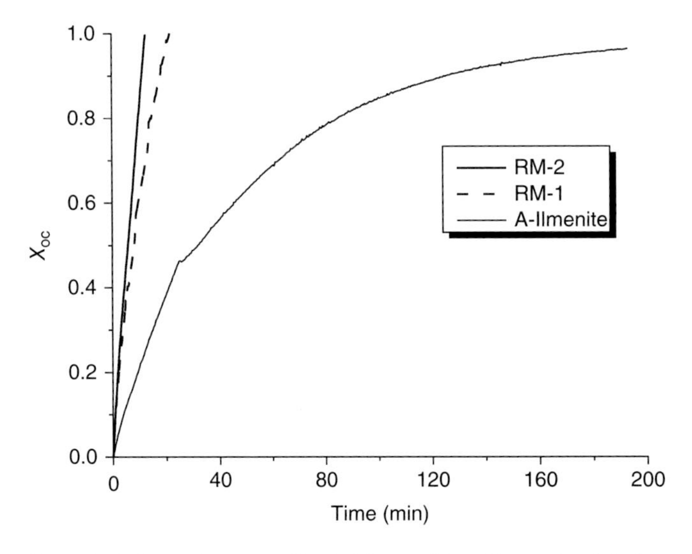

Figure 5.6 Reduction reactivity of different OCs with wet simulated syngas (reduced by mixing gas of 5% CO, 10% steam, and balanced Ar at 950 °C).

Table 5.2 Proximate and ultimate analyses of coal char.

	PRB	EKY	WKY
Proximate analysis (wt%)			
Ash	16.28	15.92	14.72
Volatiles	3.13	4.33	4.34
Fixed carbon	80.59	79.39	79.38
Ultimate analysis (wt%)			
C	77.4	78.26	76.74
H	1.38	1.67	1.84
O	1.52	1.38	2.55
N	1.72	1.75	1.67
S	1.70	1.02	2.48
Heating value (MJ kg^{-1})	27.6	28.4	27.9

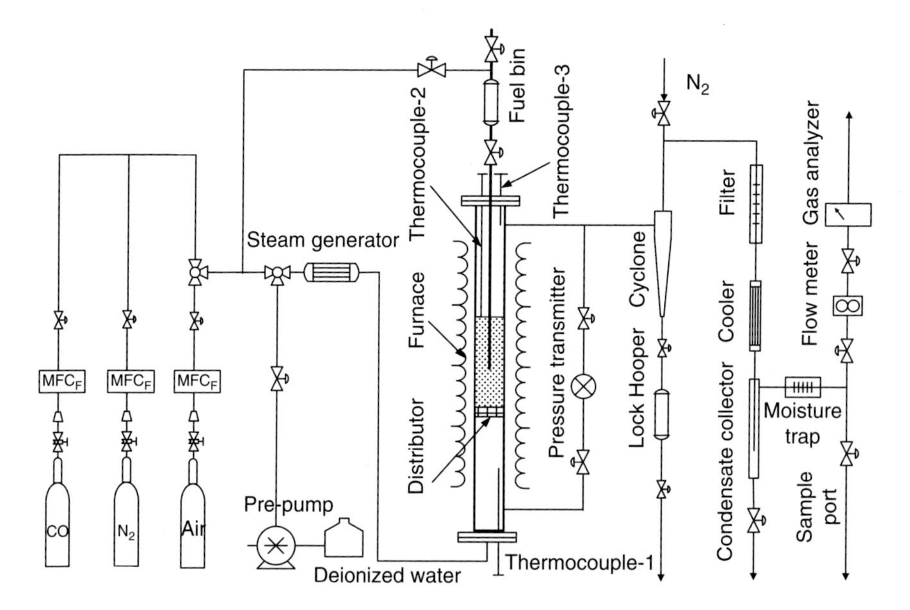

Figure 5.7 Fluidized and fixed bed reactor used for coal char-fueled PCLC experiments.

of coal char particles from the bed materials, the experiments were designed at a slow bubbling fluidization regime for the OC particles and a slightly more vigorous fluidization regime for the coal char particles. The amount of feeding gas for fluidization was enlarged linearly with operational pressure. This arrangement ensured that the superficial velocity (U_o) in the FR was fixed at 0.165 m s^{-1} at 950 °C at different operation pressures. Figure 5.8 illustrates the flow regime in the FR of various pressures, where the PRB char, RM-1, and ilmenite OCs were used as example. Experiments of external gasification were also performed in the

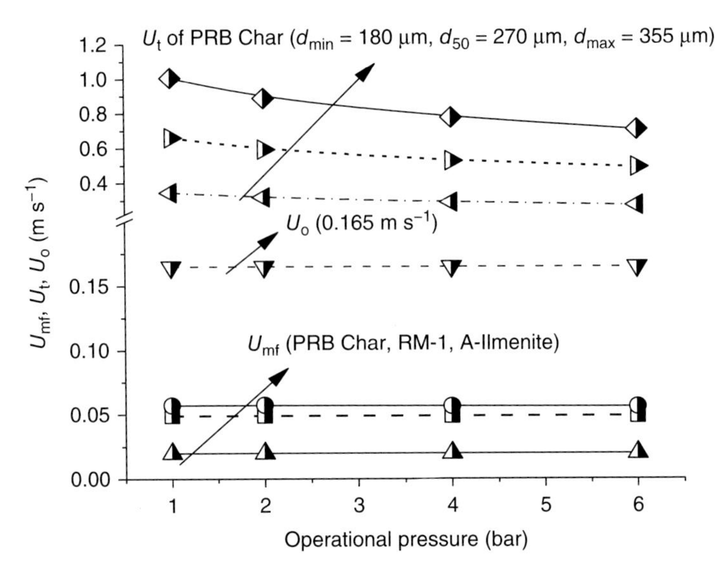

Figure 5.8 Calculated minimum velocity (U_{mf}), terminal velocity (U_t), and superficial velocity (U_o) of different solid particles involved in the PCLC experiments.

fluidized bed reactor to determine the reactivity of coal char at various pressures, which was used as the baseline to evaluate the PCLC performance. Fused Al_2O_3 particles were used as bed materials. The resulting reactivity of these coal chars are in the following order: PRB > WKY ≫ EKY.

5.3.2.3 In situ Gasification

As example, the gas concentration profiles of CLC experiments using RM-2 and PRB char at 1 and 6 bars are shown in Figure 5.9a,b. Owing to high reactivity and selectivity of RM-2, CLC experiments at various pressures consisted of high concentration of CO_2 and very low concentration of combustible gas (H_2 and CO) during the reduction period. Obviously, most of the intermediate syngas around the coal char particles were effectively removed with the presence of OC particles. The in situ gasification rate is significantly promoted as operational pressure increases.

OC reactivity and steam partial pressure are the most important factors that impact in situ gasification. The use of OCs improves coal conversion (e.g. gasification rate) through the lowered inhibition from gas products of CO and H_2. Figure 5.10 compares the instantaneous gasification rate of CLC experiment using RM-2 and that of external gasification at 1 and 6 bars. The gasification rate of the external gasification at ambient and elevated pressure linearly increased with X_C (carbon conversion under steady reaction condition). Solid conversion via in situ gasification with RM-2 increased remarkably the rate of char gasification. It is noted that both RM-2 and PRB char possess the highest reactivity among the OCs and chars tested, respectively. On the contrary, ilmenite and EKY char possess the lowest reactivity among OC and char respectively. When the pair of ilmenite and EKY char was used, char gasification exhibited a

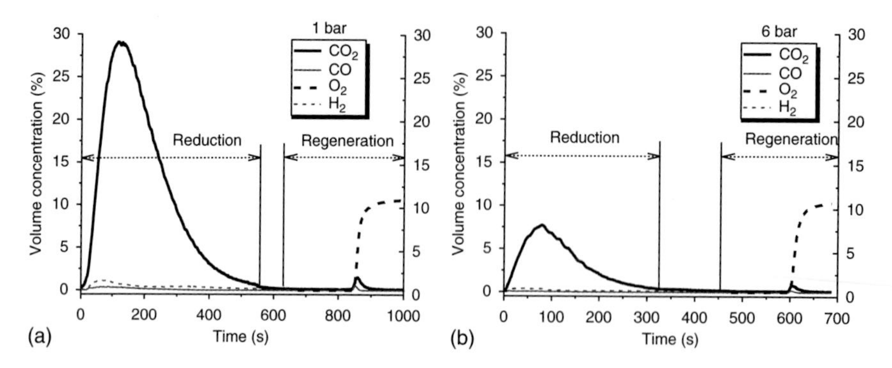

Figure 5.9 Dry gas concentration profiles of CLC at 1 bar and 6 bars when using RM-2 and PRB char.

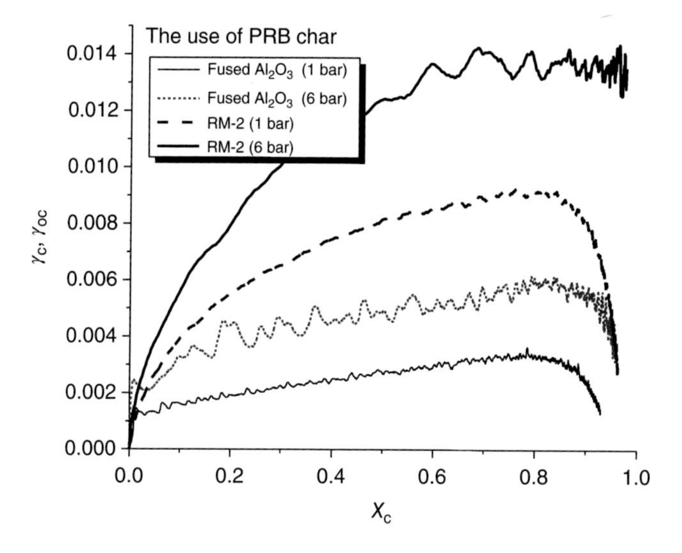

Figure 5.10 Instantaneous rates of char gasification of CLC with RM-2 and that of external gasification with fused Al_2O_3 particles when using PRB char as fuel.

different behavior, as shown in Figure 5.11. This CLC experiment resulted in little improvement in gasification, and showed similar gasification behaviors as the external gasification at various pressures.

Figure 5.12 shows the average instantaneous gasification rates of CLC and external gasification, $\overline{r_c}$ against operational pressure for different coal chars. Here, the value of $\overline{r_c}$ at steady-state reaction condition ($0.1 < X_C < 0.8$) is used. The values of $\overline{r_c}$ for both CLC and external gasification increased linearly with operational pressure despite the type of coal char. When the same OCs and pressure were used, the reaction rate of coal char from PRB (active char due to high porosity) was consistently higher than that of less active coal char from KY coal. For a given coal char and operational pressure, the value of $\overline{r_c}$ strictly followed the same order of OCs reactivity tested in TGA apparatus: RM-2 > RM-1 > ilmenite > Fused Al_2O_3.

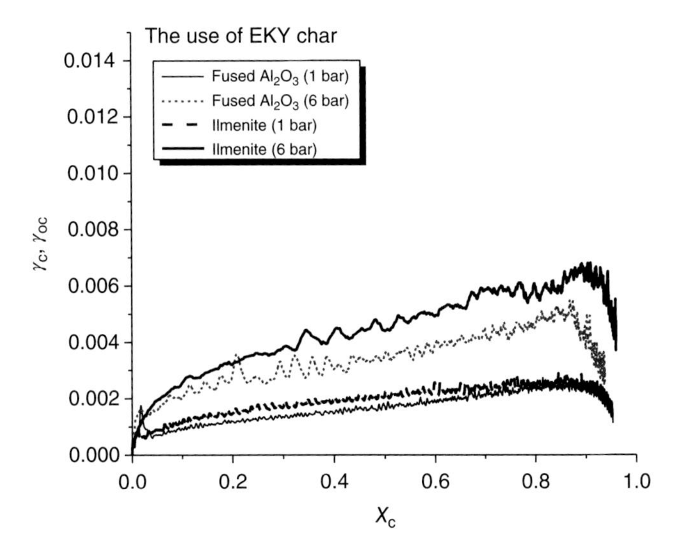

Figure 5.11 Instantaneous rates of char gasification of CLC with ilmenite OC and that of external gasification with fused Al_2O_3 when using EKY char as fuel.

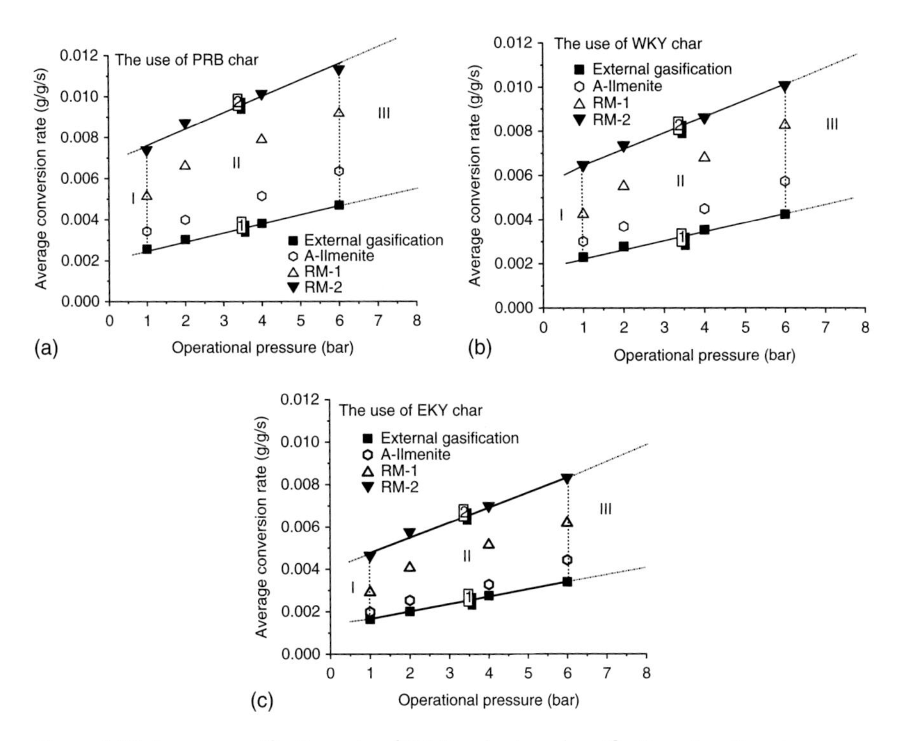

Figure 5.12 Average gasification rate of PCLC and external gasification at various pressures. (a) PRB Char; (b) Western Ky Char; and (c) Eastern Ky Char.

The gasification reaction of coal char-fueled CLC (800–1050 °C) can be classified into a kinetically controlled regime, which follows a Langmuir–Hinshelwood (LH) mechanism [8, 9].

$$C_f + H_2O \underset{k2}{\overset{k1}{\Longleftrightarrow}} C(O) + H_2 \tag{5.7}$$

$$C(O) \xrightarrow{k3} CO \tag{5.8}$$

here, C_f is a free carbon site, and C(O) is the surface complex that is converted to the gaseous phase of carbon monoxide CO via Equation (5.8). H_2 and CO are the gasification products that will inhibit any further gasification. The presence of hydrogen inhibits the steam–carbon reaction by two possible pathways:

$$C_f + H_2 \overset{k4}{\leftrightarrow} C(H_2) \tag{5.9}$$

$$C_f + 0.5H_2 \overset{k5}{\leftrightarrow} C(H) \tag{5.10}$$

According to the abovementioned regime, PCLC experiments using RM-2 and inert Al_2O_3 provide the top and bottom boundaries to analyze the roles of OCs in the complex reaction system. The in situ gasification with RM-2 was totally dominated by the reaction between steam and char (Reaction (5.7)), and almost free of inhibition from CO or H_2 because of the thermodynamic selectivity of CO_2/H_2O over CO/H_2. For the external gasification with inert Al_2O_3, the intermediate syngas was not locally consumed as it was generated, and the built-up syngas composition inhibited further char gasification. When OC reactivity fell between these two materials, the gasification process was controlled by both steam–carbon reaction and a reduced inhibition effect. Therefore, the results in Figure 5.12 qualify the limitation of using OCs and elevated steam partial pressure for the complex reaction in the FR. Line 1 (in zone II) represents the dependence of external gasification rate on the steam partial pressure (half of the total pressure), while Line 2 (in zone II) defines the dependence of the rate of "real" in situ gasification (free of gasification inhibition) on the steam partial pressure. For a given fuel char, the PCLC with OCs of different reactivity will be operated at a rate between the two boundaries. The "real" in situ gasification exhibits higher increasing rate with the steam partial pressure compared to the external gasification. Therefore, a higher potential for promoting char gasification (the differences between Line 1 and Line 2) can be achieved at a higher steam partial pressure. In addition, the potential for promoting char gasification in CLC is related to the coal char reactivity.

Reasonable operational pressure for PCLC-CC is in the range of 12–30 bars depending on the FGT available. There will be uncertainties in directly predicting the PCLC gasification behavior of higher pressure via extrapolation from the present results (such as zone III in Figure 5.12). So, reliable measurement at higher pressure is needed for more accurate understanding of the reaction model.

5.3.2.4 Combustion Efficiency

The concentrations of combustible gas (CO and H_2) in the CO_2/H_2O stream of the FR should be as low as possible. Otherwise, a purification process is needed prior to CO_2 sequestration. The following equation is used to calculate the combustion efficiency of intermediate syngas (η_{syn}) that can be oxidized by OCs in the FR [8]:

$$\eta_{syn} = 1 - \frac{0.5f_{CO} + 0.5f_{H_2}}{f_{CO_2} + 0.5f_{CO} + 0.5f_{H_2}} \tag{5.11}$$

here, f_i is the volumetric fractions of the i gas component.

Figure 5.13 shows the combustion efficiency of WKY char-fueled PCLC, where the instantaneous value of η_{syn} is plotted against OC conversion, X_{OC}. Under stable reaction conditions ($0.1 < X_C < 0.8$, X_C-carbon conversion), η_{syn} of the two RM OCs was high and stable, and unaffected by operational pressure. The values of η_{syn} for ilmenite OC decreased as the operational pressure increased. Tables 5.3–5.5 list the average combustion efficiencies ($\overline{\eta}_{syn}$) of the PCLC experiments when different coal chars were used. In the CLC experiments using RM-2, 0.97–0.98 of $\overline{\eta}_{syn}$ was achieved, and the value of $\overline{\eta}_{syn}$ was evidently independent of operational pressure and the type of coal char. Similar behaviors were observed for the CLC experiments using RM-1, but the value of $\overline{\eta}_{syn}$ was fixed between 0.93 and 0.94. The value of $\overline{\eta}_{syn}$ of ilmenite OCs declined with operational pressure, but was also independent of the type of coal char.

In Tables 5.3–5.5, the average combustion efficiency of OCs with wet syngas obtained in the fixed bed model, $\overline{S}_{syn}$, are listed for comparison. In the fixed bed experiment, the syngas was introduced from the bottom of the reactor,

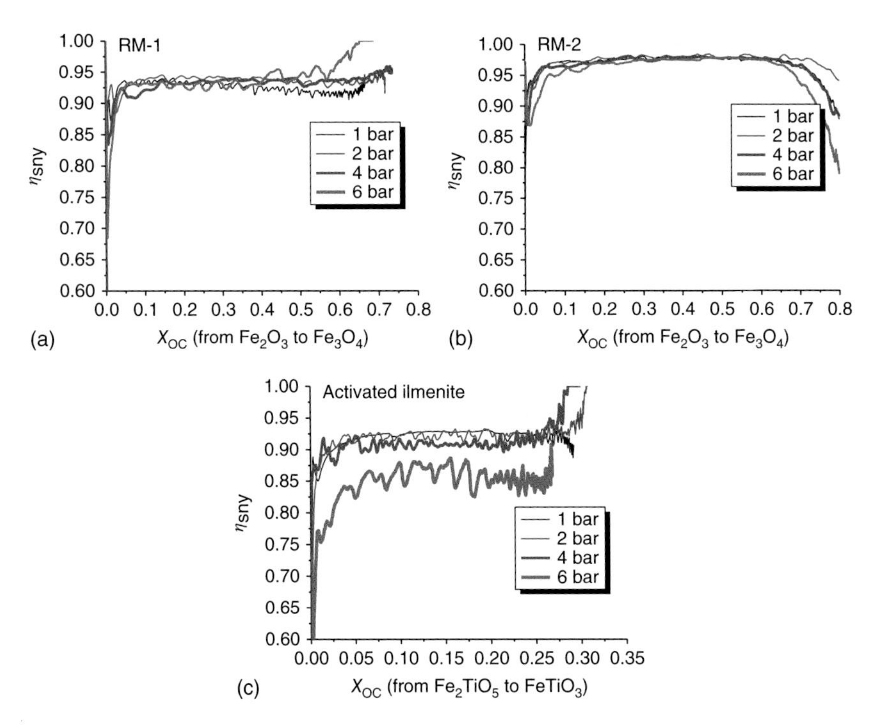

Figure 5.13 Effect of operational pressure on combustion efficiency of the WKY char-fueled PCLC. (a) RM-1 OC; (b) RM-2 OC and (c) Activated ilmenite OC.

Table 5.3 Average combustion efficiency of PCLC using RM-1.

Pressure (bar)	$\bar{S}_{syn}$	$\bar{\eta}_{syn}$ of PCLC		
		PRB char	EKY char	WKY char
1	0.996	0.928	0.935	0.924
2	0.995	0.936	0.937	0.938
4	0.994	0.928	0.941	0.930
6	0.996	0.942	0.944	0.933

Table 5.4 Average combustion efficiency of PCLC using RM-2.

Pressure (bar)	$\bar{S}_{syn}$	$\bar{\eta}_{syn}$ of PCLC		
		PRB char	EKY char	WKY char
1	0.995	0.979	0.973	0.972
2	0.994	0.980	0.975	0.973
4	0.996	0.977	0.972	0.973
6	0.995	0.975	0.969	0.963

Table 5.5 Average combustion efficiency of PCLC using Ilmenite OC.

Pressure (bar)	$\bar{S}_{syn}$	$\bar{\eta}_{syn}$ of PCLC		
		PRB char	EKY char	WKY char
1	0.971	0.931	0.923	0.927
2	0.964	0.913	0.920	0.914
4	0.948	0.911	0.918	0.900
6	0.944	0.862	0.846	0.842

and the syngas was burned out by a large amount of OCs along the bed height. So, syngas in these experiments had enough time to react with OCs, and $\bar{S}_{syn}$ is an ideal value to directly reflect the performance of OC. In the solid-fueled PCLC, the reaction between coal char and OCs occurs via the intermediate gas. The mismatch reactions between gaseous compositions and char and OCs respectively (including gasification on char particle, and combustion on OC particle and localized non-uniform gas and steam distribution) may lead to some amount of unconverted gas exiting the FR with the fluidization gas. For RM-2 OC, the incomplete combustion, expressed by the difference between $\bar{\eta}_{syn}$ and $\bar{S}_{syn}$, is minor, varying between 1.7% and 2.4% points. This suggests that the flow pattern used for the present CLC experiments could ensure char gasification taking place throughout the fluidized bed zone, and the gas compounds transferred between the two different particles (char and OC) was not the limiting step. Owing to the relatively lower OC reactivity, the value of $(\overline{S}_{syn} - \overline{\eta}_{syn})$ for RM-1 was increased to 5.6–6.4%. The value of $\overline{S}_{syn}$ for ilmenite OCs was lower than RM OCs, and evidently pressure dependent. The value of $(\overline{S}_{syn} - \overline{\eta}_{syn})$, the loss of combustion efficiency, is higher than that of RM OCs, and is strongly pressure dependent.

5.4 Direct Coal-Fueled PCLC Demonstration in Laboratory Scale

5.4.1 100 kW_{th} PCLC Facility at SEU, China

The 100 kW_{th} facility at SEU, Nanjing, China [10], is the first-in-kind to be built to validate the concept of the coal-fueled PCLC in the continuous operation of two interconnected reactors. It consists of AR/FR reactors, coal feed system, air and steam supply system, data acquisition system, and auxiliaries, as shown in Figure 5.14. The AR is of 276 mm inner diameter and 2 m in total height. The FR has an inner diameter of 80 mm and a total height of 8.5 m, with a metal distributor placed at 0.5 m from the bottom of the FR. The FR and AR were designed to be operated in fast and turbulent fluidization regime, respectively. It is believed that the fast fluidization and large height of the FR could provide the driving force of solid circulation, a favorable gas–solid contact, homogeneous mixing between OCs and coal particles, and sufficient residence time. Each reactor was equipped with an electric oven that was used to supply heat for the start-up

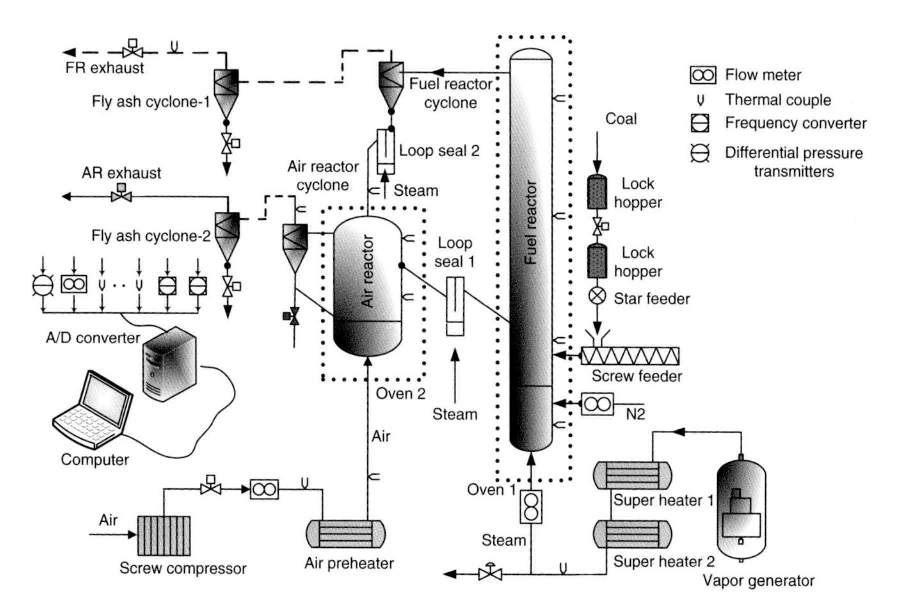

Figure 5.14 Schematic diagram of the 100 kW_{th} PCLC unit at SEU, Nanjing, China (2010–2012).

Table 5.6 Operation conditions for testing of 100 kW_{th} PCLC unit.

Parameters	Test 1	Test 2	Test 3
Operation pressure (MPa)	0.1	0.3	0.5
Coal feed rate ($kg\,h^{-1}$)	7	7	7
Steam temperature (FR, °C)	800	800	800
Air temperature (°C)	300	300	300
Bed temperature in FR (°C)	950	950	950
Bed temperature in AR (°C)	970	970	970
Hours for stable operation (h)	4.5	5.0	4

and compensate the heat loss of reactors during testing. The oven's rated electric power was 20 kW for the AR and 40 kW for the FR. Two loop seals were used to prevent gas mixing between the two reactors and to control the solid flux. One FR cyclone was mounted on the top of the AR, where the reduced OC particles were separated from the exist gas of FR and then led to AR via the loop seal. Another cyclone, named AR cyclone, was used to separate the gas–solid flow of the AR. The captured solid particles were returned to the AR.

Experiments were performed using bituminous coal as fuel and MAC iron ore as OCs at three different operational pressures (0.1, 0.3, and 0.5 MPa), and the unit was successfully operated for 19 h in total with steady coal feeding, 13.5 h of which was realized under stable operation. The operation of this unit demonstrated the major effects of operational pressure on the performance of coal-fueled PCLC. Table 5.6 summarizes the operation conditions. The coal feeding rate was 7.0 $kg\,h^{-1}$ for all tests, corresponding to a fuel thermal input of half of the design value.

The variation of bed temperature, gas concentration of FR, pressure drops of the fluidized beds, and pressure difference between the two reactors were chosen to estimate the reaction and gas–solid flow conditions in the CLC unit. The best performance was obtained under an elevated pressure of 0.5 MPa, at which the CO_2 concentration, carbon conversion, and combustion efficiency reached 97.2%, 84.7%, and 95.5%, respectively. Figure 5.15 shows the profiles of these parameters for the test under 0.5 MPa. These parameters together with the outlet gas concentrations of the FR were roughly stable during operation, indicating stable hydrodynamics of solid particles and chemical reactions. Figure 5.16 shows the average volume fraction of each gas component at the outlet of the FR. Carbon and gas conversions in the FR as well as carbon capture of the facility are shown in Figure 5.17. When the operation pressure increased from 0.1 to 0.5 MPa, the gas conversion and carbon conversion increased from 93.5% and 81.3% to 98% and 84.7%, respectively. Figure 5.18 illustrates the effects of operational pressure on heat balance of the unit, indicating that the combustion efficiency increased from 92.8% to 95.5%, and the energy loss due to unburned gas from the FR (q3) decreased from 5.2% to 2.3%. It is interesting

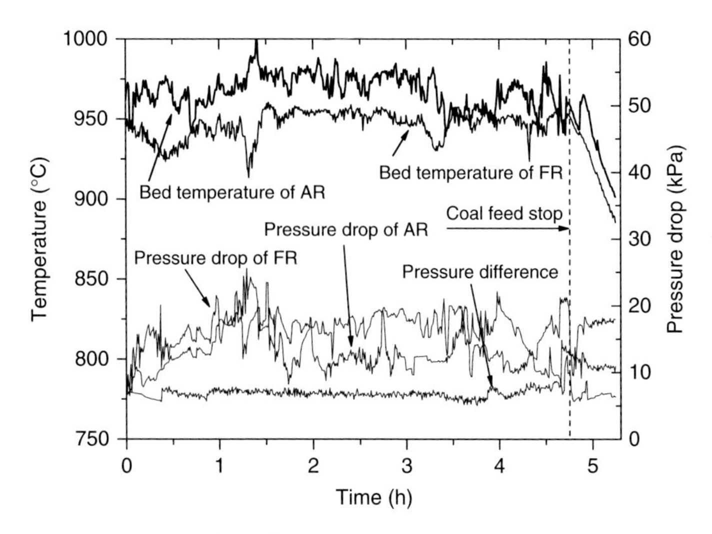

Figure 5.15 The profiles of important parameters with time for coal-fueled CLC experiment at 0.5 MPa.

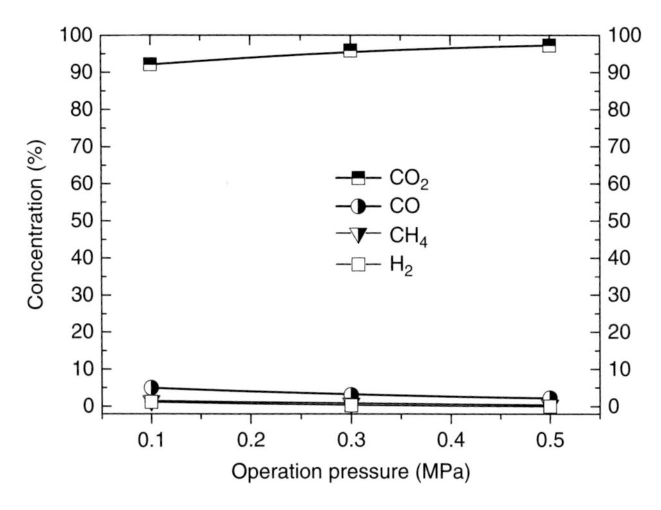

Figure 5.16 Gas concentration at the outlet of FR for coal-fueled CLC experiment at 0.5 MPa.

that the energy loss due to the unburned residual char (q4) did not change with operational pressure. The carbon conversion was relatively low, but combustion efficiency was very high. This is because most of the residual char leaving the FR was captured by the cyclone and burnt in the AR.

The losses of fine OCs were also determined for each test. The results indicated that the properties of elutriated OC particles were closely related to operational pressure. The time-averaged mass flow rates increased from 1.35 to 3.18 $kg\,h^{-1}$ when pressure increased from 0.1 to 0.5 MPa. Accordingly, the carbon concentration in the captured fly ash decreased from 14.21 to 6.45 wt%.

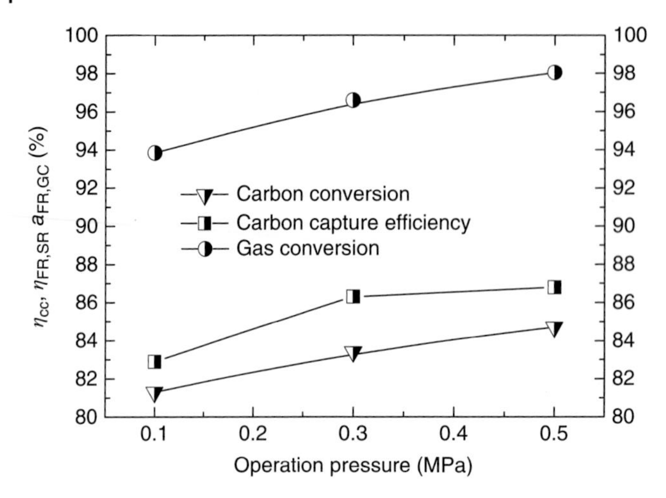

Figure 5.17 Carbon and gas conversion in the FR, and CO_2 capture of PCLC unit under different operational pressures.

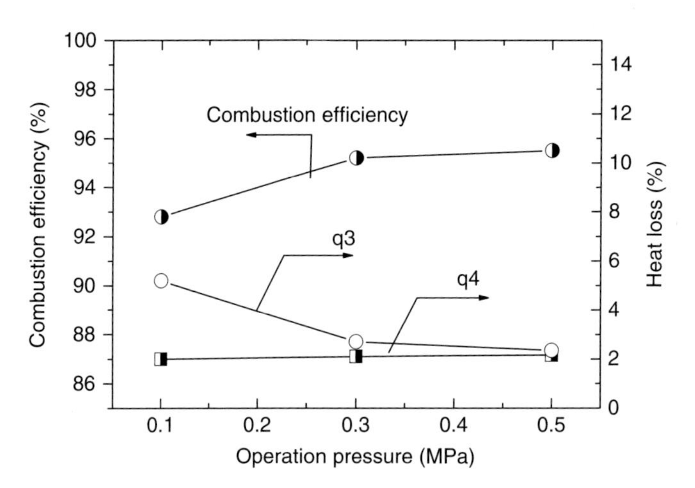

Figure 5.18 Combustion efficiency, q3 (energy loss due to unburned gaseous phase in exist gas of the FR) and q4 (energy loss due to unburned coal) under different operational pressures.

5.4.2 50 kW_{th} PCLC Unit at UK-CAER

Since 2016, UK-CAER has designed and is building a 50 kW_{th} coal-fueled PCLC facility under contract with DOE-NETL. The objective of the research program is to demonstrate a compacted CLC technology, (i) using the spouted bed reactor with a draft tube to provide strong mixing and rapid heating of the pulverized coal to avoid OC–char agglomeration, (ii) using a cost-effective Fe-based OC developed from solid waste to provide catalytic gasification and improved coal combustion rate, (iii) using the pulverized coal and elevated pressure system to enhance the overall reaction rate in the FR, and (iv) using a slow bubbling bed to serve as a device for in-line coal ash separation from the reduced OC on the basis of density and particle. The diagram of 50 kW_{th} PCLC unit is shown in Figure 5.19.

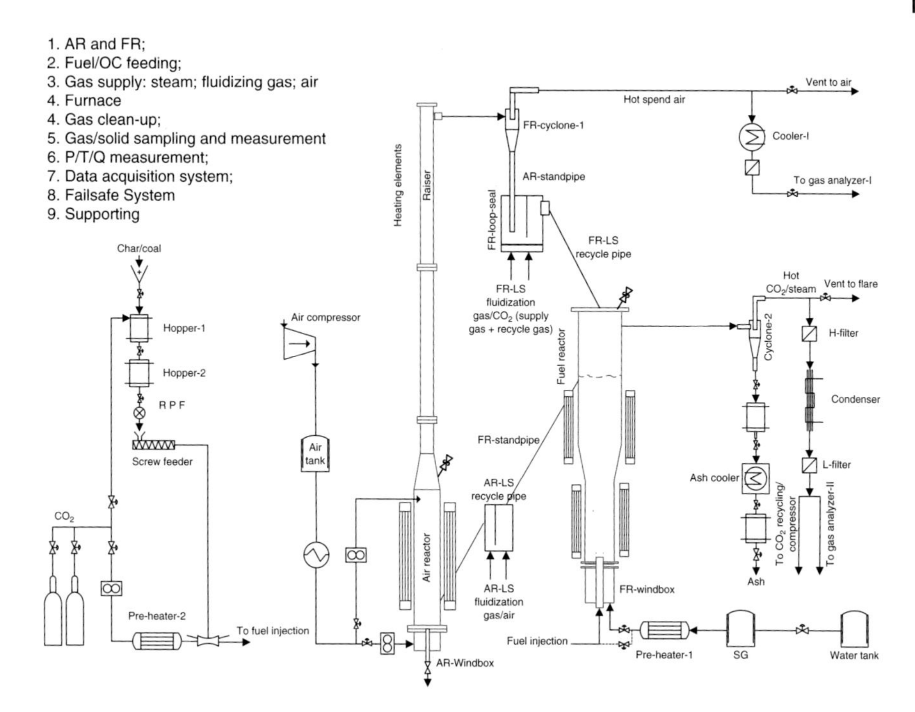

Figure 5.19 Diagram of 50 kW_{th} PCLC unit under construction at UK-CAER.

This unit will be used for the conceptual validation on the solid-fueled PCLC technology that UK-CAER has proposed, collecting various data and information to fill those technical gaps that impede the application of PCLC into solid fuel.

5.5 Tech-economic Analysis

5.5.1 Technical Performance Evaluation on the Direct Coal-Fueled PCLC-CC

Aspen Plus modeling and simulation were carried out to evaluate the plant configuration, feasibility, and performance of the direct coal-fueled PCLC-CC plant presented in Figure 5.3. The detailed plant configuration for simulation is given in Figure 5.20.

One of the key parameters for this process is the GT inlet temperature, the selection of which is limited by the melting point of OC materials. In simulation, it was assumed to be 1100 °C, based on the physical properties of the red mud OCs developed by UK-CAER. The efficiency of compressor (GC), GT, ST, generators, expander, booster, and pumps were selected according to the general performance of the components commercially available. The pressure drops of the pipelines and equipment were estimated. Sizing of key components was adjusted to meet the process design, and not picked up from the actual vendor list. In all case studies, the flow rate of air was fixed at 1000 kg h^{-1}. The reader is reminded that the following case studies are intended as a guide to the initial

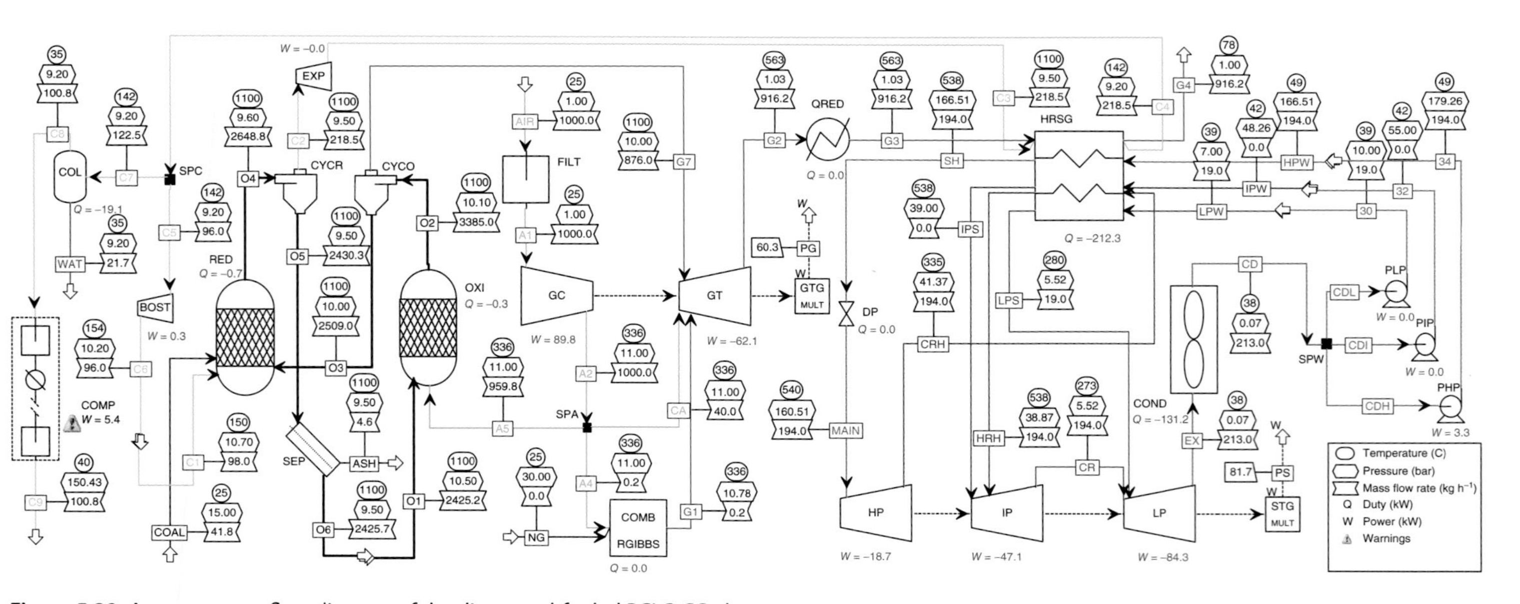

Figure 5.20 Aspen process flow diagram of the direct coal-fueled PCLC-CC plant.

consideration of deployment options, and the technical performance of the plant is little relevant to the selection of plant capacity.

5.5.1.1 Combined Cycle

The combined cycle has an FGT, HRSG, and ST. Most of the compressed air from the compressor (GC) is ducted to the AR, while the spent hot air exiting from the AR is used to drive the FGT for electric power generation. The rest of the compressed air is directly fed to the GT being used as cooling air. The heat of the exhaust gas from the FGT is recovered in HRSG to drive the bottoming ST for additional power generation. The FGT exhaust temperature is fixed at 568 °C in simulation. To improve plant efficiency, a 3-P HRSG is adopted to recover low-grade heat. 3-P HRSG is complex in terms of the pinch-type tube bank arrangement for heat recovery with different pressures. To avoid varying/iterating configuration with conditions and the available heat, a single multi-stream heat exchanger module (MHeatX) in Aspen is applied with zone analysis to manually debug the approach temperature inside HRSG by manipulating feed water flow for all 3-P pressures. The approach temperature is fixed at approximately 10 °C, and 3-P pressures are also prefixed.

5.5.1.2 PCLC Unit

The layout of PCLC unit is exactly the same as that shown in Figure 5.1. It is assumed that the spouting/circulating fluidized bed reactor is used as the respective FR and AR for heat and material balance (H & MB) calculation. Compressed air (11 bars) is used for OC regeneration and as fluidization agent, while CO_2/H_2O mixture recirculated from the exhaust gas stream of the FR is used as gasification agent. The conversion of red mud OC is specified for each reactor. The separation of solids from the gas stream is realized by cyclone, and ash separation from OC materials by a module selective electrostatic precipitator (SEP). Both are assumed to be an ideal separation process of 100% cut efficiency. In simulation, the fuel feeding rate to the FR is adjusted to ensure a conversion of oxidized OC to Fe_3O_4 state. The fuel conversion (both solid fuel and syngas product) in the FR is specified according to the results of the PCLC experiments conducted at UK-CAER. The exist gas of the FR is also ducted to another HRSG (parallel to the former HRSG) for heat recovery, and then majority of the cool CO_2/H_2O mixture is recirculated as fluidization gas of the FR. Before compression, the balanced CO_2/H_2O gas mixture is cooled down for the removal of moisture.

5.5.1.3 Physical Properties

The property method of solids in Aspen is applied for these general solids (coal, ash, and OCs). Ideal gas law is applied for the gas phase and the most recent steam table for the steam cycle. Red mud OCs developed from bauxite residual of the alumina industry by UK-CAER [11] are used as cyclic materials, and their chemical composition and heat capacity are given in Table 5.7 and Figure 5.21 respectively. In simulation, red mud OC is considered a mixture of Fe_2O_3, Fe_3O_4, Al_2O_3, SiO_2, etc. and thus its specific heat capacity (effective specific heat capacities of mixture, C_p) is the mass-weighted average of specific

Table 5.7 Chemical composition of red mud OCs (wt%).

Composition	Oxidized	Reduced
Fe_2O_3	51.56	—
Fe_3O_4	—	50.71
SiO_2	9.98	10.15
Al_2O_3	18.18	18.5
TiO_2	6.47	6.58
CaO	7.77	7.91
MgO	0.51	0.52
Na_2O	1.85	1.88
K_2O	0.18	0.18
Balance	3.5	3.56

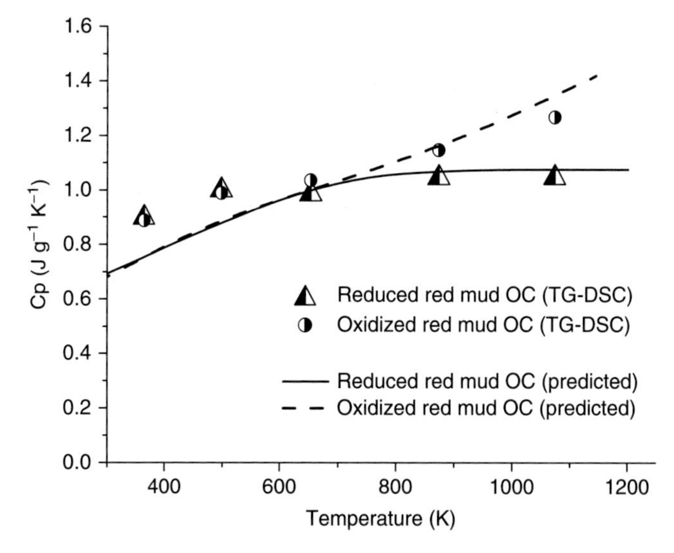

Figure 5.21 Specific heat capacity of red mud OC (RM-1). Source: Chen et al. 2017 [12]. Reproduced with permission from Elsevier.

heat capacities of the constituent substances. The predicted values of C_p for the reduced and oxidized forms of RM OCs are shown as continuous curves in Figure 5.21. The measured values (by thermogravimetric analysis-differential scanning calorimetry (TGA-DSC), shown in discrete points) are in agreement with the predicted values.

The design coal is a sub-bituminous PRB coal, and its proximate and ultimate analyses are listed in Table 5.8.

5.5.1.4 Case Study

Six case studies were performed for the direct coal-fueled PCLC-CC plant, and the results of major plant performances are listed in Table 5.9. It is worth noting

Table 5.8 Design Coal Characteristics.

Rank	**Sub-bituminous**	
Seam	PRB	
Sample location	Montana	
	AR	Dry
Proximate analysis (wt%)		
Moisture	25.77	0
Ash	8.19	11.04
Volatile matter	30.34	40.87
Fixed carbon (BD)	35.70	48.09
HHV (kJ kg^{-1})	19 920	26 787
LHV (kJ kg^{-1})	19 195	25 810
Ultimate analysis (wt%)		
Moisture	25.77	0
Carbon	50.07	67.45
Hydrogen	3.38	4.56
Nitrogen	0.71	0.96
Chlorine	0.01	0.01
Sulfur	0.73	0.98
Ash	8.19	11.03
Oxygen (BD)	11.14	15.01

that the flow rates of the recycling gas (CO_2/H_2O gas mixture) in all cases are adopted to realize a heat balance of the FR. So, in Case A and Case D, the resulting flow rates of the recycling gas are obviously out of the range for the practical operation of a fluidized bed reactor.

Case A to Case D was used to evaluate the influence of the temperature difference between the AR and FR. A net plant efficiency as high as 43.8% was obtained in Case A, which corresponds to near zero gas flowing through the FR. This results in a combination of a low circulation rate of OCs and a low temperature difference between two reactors. From Case A to Case D, it can be found that the circulation rate of red mud OCs is exactly proportional to the coal feeding rate. Thus the ratio of OC flow rate to coal feeding rate is kept constant at 143 (wt/wt). Owing to the increase in the flow rate of the recirculated CO_2/H_2O gas mixture, more heat is carried out of the PCLC unit and fed directly to HRSG to support the bottoming cycle. As a result, the net plant efficiency declines from 43.8% for Case A to 36.3% for Case D. At the same time, a change in trend can be found in that the 3-P HRSG (Case A) essentially shrinks to 1-P HRSG (Case D, no temperature pinch issues inside HRSG). Owing to the fixed air flow rate and coal feeding rate increase, O_2 concentration at the outlet of the AR declines from Case A to Case D.

Table 5.9 Performance of study cases.

Case	A	B	C	D	E	F
FR						
Temperature (°C)	1060	1030	1000	970	970	970
Coal flow rate (kg h^{-1})	36.1	45.5	60.8	91.7	50.3	50.3
Recirculated CO_2/H_2O gas (kg h^{-1})	5	193	530	1237	679	500
AR						
Temperature (°C)	1100	1100	1100	1100	1100	1100
OC circulation rate (kg h^{-1})	5144	6463	8667	13080	7181	7181
OC conversion	0.8	0.8	0.8	0.8	0.8	0.8
Air flow rate (kg h^{-1})	960	960	960	960	960	960
Temperature of compressed air (°C)	330	330	330	330	700	700
O_2 concentration at AR outlet (vol.%)	14.6	13.1	10.2	3.9	13.2	13.2
Power						
GT (kW)	66.5	62.7	58.0	48.6	61.2	61.2
ST (kW)	63.5	93.1	143.9	237.3	106	109.5
Total auxiliary load (kW)	10.9	15.0	22.1	35.2	16.8	17.3
Net output (kW)	119.1	140.8	179.8	250.7	150.4	153.4
Plant efficiency (%)						
Net	43.8	41.1	39.3	36.3	39.7	40.5

5.5.1.5 Optimization of Plant Configuration

In Case A to Case D, the temperature of the compressed air before feeding to the AR is set at 330 °C. Preheating the compressed air to high temperature level would evidently improve plant efficiency. In Case E, the compressed air stream is preheated from 330 to 700 °C by the exit gas of the FR, which is simultaneously cooled down from 970 to 564 °C before entering HRSG. Therefore, the net plant efficiency is improved from 36.3% (Case D) to 39.7% (Case E). Compared to Case D, OC circulation rate, coal feeding rate, and the flow rate of the recirculated CO_2/H_2O gas mixture of Case E decrease significantly.

Another option to further improve plant efficiency (Case F) is using an internal heat exchanger of the FR to heat up steam superheater/reheater (SH/RH) of the bottoming cycle. This would significantly reduce the flow rate of the recirculated CO_2/H_2O gas mixture. Accordingly, the net plant efficiency is boosted from 39.7% (Case E) to 40.5% (Case F).

5.5.2 Performance of the UK-CAER's PCLC-CC Plant

A conceptual cost estimate of the PCLC-CC technology has been carried out by UK-CAER and WorleyParsons Group Inc. The power plant has a nominal output of 550 MWe, in which the AR operated at 1100 °C is proposed for high combined cycle efficiency. This requires a high temperature particulate filter that has to be operated at 1100 °C, but it does not have any development track record. Fluidized bed or spouted bed reactor is suggested to be the FR, which is configured with multiple jet nozzles to feed solid fuel to reduce the system complexity and cost investment. UK-CAER performed Aspen modeling to incorporate PCLC technology into the nominally 550 MWe sub-bituminous coal-fired power generation facility to determine system performance and sizing of the major components. WorleyParsons used the information from the process modeling to identify and define the remaining balance of plant (BOP) items (such as coal feeding and drying equipment) and then developed capital costs for the facility.

The UK-CAER's PCLC-CC plant is broken down into the following subsystems:

- Fuel and sorbent handling
- Coal and sorbent preparation and feed
- Feed water and miscellaneous systems and equipment
- PCLC island
- Flue gas cleanup
- CO_2 capture (high-level)
- Air compressor–gas expander–generator
- HRSG, ducting, and stack
- ST generator and auxiliaries
- Cooling water system
- Ash/spent OC recovery and handling
- Accessory electric plant
- Instrumentation and control

Table 5.10 illustrates the overall performances of the PCLC-CC plant that is used for capital cost estimation. The DOE reference Cases 12A and 12B are also presented for comparison. The full-scale rate-based Aspen simulation with external FORTRAN sub-routine has shown that the PCLC-CC configuration will provide an overall plant net efficiency of 43.0% with CO_2 pressurized to 2215 psi, which is 3.7% points higher than that of DOE/NETL reference Case

Table 5.10 UK-CAER's PCLC-CC performance summary.

	Case 12A	Case 12B	PCLC-CC
Gross generation (terminals kWe)	582 700	673 000	534 755
Steam turbine power (kWe)	582 700	673 000	273 180
Combustion turbine power (kWe)	—	—	261 575
Total auxiliary loads (kW)	32 660	122 940	41 760
Net power (kWe)	550 040	550 060	492 995
Net plant efficiency (HHV)	39.3%	28.4%	43.0%
Net plant heat rate (kJ/kWh HHV)	9 298	13 330	8 372
Consumables			
As-received coal feed (kg h^{-1})	256 753	368 084	207 054
Thermal input (kWt)	1 420 686	2 036 717	1 146 500

S12A supercritical system without CO_2 capture (39.3%) and 14.6% points higher than Case S12B supercritical system with carbon capture system (CCS) (28.4%), respectively.

Following the Cost Estimation Methodology for NETL Assessments of Power Plant Performance, capital cost estimates have been made using various models available to WorleyParsons for conceptual level capital cost estimating, including in-house proprietary parametric models and ICARUS from Aspen Tech. ICARUS was used for developing costs for reactor vessels, cyclone, and other specialized process equipment based on the equipment size, basic design, and materials of construction information provided by UK-CAER. Factored estimates for equipment such as pumps, compressors, turbines, etc. were developed using in-house proprietary models of WorleyParsons. For components similar to those in the DOE/NETL reference report, the capital costs escalated to June 2012 dollars was used along with the developed capital costs to estimate a total plant cost. In particular, the costs of the AR and FR were developed based on the dimensions and weights; the air compressor–gas expander–generator set was assumed to have all three components mounted on the same shaft. Equipment costs were developed based on the compression requirements (gas flow rate and pressure) of the systems; the ceramic candle filter specified in the UK-CAER's PCLC-CC configuration is beyond the technical specifications for technologies currently on the market. Current particulate filters, to operate up to 1000 °C, are commercially available from Pall Corporation [13]. The technology relies on silicon carbide/alumina/mullite filtering medium. The estimated cost for the candle filter in this study is based on similar advanced filter technologies. The potential of cost savings through future advancements in particulate filtering from hot gas streams or selection of an alternate technology is considered. The results showed that the PCLC unit, particulate separation and filtration, and gas turbine/accessories are three costly items. Operation and maintenance costs estimates were based on consumable consumption rates and manpower estimates developed by UK-CAER. For general consumables such as fuels and water, costs provided by the DOE report escalated to June 2012 dollars were used. For consumables specific to the UK-CAER process, such as red mud OCs, UK-CAER has provided the costs to be used. The UK-CAER determines an alternate number of plant personnel for estimating the operating labor costs.

The comparative financial analysis was performed to compare the UK-CAER CO_2 capture technology to state-of-the-art coal-fired generation technologies (with and without CO_2 capture), as listed in Table 5.11. The analysis has estimated and compared the first year COE, the 30-year levelized cost of electricity (LCOE), and the cost of captured CO_2 emissions using the plant performance and capital and operation and maintenance costs. The UK-CAER's PCLC-CC plant has an increase in COE of 25.2% from the non-capture, Case 12A. Compared to DOE/NETL Case 12B supercritical unit with post-combustion CCS, PCLC-CC plant has the following advantages:

Table 5.11 Comparison of operating parameters and economic analysis between the DOE/NETL baseline cases and the UK-CAER's PCLC-CC (in 2012$).

	Case 12A	Case 12B	PCLC-CC
Operating parameters			
Net plant output (MWe)	550.0	550.1	493.0
Net plant heat rate (kJ/kWh HHV)	9 298	13 330	8 372
CO_2 captured (kg/MWh net)	0	1 106	770
CO_2 emitted (kg/MWh net)	858	123	9
Costs (2012$ kW^{-1})			
Risk	Low	High	High
Total plant costs	2 271	3 993	2 492
Total overnight cost	2 766	4 861	3 204
Bare erected cost	1 859	3 084	1 852
Home office expenses	169	281	199
Project contingency	244	496	101
Process contingency	0	133	341
Owners costs	495	867	712
Total overnight cost (×1 000)	1 521 592	2 673 946	2 013 286
Total as spent capital	3137	5541	4253
Annual fixed operating costs ($ yr^{-1})	44 211 196	70 779 589	60 474 591
Variable operating costs ($/MWh)	7.66	12.29	9.39
Coal price ($/ton)	16.91		
COE ($/MWh, 2012$)	70.26	128.59	87.98
CO_2 TS&M costs		5.60	3.49
Fuel costs	8.70	12.47	7.68
Variable costs	7.66	12.29	10.09
Fixed costs	10.80	17.28	16.17
Capital costs	43.10	80.95	50.56
LCOE (2012$/MWh)	89.08	163.06	111.57
Cost of CO_2 captured ($/ton CO_2)		52.74	27.62

- A lower COE by $40.61/MWh, a 31.6% reduction
- A lower LCOE by $51.49/MWh, also a 31.6% reduction
- A lower cost of CO_2 captured by 25.12 tonne^{-1} CO_2, a 47.6% reduction because of high overall plant efficiency for PCLC power plant

The total plant cost breakdown for the UK-CAER's PCLC-CC technology is listed in Tables 5.5–5.12.

Table 5.12 The Total plant cost breakdown for the UK-CAER's PCLC-CC technology.

	Client: USDOE/NETL								Report date: 31 October 13			
	Project: low rank (Western) coal baseline study											
	Total plant cost summary											
	Case: Case 1 – base case											
	Plant size: 493.0 MW, net		Estimate type: conceptual						Cost base (January) 2012 ($×1000)			
Acc. No.	Item/description	Equipment cost ($)	Material cost ($)	Labor		Sales tax ($)	Bare Erect Cost ($)	Eng'g CM HO& Fee ($)	Contingencies		Total plant cost	
				Direct ($)	Indirect ($)				Process	Project	$	$/kW
1	Coal and sorbent handling	26 674	3 335	13 958	0	0	43 966	3 642	0	8 801	56 409	114
2	Coal and sorbent prep and feed	118 822	6 511	20 372	0	0	145 705	11 413	0	31 458	188 577	383
3	Feedwater and MISC. BOP systems	8 810	7 250	7 761	0	0	23 821	2 063	0	5 992	31 876	65
4.	*Pyrolysis and CLC island*											
4.1	Pyrolysis and cracker equipment	–	–	–	0	0	0	–	–	–	–	–
4.2–4.5	CLC reactors	54 981	13 909	30 266	0	0	99 156	14 873	19 831	17 104	150 964	306
4.6–4.7	Particulate separation and filtration	60 725	17 378	34 755	0	0	112 858	13 543	10 806	15 574	152 781	310
	Subtotal 4	115 706	31 287	65 021	0	0	212 014	28 416	30 637	32 678	303 745	616

5	*Gas cleanup and compression*											
5A	GAS cleanup and piping	0	0	0	0	0 0	0	0	0	0	0	0
5B	CO_2 compression	18 093	0	10 682	0		28 775	2 561	0	6 309	37 645	76
	Subtotal 5	18 093	0	10 682	0	0	28 775	2 561	0	6 309	37 645	76
6	GAS turbine/accessories	125 000	39 933	17 778	0	0	182 711	26 784	18 271	30 879	258 645	525
7	*HRSG, ducting and stack*											
7.1	Heat recovery steam generator	37 408	0	5 418	0	0	42 826	3 764	0	4 690	51 280	104
7.2–7.9	Selective catalytic reduction (SCR) system, ductwork and stack	3 328	2 994	3 557	0	0	9 879	774	0	1 710	12 363	25
	Subtotal 7	40 736	2 994	8 975	0	0	52 705	4 538	0	6 400	63 643	129
8	*Steam turbine generator*											
8.1	Steam TG and accessories	30 317	0	5 206	0	0	35 523	3 151	0	3 893	42 567	86
8.2–8.9	Turbine plant auxiliaries and steam piping	10 678	1 081	7 701	0	0	19 460	1 637	0	4 230	25 327	51
	Subtotal 8	40 995	1 081	12 907	0	0	54 983	4 788	0	8 123	67 894	138

(Continued)

Table 5.12 (Continued)

	Client: USDOE/NETL								Report date: 31 October 13			
	Project: low rank (Western) coal baseline study											
	Total plant cost summary											
	Case: Case 1 – base case											
	Plant size: 493.0 MW, net		Estimate type: conceptual						Cost base (January) 2012 ($×1000)			
Acc. No.	Item/description	Equipment cost ($)	Material cost ($)	Labor		Sales tax ($)	Bare Erect Cost ($)	Eng'g CM HO& Fee ($)	Contingencies		Total plant cost	
				Direct ($)	Indirect ($)				Process	Project	$	$/kW
9	Cooling water system	12 322	16 363	16 567	0	0	45 252	3 880	0	10 663	59 795	121
10	ASH/SPENT sorbent handling SYS	2 383	1 420	1 257	0	0	5 060	446	0	850	6 356	13
11	Accessory electric plant	31 042	13 569	26 583	0	0	71 194	5 666	0	15 028	91 888	186
12	Instrumentation and control	9 712	1 787	6 263	0	0	17 762	1 488	888	3 378	23 516	48
13	Improvements to site	3 498	2 062	8 791	0	0	14 351	1 311	0	4 730	20 392	41
14	Buildings and structures	0	6 746	7 817	0	0	14 563	1 225	0	2 611	18 399	37
	Total cost	553 792	134 338	224 732	0	0	912 862	98 221	49 796	167 901	1 228 780	2 492

Eng'g CM HO& Fee, engineering, construction management, home office and fee.

5.6 Technical Gaps and Challenges

Some of the technical gaps and challenges are special for the development of the direct coal-fueled PCLC-CC plant.

1. *Heat integration.* Heat integration of UK-CAER's PCLC-CC plant could follow a similar method for either integrated gasification combined cycle (IGCC) or PFBC processes to obtain higher plant efficiency. However, the requirement on the HRSG configuration of PCLC-CC plant is far from that standard of IGCC or PFBC process. The special feature of the coal-fired PCLC-CC plant is the ability to generate two heat streams to drive the ST (bottoming cycle). The first one is a relatively low-quality stream of the FGT exhaust, composed of N_2 and O_2, with a temperature of 560–590 °C, and ambient pressure; the second one is high quality from the CO_2/H_2O stream of the FR with temperature up to 950–1000 °C, high pressure of 1.2–3.0 MPa, and high content of fly ash, which accounts for about 20% of the total energy for the steam cycle. To avoid the mixing of two gas streams, their heat needs to be recovered by two HRSG units to drive a supercritical, three pressure-level reheated ST, making the system much more complex compared to the typical IGCC or PFBC process. The design of HRSG and the steam cycle for the PCLC-CC plant would reconsider the effect of dual working mediums, process optimization, a thermal–economic balance, and flexibility/reliability for unit operation.
   The other special considerations for heat integration may be (i) the heat surface arrangement and water hydraulics inside HRSGs to eliminate potential heat stop points; (ii) supplementary firing of natural gas before HRSG if the use of high temperature of the AR needs to be avoided; (iii) internal heat exchanger of the FR to heat up steam SH/RH of bottoming cycle; (iv) the recycling of the CO_2/H_2O gas mixture to the FR; and (v) preheating the compressed air to the AR.
2. *Monitor and control of hydraulic flow.* Quantification of solids flow and mixing is important to ensure stable operation and maximize performance of the PCLC unit. However, direct monitoring and control of the complex multiphase fluid dynamics in the complex PCLC reactor system is not yet possible. At least, the following parameters need to be estimated: the overall solid mass flux in the chemical loop, solid residence time distribution in each component of the PCLC reactor, and the gas–solid flow pattern. In addition, monitoring the malfunction of fluidization and the pressure profile are the major concerns.
3. *Hot flue gas cleanup.* The present specifications of gas turbine manufacturers for the required flue gas quality are at a maximum particulate content of 5 mg m^{-3} STP, diameter of <5 μm, and a maximum alkali content <0.01 mg m^{-3} STP (http://netl.doe.gov/publications/proceedings/02/GasCleaning/1.08paper.pdf). No commercially available filter material has been developed to meet all the requirements for hot spent air cleanup of the coal-based PCLC-CC process. The filter material should offer the benefits of non-brittle mechanical behavior at temperatures as high as 1100 °C, and improved resistance to thermal fatigue, wear, and deformation. The material

should be light, strong, resistant to thermal shock, and relatively inexpensive. Simultaneously, the filter materials should not react with the surrounding gas phase and the metallic OC particles. In the last 30 years, as a part of the research work of PFBC/PCFBC technology, ceramic materials, FeAl and FeCr alloy, Ni-Cr-Al-Fe based alloys, and iron aluminide have been developed to form candle filter architecture for hot flue gas cleanup (https://www.netl.doe.gov/events/conference-proceedings/2002/gas-cleaning-at-high-temperatures). However, no material was successfully operated at temperatures higher than 900 °C in the long run. Furthermore, an efficient dust cake cleaning system needs to protect the filter itself from dust-induced damage; and the integrated filter with failsafe system is necessary to protect the downstream gas turbine and to prevent unscheduled shutdown of the system. The integrated filter device should be capable of limiting particle leakage with low pressure drop, and allow high system availability, commensurate with annual maintenance outages. In recent years, significant achievements in material science provide the possibility to develop advanced filter elements that could be used for the coal-fueled PCLC-CC. Technologies for flue gas cleanup at 1400 °C have been proposed with an efficient ash and alkali separation rate by several companies, including Babcock-Borsig-Power Env. GmbH (BBP Env.), E.ON Kraftwerke GmbH, SaarEnergie GmbH, Siemens AG, and Steag AG.

4. *Online ash separation.* In order to avoid coal ash accumulation in the system, ash separation from binary mixtures of OCs and solid fuel ash is needed. A simple and effective way is to develop an online ash separation device integrated with reaction system based on the difference in particle size and density between OC and coal ash. However, when operated at high pressure, the separation efficiency would drop due to the decrease in terminal velocity of coal ash. Another option is to develop a magnetic separator. Unlike nickel- and copper-based OCs, the iron-based OCs become magnetic when they are reduced to magnetite in the FR. Therefore, magnetic separation can be applied to recover the spent OCs.
5. *Retrofit of natural gas turbine.* Instead of developing a new gas turbine technology, retrofitting a natural gas turbine to FGT is the easier way for the development of the PCLC-CC plant. Some lesson can be learned for the following two cases. (i) A 15 MWe ABB STAL FGT (GT)-35P has been successfully modified and tested for the PFBC technology at Tidd in the last century. (ii) Japan tested a 75 MWe FGT on the PFBC facility at the Karita Power Station in the 2000s. Those two FGTs are different from the natural gas-based turbine. The modification/retrofit covers (i) unmatched airflow and power generation for application to PFBC – the air compressor having higher inlet flow, which requires a large open mouth and longer blades; (ii) a lower gas turbine inlet temperature between 950 and 1100 °C versus 1350 and 1400 °C for NG turbine. The blade angle and working profile across stages have to be adjusted to maintain a reasonable exhaust temperature for downstream steam generation; and (iii) blade erosion and corrosion as well as particle deposition. The gas turbine in the coal-based combined cycle is subjected to harsh conditions from coal impurities.

6. *Pollutant emission.* Little effort has been made to study the pollutant emission and mitigation for the solid-fueled CLC. The results obtained from a 100 kW_{th} CLC facility using iron-based OCs and solid fuels at Chalmers University of technology, Sweden, showed that the concentration of NO is in the range of 1000–2000 ppm and 10–20% fuel-N is oxidized to NO [14]. The sulfur-containing gas is in the form of SO_2 and H_2S in the FR, and their concentrations depend on the type of solid fuel and the content of elemental sulfur. Similar observation was obtained on a 25 kW_{th} unit at the Ohio State University [15]. The pollutants at the outlet of the FR are 600–1170 ppmv for SO_2 and 1148–1669 ppmv for NO_x. So, effective emission mitigation strategies and methods need to be developed for the coal-fueled CLC process. There are several avenues for pollutant control that can be integrated with the solid-fueled CLC, (i) using desulfurized fuels to avoid the introduction of impurities to the reaction system, (ii) in situ desulfurization in the FR using a sorbent, and (iii) using flue gas desulfurization unit prior to CO_2 sequestration.
7. *Solid waste management.* Research work is necessary to develop suitable methods for OC recovery to minimize the operating cost associated with OC makeup. Recently, some work on the solid waste management of a CLC plant was carried out by C.S.I·C Spain, in which a recovery process with 80% recovery efficiency was developed to mitigate the loss of active OCs [16]. For the UK-CAER's PCLC-CC plant, there are two spent OC streams from the PCLC island. The first stream is some amount of OC particles together with coal ash from the FR, and the second is the spent OC particles captured by the high-temperature filter from hot spent air stream. The loss of OC from the first stream is dependent on OC attrition rate and the efficiency of ash–OC separation device, while the second is dependent on the cyclone cut efficiency before the candle filter. There may be two options for solid waste management: (i) disposal of those solids to landfill in a safe manner and (ii) recovery and recycling of the spent OC materials.

References

1 Xiao, R., Song, Q., Zhang, S. et al. (2010). Pressurized chemical-looping combustion of Chinese bituminous coal: cyclic performance and characterization of iron ore-based oxygen carrier. *Energy Fuels* 24: 1449–1463.

2 Fan, Z., Chen, L., Liu, F., et al (2016). Coal based pressurized chemical looping combustion combined cycle-process development and analysis, 4th International Conference on Chemical Looping, Nanjing, China, September 26–28, 2016.

3 Adanez, J., Abad, A., Garcia-Labiano, F. et al. (2012). Progress in chemical-looping combustion and reforming technologies. *Prog. Energy Combust. Sci.* 38 (2): 215–282.

4 Schwebel, G.L., Leion, H., and Krumm, W. (2012). Comparison of natural ilmenites as oxygen carriers in chemical-looping combustion and influence

of water gas shift reaction on gas composition. *Chem. Eng. Res. Des.* 90: 1351–1360.

5 Chen, L., Bao, J., Liang, K. et al. (2016). The direct solid–solid reaction between coal char and iron-based oxygen carrier and its contribution to solid-fueled chemical looping combustion. *Appl. Energy* 184 (15): 9–18.

6 Sanders, J.P. and Gallagher, P.K. (2003). Kinetics of the oxidation of magnetite using simultaneous TG/DSC. *J. Therm. Anal. Calorim.* 72: 777–789.

7 Sanders, J.P., Gallagher, P.K., Sanders, J.P., and Gallagher, P.K. (2003). Thermomagnetometric evidence of γ-Fe_2O_3 as an intermediate in the oxidation of magnetite. *Thermochim. Acta* 406: 241–243.

8 Keller, M., Leion, H., Mattisson, T., and Lyngfelt, A. (2011). Gasification inhibition in chemical-looping combustion with solid fuel. *Combust. Flame* 158 (3): 393–400.

9 Roberts, D.G., Hodge, E.M., Harris, D.J., and Stubington, J.F. (2010). Kinetics of char gasification with CO_2 under regime II conditions: effects of temperature, reactant and total pressure. *Energy Fuels* 24 (10): 5300–5308.

10 Xiao, R., Chen, L., and Saha, C. (2012). Pressurized chemical-looping combustion of coal using an iron ore as oxygen carrier in a pilot-scale unit. *Int. J. Greenhouse Gas Control* 10: 363–373.

11 Chen, L., Zhang, Y., Liu, F., and Liu, K. (2015). Development of a cost-effective oxygen carrier from red mud for coal-fueled chemical-looping combustion. *Energy Fuels* 29 (1): 305–313.

12 Chen, L., Yang, L., Liu, F. et al. (2017). Evaluation of multi-functional iron-based carrier from bauxite residual for H_2-rich syngas production via chemical-looping gasification. *Fuel Process. Technol.* 156: 185–194.

13 Pall Corporation, (2006) Pall gas solid separation systems, advanced metal and ceramic filter systems for critical gas solid separation processes, Bulletin No. GSS-1c.

14 Markstrom, P. (2012). Design, modeling and operation of a 100 kW_{th} chemical-looping combustor for solid fuel. PhD thesis. Chalmers University of Technology.

15 Bayham, S.C., Kim, H.R., Wang, D. et al. (2013). Iron-based coal direct chemical looping combustion process: 200-h continuous operation of a 25-kW_{th} sub-pilot unit. *Energy Fuels* 27: 1347–1356.

16 García-Labiano, F., Gayán, P., Adánez, J. et al. (2007). Solid waste management of a chemical-looping combustion plant using Cu-based oxygen carriers. *Environ. Sci. Technol.* 41: 5882–5887.

Section 2

Oxygen Carriers

6

Regenerable, Economically Affordable Fe_2O_3-Based Oxygen Carrier for Chemical Looping Combustion

Hanjing Tian[1], *Ranjani Siriwardane*[2], *Esmail R. Monazam*[2], *and Ronald W. Breault*[2]

[1] *West Virginia University, Department of Chemical & Biomedical Engineering, Morgantown, WV, 26505, USA*
[2] *U.S. Department of Energy, National Energy Technology Laboratory, 3610 Collins Ferry Road, Morgantown, WV, 26507-0880, USA*

6.1 Introduction

Chemical looping combustion (CLC) produces concentrated carbon dioxide (CO_2) from the combustion of fossil fuels. Instead of air, a solid oxygen carrier (OC) supplies oxygen for combustion, which eliminates the presence of nitrogen (N_2) from the resulting flue gas. The significant advantage of CLC over conventional combustion for CO_2 capture is that CLC can produce a concentrated CO_2 stream, which minimizes the energy penalty for CO_2 separation from N_2 after a combustion process [1–3].

The development of applicable oxygen carriers is a core technology for CLC. The efficient transfer of oxygen from the lattice phase to the surface site for combustion reaction is essential for ideal oxygen carriers. So far, a tremendous number of oxide-based materials have been developed to study their reactivity/physical properties under high temperature, including iron (Fe) oxide, copper (Cu) oxide, manganese (Mn) oxide, nickel (Ni) oxide, etc. [4–22]. Besides natural ores containing those oxides, a variety of inert oxide materials are used as carrier supports/binders to synthesize practical oxygen carriers, aiming for improved reaction kinetic and stable performance, such as alumina, titania, or naturally occurring clays, such as bentonite. Meanwhile, reaction and hydrodynamic parameters, such as reaction temperature, fuel concentration, particle size, and circulation rate and residence time directly affect the selection and optimization of feasible oxygen carriers. This valuable information is critical for kinetic modeling and reactor design.

In this chapter, we will summarize the experimental results conducted on Fe_2O_3-based oxygen carriers by following the important criteria for a regenerable, economically affordable oxygen carrier [2]:

- High reactivity with fuel and air
- Low fragmentation and attrition
- Low tendency for agglomeration under high temperature

Handbook of Chemical Looping Technology, First Edition. Edited by Ronald W. Breault.

- Low carrier production cost
- Being environmentally benign
- Stability of fluidized particles under repeated reduction/oxidation cycles

6.2 Primary Oxide Selection

The oxides of the transition metals such as copper, iron, nickel, cobalt, and manganese have received much attention because they have favorable reduction/oxidation characteristics. Among those transition metal oxides, nickel oxide, iron oxide, and copper oxides are being intensively investigated for their fast reaction rates, high oxygen transfer capacities, low production cost, and stability.

Nickel oxide (NiO) has been studied in greater detail in the early stage of CLC development because it has more favorable characteristics for methane CLC. It is generally accepted that NiO-based oxygen carriers possess faster reaction rates than CuO and Fe_2O_3. Al_2O_3-supported NiO oxygen carriers have been intensively investigated with laboratory-scale instruments and bench-scale pilot plants with fluidized bed reactor facilities [6–9]. NiO/$NiAl_2O_4$ oxygen carriers were tested in a 10-kW reactor for 160 hours with approximately 99% conversion. The major drawback of NiO-based oxygen carriers is their carcinogenic characteristics, which will be a daunting issue if CLC will be deployed in pilot-scale or up to commercial deployment.

Copper oxide has several favorable features [23–25]: (i) it is highly reactive in both reduction and oxidation cycles; (ii) it is thermodynamically favored to reach complete conversion using gaseous hydrocarbon fuels (e.g. methane); (iii) reduction and oxidation reactions of Cu-based oxygen carriers are both exothermic, reducing the need for energy supply in the reduction reactor; and (iv) it is less expensive than other oxygen carrier materials, such as nickel (Ni), manganese (Mn), and cobalt (Co). However, CuO has a fairly low melting point, which results in particle agglomeration during multiple CLC reaction cycles. Inert support/binder materials such as Al_2O_3, TiO_2, SiO_2, and natural clays such as bentonite can be used to avoid particle agglomeration and enhance the characteristics of CuO and Fe_2O_3, such as reactivity, durability, and fluidizability. Overall particle size, ratio of metal oxide to inert support, porosity, oxygen carrier grain size, and geometry are some factors that affect the reaction performance of a particular oxygen carrier.

Fe_2O_3 has fair reactivity, stable reaction performance, and very little environmental concern. The reduction kinetics from Fe_2O_3 to Fe_3O_4 is fast, but the subsequent reduction to FeO is very slow for methane (CH_4) conversion to CO_2 and H_2O. Fe_2O_3/Al_2O_3 was investigated to improve reactivity and to overcome agglomeration. Samples prepared by impregnation of 80% Fe_2O_3/Al_2O_3 are reported to be stable in 20-cycle thermogravimetric analysis (TGA) tests with methane conversions between 85% and 94%. Agglomeration has been reported to be a problem with supported iron oxide materials; thus, circumventing agglomeration is important to achieve stable CLC fluidized bed operation with iron-based oxygen carriers. Although its redox characteristics are not the strongest of the metal oxides, iron oxide is widely considered an attractive option

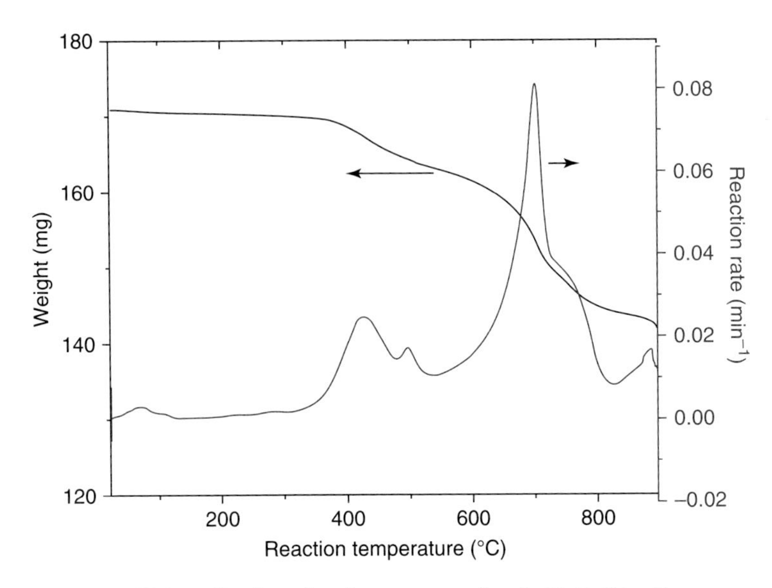

Figure 6.1 TGA profile of combustion segment of coal with CuO in nitrogen.

for CLC application because of its low cost and environmental compatibility [5, 26–28].

A systemic study of various oxides CuO, Fe_2O_3, NiO, Mn_2O_3, and Co_3O_4 has been done on direct-coal (Illinois #6 Coal) chemical looping [7]. TGA was widely applied for the measurement of CLC kinetic and weight change during coal combustion in temperature-programming mode. The weight loss profile and corresponding rate data of the CuO–coal mixture during heating in nitrogen up to 900 °C is shown in Figure 6.1. There is a continuous weight loss with the coal–CuO mixture during the heating that indicates that CuO contributes to the combustion of coal after coal volatilization. The weight loss corresponds to 100% coal combustion with CuO in the presence of N_2. The profile of reaction rates indicates the presence of two peaks at 425 and 708 °C.

The major peak at 708 °C is due to char combustion by oxygen from CuO, and the coal combustion appears to initiate around 500 °C. The rate data from TGA measurements and the outlet CO_2 profile collected from in-line mass spectrometry (MS) analysis (not shown here) are consistent. TGA data during the introduction of air after the reduction reaction with coal is shown in Figure 6.2. The oxidation rate is significantly higher than the rate of the reduction reaction. In addition, the weight gain during oxidation reaction is consistent with the amount of oxygen present in the original CuO sample, which indicates that the reduced copper can be fully reoxidized.

Combustion experiments similar to the CuO–coal system were performed with coal and NiO, Fe_2O_3, Mn_2O_3, and Co_3O_4 in the presence of both N_2 and CO_2. The results are shown in Table 6.1. It is interesting to note that the lowest combustion reaction temperature at 700 °C and the highest combustion rate were observed for the CuO–coal system. In addition, the combustion reaction is exothermic with CuO. From a practical standpoint, this could be an advantage

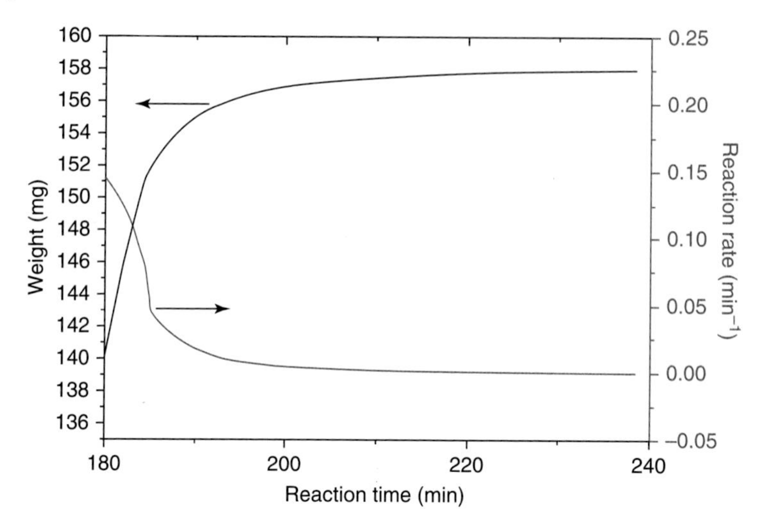

Figure 6.2 TGA profile of oxidation segment of coal with CuO in air.

in using CuO (versus other metal oxides) because both the fuel and air reactor (AR) are exothermic, avoiding the need to transfer heat from the oxidizer to the reducer. Full combustion and oxidation were also achieved for coal combusted by CuO with bigger particle size (63–173 μm), but higher reaction temperature (788 °C). The percentages of combustion and oxidation were close to 100% for the CuO, indicating that the complete coal-combustion reaction can be obtained with the CuO, and reduced copper can be completely oxidized at 900 °C.

For Fe_2O_3–coal system, the maximum combustion rate takes place at a higher temperature (973–977 °C) than that with CuO. Combustion conversion of 95% was achieved with Fe_2O_3. The percentage of oxidation was about 77% when the oxidation-state change of Fe was assumed to be from Fe(III) to Fe(II) during the CLC reaction. X-ray photoelectron spectroscopy (XPS) analysis conducted with samples after the combustion reaction also indicated that metallic iron was not present on the surface of the sample. The heat of the reduction reaction for the Fe_2O_3–coal system is endothermic. Consequently, heat transfer from the oxidizer to the reducer will be required for a CLC process with Fe_2O_3. The combustion rate for Fe_2O_3 appears to be lower than that with the CuO system, but the oxidation rate appears to be higher with Fe_2O_3.

The characteristics of the NiO–coal system were similar to the Fe_2O_3–coal system. For NiO–coal, the combustion reaction, which is also endothermic, takes place at a higher temperature of 993 °C with a lower reaction rate than with a CuO–coal system. However, the percentage of combustion with NiO–coal was lower than that observed with CuO–coal.

For Mn_2O_3–coal, the combustion reaction took place at 900 °C, showing the lowest combustion rate as compared to the four other metal oxides, but the reaction is slightly exothermic. The percentage of combustion was similar to that with NiO. For Mn_2O_3, the CLC reaction appears to occur between the oxidation states Mn(III) and Mn(II). XPS analysis data also confirmed that Mn(0) is not present in the sample after combustion. The Co_3O_4–coal system appears to be similar

Table 6.1 Thermodynamic and reaction properties of coal CLC on various metal oxides.

Samples	Gas media	Combustion				Oxidation		
		Combustion temperature (°C)	Combustion rate (min^{-1})	% Combustion	ΔH ($kJ\ mol^{-1}$)	Oxidation rate (min^{-1})	Oxygen uptake (%)	ΔH ($kJ\ mol^{-1}$)
CuO (5 μm)	CO_2	703	0.098	100	−96.5	0.172	98.6	−156
	N_2	708	0.083	100		0.175	99.2	
CuO (63 μm)	N_2	780	0.079	100	−96.5	0.174	99.2	−156
NiO (44 μm)	CO_2	993	0.061	73.05	75.2	0.84	77.5	−327.7
	N_2	1000	0.017	68.4		0.82	71.6	
Fe_2O_3 (44 μm)	CO_2	973	0.055	94.9	79.2	0.77	93.7 Fe(II)	−347.4
	N_2	977	0.05	91.6		0.78	90.6 Fe(II)	
Mn_2O_3 (44 μm)	CO_2	905	0.011	76.76	−36.1	0.42	72.2 Mn(II)	−216.4
	N_2	978	0.01	71		0.38	68.3 Mn(II)	
Co_3O_4 (70 μm)	CO_2	781	0.096	83.3	−8.6	1.74	78.2 Co(II)	−243.9
	N_2	781	0.096	83		1.74	78.0 Co(II)	

to CuO with a low combustion temperature of 781 °C, but the oxidation rate was highest for the Co_3O_4. The heat of reaction was slightly exothermic, yet the percentage of combustion was lower than that with CuO–coal. For Co_3O_4, the combustion-oxidation reaction was assumed to be between Co^{3+} and Co^{2+} oxidation states. XPS analysis verified the presence of Co^{2+} on the surface of the samples after combustion. From the five metal oxides tested, Fe_2O_3 and CuO appeared to have the best combustion characteristics. Fe_2O_3 possesses fair reactivity with the lowest cost, while the agglomeration of CuO due to a relatively lower melting point (~1000 °C) must be addressed for applicable oxygen carriers.

6.3 Supported Single Oxides

Inert supports, such as bentonite, alumina, and titania have been applied over primary oxides (Ni, Fe, Cu oxides) for improved reactivity, agglomeration resistance, and better mechanical strength. It is necessary to select consistent experimental conditions to compare the reaction performance of various oxygen carriers. A suitable screening condition should avoid carbon deposition during the CLC reaction over most oxygen carriers. Most carbon formation in CLC is due to methane decomposition between 700 and 1100 °C, which can be catalyzed by reduced metals, including metallic Ni, Cu, and Fe formed from the combustion reaction. Carbon formation is strongly affected by reaction temperatures, the nature of active species and supports, as well as other physical properties. Pre-modification of reaction temperature, methane concentration, and reaction times are necessary to identify reaction conditions that minimize carbon formation.

For the initial set of tests, 60% CuO/bentonite and 60% Fe_2O_3/bentonite were selected. The TGA profiles of 10-cycle CLC tests over 60% CuO/bentonite with methane as fuel are presented in Figure 6.3. The reduction and oxidation times were set for 30 and 60 minutes, respectively. During reduction, the sample weight decreased because the supported CuO reacted with methane and formed metallic copper, water, and CO_2. During oxidation, the sample weight gain was associated with the reduced metallic copper regenerated by air to form CuO. The extent of the reduction/oxidation reaction remained constant as the cycle number increased, also reinforcing the stability of the CuO/bentonite samples. The oxygen transfer capacity was in the range of 12–14%. No other weight change was observed, indicating no significant carbon formation for CuO oxygen carriers. Overall, 60% CuO/bentonite demonstrated excellent stability, because the reduction weight and oxidation weight remained constant for all 10-cycle tests in the temperature range of 700–900 °C.

TGA profiles of 10-cycle CLC tests over 60% Fe_2O_3/bentonite with methane as fuel are presented in Figure 6.4. Under the flow of methane, the sample weight initially decreased due to the reduction of Fe_2O_3, but then continuously increased due to carbon formation, until the methane flow was stopped. It should be noted that total sample weight gain could exceed the initial metal oxide weight due to heavy carbon deposition onto the carrier sample. Under the flow of air, the

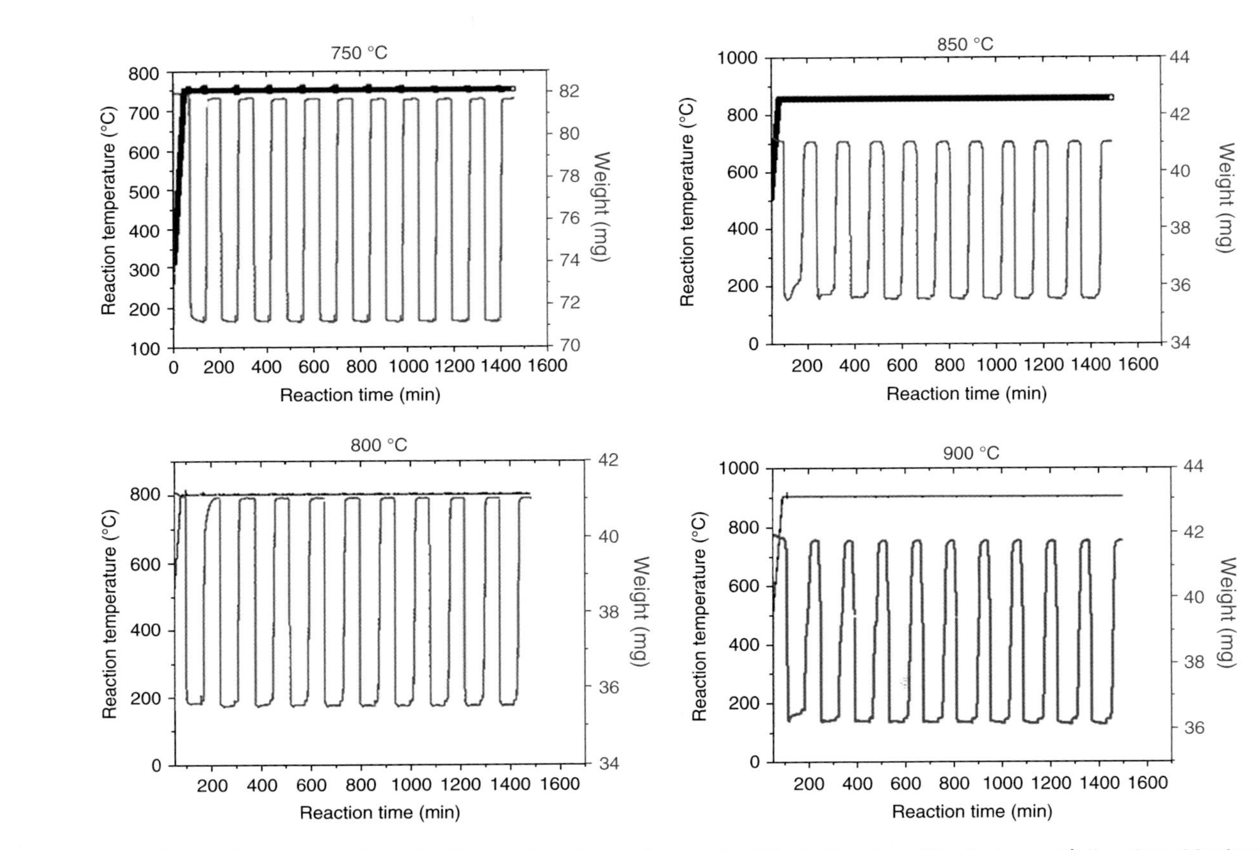

Figure 6.3 TGA profile of 10-cycle CLC tests of 60% CuO/bentonite with methane as fuel (Reduction time: 30 minutes; oxidation time: 30 minutes; pure CH_4).

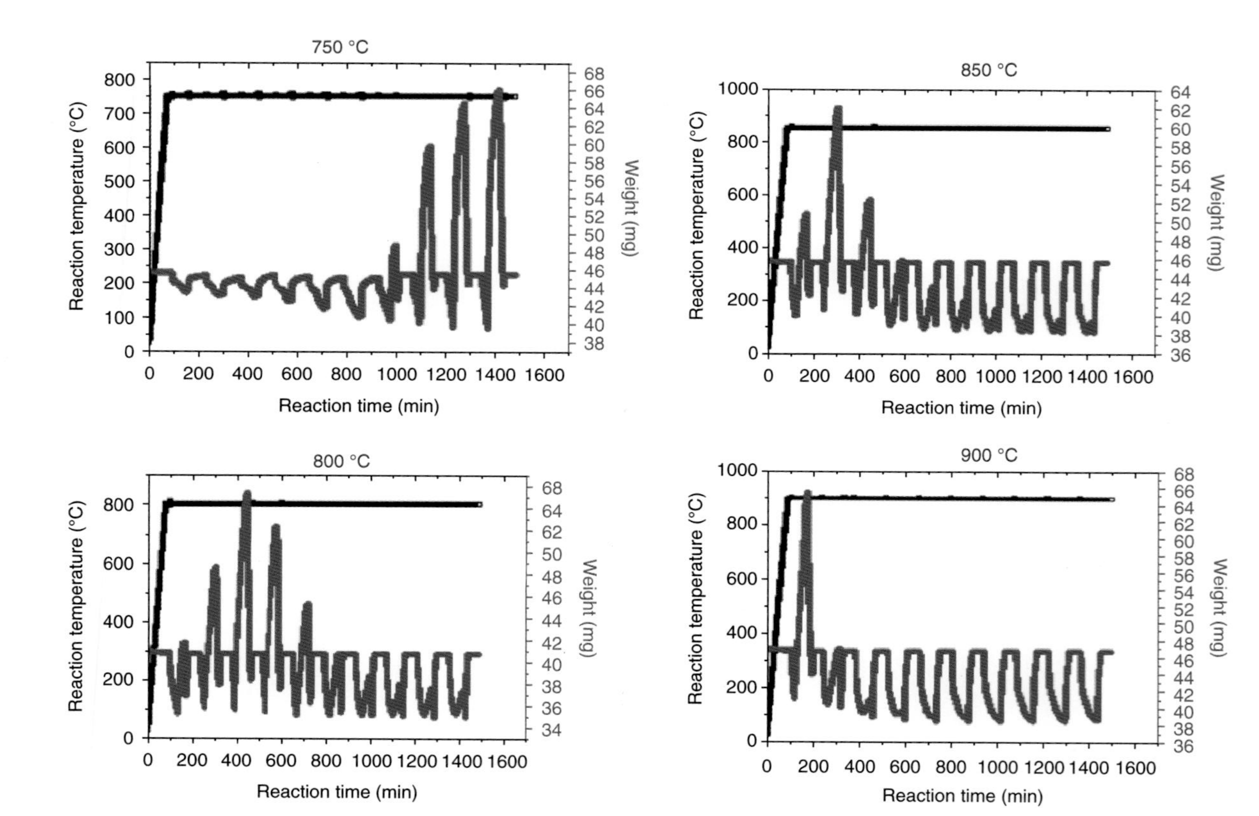

Figure 6.4 TGA profile of 10-cycle CLC tests of 60% Fe_2O_3/bentonite with pure methane as fuel (Reduction time: 30 minutes; oxidation time: 30 minutes; pure CH_4).

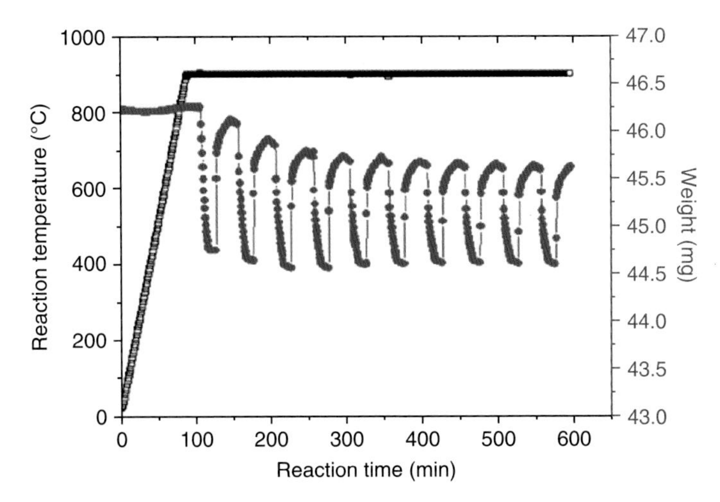

Figure 6.5 TGA profile of 10-cycle CLC tests of 60% Fe_2O_3/bentonite at 900 °C under screening conditions.

sample weight sharply decreased because of the initial oxidation and removal of carbon, and then the carrier weight continued to increase due to the reoxidation. The CO_2 produced from the removal of carbon formation is mixed with exhausted air flow and would eliminate the benefit of CO_2 separation efficiency in a large-scale CLC process. As shown in Figure 6.4, the carbon formation rate decreased at lower methane concentrations and higher reaction temperatures (700–900 °C). An effective approach for avoiding carbon formation is to dilute methane concentration or shorten reduction time before the formation of carbon deposits.

The 10-cycle TGA profile of 60% Fe_2O_3/bentonite at 800 °C with 20% CH_4/N_2 as reducing agent is shown in Figure 6.5. Using diluted CH_4, no carbon deposition was observed in reduced reduction time (10 minutes). At 800 °C, the reduction weight remained constant, but the oxidation weight decreased with increasing cycle numbers. Therefore, oxygen transfer capacity decreases from the initial value of 2.9% in the third cycle to 2.2% in the seventh cycle. The oxidation weight decreases with each test cycle during the first five cycles, suggesting that some material conditioning is required for 60% Fe_2O_3/bentonite before stable performance can be observed. After 300 minutes of activation, both reduction and oxidation weights stabilized, and the oxygen transfer capacity remained at 2.4%. Regarding the reaction activity, the reduction and oxidation rates decreased with increasing cycles at 750 °C, but slightly increased in oxygen transfer capacity with cycles observed at 900 °C. Therefore, reaction temperatures positively affect the reduction/oxidation rate, as well as oxygen transfer capacity. In this study, all oxygen carriers were compared and screened at the following conditions: 20% methane/nitrogen for reduction reaction, 750–900 °C as reaction temperatures, and 10-minute reduction time for a 10-cycle reduction/oxidation cycle.

6.4 Natural Oxide Ores

Beside synthetic materials development, some naturally occurring materials, mainly iron-based ores, have been investigated because of their low cost and wide availability [28–33]. Long-term experiments have been conducted with ilmenite, $FeTiO_3$, for approximately 25 hours at 975 °C. Scanning electron microscope (SEM) images of the sample indicate that the porosity of the ilmenite particles increased with visible cracks. The Brunauer–Emmett–Teller (BET) surface area measurements also confirmed an increase of surface area from 0.11 to 0.28 $m^2\ g^{-1}$ for used samples. Another report studied hematite for methane CLC processing. Six reduction/oxidation cycles were conducted with pure methane as fuel and air as oxidant at 950 °C. The rate of conversion of Fe_2O_3 to Fe varied from 1% to 8% per min^{-1} with the yield of methane to CO_2 ranging between 10% and 99%.

Three copper oxide ores (chryscolla, cuprite, malachite) and five iron oxide ores (hematite, ilmenite, limonite, magnetite, taconite) were purchased from Ward's Natural Science. ICP (inductively coupled plasma) analysis was conducted to measure elemental components of the samples (Ref. here). The overall reaction performances of coal/natural ores systems for coal CLC are summarized in Table 6.2.

The reactivates of iron-based ores were very low and it was not possible to obtain accurate rates by differentiating TGA data. All iron-based ores showed poor performance without steam. To obtain better reactivity with natural ores, it will be necessary to add steam in the fuel gas steam. The coal CLC reactions with natural ores were conducted in a bench-scale, fixed-bed reactor. To understand the effect of steam in CLC conversion, 20% steam was added. The maximum operational temperature for the fixed bed reactor is 800 °C, and the reduction reaction information was only obtained up to 800 °C. With the 20% steam addition, the reaction peak temperatures did not change for the coal–limonite and coal–magnetite systems, but the outlet CO_2 concentration

Table 6.2 The reaction rates and oxygen transfer capacities of natural ores in coal CLC reaction in the fifth cycle TGA tests.

	Reaction temperature (°C)	Reduction rate (min^{-1})	% Combustion	Oxidation rate (min^{-1})	% Oxidation
CuO	780	0.079	100	0.174	99.2
Chryscolla	876	0.02	35.0	0.01	67.2
Cuprite	1100	Negligible	12.9	0.11	46.1
Malachite	1100	Negligible	11.6	0.13	23.2
Fe_2O_3	977	0.05	91.6	0.78	90.6
Magnetite	1000	Negligible	10.4	0.57	70.3
Limonite	973	Negligible	15.6	0.35	65.2
Hematite	900	Negligible	12.1	0.08	34.7
Taconite	900	Negligible	26.4	0.08	35.6

Table 6.3 Analysis of extent of reduction of various ore oxygen carriers during coal CLC and methane CLC: oxygen transfer capacities were normalized by the concentration of active CuO or FeO_x species.

Coal/carbon CLC			Methane CLC		
Oxygen carriers/ reduction	**Reduction percentage**	**Species after reduction reaction**	**Oxygen carriers/ reduction reaction**	**Oxygen transfer capacity (%)**	**Species after reduction**
CuO ($CuO \rightarrow Cu$[a])	99.4	Cu	$CuO \rightarrow Cu_2O$	6	Cu_2O
			$CuO \rightarrow Cu$	12	Cu
Chryscolla ($CuO \rightarrow Cu$)	35.0	CuO (major), Cu_2O (minor)	Chryscolla ($CuO \rightarrow Cu$)	10	Cu/Cu_2O
Cuprite ($CuO \rightarrow Cu$[a])	12.9	CuO (major), Cu_2O (minor)	Cuprite ($CuO \rightarrow Cu$)	5.1	Cu_2O
Malachite ($CuO \rightarrow Cu$[a]) ($Fe_2O_3 \rightarrow FeO$[a])	11.6	CuO, Fe_2O_3 (major), Cu_2O/Fe_3O_4 (minor)	Malachite ($CuO \rightarrow Cu$) ($Fe_2O_3 \rightarrow Fe_3O_4$)	6.6	Cu_2O/Fe_3O_4
Fe_2O_3 ($Fe_2O_3 \rightarrow FeO$[a])	54	Fe_3O_4	$Fe_2O_3 \rightarrow Fe_3O_4$	3.3	Fe_3O_4
			$Fe_2O_3 \rightarrow FeO$	10	FeO
			$Fe_2O_3 \rightarrow Fe$	30	Fe
Magnetite ($Fe_3O_4 \rightarrow FeO$[a])	10.4	Fe_2O_3 (major) Fe_3O_4 (minor)	Magnetite ($Fe_3O_4 \rightarrow Fe$)	9.5	FeO/Fe
Limonite ($Fe_2O_3 \rightarrow FeO$[a])	15.6	Fe_3O_4	Limonite ($Fe_2O_3 \rightarrow Fe$)	13.1	FeO/Fe
Hematite ($Fe_2O_3 \rightarrow FeO$[a])	12.1	Fe_3O_4	Hematite ($Fe_2O_3 \rightarrow Fe$)	10.6	FeO/Fe
Taconite ($Fe_2O_3 \rightarrow FeO$[a])	26.4	Fe_3O_4	Taconite ($Fe_2O_3 \rightarrow Fe$)	9.8	FeO/Fe_3O_4

a) The reduction percentages were calculated assuming the reactions indicated in the parenthesis.

increased significantly at 800 °C after steam addition. This demonstrated better coal combustion efficiency in the presence of steam and indicated that the addition of steam is effective in improving the reaction performance of natural ores in direct coal CLC process (Table 6.3).

To understand the positive effect due to steam on coal CLC, the reaction of coal with steam was conducted in the bench-scale, fixed-bed reactor tests with 20% steam/Ar while the temperature increased to 800 °C. The outlet concentrations

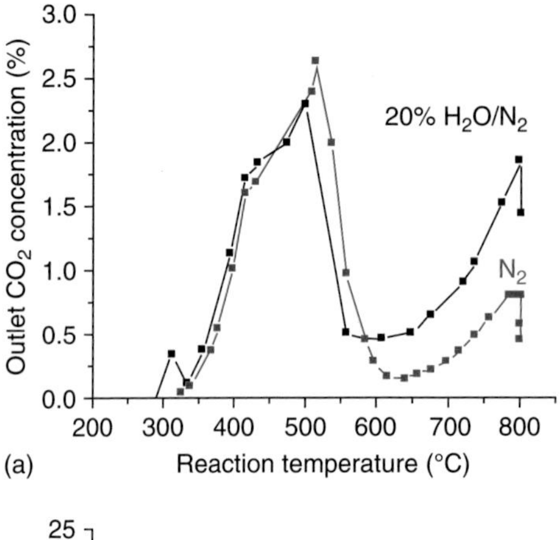

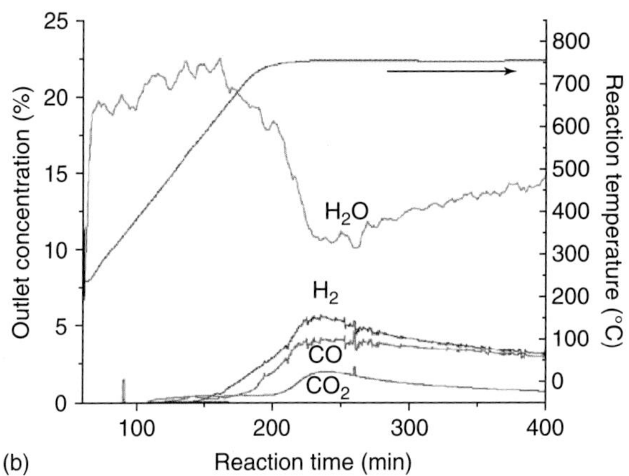

Figure 6.6 (a) Bench-scale, fixed-bed reactor test in 20% steam/Ar of coal-limonite and (b) MS profile of bench-scale fixed bed reactor test of coal gasification with 20% steam/He.

of steam gasification products are shown in Figure 6.6. The data indicates that the coal gasification reaction initiated at temperatures as low as 670 °C. Observations show that the increase in gaseous products, CO, H_2, and CO_2, corresponded with the decrease in steam concentration. While hydrogen and CO were the most abundant species in steam gasification, these gases were not observed in coal/20% steam CLC reactions when oxygen carriers were present. Therefore, it is reasonable to conclude that the oxygen carriers have reacted continuously with gaseous H_2/CO produced from steam gasification reaction initiating at a relatively low temperature (~670 °C), accelerating the combustion reaction of oxygen carrier/coal with the presence of steam. CO may have been converted to CO_2 by the oxygen carrier and H_2 may have been converted to water, which was condensed prior to mass spectrometry. Therefore, the fixed bed reactor tests data suggested that coal gasification is the major reaction when steam was introduced in the CLC process. Steam gasification of coal may be accelerated due to fast removal of gasification products, H_2 and CO, from the reaction with the oxygen carrier.

6.5 Supported Binary Oxides System

Fe_2O_3 and CuO were chosen for developing oxygen carrier materials to obtain high reactivity and stable reaction performance, and to address cost and environmental concerns. Perovskite-type compounds (such as $La_{0.8}Sr_{0.2}Co_{0.2}Fe_{0.8}O_3$) have been reported for their ability to release oxygen without major chemical or physical changes in the bulk phases, but the application of these compounds in CLC systems may not be feasible due to expensive production cost and low oxygen capacity [34]. Bimetallic compounds (such as Co–Ni/Al_2O_3 and Fe–Mn oxides) have been evaluated as oxygen carriers. It was found that Co–Ni/Al_2O_3 samples displayed high reactivity and stable behavior over multiple reduction–oxidation experiments. Ryden et al. concluded that combined oxides of iron and manganese have very favorable thermodynamic properties and could potentially be suitable for CLC applications. However, they also found that the examined material exhibited low physical stability, which could be improved by the addition of inert material. In our previous work, we investigated a supported Fe–Mn oxide mixture for chemical looping combustion using coal synthesis gas for fuel [17]. The addition of Mn oxides exhibited a positive effect on the chemical stability of Fe_2O_3 multi-cycle CLC reactions at 900 °C. We also observed that the support had a significant effect on both fractional reduction–oxidation and reaction rate. Moghtaderi et al. conducted both experimental and theoretical studies to understand the reactivity of H_2, CO, and CH_4 with binary mixtures of CuO, Fe_2O_3, and NiO [35]. It was found that the reaction parameters of the binary mixtures can be calculated directly using the kinetic parameters of the parent materials. Wang et al. have conducted bench-scale reactor tests of chemical looping combustion with a $MgAl_2O_4$ supported Fe_2O_3–CuO mixture and coke oven gas. The optimized oxide composition contained 45 wt% Fe_2O_3 and 15% CuO supported on 40% $MgAl_2O_4$.

6.5.1 Thermodynamic Analysis of CuO–Fe_2O_3 Phases

Thermodynamic calculations were conducted with Factsage 6.0 software to determine possible interactions/phases that could be formed during thermal treatment of the CuO, Fe_2O_3, and Al_2O_3 mixtures. Phase diagram of the CuO–Fe_2O_3 bimetallic system is shown in Figure 6.7. There was no interaction between CuO and Fe_2O_3 at 761 K (488 °C), but a solid solution of CuO–Fe_2O_3 could be formed with various CuO–Fe_2O_3 compositions when the mixture was heated above 761 K (488 °C). When the concentration of Fe_2O_3 was higher than that of CuO, Fe_2O_3 was also observed in addition to the CuO–Fe_2O_3 solid solution. Similarly, CuO was observed when CuO concentration was higher than that of Fe_2O_3. Therefore, at the temperature range in which CLC is conducted, the formation of Cu–Fe oxide solid solution is thermodynamically favorable.

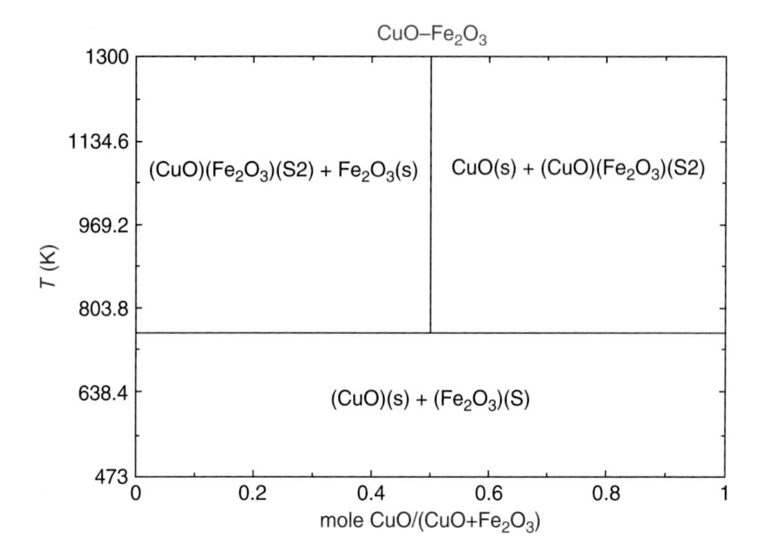

Figure 6.7 Thermodynamic analysis of interaction between $CuO–Fe_2O_3–Al_2O_3$ systems.

6.5.2 Decomposition–Oxidation Cycle of Chemical Looping Oxygen Uncoupling

Chemical looping oxygen uncoupling (CLOU) has been proposed for combustion of solid fuels, such as coal, char, or biomass, utilizing oxygen released from thermal decomposition of oxygen carriers. To determine the ability of $CuO–Fe_2O_3$ mixtures to release oxygen, 5-cycle decomposition–oxidation reaction of 60% CuO/bentonite, 60% Fe_2O_3/bentonite, 60% CuO–20% Fe_2O_3/Al_2O_3, and 40% CuO–40% Fe_2O_3/Al_2O_3 were conducted in TGA at 900 °C with nitrogen as flow gas for decomposition and air for oxidation. The plots of relative weight loss versus time of various oxygen carriers are shown in Figure 6.8. During decomposition, 60% Fe_2O_3/bentonite indicated negligible weight loss (0.33%), which may be due to partial decomposition of Fe_2O_3 to Fe_3O_4. The weight loss of 60% CuO/bentonite is 6%, which corresponds to the full decomposition of CuO to Cu_2O. The decompositions of bimetallic 60% CuO–20% Fe_2O_3/Al_2O_3 and 40% CuO–40% Fe_2O_3/Al_2O_3 oxygen carriers are 6% and 4% respectively, which are associated with the decomposition of CuO components of oxygen carriers and no significant contribution from Fe_2O_3 species. Therefore, the synergetic effect was not observed in the oxygen released during the decomposition reaction of mixed bimetallic $CuO–Fe_2O_3$ system in nitrogen.

6.5.3 Coal Chemical Looping Combustion

$CuO–Fe_2O_3$ mixtures 60% CuO/bentonite and 60% Fe_2O_3/bentonite were tested in TGA to obtain baseline data for coal chemical looping combustion, and the results are listed in Table 6.4. The combustion reaction of 60% CuO/bentonite sample showed full conversion from CuO to metallic Cu with moderate reaction rate ($0.08\ min^{-1}$) at 708 °C. In addition, the reduction of 60% Fe_2O_3/bentonite

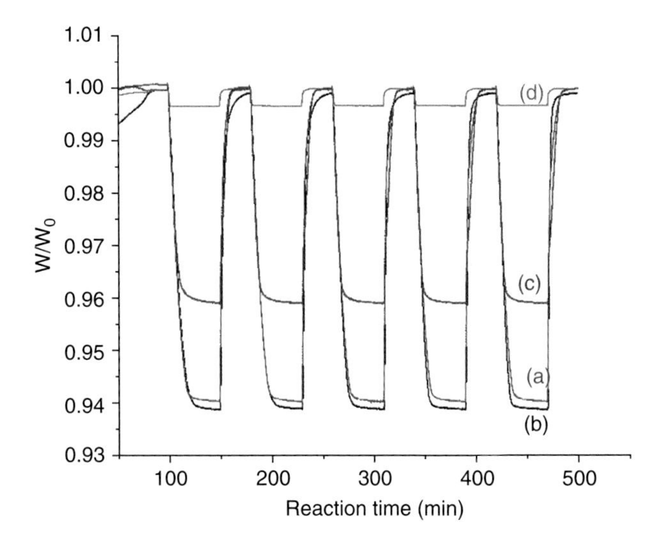

Figure 6.8 TGA profile of five-cycle decomposition in N_2/oxidation reaction with (a) 60% CuO/bentonite, (b) Fe_2O_3/bentonite, (c) 60% CuO–20% Fe_2O_3/Al_2O_3, and (d) 40% CuO–40% Fe_2O_3/Al_2O_3.

Table 6.4 Direct coal CLC reaction performance of supported copper–iron metal oxides with various compositions from TGA.

Sample (wt%)	% Combustion	Peak temperature (°C)	Combustion rate (min^{-1})
60% CuO–30% Fe_2O_3/10% bentonite	95	807	0.07
80% CuO–10% Fe_2O_3/10% bentonite	100	748, 810	0.07, 0.06
10Cu80Fe10Al	83	888	0.04
45Cu45Fe10Al	85	786	0.07
60Cu30Fe10Al	89	786	0.07
80Cu10Fe10Al	93	765	0.07
40Cu40Fe20Al	80	774	0.05
60Cu20Fe20Al	95	765	0.07
46Cu24Fe30Al	79	785	0.05
24Cu46Fe30Al	72	887	0.04
60% CuO/Bentonite	99.4 (Cu)	708	0.08
60% Fe_2O_3/Bentonite	54 (Fe(II))	979	0.04

showed lower combustion percentage from Fe_2O_3 to Fe(II) and slower reaction rate (0.04 min^{-1}) at 979 °C. Both 60% CuO/bentonite and 60% Fe_2O_3/bentonite demonstrated high oxidation rates and high percentage oxygen uptakes during the oxidation reaction. Severe agglomeration of copper particles was observed when the 60% CuO/bentonite sample was tested in a bench-scale reactor for coal

CLC reaction at 700–800 °C (not shown). Thus, bulk or supported single CuO with more than 60% CuO loading cannot be used as an oxygen carrier unless the agglomeration issue is addressed. Severe agglomeration was not observed with 60% Fe_2O_3/bentonite in bench-scale reactor tests, but its application for coal CLC may be limited due to low combustion conversion and slow reaction rates. Therefore, mixed CuO–Fe_2O_3 oxygen carriers were evaluated in this study to overcome the reactivity and physical stability issues with monometallic oxide carriers.

CLC performance of 60% CuO/bentonite, 60% Fe_2O_3/bentonite, and mixed CuO–Fe_2O_3 mixtures with coal are shown in Figure 6.9. TGA data for all supported bimetallic Fe–Cu oxygen carriers are summarized in Table 6.4. The coal/60% CuO/bentonite system peaks at low temperature (708 °C), while coal/60% Fe_2O_3/bentonite system peaks at higher temperature (979 °C). The CuO–Fe_2O_3 mixtures did not show any peaks corresponding to either CuO or Fe_2O_3. This indicated that these mixtures of bimetallic CuO–Fe_2O_3 appear to behave as one component even though they contain separate metal oxides. It is also important to note that the reaction temperatures of bimetallic oxide mixtures are lower than mixtures with Fe_2O_3, indicating that CuO has a synergetic effect on Fe_2O_3 to release oxygen at lower temperatures for the CLC reaction. As shown in Figure 6.9, when the bimetallic carriers had higher iron content, the combustion peaks were broad and the peak maximum temperature was closer to that of 60% Fe_2O_3/bentonite. For 60Cu30Fe10Al (as 60% CuO–30% Fe_2O_3/10% Al_2O_3), the reaction peak was narrow but close to the temperature of CuO. The sample with the composition of 60% CuO and 30% Fe_2O_3 appeared to act mostly like a single component material.

As shown in Table 6.4, CuO/bentonite has the highest combustion percentage while Fe_2O_3/bentonite has the lowest reaction performance. For the samples containing 10% Al_2O_3, a minimum of 60% CuO weight loading is required

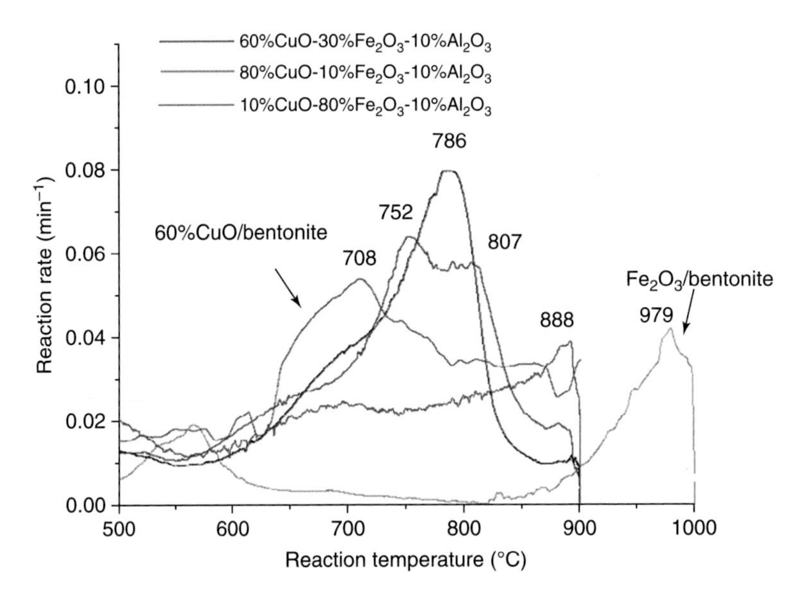

Figure 6.9 Reduction rates of various bimetallic CuO–Fe_2O_3 oxygen carriers with coal.

to obtain >90% combustion conversion. The addition of Fe_2O_3 increases the reaction temperature and decreases the reaction rate. A reasonable (~80%) combustion percentage can be obtained with samples containing as much as 80% Fe_2O_3. To achieve reasonable rates, Fe_2O_3 loading must be below 45%. In addition, the reaction rate and combustion percentages of alumina-supported mixed $CuO–Fe_2O_3$ oxygen carriers are very sensitive to Al_2O_3 loading; with 30% inert supports, the combustion percentage of $CuO–Fe_2O_3$ oxygen carriers was below 80%. Combustion percentages greater than 80% can be obtained when the support concentration is less than 20%. Both 60Cu20Fe20Al and 40Cu40Fe20Al showed a reasonable combustion percentage, lower reaction peak temperatures, and moderate reaction rates. To obtain reasonable performance at temperatures below 900 °C, mixed Cu–Fe oxide carriers should contain at least 40% CuO, but less than 45% Fe_2O_3 and less than 30% inert support, such as Al_2O_3. In addition, bentonite-supported oxygen carriers generally showed better reaction performance than alumina-supported samples. Assuming CuO to be fully reduced to metallic Cu and Fe_2O_3 to Fe(II)O, the combustion percentages of various mixed $CuO–Fe_2O_3$ oxygen carriers were calculated and are summarized in Table 6.4. Since the sample containing 60Cu20Fe20Al showed good performance during TGA tests, that composition was chosen for bench-scale reaction tests and evaluation of different preparation methods.

The samples prepared with the physical mixing method, co-precipitation method, and direct decomposition have similar reduction rates and combustion percentages. The sample prepared by the direct decomposition method exhibits slightly higher combustion temperature due to homogenous mixing of chemical components. However, the physical mixing method and co-precipitation were the preferred preparation methods because they were simple and the samples exhibited good performance.

6.5.4 Chemical Looping Combustion with Methane as Fuel

The 10-cycle TGA profiles of methane CLC reaction over 60% CuO/bentonite, 60% Fe_2O_3/bentonite, 60Cu20Fe20Al, and 40Cu40Fe20Al at 800 °C are shown in Figure 6.10a–d, and the reaction performance including reduction/oxidation rates and oxygen transfer capacity at various temperatures are summarized in Table 6.5. 60% CuO/bentonite demonstrated fast reduction and reoxidation. The oxygen transfer capacity obtained in TGA was about 12%, indicating that CuO was fully reduced to a metallic Cu species even with a shortened reduction time. The final reduced weight of 60% Fe_2O_3/bentonite remained constant, but the final oxidation weight decreased with increasing cycles in the 10-cycle CH_4 CLC reaction in TGA at 800 °C. The oxygen transfer capacity continuously decreased with increasing cycles as shown in Table 6.5. Similar behavior was observed at 900 °C, but the increase in reaction temperatures had a positive effect on the reduction/oxidation rate for 60% Fe_2O_3/bentonite. The oxygen transfer capacity of Fe_2O_3/bentonite is significantly lower than that of CuO-based oxygen carriers for methane CLC.

The samples of 60Cu20Fe20Al and 40Cu40Fe20Al prepared with the physical mixing method showed constant reduction/oxidation weight during 10-cycle

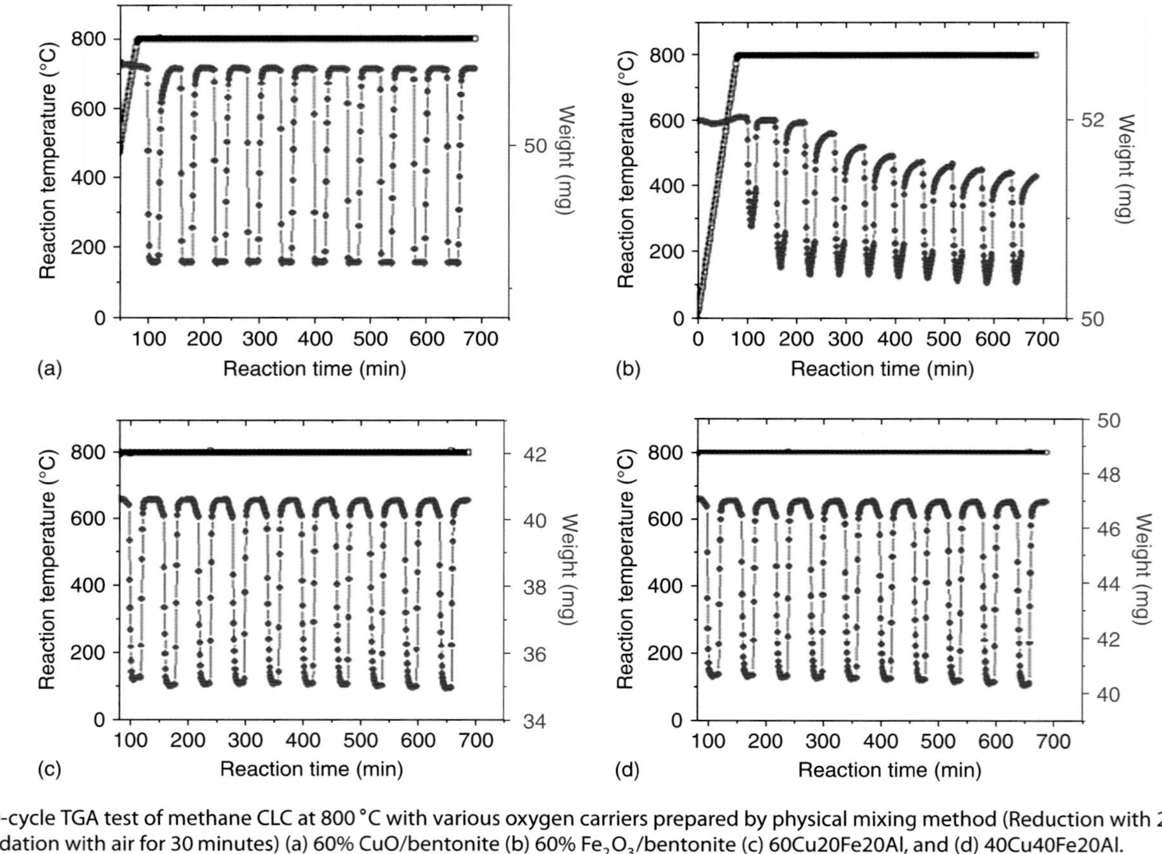

Figure 6.10 10-cycle TGA test of methane CLC at 800 °C with various oxygen carriers prepared by physical mixing method (Reduction with 20% CH_4/N_2 for 10 minutes; oxidation with air for 30 minutes) (a) 60% CuO/bentonite (b) 60% Fe_2O_3/bentonite (c) 60Cu20Fe20Al, and (d) 40Cu40Fe20Al.

Table 6.5 CLC reaction performance of various oxygen carrier materials in the 3rd, 5th, 7th, and 9th cycle TGA tests with methane.

		Cycle 3			Cycle 5			Cycle 7			Cycle 9		
Sample	***T* (°C)**	R_{red}	R_{oc}	R_{ox}	R_{red}	R_{oc}	R_{ox}	R_{red}	R_{oc}	R_{ox}	R_{red}	R_{oc}	R_{ox}
60% CuO/ bentonite	750	0.37	12.85	0.25	0.38	12.85	0.25	0.38	12.85	0.25	0.38	12.85	0.25
	800	0.68	12.92	0.24	0.69	12.92	0.24	0.69	12.92	0.24	0.69	12.92	0.24
	850	0.78	13.02	0.22	0.79	13.01	0.22	0.79	13.02	0.22	0.79	13.02	0.22
	900	0.94	13	0.39	0.94	13	0.39	0.95	13	0.39	0.94	13	0.39
	750	0.27	1.37	1.16	0.23	1.37	1	0.23	1.53	1.04	0.23	1.56	1.01
60% Fe_2O_3/ bentonite	800	0.36	2.7	0.83	0.32	2.7	0.96	0.32	2.24	1	0.31	2.1	1.07
	900	2.61	2.98	1.69	2.14	2.98	1.99	2.14	2.4	2.1	2.14	2.38	2.12
60Cu20Fe20Al	800	1.09	13.63	0.38	1.1	13.63	0.39	1.09	13.68	0.38	1.07	13.84	0.38
	900	1.26	13.63	0.38	1.21	13.67	0.37	1.176	13.71	0.36	1.162	14.1	0.36
40Cu40Fe20Al	800	0.92	13.86	0.35	0.92	13.86	0.35	0.91	14.13	0.35	0.9	14.31	0.34
	900	1.49	16.5	0.35	1.378	17.9	0.292	1.295	17	0.27	1.3	17.9	0.27

R_{red}, rate of reduction (min^{-1}); R_{ox}, rate of oxidation (min^{-1}); R_{oc}, oxygen transfer capacity (%).

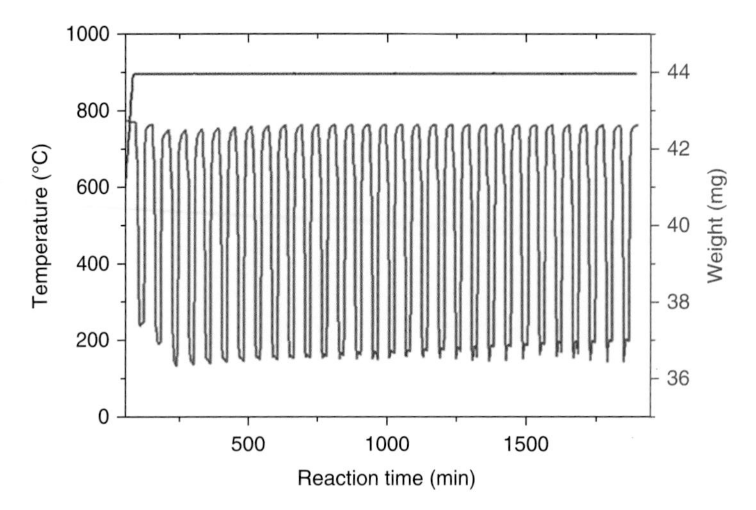

Figure 6.11 30-Cycle TGA test of methane CLC at 900 °C with 60Cu20Fe20Al prepared by physical mixing method (Reduction with 20% CH_4/N_2 for 10 minutes; oxidation with air for 30 minutes).

TGA tests. The reduction rate of 60Cu20Fe20Al oxygen carrier was about 1.1 min^{-1}, while the oxidation reaction rates are about 0.4 min^{-1}. The rates remained constant during 10-cycle tests. The reaction rates of 40Cu40Fe20Al were slightly lower than those of 60Cu20Fe20Al, but the oxidation rates were similar. It is interesting to note that the reduction rates are higher with $CuO–Fe_2O_3$ mixture than that with either CuO or Fe_2O_3 alone. This indicates a synergetic effect of the $CuO–Fe_2O_3$ system in improving reduction rates. This synergetic effect of the bimetallic $CuO–Fe_2O_3$ oxygen carrier system was also observed with oxygen transfer capacities.

TGA profiles of 60Cu20Fe20Al oxygen carrier materials for 30-cycle (2000 minutes) long-term TGA tests are shown in Figure 6.11. For 60Cu20Fe20Al, a reduction rate increase of 22%, an oxidation rate increase of 7%, and an oxygen transfer capacity increase of 1% were observed from the 5th cycle to the 30th cycle.

6.5.5 Bulk Phase and Oxidation State Analysis of Mixed $CuO–Fe_2O_3$ System

To understand the reduction behavior of the $CuO–Fe_2O_3$ system, temperature programmed reduction (TPR) experiments were conducted with TGA. The oxygen carriers were heated in 5% methane/N_2 flow at a heating rate of 5 °C min^{-1}. The relative weight loss rates ($d(w/w_0)/dt$) versus reaction temperature are shown in Figure 6.12. TPR plot of 60% CuO/bentonite indicated a single reduction peak starting from 700 °C with maximum value at 870 °C, which corresponds to the reduction of CuO to Cu_2O. The starting temperature of the reduction reaction of 60% Fe_2O_3/bentonite is quite lower (650 °C), and the first reduction peak

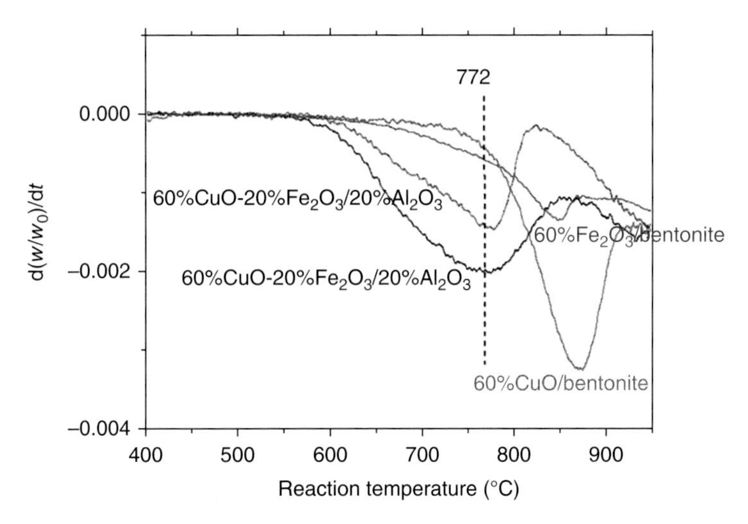

Figure 6.12 TPR of (a) 60% CuO/bentonite, (b) 60% Fe_2O_3/bentonite (c) 60Cu20Fe20Al, and (d) 40Cu40Fe20Al with 5% CH_4/N_2 up to 950 °C.

temperature is observed at 850 °C, which proved the fast reduction of Fe_2O_3 to Fe_3O_4. The intensities of TPR peaks of 60% Fe_2O_3/bentonite are much lower than those of other oxygen carriers. 60Cu20Fe20Al and 40Cu40Fe20Al start to reduce around 580 °C, a temperature which is significantly lower than that of both 60% CuO/bentonite and 60% Fe_2O_3/bentonite, demonstrating improved oxygen release for methane combustion. In addition, all three CuO–Fe_2O_3 oxygen carriers showed the reduction peak maximum at 772 °C, which is significantly lower than the reduction peaks of either 60% CuO/bentonite or 60% Fe_2O_3/bentonite. This single reduction peak at lower temperature (772 °C) indicates that CuO–Fe_2O_3 acts as a single component even though it contains two separate metal oxides. This faster oxygen release in the bimetallic CuO–Fe_2O_3 system versus the single CuO or Fe_2O_3 species may be attributed to the formation of solid solution of CuO and Fe_2O_3.

X-ray diffraction (XRD) analysis has been performed to validate the formation of solid solution of the CuO and Fe_2O_3 system: For the physical mixture of 60% CuO, 20% Fe_2O_3, and 20% Al_2O_3 before calcinations at 900 °C, crystalline CuO and Fe_2O_3 were observed with non-crystalline alumina state. After calcinations at 900 °C, only extra CuO species were present, indicating that crystalline Fe_2O_3 may have dispersed in the CuO phase to form $CuFe_2O_4$, which is consistent with thermodynamic calculations and temperature-programming reduction-thermogravimetric analysis (TPR-TGA) results. Particle agglomeration is a major concern with CuO-containing oxygen carriers. Discrete particles were observed in the SEM images (similar magnification) of 60Cu20Fe20Al after 10-cycle bench-scale fixed bed reactor tests with carbon, as shown in Figure 6.13. This clearly demonstrated the improved physical stability of bimetallic CuO–Fe_2O_3 oxygen carriers in cyclic carbon CLC tests as compared to CuO.

Figure 6.13 Photomicrographs of 60Cu20Fe20Al: Fresh, after reduction with methane, and after reoxidation (elemental mapping: copper-green, iron-blue, and aluminum-red).

6.5.6 Synergetic Reactivity–Structure of $CuO–Fe_2O_3$ Oxygen Carriers

Based on theoretical oxygen capacities for possible reactions, the final reduced oxide species were identified and are summarized in Table 6.6. For the coal/carbon CLC reaction, the reduction percentage based on TGA weight loss of 60% CuO/bentonite compares favorably with the reaction weight loss corresponding to full reduction of CuO to metallic copper. For 60% Fe_2O_3/bentonite, there was only partial (54%) conversion of Fe_2O_3 to FeO during coal CLC reaction and the major components after reduction may correspond to FeO/Fe_2O_3. The bimetallic $CuO–Fe_2O_3$ system showed very high combustion conversions: 95% for 60Cu20Fe20Al and 80% for 40Cu40Fe20Al. The presence of CuO species enhanced the extent of reduction of Fe_2O_3 to FeO in coal CLC reaction, and the major components for reduced bimetallic $CuO–Fe_2O_3$ are Cu and FeO. This demonstrated the synergetic effect of CuO in increasing the extent of reduction of Fe_2O_3.

In addition, CuO species in 60% CuO/bentonite was reduced to metallic Cu in methane, which is similar to what was observed in coal CLC. During the methane CLC reaction, 60% Fe_2O_3/bentonite showed an increase in oxygen transfer capacity from 1.37% to 3% when the reaction temperature was increased

Table 6.6 Analysis of extent of reduction of 60% CuO/bentonite, 60% Fe_2O_3/bentonite, 60Cu20Fe20Al, and 40Cu40Fe20Al during coal CLC and methane CLC.

	Coal/carbon CLC			Methane CLC			
Samples	Reduction reaction	Reduction percentage (%)	Reduced species	Reduction reaction	Theoretical oxygen transfer capacity (%)	Actual oxygen transfer capacity (%)	Reduced species
60% CuO/bentonite	$CuO \rightarrow Cu$	99.4	Cu	$CuO \rightarrow Cu_2O$	6	12	Cu
				$CuO \rightarrow Cu$	12		
60% Fe_2O_3/bentonite	$Fe_2O_3 \rightarrow FeO$	54	Fe_3O_4	$Fe_2O_3 \rightarrow Fe_3O_4$	2		Fe_3O_4/FeO
				$Fe_2O_3 \rightarrow FeO$	4.8	1.3–3	
				$Fe_2O_3 \rightarrow Fe$	18		
60Cu20Fe20Al	$CuO \rightarrow Cu$ $Fe_2O_3 \rightarrow FeO$	95	Cu, FeO	$CuO \rightarrow Cu$			
				$Fe_2O_3 \rightarrow FeO$	13.6		
				$CuO \rightarrow Cu$		13.63–14	Cu, FeO/Fe
				$Fe_2O_3 \rightarrow Fe$	18		
40Cu40Fe20Al				$CuO \rightarrow Cu$			
	$CuO \rightarrow Cu$			$Fe_2O_3 \rightarrow FeO$	12.2		
	$Fe_2O_3 \rightarrow FeO$	80	Cu, FeO	$CuO \rightarrow Cu$		13.86–18	Cu, FeO/Fe
				$Fe_2O_3 \rightarrow Fe$	20		

from 750 to 900 °C. At temperatures below 900 °C, this reduction weight loss corresponds to Fe_3O_4 formation, while at 900 °C it corresponds to the formation of both Fe_3O_4 and FeO. The higher reaction temperature improved the reduction conversion of Fe_2O_3 to FeO. Excellent reaction rate and oxygen transfer capacities were observed during the reaction of methane with bimetallic CuO–Fe_2O_3 oxygen carriers. As shown in Table 6.6, CuO–Fe_2O_3 carriers were reduced to metallic Fe and Cu in the methane CLC system at 900 °C during the same reduction time. This shows the synergetic effect of the CuO–Fe_2O_3 system in increasing the reduction rate of Fe_2O_3 to Fe, which could not have been achieved with Fe_2O_3 alone in methane.

The reduction ability of CuO–Fe_2O_3 mixtures was superior to that of Fe_2O_3 and CuO with both methane and carbon/coal. Kang et al. observed better reduction of $CuFe_2O_4$ as compared to Fe_2O_3 with methane during their studies related to methane reforming reactions [36]. $CuFe_2O_4$ has a tetragonally distorted spinel structure with the structural formula $Fe^{3+}(Fe^{3+}Cu^{2+})O_4^{2-}$, in which the octahedral cations are placed inside the square bracket. Thermal treatment is known to place some of the Cu^{2+} ions in tetrahedral sites, dispersing both Fe^{3+} and Cu^{2+} uniformly throughout the $CuFe_2O_4$ structure. During reduction, Cu metal segregates from $CuFe_2O_4$ and simultaneously Fe_2O_3 further reduces to FeO. The segregation of Cu from the spinel structure could promote defective structure formation and cracking in the micro structure. This crack growth can enhance the methane gas transfer and facilitate Fe_2O_3 reduction. This may explain why the extent of reduction with $CuFe_2O_4$ in methane is faster and better than with Fe_2O_3 as observed in the present study.

For pure CuO, the Cu–O bond is reported to be about 1.73 and 1.738–1.819 Å. For Cu_2O, the reported Cu–O distance is about 1.75 Å. The Fe–O bond distances reported for Fe_2O_3 are 1.95 and 2.08 Å, those for Fe_3O_4 are 1.887 and 2.059 Å, and that for FeO is 2.15 Å. In $CuFe_2O_4$, copper ions occupy the octahedral interstices in the spinel lattice, wherein the copper atoms form four short coplanar covalent bonds and two long ionic bonds with the six neighboring oxygen atoms. The corresponding Cu–O distances have been reported to be 1.97 and 2.21 Å. The Fe–O distances of tetrahedrally and octahedrally coordinated iron atoms are 1.74 and 2.14 Å respectively. The metal–oxygen bond distances in $CuFe_2O_4$ are different from Fe_2O_3 and CuO. Cu–O distances are longer in $CuFe_2O_4$ as compared to that of CuO. Better reduction ability of $CuFe_2O_4$ may also be due to the differences in metal–oxygen bond energies in $CuFe_2O_4$ as compared to single metal oxides, in addition to the contributions from better gas diffusion and better elemental distribution.

Kameoka et al. studied the reduction of $CuFe_2O_4$ with hydrogen as it relates to work on steam reforming of methane [37]. These studies showed that the sintering of Cu particles over $CuFe_2O_4/SiO_2$ during reduction with H_2 was significantly lower than with CuO/SiO_2. The crystallite particle size of reduced copper particles on $CuFe_2O_4/SiO_2$ was significantly smaller than with CuO/SiO_2. The even dispersion of these fine copper particles in the $CuFe_2O_4$ inhibited sintering of copper even at high temperature. They also observed that the presence of copper promoted the reduction of Fe^{3+} by H_2, suggesting that dissociated H_2

may migrate from Cu to iron oxide and contribute to faster reduction of Fe_2O_3. Kamoeka et al. also found that Cu(0) and Fe_3O_4 are immiscible, which keeps Cu highly dispersible while maintaining small particle size. During oxidation, Cu also reforms $CuFe_2O_4$ quickly because of the well-dispersed Cu and Fe in the structure.

In order to form 1 mol of $CuFe_2O_4$, it is necessary to have 1 mol of CuO and 1 mol of Fe_2O_3. This corresponds to a 1 : 2 weight ratio of CuO:Fe_2O_3. In the 40% CuO–40% Fe_2O_3 formulation, if all Fe_2O_3 is consumed for $CuFe_2O_4$ formation, 20% of CuO may be present in a monometallic state. Some of that may also be consumed for $CuAl_2O_4$ formation with the support, leaving only a small amount of CuO present in the formulation and that may be well distributed in the structure. Since copper enhances the faster and deeper reduction of Fe_2O_3 during CLC with methane, the formulation with higher Fe_2O_3 showed a better oxygen capacity. Even distribution of fine Cu and Fe oxide particles in the $CuFe_2O_4$ and better gas diffusion in the micro-cracks formed during reduction may have contributed to the improved reactivity performances that were observed in the present work. Even distribution of Cu particles in the structure may also have contributed to the minimal agglomeration problems observed in the present work with bimetallic Cu–Fe carriers.

6.6 Kinetic Networks of Fe_2O_3-based Oxygen Carriers

Research on oxygen carrier development for CLC has so far mainly been focused on the selection of suitable oxygen carrier materials and the implementation of this concept in pilot-scale tests. However, few studies reported have been focused on the determination of kinetic parameters in particle scale. The knowledge of the kinetics data is of great importance in the design of a CLC process, because it determines the solid inventory necessary in the fuel and air reactors, as well as the recirculation rate of oxygen carriers between the reactors.

One of the most important parameters of the reduction reaction is the apparent activation energy as it defines the reactor dimensions and the energy consumption, which depends on many factors such as chemical structure of the starting raw material, nature of reducing gas, temperature range, reaction step, presence of water vapor, and other gases in the gas mixture. The literature data suggests that activation energy obtained by reducing iron oxides with methane varies from 49 to 271 $kJ\ mol^{-1}$ [27, 35]. This wide range is due to the variation in the rate-controlling step assumed in developing the kinetic rates. Moreover, the rate-controlling mechanism is related to many factors such as the chemical and physical nature of the starting raw material, gas composition, and temperature range. To determine the kinetic parameters of the hematite reduction reaction with methane, several experiments were conducted at different temperatures (700–825 °C) and concentrations (15%, 20%, and 35%).

Consider the following series of reactions for hematite reduction:

$$Fe_2O_3 \rightarrow Fe_3O_4 \rightarrow FeO \rightarrow Fe \tag{6.1}$$

The theoretical weight changes in accordance to reaction stoichiometry for CH_4 gas reduction of hematite were calculated according to the following reactions:

$$12Fe_2O_3 + CH_4 \rightarrow 8Fe_3O_4 + CO_2 + 2H_2O \quad (6.2)$$

$$4Fe_2O_3 + CH_4 \rightarrow 8FeO + CO_2 + 2H_2O \quad (6.3)$$

$$4Fe_2O_3 + 3CH_4 \rightarrow 8Fe + 3CO_2 + 6H_2O \quad (6.4)$$

It was thus determined that the theoretical weight decrease corresponding to transformation of Fe_2O_3 into Fe_3O_4 is 3.3 wt%. Transformation of Fe_2O_3 into FeO and Fe corresponds to weight decrease of 10 and 30 wt% respectively. In this chapter, m_f is considered as the mass of FeO and 100% conversion means that the hematite was all converted to FeO (wüsite).

Kinetic analysis of thermally stimulated reactions is traditionally expected to produce an adequate kinetic description of the process in terms of the reaction model and of the Arrhenius parameters using a single-step kinetic equation

$$\frac{dX}{dt} = k(T)f(X) \quad (6.5)$$

where t is the time, T is the temperature, X is the extent of conversion, and $f(X)$ is the reaction model. In Eq. (6.9), $k(T)$ is the Arrhenius rate constant, which is given as

$$k(T) = A \exp\left(\frac{-E}{RT}\right) \quad (6.6)$$

where R is the gas constant, and A and E are Arrhenius parameters, the pre-exponential factor and the activation energy, respectively. For reaction kinetics under isothermal conditions, Eq. (6.7) can be analytically integrated to yield

$$g(X) = \int_0^X \frac{dX}{f(X)} = k(T)t \quad (6.7)$$

where $g(X)$ is the integral form of the reaction model.

To apply the model-fitting method, the cited mathematical integral expressions $g(X)$ (Table 6.7) together with the experimental X and t values for a given temperature were inserted in Eqs. (6.6) and (6.7). If one of the reaction models presented in Table 6.7 fits the experimental data, it will result in a straight line with the slope of k. That is, the values of kinetic rate constant, k, can be determined at different temperatures from the slope of the straight line obtained by plotting $g(X)$ against time. This value can be subsequently inserted in the Arrhenius equation together with the corresponding temperature value to yield activation energy and pre-exponential factor values from the slope and intercept of the regression straight line.

We have chosen a simple multi-step process that involves two parallel or series independent reactions.

$$A \xrightarrow{k_1(T)} B$$

$$C \xrightarrow{k_2(T)} D \quad (6.8)$$

Table 6.7 Kinetic models for solid-state reactions.

Kinetic model	Kinetic mechanism	f(X)	g(X)
Kinetics-order models	1st order	$(1-X)$	$-\ln(1-X)$
	2nd order	$(1-X)^2$	$(1-X)^{-1}-1$
	3rd order	$(1-X)^3$	$(1/2)((1-X)^{-2}-1)$
Diffusion model	1-D	$1/(2X)$	X^2
	2-D	$1/(-\ln(1-X))$	$(1-X)\ln(1-X)+X$
	3-D	$(3/2)(1-X)^{2/3}(1-(1-X)^{1/3})$	$(1-(1-X)^{1/3})^2$
Contraction model	2-D	$2(1-X)^{1/2}$	$(1-(1-X)^{1/2})$
	3-D	$3(1-X)^{2/3}$	$(1-(1-X)^{1/3})$
Nucleation model	variable n	$n(1-X)(-\ln(1-X))^{(1-1/n)}$	$(-\ln(1-X))^{(1/n)}$

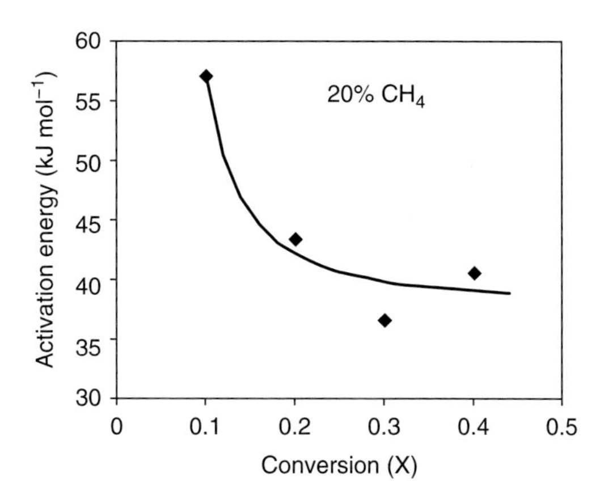

Figure 6.14 Activation energy values as a function of X obtained by an isothermal operation.

Although simple, this mechanism may reasonably approximate the process of conversion of a substance that exists in two isomeric forms, or a conversion of a reactant that simultaneously exists in two phases or a conversion of a solid by two separate paths to different products. As Figure 6.14 indicates, the activation energy decreases as conversion increases. The decreasing trend of activation energy in the multi steps of the reduction can be explain in terms of a nucleation and growth model of reduction in which the activation energy required during nucleation is higher than that needed in the growth stage [38]. The equations involved in isothermal processes are fitting nucleation and kinetic control in parallel:

$$\frac{X_t}{X_\infty} = w_1(1 - e^{-a_1 t^{n_1}}) + w_2(1 - e^{-a_2 t^{n_2}}) \tag{6.9}$$

where

a_1	=	nucleation rate constant for the first mechanism (min.-n_1),
a_2	=	nucleation rate constant for the second mechanism (min.-n_2),
n_1	=	shape parameter for the first mechanism,
n_2	=	shape parameter for the second mechanism,
t	=	time (min),
w_1	=	weight factor for the first mechanism,
w_2	=	weight factor the second mechanism,
X_t	=	total conversion at any time t,
X_∞	=	equilibrium conversion.

For a given temperature, values of X_∞, w_2, a_1, a_2, n_1, and n_2 were determined by curve fitting the raw TGA data with the parameters in Eq. (6.9) using TABLE-CURVE available from Statistical Package for the Social Sciences. The values determined for the shape parameters, n_1, ranged from 0.8 to 1.4 for all the temperatures and all the CH_4 concentrations; the average value of n_1 was 1.01 ± 0.2 (95% CL). The values determined for the shape parameters, n_2, ranged from 1.57 to 3.1 for all the temperatures and all the CH_4 concentrations. The observed value of $n_1 = 1.01$ was very close to the value of $n_1 = 1$ that defines the pseudo-first order rate expression. To simplify the analysis, value of $n_1 = 1$ was used. The values of X_∞, w_2, a_1, a_2, and n_2 were recalculated based on the approximation of $n_1 = 1$ for every set of conversion data at different temperatures and CH_4 concentrations. The values determined for the shape parameters, n_2, ranged from 1.37 to 2.8 for all the temperatures and CH_4 concentrations with an average value of 2.17 ± 0.44. The value of $n_2 = 2.17$ is relatively close to the value of $n_2 = 2$, which defines the Rayleigh distribution. The Rayleigh distribution is a special case for modeling the lifetime of a device that has a linearly increasing instantaneous failure rate. In order to simplify further analysis, a value of $n_2 = 2$ was used. The values of X_∞, w_2, a_1, and a_2, were recalculated based on the approximation of $n_1 = 1$ and $n_2 = 2$ for every set of conversion data taken at different temperatures and CH_4 concentrations.

In order to determine the rate-controlling mechanism, the value of apparent activation energy was calculated from Arrhenius equation as a function of Avrami kinetic constant (a):

$$k = a^{1/n} = k_0 e^{\frac{-E}{RT}} \tag{6.10}$$

where k is the reduction rate constant, k_0 is the frequency factor, R is the gas constant, and T is the absolute temperature. The linear regression of the experimental data of ln k against $1/T$ determines E/R. A plot of ln k versus $1/T$ for reduction of hematite for both R_1 and R_2 is shown in Figure 6.11 at different reaction temperatures for all inlet CH_4 concentrations (15%, 20%, and 35%). The error bars lengths are defined by the range of the data at each temperature. The pre-exponential factor, k_0, and activation energy, E, were obtained from the intercept and slope of the straight line of ln k versus $1/T$ for a given CH_4 concentration. The apparent activation energies for both reactions R_1 and R_2 were estimated to be 34.4 ± 0.5 and 39.3 ± 1.5 kJ mol^{-1} respectively (Table 6.8).

Table 6.8 Parameters obtained by fitting Equation (6.15) to experimental data for different temperatures and CH_4 concentrations. R^2 for all curve fits was greater than 0.999.

	$CH_4 = 35\%$				$CH_4 = 20\%$				$CH_4 = 15\%$		
T (°C)	700	750	800	825	700	750	800	825	750	800	825
X_∞	0.524	0.583	0.808	1.000	0.311	0.565	0.638	0.746	0.330	0.489	0.623
w_1	0.492	0.565	0.625	0.638	0.563	0.577	0.621	0.679	0.582	0.581	0.646
a_1	0.068	0.071	0.085	0.085	0.039	0.049	0.042	0.051	0.020	0.023	0.025
n_1	0.831	0.802	0.796	0.791	1.132	1.012	0.937	0.923	1.248	1.133	1.111
a_2	0.057	0.109	0.292	0.325	0.036	0.038	0.122	0.134	0.016	0.039	0.048
n_2	2.518	2.346	1.788	1.568	2.319	2.441	1.946	1.962	2.392	2.102	2.078
					$n_1 = 1$						
X_∞	0.506	0.552	0.772	0.955	0.315	0.565	0.624	0.730	0.349	0.506	0.642
w_1	0.569	0.513	0.465	0.458	0.380	0.417	0.404	0.354	0.329	0.369	0.310
a_1	0.073	0.140	0.331	0.360	0.023	0.036	0.131	0.151	0.007	0.030	0.037
a_2	2.312	2.081	1.555	1.374	2.582	2.472	1.881	1.844	2.800	2.264	2.257
n_2	0.041	0.040	0.046	0.045	0.060	0.051	0.035	0.041	0.042	0.034	0.035
					$n_1 = 1$ $n_2 = 2$						
X_∞	0.520	0.555	0.757	0.926	0.323	0.581	0.618	0.724	0.385	0.527	0.658
w_1	0.576	0.515	0.446	0.425	0.435	0.448	0.401	0.348	0.385	0.380	0.321
a_1	0.099	0.149	0.273	0.280	0.048	0.062	0.118	0.133	0.027	0.043	0.051
a_2	0.035	0.039	0.052	0.054	0.025	0.035	0.037	0.042	0.024	0.029	0.032

One of the goals of kinetic studies is to deduce the reaction mechanism. As apparent from Figure 6.14, activation energy decreases with increasing conversion. In our case, the reduction of hematite to magnetite probably proceeds through the exothermic stage at low conversion, which results in high activation energy. The initial high activation energy represents the sum of the enthalpy of reversible process and of the activation energy of the irreversible process. On the other hand, lower value of the activation energy at higher conversion is a characteristic of the process proceeding through a reversible endothermic process.

The possible reaction mechanism is summarized below.

Methane decomposition and reduction of hematite to Fe_3O_4:

$$3CH_4 \rightarrow 3C + 6H_2 \quad \Delta H^{\circ}_{r,800\,°C} = 268.34\ \text{kJ mol}^{-1} \tag{6.11}$$

$$CH_4 + 12Fe_2O_3 \rightarrow 8Fe_3O_4 + CO_2 + 2H_2O \quad \Delta H^{\circ}_{r,800\,°C} = 164.1\ \text{kJ mol}^{-1} \tag{6.12}$$

$$3C + 12Fe_2O_3 \rightarrow 8Fe_3O_4 + CO_2 + 2CO \quad \Delta H^{\circ}_{r,800\,°C} = 346.2\ \text{kJ mol}^{-1} \tag{6.13}$$

$$H_2 + 3Fe_2O_3 \rightarrow 2Fe_3O_4 + H_2O \quad \Delta H^{\circ}_{r,800\,°C} = -6.75\ \text{kJ mol}^{-1} \tag{6.14}$$

$$4CH_4 + 27Fe_2O_3 \rightarrow 18Fe_3O_4 + 2CO_2 + 2CO + 3H_2O + 5H_2 \quad \Delta H^{\circ}_{r,800\,°C} = 771.86\ \text{kJ mol}^{-1} \tag{6.15}$$

Methane decomposition and reduction of Fe_3O_4 to FeO:

$$2CH_4 \rightarrow 2C + 4H_2 \quad \Delta H^{\circ}_{r,800\,°C} = 178.9\ \text{kJ mol}^{-1} \tag{6.16}$$

$$CH_4 + 4Fe_3O_4 \rightarrow 3FeO + 2H_2O + CO_2 \quad \Delta H^{\circ}_{r,800\,°C} = 377.9\ \text{kJ mol}^{-1} \tag{6.17}$$

$$2C + 3Fe_3O_4 \rightarrow 9FeO + CO_2 + CO \quad \Delta H^{\circ}_{r,800\,°C} = 377.5\ \text{kJ mol}^{-1} \tag{6.18}$$

$$H_2 + Fe_3O_4 \rightarrow 3FeO + H_2O \quad \Delta H^{\circ}_{r,800\,°C} = 46\ \text{kJ mol}^{-1} \tag{6.19}$$

$$3CH_4 + 8Fe_3O_4 \rightarrow 24FeO + 2CO_2 + CO + 3H_2O + 3H_2 \quad \Delta H^{\circ}_{r,800\,°C} = 981\ \text{kJ mol}^{-1} \tag{6.20}$$

$$7CH_4 + 27Fe_2O_3 \rightarrow 10Fe_3O_4 + 24FeO + 4CO_2 + 3CO + 6H_2O + 8H_2 \quad \Delta H^{\circ}_{r,800\,°C} = 1752.9\ \text{kJ mol}^{-1} \tag{6.21}$$

These results suggest that the reduction of hematite (Fe_2O_3) is initially controlled by the topochemical process. First, a thin layer of lower iron oxides (wüsite) is formed on the surface; then the mechanism shifts to two simultaneous reaction mechanisms (Fe_2O_3 to FeO and production of CO and H_2, intrinsic topochemical kinetics; and Fe_2O_3 to Fe_3O_4 and production of CO_2 and H_2O, nucleation and growth kinetics). The apparent activation energy was obtained as 206 kJ mol^{-1} in the temperature range of 800–950 °C and 105.7 kJ mol^{-1} in the temperature range of 950–1025 °C for Reaction (6.1). Furthermore, they claimed that most of this apparent value was due to increase in concentration of hydrogen with increase in temperature due to enhanced cracking of CH_4.

6.7 50-kW$_{th}$ Methane/Air Chemical Looping Combustion Tests

The chemical looping facility is designed for 50 kW$_{th}$ of natural gas and consists of a 20-cm diameter bubbling bed fuel reactor (FR) and a 15-cm diameter turbulent bed air reactor. A schematic of the chemical looping system components is shown in Figure 6.15. Solids are conveyed from the fuel reactor to the air reactor using an L-valve and conveyed from the air reactor to a seal pot through a 5-cm riser. The solids pass through a cyclone and enter a fluidized bed seal pot, which helps to keep the air and fuel streams separate at the top of the unit. The solids circulation rate is controlled with the L-valve, which also acts to seal gases between the air reactor and the fuel reactor. The fuel reactor, air reactor, and seal pot gas exit streams contain secondary cyclones to separate any coarse solids that are entrained from the reactors. The oxygen carrier materials separated in these cyclones are drained and weighed through the test period. Additionally, sintered metal filters are used to capture fines that are not collected from the secondary cyclones. The pressures in the reactors are controlled using back-pressure control

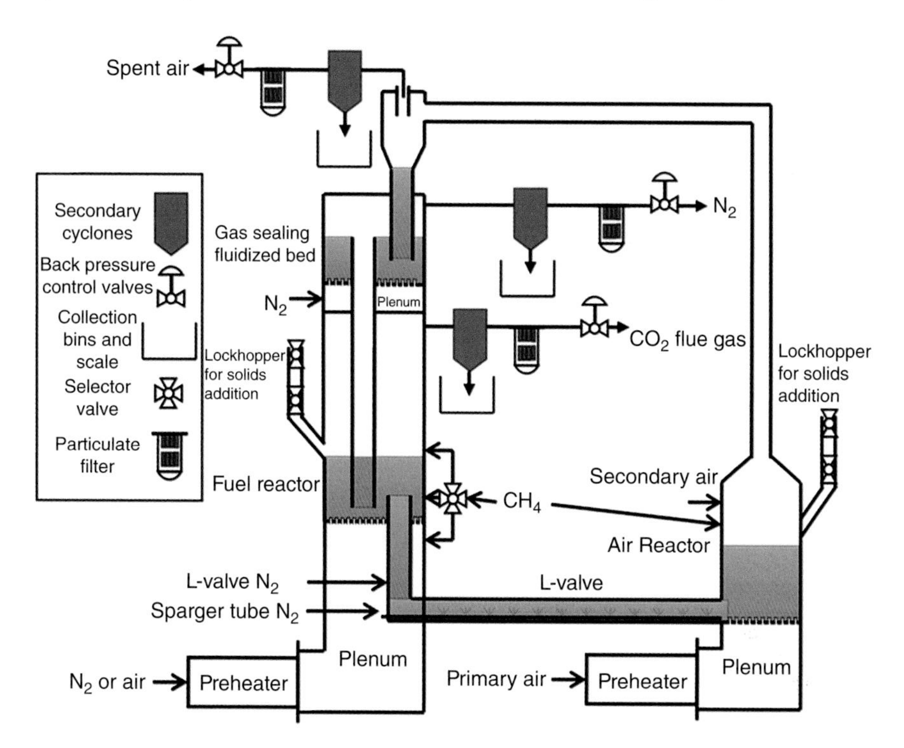

Figure 6.15 Schematic of NETL's 50-kW$_{th}$ chemical looping reactor.

Table 6.9 Hydrodynamic and physical properties of commercially prepared CuO–Fe_2O_3–alumina oxygen carrier (200–600 μm, tumbling method).

Property	Particles from tumbling method
Particle density ($g\,cm^{-3}$)	2.9
Skeletal density ($g\,cm^{-3}$)	4.73
Bulk density ($g\,cm^{-3}$)	1.85
Heat capacity ($J\,kg^{-1}\,°C$)	2.9
Surface area ($m^2\,g^{-1}$)	1
Minimum fluidization velocity (U_{mf}) at 800 °C ($m\,s^{-1}$)	0.135
Particle size (μm)	200–600

valves. Make-up solids can be added via a lock hopper system, and the operator can choose to add make-up oxygen carrier material to either the air reactor or the fuel reactor. To measure reactor performance, such as methane conversion, carbon balance, and CO_2 purity, gases from the fuel reactor are characterized using a bank of non-disperse infrared analyzers (NDIRs) and a gas chromatograph. Compared to other prototypes of this scale, the Nation Energy Technology Laboratory (NETL) 50-kW_{th} unit is unique because it is refractory-lined and no heat can be added directly to the solids flow path. This allows the oxygen carriers tested in the unit to experience conditions that more represent a potential commercial unit.

Commercial-scale batches of the oxygen carrier were also prepared to obtain particle sizes in the range 200–600 μm using the solid-state mixing/tumbling method. The composition of the oxygen carrier was 30 wt% CuO, 30 wt% Fe_2O_3, and 40 wt% alumina. Hydrodynamic and other physical properties of the oxygen carriers are listed in Table 6.9.

The overall operational summary is shown in Figure 6.16, which shows the temperature profiles for the fuel reactor, air reactor, and seal pot, the pressures in the air and fuel reactors, and the total solids inventory. The second-generation carrier particles circulated at the target temperature for 75 hours. As indicated by the green bars in Figure 6.16, the carrier underwent chemical looping conditions for 40 hours, which ranged from 7 to 50 kW_{th} of fuel input. Around 5.4 hours of chemical looping, trials were performed without natural gas addition to the air reactor, and 1.6 hours of "autothermal" operation was performed. "Autothermal" is a term that is used to describe operation without any gas preheating or supplemental heat addition to the system. Further details of the autothermal operation period showing flow rates, methane conversion, and gas mole fractions in the fuel reactor flue gas are presented in Figure 6.17, with the autothermal period shaded red. During this time, the fuel reactor was fed with natural gas at a rate of 45. 7–50.0 kW_{th} diluted at 40–42 mol% using nitrogen as the diluent; the carrier was able to produce a natural gas conversion between 40% and 65%, with a circulation rate estimated between 146 and 194 $kg\,h^{-1}$ from L-valve cutoff tests.

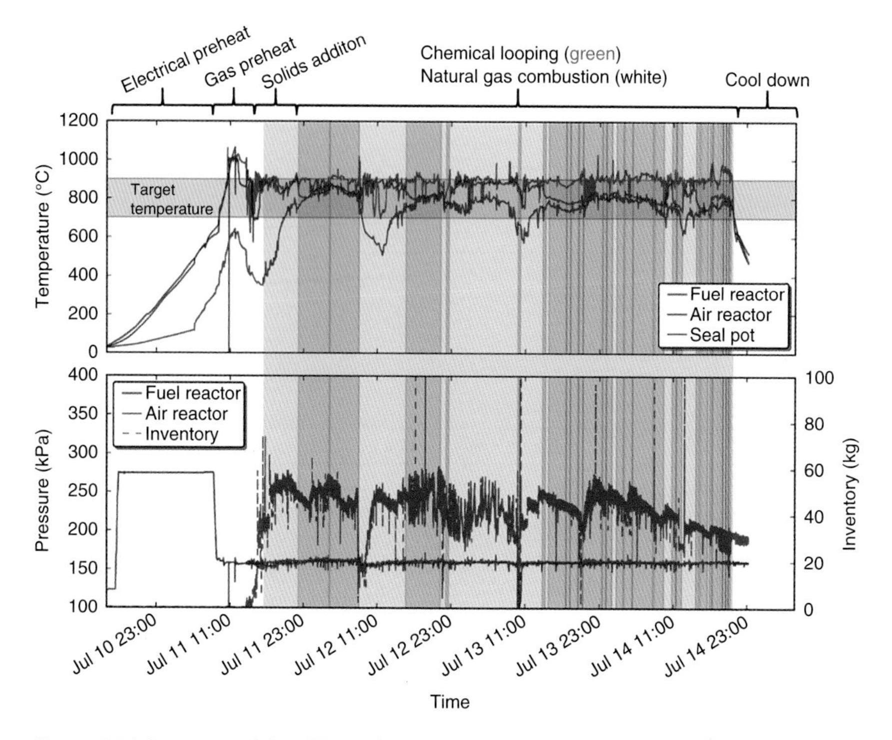

Figure 6.16 Summary of the NETL 50-kW$_{th}$ operation with oxygen carrier manufactured using the tumbling method. Source: Mattisson et al. 2001 [5]. Reproduced with permission of Elsevier.

The fuel reactor was operated around 2 U_{mf}. Thus, it is estimated that the oxygen carrier utilization, i.e. the moles of oxygen consumed from the carrier divided by the total available oxygen in the carrier ranged from 25.4% to 32.0%.

The particle losses from the reactor with the oxygen carrier prepared with the tumbling method (200–600 μm) were significantly low. The losses are due to elutriation, attrition, system upsets, and transport disintegration due to the design of the system. Particle size distributions of fresh oxygen carrier, collected from the air reactor and from the fuel reactor, are shown in Figure 6.18. Particle size distribution of the oxygen carrier collected from the reactor beds after the test was very similar to that of the fresh carrier. A slight decrease in the particle size was observed with OC collected from the fuel reactor.

Tests in NETL's 50-kW$_{th}$ chemical looping circulating fluidized bed combustor unit with the first-generation oxygen carrier (100–200 μm) showed promising results. It is remarkable that there was no agglomeration of solid during the week-long tests at 900 °C even though the oxygen carrier had a high CuO concentration (30%). The average fuel reactor temperatures ranged from 760 to 815 °C for these test periods, and the average air reactor temperature ranged from 840 to 915 °C. The fuel conversion from natural gas to CO_2 ranged from 50% to 80%. Approximately 1.6 hours of continuous operation was achieved with no gas preheat and no augmented natural gas combustion.

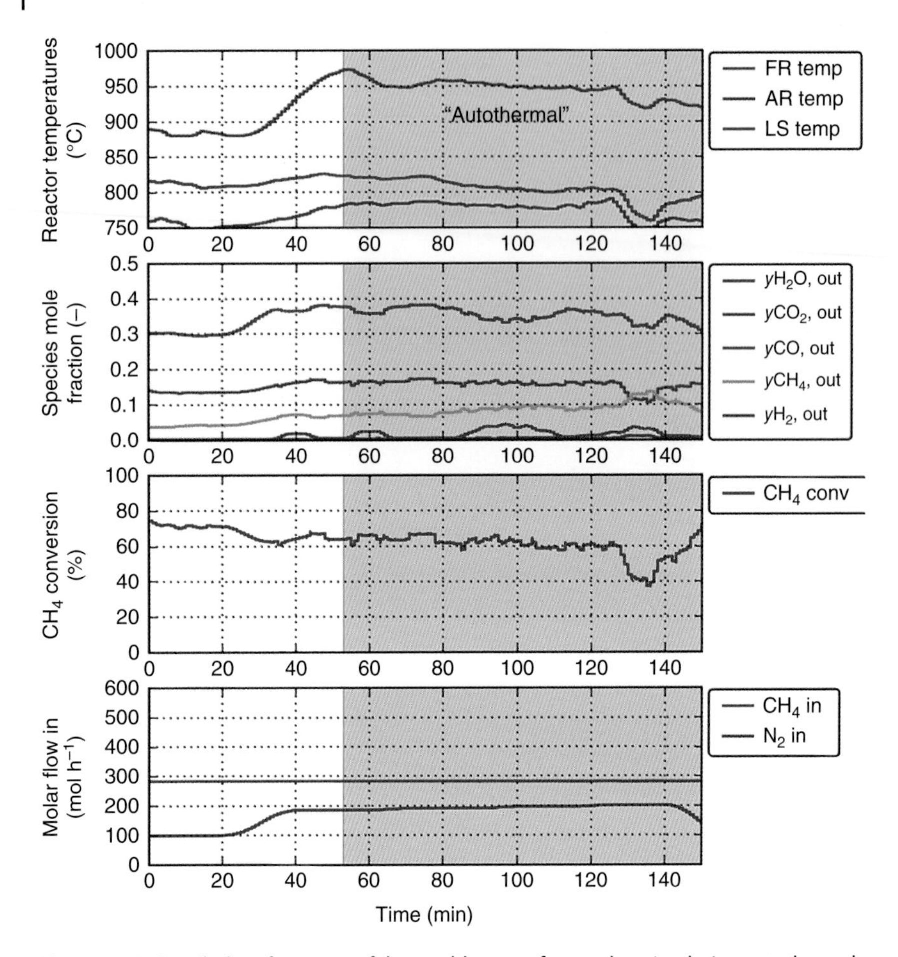

Figure 6.17 Detailed performance of the tumble-manufactured carrier during autothermal operation. Source: Mattisson et al. 2001 [5]. Reproduced with permission of Elsevier.

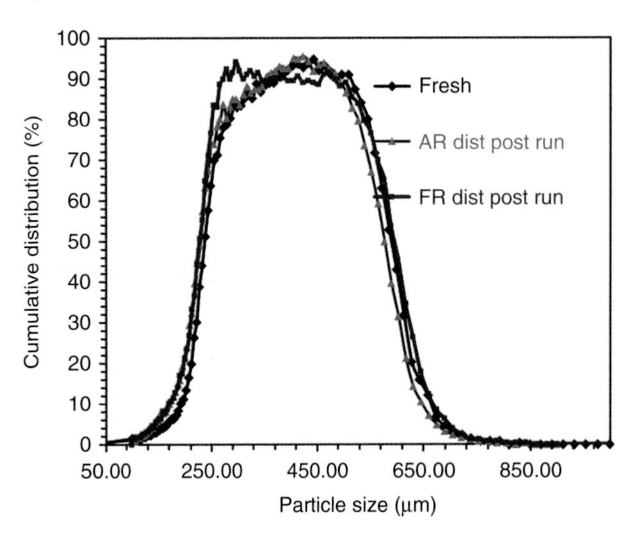

Figure 6.18 Particle size distribution of fresh oxygen carrier (OC), OC collected from the air reactor (AR), and OC from the fuel reactor (FR).

References

1 Hossain, M.M. and de Lasa, H.I. (2008). Chemical-looping combustion (CLC) for inherent CO_2 separations – a review. *Chem. Eng. Sci.* 63: 4433–4451.
2 Jerndal, E., Mattisson, T., and Lyngfelt, A. (2009). Investigation of different $NiO/NiAl_2O_4$ particles as oxygen carriers for chemical-looping combustion. *Energy Fuels* 23: 665–676.
3 Abu-Khader, M.M. (2006). Recent progress in CO_2 capture/sequestration: a review. *Energy Sources Part A* 28: 1261–1279.
4 Adanez, J., Abad, A.F., Garcia-Labiano, P., and de Diego, L.F. (2012). Progress in chemical-looping combustion and reforming technologies. *Prog. Energy Combust. Sci.* 38: 215–282.
5 Mattisson, T., Lyngfelt, A., and Cho, P. (2001). The use of iron oxide as an oxygen carrier in chemical-looping combustion of methane with inherent separation of CO_2. *Fuel* 80: 1953–1962.
6 Hossain, M.M., Sedor, K.E., and de Lasa, H.I. (2007). Co-Ni/Al_2O_3 oxygen carrier for fluidized bed chemical-looping combustion: desorption kinetics and metal-support interaction. *Chem. Eng. Sci.* 62: 5464–5472.
7 Siriwardane, R., Tian, H.J., Richards, G. et al. (2009). Chemical-looping combustion of coal with metal oxide oxygen carriers. *Energy Fuels* 23: 3885–3892.
8 Zafar, Q., Abad, A., Mattisson, T. et al. (2007). Reduction and oxidation kinetics of Mn_3O_4/Mg-ZrO_2 oxygen carrier particles for chemical-looping combustion. *Chem. Eng. Sci.* 62: 6556–6567.
9 Shen, L.H., Gao, Z.P., Wu, J.H., and Xiao, J. (2010). Sulfur behavior in chemical looping combustion with NiO/Al_2O_3 oxygen carrier. *Combust. Flame* 157: 853–863.
10 Jerndal, E., Leion, H., Axelsson, L. et al. (2011). Using low-cost iron-based materials as oxygen carriers for chemical looping combustion. *Oil Gas Sci. Technol.* 66: 235–248.
11 Dueso, C., Abad, A., Garcia-Labiano, F. et al. (2010). Reactivity of a NiO/Al_2O_3 oxygen carrier prepared by impregnation for chemical-looping combustion. *Fuel* 89: 3399–3409.
12 Petrakopoulou, F., Boyano, A., Cabrera, M., and Tsatsaronis, G. (2011). Exergoeconomic and exergoenvironmental analyses of a combined cycle power plant with chemical looping technology. *Int. J. Greenhouse Gas Control* 5: 475–482.
13 Sedor, K.E., Hossain, M.M., and de Lasa, H.I. (2008). Reduction kinetics of a fluidizable nickel-alumina oxygen carrier for chemical-looping combustion. *Can. J. Chem. Eng.* 86: 323–334.
14 Tian, H.J., Guo, Q.J., and Chang, J. (2008). Investigation into decomposition behavior of $CaSO_4$ in chemical-looping combustion. *Energy Fuels* 22: 3915–3921.
15 Zafar, Q., Abad, A., Mattisson, T., and Gevert, B. (2007). Reaction kinetics of freeze-granulated $NiO/MgAl_2O_4$ oxygen carrier particles for chemical-looping combustion. *Energy Fuels* 21: 610–618.

16 Johansson, E., Mattisson, T., Lyngfelt, A., and Thunman, H.A. (2006). 300 W laboratory reactor system for chemical-looping combustion with particle circulation. *Fuel* 85: 1428–1438.
17 Ryden, M. and Lyngfelt, A. (2006). Using steam reforming to produce hydrogen with carbon dioxide capture by chemical-looping combustion. *Int. J. Hydrogen Energy* 31: 271–1283.
18 Jin, H.G. and Ishida, M. (2001). Reactivity study on a novel hydrogen fueled chemical-looping combustion. *Int. J. Hydrogen Energy* 26: 889–894.
19 Chakravarthy, V.K., Daw, C.S., and Pihl, J.A. (2011). Thermodynamic analysis of alternative approaches to chemical looping combustion. *Energy Fuels* 25: 656–669.
20 Siriwardane, R., Tian, H.J., Chaudhari, K. et al. (2008). Chemical-looping combustion of coal-derived synthesis gas over copper oxide oxygen carriers. *Energy Fuels* 22: 3744–3755.
21 Siriwardane, R., Tian, H.J., Miller, D. et al. (2010). Evaluation of reaction mechanism of coal-metal oxide interactions in chemical-looping combustion. *Combust. Flame* 57: 2198–2208.
22 Song, K.S., Seo, Y.S., Yoon, H.K., and Cho, S.J. (2003). Characteristics of the NiO/hexaaluminate for chemical looping combustion. *Korean J. Chem. Eng.* 20: 471–475.
23 Corbella, B.M., de Diego, L., Garcia-Labiano, F. et al. (2006). Characterization and performance in a multicycle test in a fixed-bed reactor of silica-supported copper oxide as oxygen carrier for chemical-looping combustion of methane. *Energy Fuels* 20 (1): 148–154.
24 Adanez, J., Gayan, P., Celaya, J. et al. (2006). Chemical looping combustion in a 10 kW(th) prototype using a CuO/Al_2O_3 oxygen carrier: effect of operating conditions on methane combustion. *Ind. Eng. Chem. Res.* 45: 6075–6080.
25 de Diego, L.F., Garcia-Labiano, F., Adanez, J. et al. (2004). Development of Cu-based oxygen carriers for chemical-looping combustion. *Fuel* (13): 1749–1757.
26 Mattisson, T., Johansson, M., and Lyngfelt, A. (2004). Multicycle reduction and oxidation of different types of iron oxide particles – application to chemical-looping combustion. *Energy Fuels* 18: 628–637.
27 Abad, A., Adanez, J., Garcia-Labiano, F. et al. (2007). Mapping of the range of operational conditions for Cu-, Fe-, and Ni-based oxygen carriers in chemical-looping combustion. *Chem. Eng. Sci.* 62: 533–549.
28 Cho, P., Mattisson, T., and Lyngfelt, A. (2006). Defluidization conditions for a fluidized bed of iron oxide-, nickel oxide-, and manganese oxide-containing oxygen carriers for chemical-looping combustion. *Ind. Eng. Chem. Res.* 45: 968–977.
29 Leion, H., Jerndal, E., Steenari, B.M. et al. (2009). Solid fuels in chemical-looping combustion using oxide scale and unprocessed iron ore as oxygen carriers. *Fuel* 88: 1945–1954.
30 Leion, H., Lyngfelt, A., Johansson, M. et al. (2008). The use of ilmenite as an oxygen carrier in chemical-looping combustion. *Chem. Eng. Res. Des.* 86: 1017–1026.

31 Ryden, M., Johansson, M., Cleverstam, E. et al. (2010). Ilmenite with addition of NiO as oxygen carrier for chemical-looping combustion. *Fuel* 89: 3523–3533.

32 Johansson, M., Mattisson, T., and Lyngfelt, A. (2006). Investigation of Mn_3O_4 with stabilized ZrO_2 for chemical-looping combustion. *Chem. Eng. Res. Des.* 84: 807–818.

33 Fossdal, A., Bakken, E., Øye, B.A., and Schøning, C. (2006). Study of inexpensive oxygen carriers for chemical looping combustion. *Int. J. Greenhouse Gas Control* 5: 483–488.

34 Readman, J.E., Olafsen, A., Larring, Y., and Blom, R. (2005). $La_{0.8}Sr_{0.2}Co_{0.2}Fe_{0.8}O_{3-\delta}$ as a potential oxygen carrier in a chemical looping type reactor, an in-situ powder X-ray diffraction study. *J. Mater. Chem.* 15: 1931–1937.

35 Moghtaderi, B. and Song, H. (2010). Reduction properties of physically mixed metallic oxide oxygen carriers in chemical looping combustion. *Energy Fuels* 24: 5359–5368.

36 Kang, K.S., Kim, C.H., Cho, W.C. et al. (2008). Reduction characteristics of $CuFe_2O_4$ and Fe_3O_4 by methane; $CuFe_2O_4$ as an oxidant for two-step thermochemical methane reforming. *Int. J. Hydrogen Energy* 33: 4560–4568.

37 Kameoka, S., Tanabe, T., and Tsai, A.P. (2005). Spinel $CuFe_2O_4$: a precursor for copper catalyst with high thermal stability and activity. *Catal. Lett.* 100: 89–93.

38 Pourghahramani, P. and Forssberg, E. (2007). Reduction kinetics of mechanically activated hematite concentrate with hydrogen gas using nonisothermal methods. *Thermochim. Acta* 454: 69–77.

7

Oxygen Carriers for Chemical-Looping with Oxygen Uncoupling (CLOU)

Tobias Mattisson[1] and Kevin J. Whitty[2]

[1] *Chalmers University of Technology, Department of Energy and Environment, Division of Energy Technology, Hörsalsvägen 7B, Göteborg, 41296, Sweden*
[2] *University of Utah, Department of Chemical Engineering, 50 S. Central Campus Drive, Room 3290, Salt Lake City, UT 84112, USA*

7.1 Introduction

As early as the 1950s Lewis and Gilliland proposed a method for production of syngas or carbon dioxide from a carbonaceous fuel using iron- and copper-based oxygen carriers [1, 2]. Although the aim of the proposed methodology had little to do with sustainable energy conversion, the concept is in principle the same as chemical-looping combustion (CLC), as coined by Ishida et al. [3], and described in more detail in other chapters of this book. There was only limited research in the 1990s on the process, most of it performed by Ishida et al., e.g. [4–6]. It was in 1994 that the technology was first proposed as a combustion option for efficient CO_2 capture. At the start of the millennium, research around CLC accelerated, spurred by increased interest in carbon capture, where this emerging technology had a significant advantage with respect to efficiencies and costs in comparison to other technologies. Since then, a large number of materials have been investigated as oxygen carriers, and the technology has been demonstrated at a scale up to 3 MW [7, 8]. The cornerstone of the CLC technology is finding oxygen carrier particles that are sufficiently reactive with fuel, while at the same time having high resistance toward attrition over many redox cycles. This has been described more in detail in other chapters of this book. The initial focus was on monometallic oxide systems based on the transition metals Ni, Fe, Cu, and Mn. These metals have oxide systems that enable a rapid bulk transfer of oxygen at conditions relevant for combustion. As the metal oxides are often supported on an inert material, such as Al_2O_3, $MgAl_2O_4$, or SiO_2, there have been numerous different compositions investigated in the last two decades [9–11].

The CLC process is highly fuel flexible. The utilization of gaseous fuels is straightforward, and the gas is actually the fluidizing medium, e.g. [12]. Liquid fuels could also be injected directly to the bed together with steam, although this likely depends on the viscosity and form of the fuel [13, 14]. For solid fuels, the fuel conversion process is a little different. One way to process solid

Handbook of Chemical Looping Technology, First Edition. Edited by Ronald W. Breault.

fuel via CLC is to first perform an external gasification of the fuel to syngas, and then process the syngas using gaseous CLC [15, 16]. As this would entail an external gasification unit and also an air separation unit (ASU) for oxygen production for the partial oxidation, it is doubtful that this is an economically viable alternative. Instead, the fuel can be directly introduced into the fuel reactor together with the metal oxide particles [17–20]. The fuel in this case devolatilizes to a gaseous fraction, i.e. volatiles, and solid char part; see Figure 4.1. The volatile part, composed mainly of CO, H_2, and CH_4, reacts directly with the oxygen carrier particles, here denoted by Me_xO_y, according to Reactions (7.1)–(7.3).

$$C_nH_{2m} + (2n+m)Me_xO_y \rightarrow (2n+m)Me_xO_{y-1} + nCO_2 + mH_2O \quad (7.1)$$

$$CO + Me_xO_y \rightarrow CO_2 + Me_xO_{y-1} \quad (7.2)$$

$$H_2 + Me_xO_y \rightarrow H_2O + Me_xO_{y-1} \quad (7.3)$$

The char component is gasified in the presence of steam and carbon dioxide, producing CO and H_2 according to Reactions (7.4) and (7.5):

$$C_nH_{2m}\text{ (mainly C)} + nH_2O \rightarrow nCO + (n+m)H_2 \quad (7.4)$$

$$C_nH_{2m}\text{ (mainly C)} + nCO_2 \rightarrow 2nCO + mH_2 \quad (7.5)$$

The CO and H_2 from these reactions will then need to react with the oxygen carrier to fully oxidized products, as in Reactions (7.2) and (7.3). Clearly, Reactions (7.4) and (7.5) do not directly involve the oxygen carrier particles. Still, the rate of these reactions determines how well the solid fuel is converted to gas in the fuel reactor, and governs the residence time of solids needed in the fuel reactor for high char conversion [18]. When work was initiated around CLC with solid fuels, Fe-based materials such as ilmenite or iron oxide were often employed [17, 18, 21]. A problem identified early on was the slow rate of Reactions (7.4) and (7.5), which meant significant residence times and solids inventories for complete carbon burnout in the fuel reactor. Strategies were identified to limit the bed inventory, including raising the temperature for enhanced gasification rates, or separating and recirculating unburnt carbon via a carbon stripper. Another alternative is the use of chemical-looping with oxygen uncoupling (CLOU), which is the main topic of this chapter as well as Chapter 4.

In CLOU, combustion is completed through three overall steps, Figure 7.1. First, the oxygen carrier particles release oxygen to the gas phase in the fuel reactor, Reaction (7.6). In the following step, the released oxygen reacts directly with the fuel through combustion, Reaction (7.7).

$$2Me_xO_y \leftrightarrow 2Me_xO_{y-1} + O_2 \quad (7.6)$$

$$C_nH_{2m} + \left(n + \frac{m}{2}\right) O_2 \rightarrow nCO_2 + mH_2O \quad (7.7)$$

In the third and final steps, the oxygen carriers need to be oxidized back to the fully oxidized state, i.e. reverse of Reaction (7.6). CLOU particles need to fulfill many of the same criteria as normal CLC oxygen carriers. The particles need to (i) have a high resistance toward attrition and fragmentation in a fluidized bed system, (ii) be produced at low cost for an economical carbon capture process,

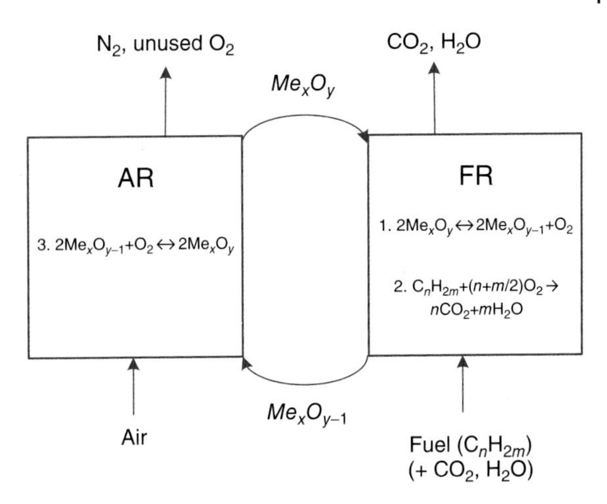

Figure 7.1 Principal layout of chemical-looping with oxygen uncoupling (CLOU). The oxygen carrier is denoted by Me_xO_y and Me_xO_{y-1}, where Me_xO_y is a metal oxide and Me_xO_{y-1} is a metal or a metal oxide with lower oxygen content compared to Me_xO_y. The fuel is here carbon (C). A fluidization gas, e.g. recirculated CO_2 or steam, is most likely needed for the case of solid fuel.

(iii) be nontoxic for environment and human health. In addition, CLOU oxygen carriers need to have special thermodynamic and kinetic requirements in comparison to oxygen carriers used in normal CLC. More specifically, (iv) the metal oxide needs to have a sufficient equilibrium partial pressure of O_2 as to create a significant driving force for reaction in the fuel reactor; (v) the oxide system needs to have an equilibrium partial pressure low enough for oxidation to be possible at the oxygen concentration level in the outlet of the air reactor; and (vi) the reactions in the fuel reactor, including O_2 release by the carrier and combustion of the fuel, should be overall exothermic. This will result in a temperature increase in an adiabatic fuel reactor, which will promote the release rate of gas-phase oxygen from the particle to the bulk gas.

There are several advantages with utilizing oxygen carriers with the so-called uncoupling properties:

1. The combination of uncoupling and combustion of solid fuels means that the rather slow char gasification reactions, i.e. Reactions (7.4) and (7.5), could be partially or completely eliminated.
2. Oxygen release from particles in the freeboard could help convert combustible components not oxidized in the bottom region of a fluidized bed.
3. The oxygen release could compensate for imperfect mixing in a fluidized bed as direct contact between the fuel and oxygen carrier is not necessary. This could make CLOU highly applicable also for gaseous and liquid fuel chemical-looping.

Since the technology was proposed in 2009, there has been significant research around CLOU. Aspects related to oxygen carrier development and testing, kinetic modeling as well as process simulations have been studied, and also reviewed in several articles [22, 23]. The focus of this chapter will be on work related to oxygen carrier development, with emphasis on the research conducted in the last few years. The chapter begins with a brief consideration of important thermodynamic criteria for these types of oxygen carriers. This is then followed by a review of literature around oxygen carrier development.

7.2 Thermodynamics of CLOU

Oxygen carriers for CLOU need to fulfill many of the same criteria as normal CLC oxygen carriers, e.g. the particles need to be mechanically stable, have limited tendencies for agglomeration, and be inexpensive. In addition, for the CLOU process it is important to consider the interplay between oxygen partial pressure of the oxygen carrier system, the heats of reaction, and the temperature levels in the air and fuel reactors.

7.2.1 Equilibrium Partial Pressure of O_2

In a thermodynamic analysis of oxygen carriers for CLC, Jerndal et al. [24] showed that it was not possible to use the oxide pairs $CuO–Cu_2O$, $Mn_2O_3–Mn_3O_4$, and $Co_3O_4–CoO$ as oxygen carriers for CLC at higher temperatures due to decomposition of the most oxidized phase. Later, Mattisson et al. identified this aspect as a possible advantage, and suggested these systems as possible candidates for chemical-looping. The conditions for applications were outlined in two papers [25, 26]. The mono-metallic oxides of Cu, Mn, and Co release oxygen in the fuel reactor through Reactions (7.8)–(7.10):

$$4CuO \leftrightarrow 2Cu_2O + O_2(g) \quad \Delta H_{850} = 263.2\,\text{kJ}\,\text{mol}^{-1}\,O_2 \tag{7.8}$$

$$6Mn_2O_3 \leftrightarrow 4Mn_3O_4 + O_2(g) \quad \Delta H_{850} = 193.9\,\text{kJ}\,\text{mol}^{-1}\,O_2 \tag{7.9}$$

$$2Co_3O_4 \leftrightarrow 6CoO + O_2(g) \quad \Delta H_{850} = 408.2\,\text{kJ}\,\text{mol}^{-1}\,O_2 \tag{7.10}$$

The equilibrium partial pressure of O_2 for the above reactions can easily be calculated. It will be assumed that the condensed phases are pure species and that standard Gibbs free energy of the gas phase components is much greater than the Gibbs free energy of the solid phase. The equilibrium constant and also equilibrium partial pressure of oxygen for the above reactions can then be calculated from the change in the Gibbs free energy of Reactions (7.8)–(7.10):

$$\Delta G^{\circ} = -RT\ln(p_{O_2,eq}) = -RT\ln(K) \tag{7.11}$$

where $p_{O_2,eq}$ is the equilibrium partial pressure of oxygen and K is the equilibrium constant for the above reactions. Normally, ΔG° can be approximated as a linear function of temperature:

$$\Delta G^{\circ} = A + BT \tag{7.12}$$

where A and B are constants, and T is the temperature. Figure 7.2 shows the equilibrium partial pressure of O_2 as a function of temperature for the above reactions involving pure condensed species. From this, it can be seen that CuO releases oxygen in air ($p_{O_2} = 0.21$ bar) at temperatures above 1028 °C, Mn_2O_3 at temperatures above 899 °C, and Co_3O_4 above 889 °C. It is expected that the temperature in the air reactor would be lower since an efficient combustion process will be operated with a lower partial pressure of O_2 from the outlet of the air reactor, a value which is dependent upon the excess air ratio [26].

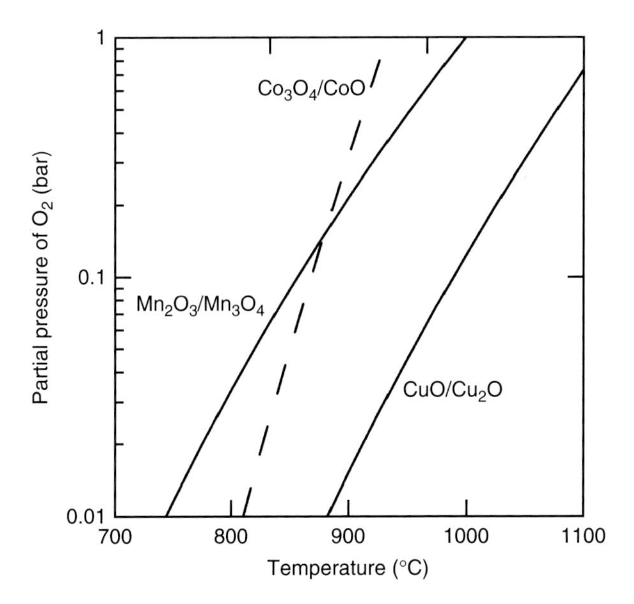

Figure 7.2 The partial pressure of gas phase O_2 over the metal oxide systems CuO/Cu_2O, Co_3O_4/CoO, and Mn_2O_3/Mn_3O_4 as a function of temperature.

Several combined or mixed oxides systems containing more than one metal in the lattice have been studied for CLOU (Table 7.1). Of special interest are combined oxides based on Mn, as doping with various metals can tweak the thermodynamic properties for enhanced performance. One of the more extensively studied systems is the Mn–Fe–O system, which has shown great promise for the CLOU application. Figure 7.3 shows a phase diagram for this binary system calculated at an equilibrium partial pressure of 0.05 atm O_2, as presented by Azimi [64]. It is evident that the temperatures for the different phase transitions are highly dependent upon the ratio of Mn/Fe. Rydén et al. [65] reviewed the thermodynamics of this system in addition to a number of other such combined oxide systems, most of which have now been experimentally investigated, e.g. [41, 66–69]. The decomposition reaction for a combined oxide system of $(Mn_yFe_{1-y})O_x$, which has shown promise as a CLOU material both thermodynamically and experimentally, can be written as

$$6(Mn_{0.8}Fe_{0.2})_2O_3 \leftrightarrow 4(Mn_{0.8}Fe_{0.2})_3O_4 + O_2(g) \quad \Delta H_{850} = 254 \text{ kJ mol}^{-1}\ O_2 \tag{7.13}$$

Figure 7.4 shows the partial pressure of O_2 for three such mixed oxide systems, where critically evaluated thermodynamic data are available, i.e. for Mn–Fe–O and Mn–Si–O. For reference, the pure Mn_2O_3/Mn_3O_4 is also included. It is clear that the equilibrium temperature is increased for the combined oxides in comparison to the pure system. This could be an advantage as the concentration driving force, i.e. $(p_{O_2,eq} - p_{O_2})$ for uncoupling will increase in the fuel reactor at a given

Table 7.1 Overview of oxygen carriers investigated for chemical-looping with oxygen uncoupling (CLOU) in the last few years.

Authors	References	Oxygen carrier (OC) (Me_xO_y/support)	Preparation method (OC)	Fuel	T_{CLOU} (°C)[a)]	Apparatus	Notes
Arjmand et al. (2014)	[27]	Mn_3O_4/SiO_2	SD	CH_4, biomass char	900–1100	FzB	a, b, d
Adánez-Rubio et al. (2014)	[28]	$CuO/MgAl_2O_4$	SD	Lignite	900–935	CFzB	d
Hallberg et al. (2014)	[29]	$CaMn_yMg_{(1-y)}O_{3-\delta}$, $CaMn_{0.775}Ti_{0.125}Mg_{0.1}O_{3-\delta}$	SD	CH_4	900	CFzB	a
Jing et al. (2014)	[30]	Mn_3O_4/SiO_2	SD	CH_4	900–1100	FzB	a
Adánez-Rubio et al. (2014)	[31]	$CuO/MgAl_2O_4$	SD	Biomass	860–930	CFzB	
Ksepko et al. (2014)	[32]	Fe_2O_3-CuO/Al_2O_3	MM	H_2	600–950	TGA	b
Arjmand et al. (2014)	[33]	$Ca_xMn_{1-y}M_yO_{3-\delta}$ (M = Mg, Ti)	SD	CH_4	900–1050	FzB	d
Clayton et al. (2014)	[34]	CuO/TiO_2, CuO/ZrO_2	FG, MM	—	600–950	TGA	a, c
Clayton and Whitty (2014)	[35]	CuO/SiO_2, CuO/TiO_2, CuO/ZrO_2	FG, MM, I	—	750–950	TGA	a, b
Zhao et al. (2014)	[36]	Cu-ore	Conc.	Anthracite	850–1000	FzB	
Imtiaz et al. (2014)	[37]	$CuO/MgAl_2O_4$, CuO/CeO_2, CuO/Al_2O_3,	CP	—	950	FzB, TGA	a
Rydén et al. (2014)	[38]	CuO/ZrO_2, $CuO/ZrO_2/La_2O_3$, $CuO/Y-ZrO_2$, $CuO/MgAl_2O_4$, $CuO/MgAl_2O_4/La_2O_3$, CuO/CeO_2, $CuO/CeO_2/La_2O_3$, CuO/Fe_2O_3, $CuO/Fe_2O_3/Mn_2O_3$, $CuO/Fe_2O_3/MgO$, $CuO/Fe_2O_3/Al_2O_3$	SD	CH_4	800–925	FzB, CFzB	
Azimi et al. (2014)	[39]	$(Mn_{0.75}Fe_{0.25})_2O_3/MgAl_2O_4$, $(Mn_{0.75}Fe_{0.25})_2O_3/CeO_2$, $(Mn_{0.75}Fe_{0.25})_2O_3/Y_2O_3-ZrO_2$, $(Mn_{0.75}Fe_{0.25})_2O_3/ZrO_2$	SD	CH_4, syngas	850	FzB	
Pour et al. (2014)	[40]	Mn-ore/Fe_2O_3	MM	CH_4, syngas, petcoke	900–950	FzB	e
Shafiefarhood et al. (2015)	[41]	$(Co_xFe_{1-x})_3O_4$, $(Mn_xFe_{1-x})_2O_3$, $(Co_xFe_{1-x})_3O_4/La_{0.8}Sr_{0.2}Co_xFe_{1-x}O_{3-\delta}$, $(Mn_xFe_{1-x})_2O_3/La_{0.8}Sr_{0.2}Mn_xFe_{1-x}O_{3-\delta}$	MM, SG	CH_4, H_2	850–950	TGA	a, f

Imtiaz et al. (2015)	[42]	CuO/CeO_2, CuO/Al_2O_3	CP	CH_4	900–950	TGA, FzB	a, f
Adánez-Rubio (2014)	[43]	$CuO/MgAl_2O_4$	SD		900–1000	TGA	a, b
Arjmand et al. (2015)	[44]	$(Mn_xFe_{1-x}Si_y)_2O_5$	SD	CH_4, syngas	850–1000	FzB	a
Wang et al. (2015)	[45]	CuO/ZrO_2	MM		1065	TGA	a, f
Mayer et al. (2015)	[46]	$CaMn_{0.9}Mg_{0.1}O_{3-\delta}$, $CaMn_{0.775}Ti_{0.125}Mg_{0.1}O_{3-\delta}$	SD	Natural gas	900–960	CFzB	
Mei et al. (2015)	[47]	$CuO/CuAl_2O_4$	SG	CH_4, H_2	900–1000	TGA	
Galinsky et al. (2015)	[48]	$Ca_{1-x}A_xMnO_3$ (A = Sr, Ba)	SG	Coal char	650–950	TGA, FzB	a, f
Azimi et al. (2015)	[49]	$(Mn_xFe_{1-x}Al_y)_2O_5$	SD	CH_4, syngas, wood char	850–1000	FzB	a
Zhang et al. (2015)	[50]	CuO, $CuO/CuAl_2O_4$, $CuAl_2O_4$	SG		900–1000	TGA	b
Galinsky et al. (2016)	[51]	$CaMn_{1-x}B_xO_{3-\delta}$ (B = Al, V, Fe, Co, Ni)	SG	Coal char	850	TGA, FzB	a
Hanning et al. (2016)	[52]	Mn_3O_4/SiO_2, $Mn_3O_4/SiO_2/TiO_2$	SD	CH_4, syngas, wood char	800–1050	FzB, CFzB	b
Schmitz and Linderholm (2016)	[53]	$CaMn_{0.775}Ti_{0.125}Mg_{0.1}O_{3-\delta}$	SD	Wood char	800–950	CFzB	
Ku et al. (2017)	[54]	CuO/ZrO_2	MM	Syngas	950	TGA	a
Mattisson et al. (2016)	[55]	Mn-ores, $CaMnO_{3-\delta}$	SD	CH_4, syngas	900–1050	FzB	a
Pishahang et al. (2016)	[56]	$CaTi_xMn_{0.9-x}Mg_{0.1}O_{3-\delta}$	SD	H_2, CH_4	800–1000	TGA	d
Mendiara et al. (2016)	[57]	$CuO/MgAl_2O_4$	SD	Pine sawdust	900–935	CFzB	
Zhao et al. (2017)	[58]	CuO/TiO_2, CuO/ZrO_2, $CuO/CuAl_2O_4$, $CuO/MgAl_2O_4$	I		900–950	TGA	a, b
Abian et al. (2017)	[59]	$(Mn_xFe_{1-x}Ti_y)Oz$	E	CO, H_2, CH_4	850–950	TGA	a
Mungse et al. (2017)	[60]	$FeMnO_3$	MM	—		FzB	f
Perez-Vega et al. (2017)	[61]	$(Mn_{0.77}Fe_{0.23})_2O_3$	MM	CO, H_2, CH_4	900–950	TGA, FzB	a
Tian et al. (2017)	[62]	Cu-ore	—	Coal	870–930	FzB	c
Sundqvist et al. (2017)	[63]	Mn-ores (8)	—	CH_4, syngas	850–1000	FzB	a

Abbreviations: SD, Spray-drying; FG, Freeze granulation; MM, Mechanical mixing; SG, Sol–gel; E, Extrusion; CP, Co-precipitation; I, Impregnation; SP, Spray pyrolysis; Conc., Concentrating; NG, Natural gas; FzB, Batch fluidized bed reactor; CFzB, Continuous interconnected fluidized bed reactor; FxB, Fixed bed reactor; TGA, Thermogravimetric analyzer.

Notes: (a) utilized inert with or without O_2 to study decomposition reactions; (b) kinetic/mechanistic investigation of decomposition; (c) kinetic/reactivity of oxidation; (d) investigated effect of sulfur; (e) investigated different manganese ores; and (f) conducted TPR/TPO.

a) The temperature generally refers to the temperature used during isothermal decomposition (uncoupling) reaction or the temperature in the fuel reactor, the exception being in studies with the specific aim of studying the oxidation reaction.

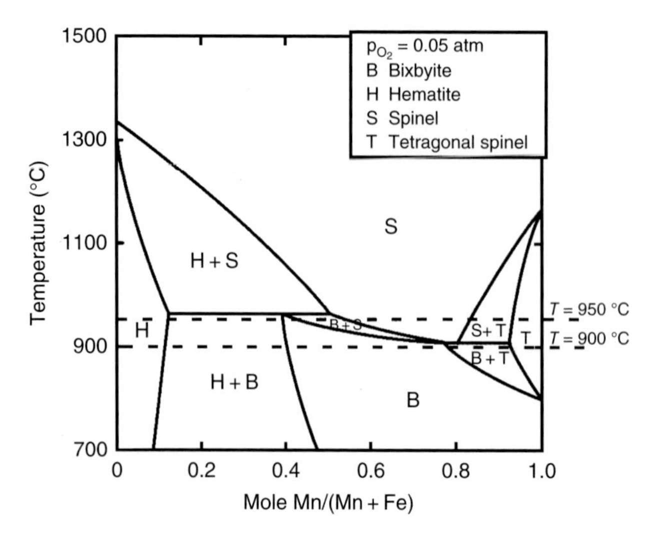

Figure 7.3 Phase diagram of $(Mn_yFe_{1-y})O_x$ in an atmosphere with an O_2 partial pressure of 0.05 atm. Source: Azimi 2014 [64]. Reproduced with permission of American Chemical Society.

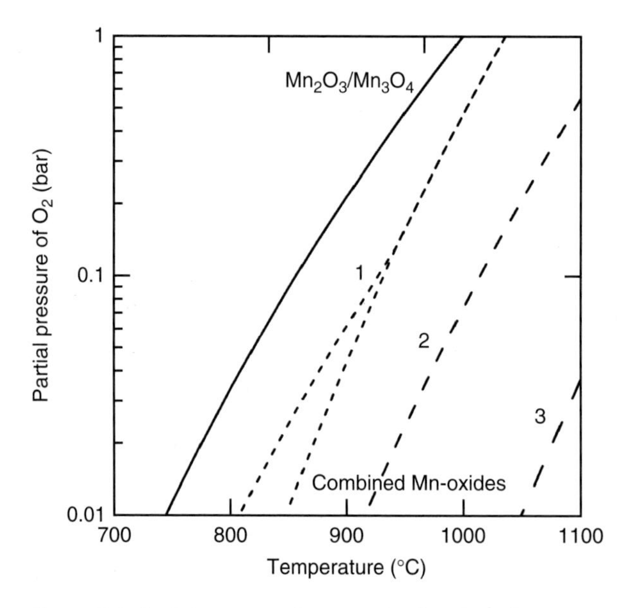

Figure 7.4 Partial pressure of oxygen as a function of temperature for a series of combined Mn-oxides as well as pure Mn_2O_3/Mn_3O_4. Partial pressure of O_2 is calculated from the reactions (1) $6(Mn_{0.8}Fe_{0.2})_2O_3 \leftrightarrow 4(Mn_{0.8}Fe_{0.2})_3O_4 + O_2$ (g), (2) $(^2/_3)Mn_7SiO_{12} + 4SiO_2 \leftrightarrow (^{14}/_3)MnSiO_3 + O_2$ (g), (3) $(^{10}/_3)MnSiO_3 + (^2/_3)Mn_7SiO_{12} \leftrightarrow 4Mn_2SiO_4 + O_2$ (g).

temperature. For the combined oxide system of $(Mn_{0.8}Fe_{0.2})_2O_3$ there are two parallel curves that converge at higher temperatures, which are due to the presence of a two-phase region where both the oxidized and reduced oxide exist, and this is clearly seen in the phase diagram in Figure 7.3.

Combined oxides with perovskite structure have also been explored as oxygen carriers for CLOU. These types of materials have the general formula ABO_3, where A is a large cation and B is a smaller cation. Perovskite materials exhibit interesting redox properties and have been investigated for both CLC and CLOU [70–74]. Oxygen carrier particles can exhibit oxygen non-stoichiometry and can release substantial amounts of oxygen via

$$ABO_{(3-\delta)_{AR}} \leftrightarrow ABO_{(3-\delta)_{FR}} + \frac{1}{2}(\delta_{FR} - \delta_{AR})O_2 \tag{7.14}$$

The oxygen released in the gas phase, given by $(\delta_{FR} - \delta_{AR})$, can be used during CLOU process to burn gas, liquid, or solid fuel. The equilibrium value for oxygen in the gas phase is thus given by the temperature and the degree of solids conversion, which is a little bit different compared to the other systems explored above. The thermodynamics of the most common material $CaMnO_3$ has been explored by several researchers, e.g. [75, 76].

When the oxygen carrier particles are transferred to the fuel reactor, it is of course an advantage if the temperature is as high as possible, as this promotes the rate of oxygen release. This depends upon the thermal nature of the overall reaction in the fuel reactor, and this will be explored in the following section.

7.2.2 Thermal Considerations

When the hot oxygen carrier material from the air reactor is transferred to the fuel reactor, the oxygen carrier should release oxygen, which will react with the fuel. Considering that the total heat evolved from a CLOU system will be that of normal combustion, the sum of the heats of reaction in the fuel and air reactor will be that of normal combustion. The enthalpies of the uncoupling reactions for several metal oxide systems are shown in Figure 7.5 [65]. Also shown is the reaction enthalpy of fuels, carbon (representing coal) and methane, the major component in natural gas. Systems with an oxygen carrier reaction enthalpy greater than the reaction enthalpy of the fuel would have an overall endothermic reaction in the fuel reactor. This would effectively mean a temperature drop with respect to the air reactor. From the figure it can be observed that utilization of CuO, Mn_2O_3, and combined Mn-oxides would always result in an overall exothermic reaction in the fuel reactor, which would result in a temperature increase in an adiabatic fuel reactor. For Co_3O_4 the reaction is slightly endothermic when utilizing CH_4, and thus there will be a temperature decrease for Co_3O_4 using such fuels. Although the reaction with Co_3O_4 is slightly exothermic for carbon, considering heat losses and that heat will be needed to heat the fluidization gas, there will likely be a temperature drop also for a Co-based oxygen carrier. The fact that a temperature increase is possible for most oxide systems is significant and of considerable advantage for the following reasons: (i) the higher temperature will promote the oxygen release rate from the metal oxide particles and (ii) the recirculation rate of particles between the air and fuel reactor can be kept low, as the

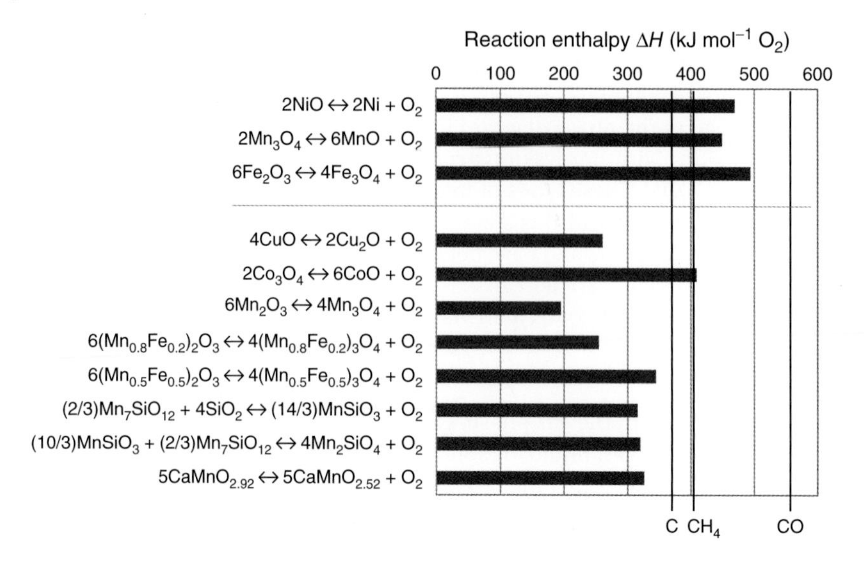

Figure 7.5 Heat or reaction, ΔH, for O_2 release for different oxygen carrier materials. Source: Rydén et al. 2013 [65]. Reproduced with permission of Elsevier.

recirculating particles do not need to provide heat for an endothermic reaction, which is the case for CLC with solid fuels [18]. The temperature drop in the fuel reactor for the cobalt oxide in combination with its high cost and toxicity likely makes this oxygen carrier unsuitable for a CLOU system.

7.3 Overview of Experimental Investigations of CLOU Materials

In the last eight years, a significant number of articles have been published with experimental investigations of different oxygen carrier materials. There have also been several reviews that are entirely or partially devoted to CLOU, including reviews of oxygen carrier materials [9, 22, 23]. Table 7.1 contains an overview of the experimental investigations of CLOU materials in the last few years. A similar table was presented by Mattisson for work conducted prior to 2014 [23]. The table is not meant to be a complete and detailed analysis, but rather is included to give a broad overview of the type of materials that are currently being investigated for CLOU. Further, it only contains research published in peer-reviewed journals in English. It should be noted that only articles that have the aim of utilizing the oxygen uncoupling behavior of metal oxide particles are included. Most of the investigations related to CLOU have been carried out at Chalmers University of Technology in Sweden and at ICB-CSIC in Zaragoza, Spain, but there are several other groups working with development of the CLOU process, including research groups at the University of Utah in the United States, Tsinghua University in China, Swiss Federal Institute of Technology in Switzerland, and North Carolina State University in the United States among others. From the table, the investigations can be divided into two major classes of (i) Cu-based materials and (ii) combined/mixed oxides.

7.3.1 Copper Oxide

Prior to 2014, most investigations of CuO materials utilized particles with Al_2O_3, $MgAl_2O_4$, or ZrO_2 as support, but there had also been studies of Cu-materials for CLOU utilizing refined natural materials [77] as well as other supports [37, 78–80]. For instance, Hedayati et al. used CuO with CeO_2 and $Ce_{0.9}Gd_{0.1}O_{1.9}$ prepared by extrusion [78]. The particles were investigated in a batch fluidized bed reactor with respect to oxygen release behavior and reactivity with CH_4 and syngas. Most work prior to 2014 was performed with gaseous fuels, but solid and liquid fuels had also been investigated [81, 82].

From Table 7.1 it is clear that significant amount of research has been conducted with Cu-based materials also in the last several years, including the use of various supports and preparation methods, including the use of Cu as part of mixed oxides oxygen carriers [83]. A variety of reactors have also been employed, from thermogravimetric analyzers using only small amounts of particles to continuous circulating units using kilograms of particles. Most of the studies utilized some type of fuel, but most investigations also evaluated the oxygen carriers' decomposition in an inert atmosphere with no or some oxygen in the gas phase. Here a thermogravimetric analysis (TGA) is highly applicable as the differential conditions facilitate the use of a precise oxygen partial pressure in the gas phase, for both oxidation and reduction [34, 35]. A number of studies have used this methodology to establish kinetic constants and solid-state reaction models, as will be described in Section 7.4. In addition, several more fundamental studies have been conducted recently with respect to Cu-based oxygen carriers, including first-principles investigation of oxygen transport activation barriers [50].

With respect to continuously operated chemical-looping investigations, Rydén et al. conducted a major screening study of nine different Cu-based oxygen carriers in batch and continuous operation [38]. Although the reactivity of many of these was very high, the carriers tested in continuous operation always suffered degradation and fragmentation. Adánez-Rubio et al. conducted the first successful tests with biomass in continuous operation [31]. Here, an oxygen carrier of $CuO/MgAl_2O_4$ was employed, showing a combination of high combustion efficiency and carbon capture, i.e. char burnout in the fuel reactor. However, only 10 hours of operation was performed, and thus the mechanical stability over a significant period was not determined. Mendiara et al. compared the combustion efficiency and carbon capture efficiency of the same Cu-based CLOU oxygen carrier with an Fe-based material using sawdust as fuel [57]. The Cu-based material showed significantly better behavior with respect to the combustion efficiency, and somewhat increased carbon capture. Also, the tar content in the off-gases was non-existent using the CLOU material, which was not the case for the iron oxide. Prior to the work with biomass, the same group of researchers had conducted successful tests with coal using the same oxygen carrier and circulating reactor system [84].

7.3.2 Combined and Mixed Oxides

There are several disadvantages with utilizing Cu-material for CLOU, including questionable stability of the particles and the relatively high cost of Cu in addition to the low melting temperature of copper, 1089 °C. Although the melting temperatures of the relevant oxides, CuO and Cu_2O, are higher compared to metallic Cu, it is expected that some Cu will be formed in a fuel reactor, owing to reduction of copper oxides with volatiles or gaseous fuel components. It is also clear that there could be problems related to attrition and fragmentation of Cu-based oxygen carriers. Manganese oxides, on the other hand, suffer from the apparent slow kinetics of oxidizing Mn_3O_4 to Mn_2O_3 at the low temperature needed in order to oxidize the material fully, Figure 7.2 [78, 85]. The use of combined Mn-oxides could be a way of overcoming the low equilibrium temperatures. Shulman et al. investigated a number of combined oxides of Mn for the first time with respect to CLOU, and found that the incorporation of certain metals into the Mn_3O_4 matrix could result in an enhanced reaction rate of the manganese oxide [68]. This was followed by a number of articles where combined oxides were utilized as oxygen carriers for CLOU, e.g. [69, 86, 87]. The common denominator for all of these investigated materials is that they contain manganese in the structure, but the Mn is exchanged partly with another metal in the lattice. These combined oxides have thermodynamic and/or kinetic properties that differ from pure Mn_2O_3, and which enable decomposition of the oxide with release of gaseous oxygen at conditions suitable for CLOU, Figures 7.3 and 7.4. Figure 7.6 summarizes the compositions evaluated in the literature in a number of ternary diagrams, which clearly shows that the ternary and binary systems of Mn with Mg, Si, Fe, and Ca are very well explored.

Of the combined manganese oxygen carriers studied for CLOU, most research has been conducted around the $CaMnO_3$-type of perovskite, Reaction (7.14). It has been found that such material can exhibit a combination of high mechanical stability, good fluidization properties as well as good reactivity with gaseous fuels. The fact that this type of material can be produced with cheap raw materials is an added advantage. Several works had been performed prior to 2014, mostly utilizing gaseous fuels in smaller reactor units. Leion et al. [93] investigated the CLOU behavior and reactivity of a series of materials of general composition $CaMn_xTi_{1-x}O_3$ with CH_4 in a TGA. Based on the oxygen capacity and the reactivity, the oxygen carrier of $CaMn_{0.875}Ti_{0.125}O_3$ was chosen for further testing with CH_4 and petroleum coke. The reactivity with petroleum coke at 900 and 950 °C was slower compared to that in previous experiments with Cu-based oxygen carrier [25], which was explained by the slower rate of oxygen uncoupling in this material. Rydén et al. investigated the same material in a bench-scale continuously circulating reactor using natural gas [71]. The material exhibited very smooth fluidization behavior for over 70 h at hot conditions. The uncoupling effect was dependent upon the oxygen concentration in the air reactor, with a higher concentration of O_2 from the fuel reactor as the oxygen concentration was increased in the air reactor. It was possible to achieve complete fuel conversion at 950 °C. Further, Källén et al. investigated the performance of $CaMn_{0.9}Mg_{0.1}O_{3-\delta}$ in a 10 kW natural gas-fired CLC unit [94]. The results showed high rates of

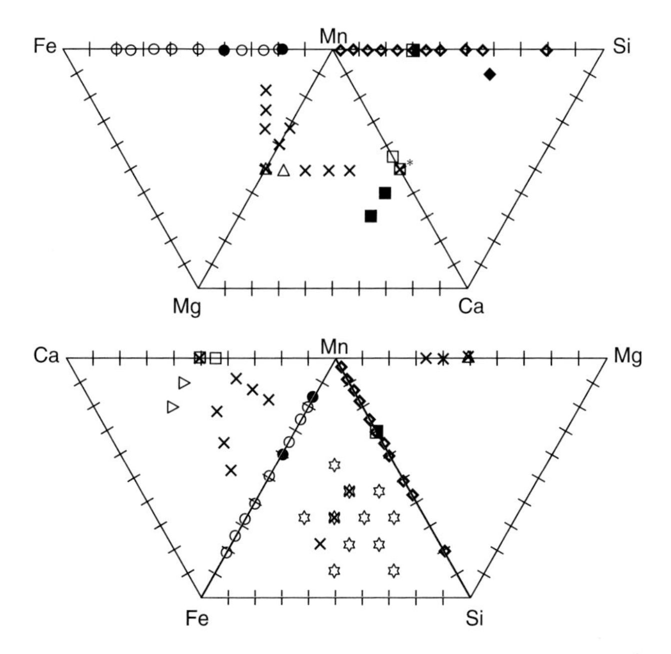

Figure 7.6 Molar fractions of metals/metalloids of investigated combined manganese oxides. Symbols indicate compositions that have been investigated in the literature according to Mattisson et al. [67] (×), Azimi et al. [88] (○), Shulman et al. [68] (•), Shulman et al. [69] (△), Jing et al. [30] (◇), Frick et al. [89] (♦), Arjmand et al. [33] (□), Jing et al. [90] (■), Hallberg et al. [91] (▷), Arjmand et al. [44] (✡). *Several works have studied particles in the Ca–Mn binary with a similar amount of Ca and Mn, often doped with additional metals, e.g. [46, 51, 56, 92].

oxygen release and full conversion of natural gas – a behavior superior to that of NiO-based oxygen carrier in the same unit [95, 96]. As is evident from Table 7.1, significant research has been conducted around this type of oxygen carrier also in the most recent years. Most of the research has been performed in Europe, within the scope of EU-financed projects [97]. This includes testing in different types of batch reactors [51, 56, 98], continuous units with gas and solid fuels [29, 46, 53], in addition to investigations of kinetics and effect of impurities [33, 56, 92]. For example, Figure 7.7 shows the gas conversion from natural gas obtained in a 10 kW pilot CLC unit (Figure 7.7a) using different types of calcium manganites, most of which were produced in the EU-financed SUCCESS project. Here, the idea was to up-scale the production of $CaMn_{0.775}Ti_{0.125}Mg_{0.1}O_{3-\delta}$ to multi-ton scale [99]. It is evident that full conversion of gas could be obtained in the fuel reactor, although the reactivity was dependent on the actual material used. The CLOU effect of these materials was also clearly seen during operation with these types of materials. This is illustrated in Figure 7.8 where the outlet oxygen concentration is shown as a function of temperature using a small bench-scale unit [99]. Here, no fuel is added to the fuel reactor, but it is simply fluidized by nitrogen. The oxygen carrier is oxidized in the air reactor, in a similar manner as when fuel is added. Evidently, the measured outlet volume fraction from the fuel reactor can approach 13 vol%. Based on these promising results, an

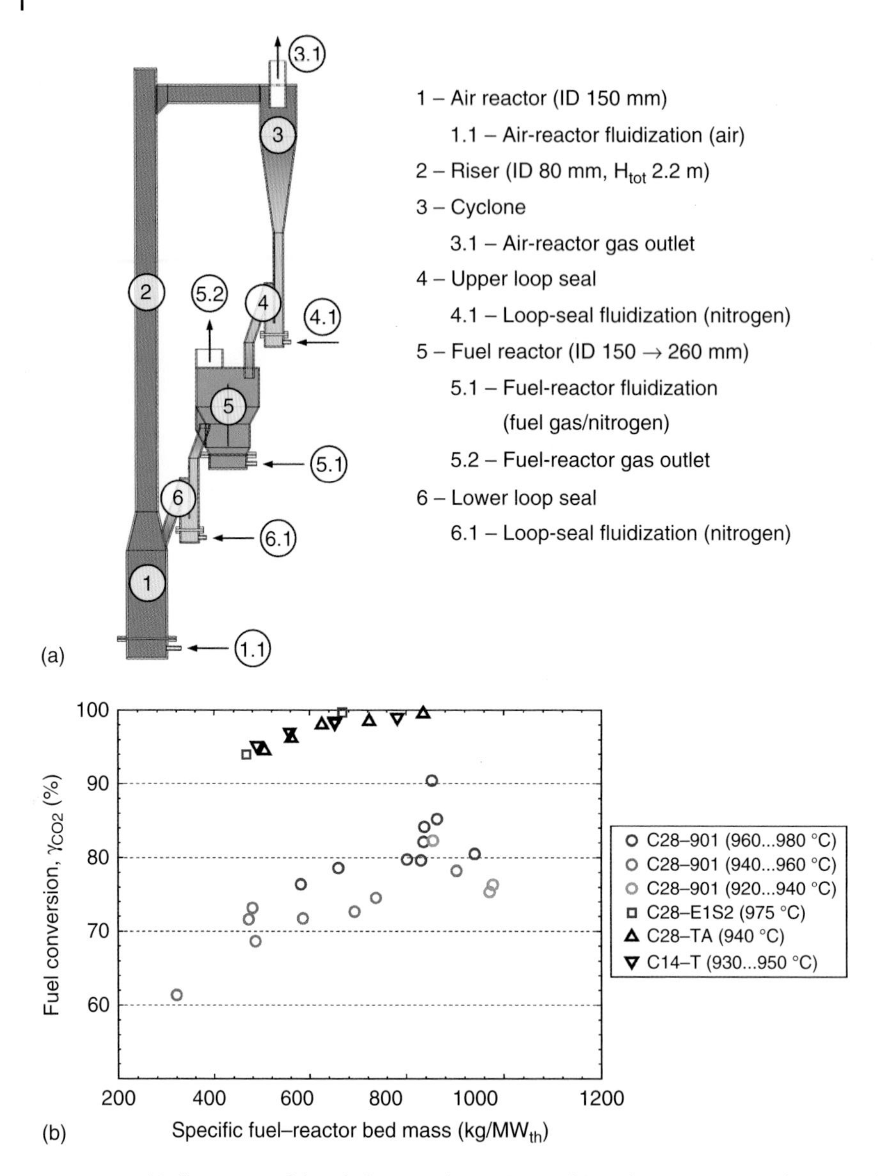

Figure 7.7 (a) Illustration of the Chalmers 10 kW CLC unit. (b) Fuel conversion as a function of specific fuel-reactor bed mass in the 10 kW unit using methane with different calcium manganite oxygen carriers. The temperatures given in the legend are for the fuel reactor. Source: Taken from Moldenhauer et al. 2017 [99].

oxygen carrier of $CaMn_{0.775}Ti_{0.125}Mg_{0.1}O_3$ was scaled up to multi-ton production and tested successfully at 1 MW scale using natural gas [100]. A similar type of material has also been tested using solid fuels. Schmitz and Linderholm tested $CaMn_{0.9}Mg_{0.1}O_{3-\delta}$ using petroleum coke and wood char using a pilot similar to the one shown in Figure 7.7a, but designed for in-bed feeding of fuel as well as with a carbon stripper [53]. The oxygen demand (OD) of the outlet gases from the

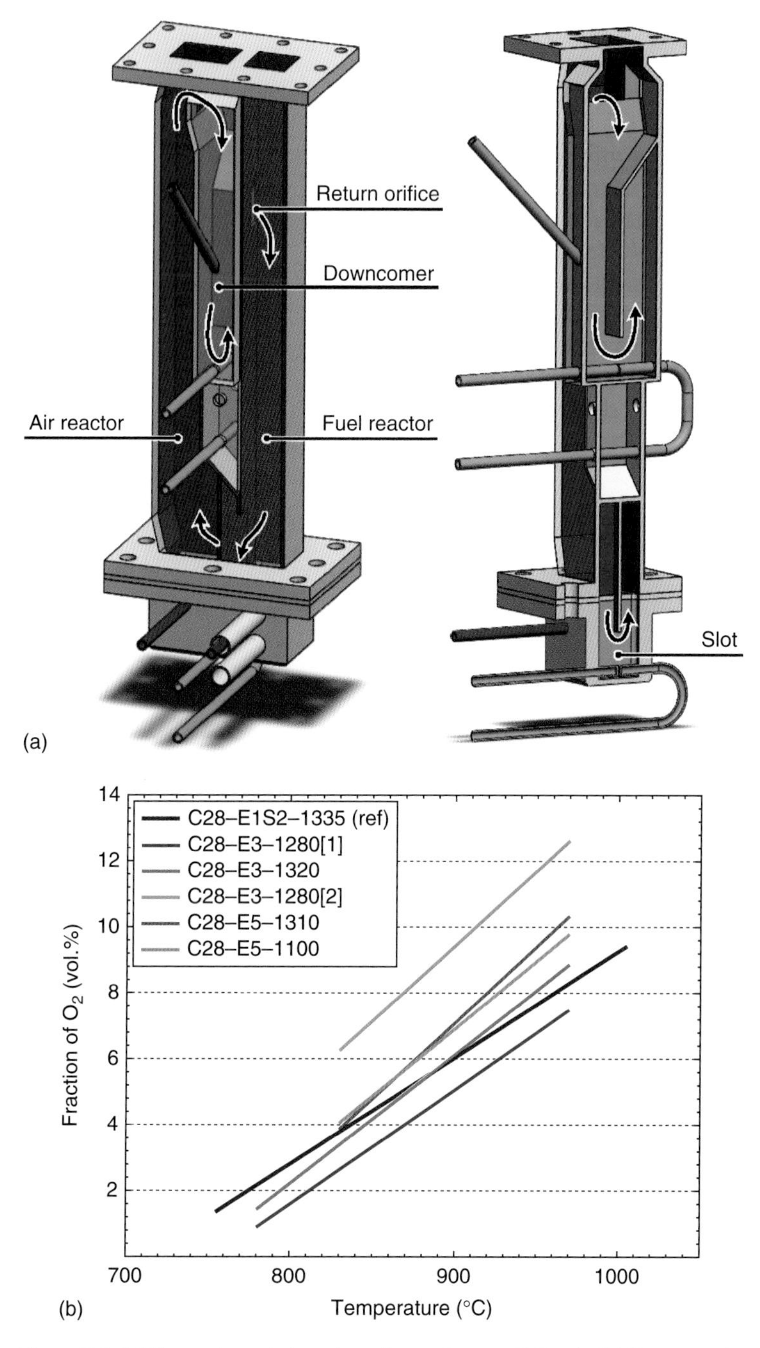

Figure 7.8 (a) Illustration of a bench-scale continuous CLC unit. (b) Oxygen volume fraction from the fuel reactor as a function of the temperature using different calcium manganite oxygen carriers. Here, the air reactor is fluidized with air and the fuel reactor with argon. Source: Taken from Moldenhauer et al. 2017 [99].

fuel reactor is shown as a function of the circulation index in Figure 7.9a. The oxygen demand is below 10% for both fuels, but lowest for the low volatile char. Also, here the CLOU effect was determined by measuring oxygen release during inert fluidization of the fuel reactor, Figure 7.9b. All in all, from the studies performed with calcium manganite it can be concluded that very high gas conversions can be obtained with gas and solid fuels while the mechanical integrity seems reasonable and indicates that lifetimes of greater than 1000 hours should be possible, at least using gaseous fuels. However, the material does seem to react with sulfur components, such as H_2S and SO_2, which likely affects the reactivity negatively. However, several studies have shown that this effect is very dependent upon process conditions, such as temperature and time under oxidation and reduction [33, 56].

Calcium manganate can be produced with very cheap material, i.e. CaO and manganese ore [102]. This is likely a very important aspect for solid fuel

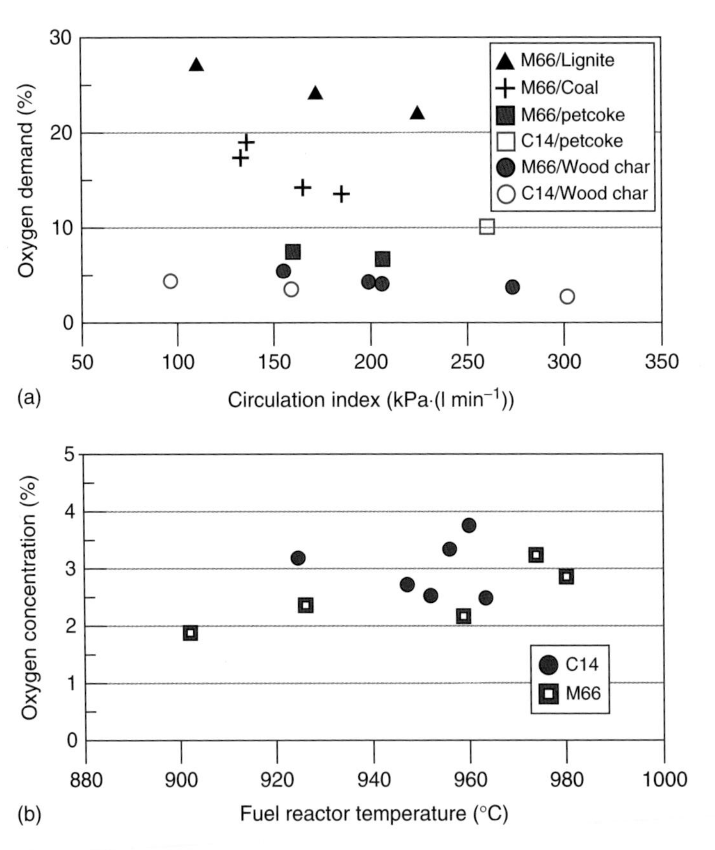

Figure 7.9 (a) Oxygen demand as a function of the circulation index of solids from the air reactor to the fuel reactor using two combined oxides of Ca–Mn–O and Mn–Si–Ti–O and different solid fuels in the temperature range 900–950 °C. The oxygen demand is defined as the fraction of oxygen needed for complete combustion of the gases generated in the fuel reactor. (b) The volume fraction of oxygen from the fuel reactor as a function of fuel reactor temperature. Here, nitrogen is used to fluidize the fuel reactor. Source: Data from Schmitz and Linderholm 2016 [53] and Schmitz et al. 2017 [101].

combustion, as the lifetime of particles in a solid fuel combustor is likely to be much lower than in a gas-fired unit due to deactivation with ash as well as loss of material with the ash flows. The highly promising results from testing of these materials clearly make this one of the more promising combined oxides.

Another well-investigated combined oxide is the system composed of Mn–Fe–O. Oxygen carriers composed of Mn–Fe have earlier been tested by Lambert et al. [103], but the CLOU effect was not observed in this investigation. In addition to the work of Shulman et al. described earlier, Azimi et al. [66, 86, 88, 104] conducted extensive investigations of this system using spray-dried oxygen carriers and using both solid and gaseous fuels. In one investigation [88], an extensive number of Fe/Mn particles were tested with Mn/(Mn + Fe) from 0.2 to 0.8 using methane and syngas. The oxygen uncoupling effect was dependent on the Mn/Fe ratio and the temperature of oxidation and reduction. Oxygen carrier particles with a Mn/(Mn + Fe) ratio of 0.2–0.4 release oxygen at 900 °C, whereas materials with a higher Mn content showed no oxygen uncoupling. On the other hand, at the lower temperature of 850 °C materials with a Mn/(Mn + Fe) ratio greater than 0.5 released considerable amounts of oxygen. The behavior of these materials agreed well with what is expected based on the calculated phase diagram of the Fe–Mn–O system, seen in Figure 7.3.

From Table 7.1, it can be seen that a number of papers have focused on the use of Mn–Fe combined with other active or inert supports. Both Shulman et al. [68] and Rydén et al. [105] have earlier reported on problems with respect to mechanical stability, which may be the driving factor for many of these investigations [68]. For instance, Azimi et al. investigated the use of $(Fe_{0.2}Mn_{0.8})_2O_3$ on inert supports of CeO_2, $MgAl_2O_4$, and ZrO_2 [39].

Another binary system of interest for CLOU is that of Mn–Si–O, Figure 7.4. Shulman et al. [68] prepared particles of 80 wt% Mn_3O_4 with SiO_2 by freeze granulation. The particles released oxygen in the gas phase in the temperature interval 810–900 °C. Interestingly, from X-ray diffraction (XRD) analysis of the samples calcined at 950 and 1100 °C, there did not seem to be any interaction between Mn_3O_4 with SiO_2. However, in a thermal evaluation, Rydén et al. presented a relevant phase diagram for the system and identified several possible CLOU transformations [65]. Jing et al. investigated a significant number of spray-dried particles of Mn_3O_4/SiO_2 with varying contents of Mn using a batch fluidized bed reactor [30]. Here, the SiO_2 content was varied between 2 and 75 wt%. The propensity to uncouple oxygen to the gas phase and the reactivity with methane were varied in the temperature interval 900–1100 °C. The experiments showed that the reactivity was highly dependent upon the SiO_2 content in addition to the temperature, with very low reactivity for oxygen carriers with a SiO_2 content exceeding 45 wt%. But for materials with lower SiO_2 fractions, full CH_4 conversion could be obtained together with a high oxygen release rate provided that the temperature was above 950 °C. For these materials, it seems as if the main oxygen releasing phase was braunite, Mn_7SiO_{12}, which can be reduced to both $MnSiO_3$ and Mn_2SiO_4 in the fuel reactor. This is in conformity with the phase diagram presented by Rydén et al. [65]. Hanning et al. conducted experiments with two optimized oxygen carriers of Mn–Si and Mn–Si–Ti in both batch and continuous fluidized bed reactors [52]. It was found that although the inclusion of some Ti in the structure resulted in lower reactivity, the attrition resistance increased substantially. The latter materials were also investigated successfully by Schmitz et al. in a 10 kW CLC pilot, Figure 7.9 [101]. The material, denoted here as M66, showed a combination of high propensity for CLOU, high rate of reaction with fuels of low volatile content, and good stability. Using wood char, an oxygen demand of only 2% was achieved in this study. Finally, Arjmand evaluated a matrix of materials in the Fe–Mn–Si–O system and found that particles with a combination of high reactivity and low attrition could be obtained [44].

7.3.3 Naturally Occurring Oxygen Carriers

Almost all investigated oxygen carriers for CLOU have been prepared using some type of production method, such as impregnation or spray-drying. Although many such processes are commercial and available for multi-ton production of particles, the production cost could be substantial and be a significant added cost for the carbon capture process. The cost of oxygen carrier is related to the specific solids inventory (kg/MW), the cost of manufacturing of the particles ($/kg), and the lifetime (h). With respect to the latter, this may be particularly challenging for solid fuels, where the lifetime may be limited due to flow of particles out of the process together with ash as well as deactivation due to ash–oxygen carrier interaction. Cheaper natural materials, such as iron ore and ilmenite, are commonly employed for normal CLC with solid fuels. One interesting development with respect to CLOU in the last few years is the use of cheaper natural ores of Cu and Mn as oxygen carriers. The Cu content in natural ores is very low, and thus concentrating and refining is necessary prior to use in order to be applicable as oxygen carriers. Manganese ores, on the other hand, have Mn-oxide as the main phase. In addition, they often contain high fractions of Fe, Si, and Ca, all of which could actually be beneficial for the CLOU process, as discussed earlier. Arjmand et al. [106, 107] found a small CLOU effect for several manganese ores at 950–970 °C. Further, Sundqvist et al. evaluated a significant number of manganese ores both with respect to oxygen uncoupling and reactivity with methane and syngas using a small fluidized bed reactor [63, 108, 109]. Although the uncoupling rate was relatively low when using nitrogen as fluidizing gas, a small oxygen release was noted for most materials. Further, when evaluating the rate using devolatilized char in the fluidized bed, several ores showed high rates of oxygen transfer at temperatures of 950 °C and above [109]. Figure 7.10 shows the mass-based conversion, ω, as a function of time for one such manganese ore, Gloria (from South Africa). At 1000 °C almost 1 wt% of oxygen is removed from the particles by the underlying uncoupling reactions [109]. As the fluidized bed in the evaluations was fluidized with nitrogen, the only source of oxygen for wood combustion was from the oxygen carrier. The rate of release was judged to be of such a magnitude that it could certainly be of relevance in the fuel conversion process at high temperatures.

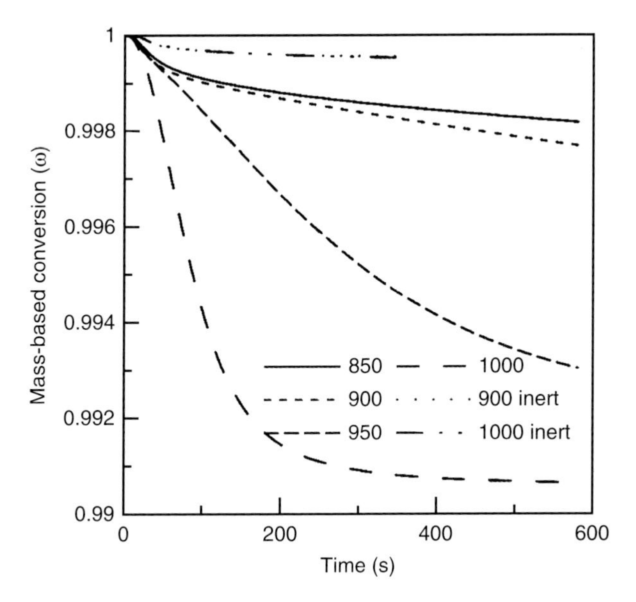

Figure 7.10 The mass-based conversion of oxygen carrier as a function of time when manganese ore (Gloria) reacts with devolatilized wood char in a batch fluidized bed reactor. The fluidizing gas was pure nitrogen. For comparison, curves are included for when the particles are exposed only to nitrogen gas. Source: Sundqvist and Mattisson 2016 [109]. Reproduced with permission of Elsevier.

7.4 Kinetics of Oxidation and Reduction of Oxygen Carriers in CLOU

The rate for the oxygen uncoupling, Reaction (7.6), is a function of an effective rate constant, the conversion level of the solids, and the partial pressure of O_2. In general, the rate of the oxygen release can be expressed as

$$r_{red} = k_r(p_{O_2,eq} - p_{O_2})^n f(\omega) \tag{7.15}$$

Here, k_r is an effective reaction rate constant, p_{O_2} is the bulk partial pressure of oxygen in the gas phase, and $f(\omega)$ is a function that describes the rate dependence on the conversion of the solid oxygen carrier. In a similar way, the oxidation reaction, reverse of Reaction (7.6), can be expressed as

$$r_{ox} = k_o(p_{O_2} - p_{O_2,eq})^n f(\omega) \tag{7.16}$$

From Equation (7.11) it is clear that the equilibrium partial pressure of O_2 varies exponentially with temperature. This means that not only the intrinsic rate constant is a function of temperature but also the driving force in the gas phase, given by $(p_{O_2,eq} - p_{O_2})$ in Equation (7.15). To generalize, research around kinetics has focused on determining (i) the activation energy and pre-exponential factor (from Arrhenius equation) and (ii) the form of $f(\omega)$.Until 2013, several studies had been performed with respect to determination of kinetic parameters of the CLOU reactions, e.g. [80, 110–112]. There had also been several relevant studies related to the process known as chemical-looping air separation (CLAS), which utilizes the same underlying reactions as CLOU, e.g. [113]. The investigations up to this time were almost exclusively conducted on Cu-based oxygen carriers, with rate of reaction and kinetic parameters determined for both pure CuO and with CuO on several types of inert support materials. Adánez-Rubio et al. provided a review of the results of these investigations [43]. The methodology and conditions for kinetic determinations varied in a wide range, as did the calculated activation energies and the conversion form, $f(\omega)$. For instance, Arjmand et al. [110] introduced an excess of devolatilized wood char to a bed of freeze-granulated $CuO/MgAl_2O_4$ particles, and in this way were able to calculate the rate of decomposition, Reaction (7.6), in the temperature interval 850–900 °C. The decomposition reaction was modeled using the Avrami–Erofeev mechanism and the kinetic data were obtained using an Arrhenius expression. From the analysis, an overall activation energy of 139.3 kJ mol^{-1} was calculated for the decomposition reaction, Reaction (7.6).

Whitty and Clayton [80] performed decomposition experiments of CuO on SiO_2, TiO_2, and ZrO_2 in three types of reactor systems using N_2 and found that the rates of decomposition were similarly independent of reactor type, and also very similar for the three Cu-based oxygen carriers evaluated, with activation energies of 147–180 kJ mol^{-1} in the temperature range 800–1000 °C. However, it was also found that the activation energy was dependent on the temperature range investigated. A considerably higher activation energy of 281 kJ mol^{-1} was obtained by Sahir et al. [114] for the decomposition reaction of a freeze-granulated CuO/ZrO_2 oxygen carrier. Peterson et al. investigated the use of impregnated CuO on SiC support [112]. An activation energy of the uncoupling of 220 kJ mol^{-1} was determined in the temperature range 850–950 °C.

From Table 7.1 it is evident that there have been a number of newer investigations with respect to kinetics of Cu-based materials [35, 43]. Clayton and Whitty conducted a follow-up investigation on earlier investigations of CuO/TiO_2 and CuO/ZrO_2, and decoupled the reaction rate constant from the oxygen driving force [35]. In this way two separate activation energies were determined, i.e. one global and one kinetic. Adánez-Rubio et al. investigated the reduction reactions of a spray-dried oxygen carrier of $CuO/MgAl_2O_4$ using a TGA at different O_2 concentrations and at varying temperatures [43]. A combined Langmuir–Hinshelwood and nucleation model was employed to describe the surface and bulk reactions, and an activation energy of 270 kJ mol^{-1} was determined.

Most of the investigations mentioned above also evaluated the kinetics of the oxidation reaction, i.e. reverse of Reaction (7.6). From Equation (7.16) it is clear that a higher temperature would reduce the driving force for the reaction, i.e. $p_{O_2} - p_{O_2,eq}$, and thus off-set the expected kinetic increase (given by the reaction rate constant k_o in Equation (7.16)). Whitty and Clayton investigated the rate of oxidation for an oxygen carrier of CuO/ZrO_2 as a function of temperature and oxygen driving force. Although an increase in oxidation was seen when the oxygen partial pressure was increased, the increase in rate was not as high as expected, indicating that there were other mechanisms restricting oxidation, which is an observation in conformity with previous studies [80]. An activation energy of the oxidation of 202 kJ mol^{-1} was found for this oxygen carrier. A considerably lower activation energy was found in a later study of $CuO/MgAl_2O_4$, where an activation energy of 30 kJ mol^{-1} was determined from experiments in a TGA [43].

Studies on the kinetics of the CLOU reactions for other metal oxide systems, such as Mn-based, are few. Arjmand et al. [115], used a batch fluidized bed reactor to determine the rate of oxygen release for a series of perovskite materials of general formula $Ca_xLa_{1-x}Mn_{1-y}M_yO_3$. The maximum rate of uncoupling was approximately 0.0003–0.0004 kg O_2/(kg OCs) at 950 °C, which is about 15 times slower than for the earlier investigated Cu-based oxygen carrier at 900 °C [110]. A similar perovskite material of $CaMn_{0.775}Ti_{0.125}Mg_{0.1}O_3$ was investigated by de Diego et al. using a TGA with different fuel components [92]. Both the CLC and CLOU reaction kinetics were determined and modeled using a shrinking-core type of mechanism. With respect to other Mn-combined oxides, the kinetic investigations are in principle non-existent. In one such study directly related to the CLOU concept, Arjmand et al. determined the apparent kinetics of a few oxygen carriers of Mn_2O_3/SiO_2 [27]. Three spray-dried oxygen carriers were exposed to wood char at 975–1100 °C, and the kinetics of release were determined from the measured outlet CO_2 concentration. Activation energies of between 229 and 300 kJ mol^{-1} were calculated for the oxygen release. At temperatures above 1050 °C, rates were very high, translating into very low solids inventory in the fuel reactor, i.e. <50 kg/MW_{th}.

7.5 Conclusions

CLOU is a relatively new and promising variation of normal chemical-looping. The difference between CLC and CLOU is the mechanism for oxidation of the fuel in the fuel reactor. In CLOU the oxygen carrier releases oxygen in the gas phase, which then reacts with the fuel through normal combustion, in contrast to CLC, where the oxygen carrier reacts directly or indirectly with the fuel through a non-catalytic gas–solid reaction. This chapter surveys the development of CLOU materials and necessary requirements of these, and provides an overview of the experimental investigations that have been performed in the last couple of years. Oxygen carrier development has to a large extent focused on Cu-based material, although there has been an increase in the development of combined oxides, specifically oxides based on Mn. The process has been tested in continuous working CLOU reactors for gaseous, liquid, and solid fuels, with very promising results with respect to rates of fuel conversion. Although it is likely that the biggest advantages with CLOU can be obtained for solid fuels, most experiments have been performed with gaseous fuels. Here, oxygen carriers with CLOU properties have been scaled-up to multi-ton production and investigated at MW scale.

Acknowledgment

The authors wish to acknowledge Chalmers Area of Advance for their financial support of this work.

References

1 Lewis, W.K. and Gilliland, E.R. (1954). Production of pure carbon dioxide. US Patent 2,665,972.

2 Lewis, W.K., Gilliland, E.R., and Sweeney, M.P. (1951). Gasification of carbon: metal oxides in a fluidized powder bed. *Chem. Eng. Prog.* 47: 251–256.

3 Ishida, M., Zheng, D., and Akehata, T. (1987). Evaluation of a chemical-looping-combustion power-generation system by graphic exergy analysis. *Energy* 12: 147–154.

4 Ishida, M. and Jin, H. (1994). A novel combustor based on chemical-looping reactions and its reaction kinetics. *J. Chem. Eng. Jpn.* 27: 296–301.

5 Ishida, M. and Jin, H. (1996). A novel chemical-looping combustor without NOx formation. *Ind. Eng. Chem. Res.* 35: 2469–2472.

6 Ishida, M., Jin, H., and Okamoto, T. (1996). A fundamental study of a new kind of medium material for chemical-looping combustion. *Energy Fuels* 10: 958–963.

7 Abanades, J., Arias, B., Lyngfelt, A. et al. (2015). Emerging CO_2 capture systems. *Int. J. Greenhouse Gas Control* 40: 126–166.

8 Berdugo Vilches, T., Lind, F., Rydén, M., and Thunman, H. (2016). Experience of more than 1000 h of operation with oxygen carriers and solid biomass at large scale. 4th International Conference on Chemical Looping, Nanjing, China.

9 Adánez, J., Abad, A., Garcia-Labiano, F. et al. (2011). Progress in chemical-looping combustion and reforming technologies. *Prog. Energy Combust. Sci.* 38: 215–282.

10 Luo, S., Zeng, L., and Fan, L. (2015). Chemical looping technology: oxygen carrier characteristics. *Annu. Rev. Chem. Biomol. Eng.* 6: 53–75.

11 Lyngfelt, A., Mattisson, T., Linderholm, C., and Rydén, M. (2016). Chemical-looping combustion of solid fuels – What is needed to reach full scale? 4th International Conference on Chemical Looping, Nanjing.

12 Kolbitsch, P., Bolhar-Nordenkampf, J., Pröll, T., and Hofbauer, H. (2009). Comparison of two Ni-based oxygen carriers for chemical looping combustion of natural gas in 140 kW continuous looping operation. *Ind. Eng. Chem. Res.* 48: 5542–5547.

13 de Diego, L., Serrano, A., García-Labiano, F. et al. (2016). Bioethanol combustion with CO_2 capture in a 1 kW_{th} chemical looping combustion prototype: suitability of the oxygen carrier. *Chem. Eng. J.* 283: 1405–1413.

14 Moldenhauer, P., Rydén, M., Mattisson, T. et al. (2014). Chemical-looping combustion with fuel oil in a 10 kW pilot plant. *Energy Fuels* 28: 5978–5987.

15 Mattisson, T., Johansson, M., and Lyngfelt, A. (2006). CO_2 capture from coal combustion using chemical-looping combustion – reactivity investigation of Fe, Ni and Mn based oxygen carriers using syngas. The Clearwater Coal Conference, Clearwater, FL.

16 Mattisson, T., García-Labiano, F., Kronberger, B. et al. (2007). Chemical-looping combustion using syngas as fuel. *Int. J. Greenhouse Gas Control* 1: 158–169.
17 Berguerand, N. and Lyngfelt, A. (2008). Design and operation of a 10 kW chemical-looping combustor for solid fuels – testing with south African coal. *Fuel* 87: 2713–2726.
18 Leion, H., Mattisson, T., and Lyngfelt, A. (2007). The use of petroleum coke as fuel in chemical-looping combustion. *Fuel* 86: 1947–1958.
19 Lyon, R. and Cole, J. (2000). Unmixed combustion: an alternative to fire. *Combust. Flame* 121: 249–261.
20 Scott, S., Dennis, J., Hayhurst, A., and Brown, T. (2006). In situ gasification solid fuel and CO_2 separation using chemical looping. *AIChE J.* 52: 3325–3328.
21 Dennis, J., Scott, S., and Hayhurst, A. (2006). In situ gasification of coal using steam with chemical looping: a technique for isolating CO_2 from burning solid fuel. *J. Energy Inst.* 79: 187–190.
22 Imtiaz, Q., Hosseini, D., and Muller, C. (2013). Review of oxygen carriers for chemical looping with oxygen uncoupling (CLOU): thermodynamics, material development, and synthesis. *Energy Technol.* 1: 633–647.
23 Mattisson, T. (2013, 2013). Materials for chemical-looping with oxygen uncoupling. *ISRN Chem. Eng.* 1–19. doi: 10.1155/2013/526375.
24 Jerndal, E., Mattisson, T., and Lyngfelt, A. (2006). Thermal analysis of chemical-looping combustion. *Chem. Eng. Res. Des.* 84: 795–806.
25 Mattisson, T., Leion, H., and Lyngfelt, A. (2009). Chemical-looping oxygen uncoupling using CuO/ZrO_2 with petroleum coke. *Fuel* 88: 683–690.
26 Mattisson, T., Lyngfelt, A., and Leion, H. (2009). Chemical-looping oxygen uncoupling for combustion of solid fuels. *Int. J. Greenhouse Gas Control* 3: 11–19.
27 Arjmand, M., Rydén, M., Leion, H. et al. (2014). Sulfur tolerance and rate of oxygen release of combined Mn–Si oxygen carriers in chemical-looping with oxygen uncoupling (CLOU). *Ind. Eng. Chem. Res.* 53: 19488–19497.
28 Adánez-Rubio, I., Abad, A., Gayan, P. et al. (2014). The fate of sulphur in the Cu-based chemical looping with oxygen uncoupling (CLOU) process. *Appl. Energy* 113: 1855–1862.
29 Hallberg, P., Källén, M., Jing, D. et al. (2014, 2014). Experimental investigation of $CaMnO_{3-\delta}$ based oxygen carriers used in continuous chemical-looping combustion. *Int. J. Chem. Eng.* 1–9. doi: 10.1155/2014/412517.
30 Jing, D., Arjmand, M., Mattisson, T. et al. (2014). Examination of oxygen uncoupling behaviour and reactivity towards methane for manganese silicate oxygen carriers in chemical-looping combustion. *Int. J. Greenhouse Gas Control* 29: 70–81.
31 Adánez-Rubio, I., Abad, A., Gayán, P. et al. (2014). Biomass combustion with CO_2 capture by chemical looping with oxygen uncoupling (CLOU). *Fuel Process. Technol.* 124: 104–114.

32 Ksepko, E., Sciazko, M., and Babinski, P. (2014). Studies on the redox reaction kinetics of Fe_2O_3–CuO/Al_2O_3 and Fe_2O_3/TiO_2 oxygen carriers. *Appl. Energy* 115: 374–383.

33 Arjmand, M., Kooiman, R.F., Rydén, M. et al. (2014). Sulfur tolerance of $Ca_xMn_{1-y}M_yO_{3-\delta}$ (M = Mg, Ti) perovskite-type oxygen carriers in chemical-looping with oxygen uncoupling (CLOU). *Energy Fuels* 28: 1312–1324.

34 Clayton, C., Sohn, H., and Whitty, K. (2014). Oxidation kinetics of Cu_2O in oxygen carriers for chemical looping with oxygen uncoupling. *Ind. Eng. Chem Res.* 53: 2976–2986.

35 Clayton, C. and Whitty, K. (2014). Measurement and modeling of decomposition kinetics for copper oxide-based chemical looping with oxygen uncoupling. *Appl. Energy* 116: 416–423.

36 Zhao, H., Wand, K., Fang, Y. et al. (2014). Characterization of natural copper ore as oxygen carrier in chemical-looping with oxygen uncoupling of anthracite. *Int. J. Greenhouse Gas Control* 22: 154–164.

37 Imtiaz, Q., Kierzkowska, A., Broda, M., and Muller, C. (2012). Synthesis and performance of Al_2O_3, CeO_2 or $MgAl_2O_4$-supported CuO-based materials for chemical looping with oxygen uncoupling. 2nd International Conference on Chemical Looping, Darmstadt, Germany.

38 Rydén, M., Jing, D., Källén, M. et al. (2014). CuO-based oxygen-carrier particles for chemical-looping with oxygen uncoupling-experiments in batch reactor and in continuous operation. *Ind. Eng. Chem. Res.* 53: 6255–6267.

39 Azimi, G., Leion, H., Mattisson, T. et al. (2014). Mn–Fe oxides with support of $MgAl_2O_4$, CeO_2, ZrO_2 and Y_2O_3–ZrO_2 for chemical-looping combustion and chemical-looping oxygen uncoupling. *Ind. Eng. Chem Res.* 53: 10358–10365.

40 Pour, N., Azimi, G., Leion, H. et al. (2014). Investigation of manganese-iron oxide materials based on manganese ores as oxygen carriers for chemical looping with oxygen uncoupling (CLOU). *Energy Technol.* 2: 469–479.

41 Shafiefarhood, A., Stewart, A., and Li, F. (2015). Iron-containing mixed-oxide composites as oxygen carriers for chemical looping with oxygen uncoupling (CLOU). *Fuel* 139: 1–10.

42 Imtiaz, Q., Kurlov, A., Rupp, J., and Muller, C. (2015). Highly efficient oxygen-storage material with intrinsic coke resistance for chemical looping combustion-based CO_2 capture. *ChemSusChem* 8: 2055–2065.

43 Adánez-Rubio, I., Gayan, P., Abad, A. et al. (2014). Kinetic analysis of a Cu-based oxygen carrier: relevance of temperature and oxygen partial pressure on reduction and oxidation reactions rates in chemical looping with oxygen uncoupling (CLOU). *Chem. Eng. J.* 256: 69–84.

44 Arjmand, M., Frick, V., Rydén, M. et al. (2015). Screening of combined Mn–Fe–Si oxygen carriers for chemical looping with oxygen uncoupling (CLOU). *Energy Fuels* 29: 1868–1880.

45 Wang, K., Yu, Q., Qin, Q., and Zuo, Z. (2015). Analysis of oxygen releasing rate of Cu-based oxygen carrier in N_2–O_2 atmosphere. *J. Therm. Anal. Calorim.* 119: 2221–2227.

46 Mayer, K., Penthor, S., Pröll, T., and Hofbauer, H. (2015). The different demands of oxygen carriers on the reactor system of a CLC plant – results of oxygen carrier testing in a 120 kW_{th} pilot plant. *Appl. Energy* 157: 323–329.

47 Mei, D., Abad, A., Zhao, H., and Adánez, J. (2015). Characterization of a sol-gel derived $CuO/CuAl_2O_4$ oxygen carrier for chemical looping combustion (CLC) of gaseous fuels: relevance of gas–solid and oxygen uncoupling reactions. *Fuel Process. Technol.* 133: 210–219.

48 Galinsky, N., Mishra, A., Zhang, J., and Li, F. (2015). $Ca_{1-x}A_xMnO_3$ (a = Sr and Ba) perovskite based oxygen carriers for chemical looping with oxygen uncoupling (CLOU). *Appl. Energy* 157: 358–367.

49 Azimi, G., Mattisson, T., Leion, H. et al. (2015). Comprehensive study of Mn-Fe-Al oxygen-carriers for chemical-looping with oxygen uncoupling (CLOU). *Int. J. Greenhouse Gas Control* 34: 12–24.

50 Zhang, Y., Zhao, H., Guo, L., and Zheng, C. (2015). Decomposition mechanisms of Cu-based oxygen carriers for chemical looping with oxygen uncoupling based on density functional theory. *Combust. Flame* 162: 1265–1274.

51 Galinsky, N., Sendi, M., Bowers, L., and Li, F. (2016). $CaMn_{1-x}B_xO_{3-\delta}$ (B = Al, V, Fe, co, and Ni) perovskite based oxygen carriers for chemical looping with oxygen uncoupling. *Appl. Energy* 174: 80–87.

52 Hanning, M., Frick, V., Mattisson, T. et al. (2016). Performance of combined manganese-silicon oxygen carriers and effects of including titanium. *Energy Fuels* 30: 1171–1182.

53 Schmitz, M. and Linderholm, C. (2016). Performance of calcium manganate as oxygen carrier in chemical looping combustion of biochar in a 10 kW pilot. *Appl. Energy* 169: 729–737.

54 Ku, Y., Shiu, S., Liu, Y. et al. (2017). Liquid sintering behavior of Cu-based oxygen carriers for chemical looping process. *Catal. Commun.* 92: 70–74.

55 Mattisson, T., Linderholm, C., Jerndal, E., and Lyngfelt, A. (2016). Enhanced performance of manganese ore as oxygen carrier for chemical-looping with oxygen uncoupling (CLOU) by combination with $Ca(OH)_2$ through spray-drying. *J. Environ. Chem. Eng.* 4: 3707–3717.

56 Pishahang, M., Larring, Y., Sunding, M. et al. (2016). Performance of perovskite-type oxides as oxygen-carrier materials for chemical looping combustion in the presence of H_2S. *Energy Technol.* 4: 1305–1316.

57 Mendiara, T., Adánez-Rubio, I., Gayan, P. et al. (2016). Process comparison for biomass combustion: in situ gasification chemical looping combustion (iG-CLC) versus chemical looping with oxygen uncoupling (CLOU). *Energy Technol.* 4: 1130–1136.

58 Zhao, H., Zhang, Y., Wei, Y., and Gui, J. (2017). Understand CuO-support interaction in Cu-based oxygen carriers at a microcosmic level. *Proc. Combust. Inst.* 36: 4069–4077.

59 Abian, M., Abad, A., Izquierdo, M. et al. (2017). Titanium substituted manganese-ferrite as an oxygen carrier with titanium substituted manganese-ferrite for chemical looping combustion of solid fuels. *Fuel* 195: 38–48.

60 Mungse, P., Saravanan, G., Nishibori, M. et al. (2017). Solvent-free, improved synthesis of pure bixbyite phase of iron and manganese mixed oxides as low-cost, potential oxygen carrier for chemical looping with oxygen uncoupling. *Pure Appl. Chem.* 89: 511–521.

61 Perez-Vega, R., Abad, A., Gayan, P. et al. (2017). Development of $(Mn_{0.77}Fe_{0.23})_2O_3$ particles as an oxygen carrier for coal combustion with CO_2 capture via in-situ gasification chemical-looping combustion (iG-CLC) aided by oxygen uncoupling (CLOU). *Fuel Process. Technol.* 164: 69–79.

62 Tian, X., Wang, K., Zhao, H., and Su, M. (2017). Chemical looping with oxygen uncoupling of high-sulfur coal using copper ore as oxygen carrier. *Proc. Combust. Inst.* 36: 3381–3388.

63 Sundqvist, S., Khalilian, N., Leion, H. et al. (2017). Manganese ores as oxygen carriers for chemical-looping combustion (CLC) and chemical-looping with oxygen uncoupling (CLOU). *J. Environ. Chem. Eng.* 5: 2552–2563.

64 Azimi, G. (2014). *Chemical-Looping Combustion and Chemical-Looping with Oxygen Uncoupling – Use of Combined Manganese and Iron Oxides for Oxygen Transfer*. Göteborg: Department of Chemical and Biological Engineering, Chalmers University of Technology.

65 Rydén, M., Leion, H., Mattisson, T., and Lyngfelt, A. (2013). Combined oxides as oxygen-carrier material for chemical-looping with oxygen uncoupling. *Appl. Energy* 113: 1924–1932.

66 Azimi, G., Leion, H., Mattisson, T., and Lyngfelt, A. (2011). Chemical-looping with oxygen uncoupling using combined Mn–Fe oxides-testing in batch fluidized bed. *Energy Procedia* 4: 370–377.

67 Mattisson, T., Jing, D., Lyngfelt, A., and Rydén, M. (2016). Experimental investigation of binary and ternary combined manganese oxides for chemical-looping with oxygen uncoupling (CLOU). *Fuel* 164: 228–236.

68 Shulman, A., Cleverstam, E., Mattisson, T., and Lyngfelt, A. (2009). Manganese/iron, manganese/nickel, and manganese/silicon oxides used in chemical-looping with oxygen uncoupling (CLOU) for combustion with methane. *Energy Fuels* 24: 5269–5275.

69 Shulman, A., Cleverstam, E., Mattisson, T., and Lyngfelt, A. (2011). Chemical – looping with oxygen uncoupling using Mn/Mg-based oxygen carriers – oxygen release and reactivity with methane. *Fuel* 90: 941–950.

70 Readman, J., Olafsen, A., Larring, Y., and Blom, R. (2005). $La_{0.8}Sr_{0.2}Fe_{0.8}O_{3-\delta}$ as a potential oxygen carrier in a chemical looping type reactor, an in-situ powder X-ray diffraction study. *J. Mater. Chem.* 15: 1931–1937.

71 Rydén, M., Lyngfelt, A., and Mattisson, T. (2011). $CaMn_{0.875}Ti_{0.125}O_3$ as oxygen carrier for chemical-looping combustion with oxygen uncoupling (CLOU)-experiments in a continuously operating fluidized-bed reactor system. *Int. J. Greenhouse Gas Control* 5: 356–366.

72 Rydén, M., Lyngfelt, A., Mattisson, T. et al. (2008). Novel oxygen-carrier materials for chemical-looping combustion and chemical-looping reforming; $La_xSr_{1-x}Fe_yCo_{1-y}O_{3-\delta}$ perovskites and mixed-metal oxides of NiO, Fe_2O_3 and Mn_3O_4. *Int. J. Greenhouse Gas Control* 2: 21–36.

73 Sarshar, Z., Kleitz, F., and Kaliaguine, S. (2011). Novel oxygen carriers for chemical looping combustion: $La_{1-x}Ce_xBO_3$ (B = Co, Mn) perovskites synthesized by reactive grinding and nanocasting. *Energy Environ. Sci.* 4: 4258–4269.

74 Sarshar, Z., Sun, Z., Zhao, D., and Kaliaguine, S. (2012). Development of sinter-resistant core–shell $LaMn_xFe_{1-x}O_3$@$mSiO_2$ oxygen carriers for chemical looping combustion. *Energy Fuels* 26: 3091–3103.

75 Rörmark, L., Mörch, A., Wiik, K. et al. (2001). Enthapies of oxidation of $CaMnO_{3-\delta}$, $Ca_2MnO_{4-\delta}$ and $SrMnO_{3-\delta}$ – deduced redox properties. *Chem. Mater.* 13: 4005–4013.

76 Rörmark, L., Wiik, K., Stölen, S., and Grande, T. (2002). Oxygen stoichiometry and structural properties of $La_{1-x}A_xMnO_{3-\delta}$ (A = Ca or Sr and $0 < x < 1$). *J. Mater. Chem.* 12: 1058–1067.

77 Wen, Y., Li, Z., Xu, L., and Cai, N. (2012). Experimental study of natural Cu ore particles as oxygen carriers in chemical looping with oxygen uncoupling (CLOU). *Energy Fuels* 26: 3919–3927.

78 Hedayati, A., Azad, A.-M., Rydén, M. et al. (2012). Evaluation of novel ceria-supported metal oxides as oxygen carriers for chemical-looping combustion. *Ind. Eng. Chem. Res.* 51: 12796–12806.

79 Jing, D., Mattisson, T., Rydén, M. et al. (2013). Innovative oxygen carrier materials for chemical-looping combustion. *Energy Procedia* 37: 645–653.

80 Whitty, K., and Clayton, C. (2012). Measurement and modeling of kinetics for copper-based chemical looping with oxygen uncoupling. 2nd International Conference on Chemical Looping, Darmstadt, Germany.

81 Adánez-Rubio, I., Abad, A., Gayan, P. et al. (2013). Performance of CLOU process in the combustion of different types of coal with CO_2 capture. *Int. J. Greenhouse Gas Control* 12: 430–440.

82 Moldenhauer, P., Rydén, M., Mattisson, T., and Lyngfelt, A. (2012). Chemical-looping combustion and chemical-looping reforming of kerosene in a circulating fluidized-bed 300 W laboratory reactor. *Int. J. Greenhouse Gas Control* 9: 1–9.

83 Ksepko, E., Talik, E., and Figa, J. (2008). Preparation and characterization of $Sr(Mn_{1-x}Ni_x)O_3$ solid solution in relation to their use in chemical looping oxygen transfer. 25th Annual International Pittsburgh Coal Conference, Pittsburgh, PA.

84 Abad, A., Adánez-Rubio, I., Gayan, P. et al. (2012). Demonstration of chemical-looping with oxygen uncoupling (CLOU) process in a 1.5 kW_{th} continuously operating unit using a Cu-based oxygen-carrier. *Int. J. Greenhouse Gas Control* 6: 189–200.

85 Zafar, Q., Abad, A., Mattisson, T. et al. (2007). Reduction and oxidation kinetics of Mn_3O_4/mg-ZrO_2 oxygen carrier particles for chemical-looping combustion. *Chem. Eng. Sci.* 62: 6556–6567.

86 Azimi, G., Rydén, M., Leion, H. et al. (2013). $(Mn_yFe_{1-y})O_x$ combined oxides as oxygen carrier for chemical-looping with oxygen uncoupling (CLOU). *AIChE J.* 59: 582–588.

87 Jing, D., Hermans, E., Leion, H. et al. (2012). Manganese-silica combined oxides as oxygen carrier for chemical-looping combustion. 2nd International Conference on Chemical Looping, Darmstadt, Germany.
88 Azimi, G., Leion, H., Rydén, M. et al. (2013). Investigation of different Mn–Fe oxides as oxygen carrier for chemical-looping with oxygen uncoupling (CLOU). *Energy Fuels* 27: 367–377.
89 Frick, V., Rydén, M., Leion, H. et al. (2015). Screening of supported and unsupported Mn–Si oxygen carriers for chemical-looping with oxygen uncoupling (CLOU). *Energy* 93: 544–554.
90 Jing, D., Mattisson, T., Leion, H. et al. (2013). Examination of perovskite structure $CaMnO_{3-\delta}$ with MgO addition as oxygen carrier for chemical looping with oxygen uncoupling using methane and syngas. *Int. J. Chem. Eng.* 2013: 1–16. doi: 10.1155/2013/679560.
91 Hallberg, P., Jing, D., Rydén, M. et al. (2013). Chemical looping combustion and chemical looping with oxygen uncoupling experiments in a batch reactor using spray-dried $CaMn_{1-x}M_xO_{3-\delta}$ (M = Ti, Fe, mg) particles as oxygen carriers. *Energy Fuels* 27: 1473–1481.
92 de Diego, L., Abad, A., Cabello, A. et al. (2014). Reduction and oxidation kinetics of a $CaMn_{0.9}Mg_{0.1}O_{3-\delta}$ oxygen carrier for chemical-looping combustion. *Ind. Eng. Chem. Res.* 53: 87–103.
93 Leion, H., Larring, Y., Bakken, E. et al. (2009). Use of $CaMn_{0.875}Ti_{0.125}O_3$ as oxygen carrier in chemical-looping with oxygen uncoupling. *Energy Fuels* 23: 5276–5283.
94 Källén, M., Rydén, M., Dueso, C. et al. (2013). $CaMn_{0.9}Mg_{0.1}O_{3-\delta}$ as oxygen carrier in a gas-fired 10 kW_{th} chemical-looping combustion unit. *Ind. Eng. Chem. Res.* 52: 6923–6932.
95 Linderholm, C., Abad, A., Mattisson, T., and Lyngfelt, A. (2008). 160 Hours of chemical-looping combustion in a 10 kW reactor system with a NiO-based oxygen carrier. *Int. J. Greenhouse Gas Control* 2: 520–530.
96 Linderholm, C., Mattisson, T., and Lyngfelt, A. (2009). Long-term integrity testing of spray-dried particles in a 10-kW chemical-looping combustor using natural gas as fuel. *Fuel* 88: 2083–2096.
97 Mattisson, T., Adánez, J., Mayer, K. et al. (2014). Innovative oxygen carriers uplifting chemical-looping combustion. *Energy Procedia* 63: 113–130.
98 Jing, D., Jacobs, M., Hallberg, P. et al. (2016). Development of $CaMn_{0.775}Mg_{0.1}Ti_{0.125}O_{3-\delta}$ oxygen carriers produced from different Mn and Ti sources. *Mater. Des.* 89: 527–542.
99 Moldenhauer, P., Hallberg, P., Biermann, M. et al. (2017). Oxygen carrier development of calcium manganite-based materials with perovskite structure for chemical looping combustion of methane. 42nd International Technical Conference on Clean Energy, Clearwater, FL.
100 Ohlemuller, P., Reitz, M., Ströhle, J., and Epple, B. (2017). Operation of a 1 MW_{th} chemical looping pilot plant with natural gas. TCCS-9, Trondheim, Norway.
101 Schmitz, M., Linderholm, C.J., and Lyngfelt, A. (2017). Chemical looping combustion of four different solid fuels using a manganese-silicon-titanium oxygen carrier. *Int. J. Greenhouse Gas Control* 70: 88–96.

102 Fossdal, A., Bakken, E., Öye, B. et al. (2011). Study of inexpensive oxygen carriers for chemical looping combustion. *Int. J. Greenhouse Gas Control* 5: 483–488.

103 Lambert, A., Delquie, C., Clemencon, I. et al. (2009). Synthesis and characterization of bimetallic Fe/Mn oxides for chemical looping combustion. *Energy Procedia* 1: 375–381.

104 Azimi, G., Leion, H., Rydén, M. et al. (2012). Solid fuel conversion of iron manganese oxide as oxygen carrier for chemical-looping with oxygen uncoupling (CLOU). 2nd International Conference on Chemical Looping, Darmstadt, Germany.

105 Rydén, M., Lyngfelt, A., and Mattisson, T. (2011). Combined manganese/iron oxides as oxygen carrier for chemical-looping combustion with oxygen uncoupling (CLOU) in a circulating fluidized bed reactor system. *Energy Procedia* 4: 341–348.

106 Arjmand, M., Leion, H., Lyngfelt, A., and Mattisson, T. (2012). Use of manganese ore in chemical-looping combustion (CLC)-effect on steam gasification. *Int. J. Greenhouse Gas Control* 8: 56–60.

107 Arjmand, M., Leion, H., Mattisson, M., and Lyngfelt, A. (2012). Evaluation of different manganese ores as oxygen carrier in chemical-looping combustion (CLC) for solid fuels. 2nd International Conference on Chemical Looping, Darmstadt, Germany.

108 Sundqvist, S., Arjmand, M., Mattisson, T. et al. (2016). Screening of different manganese ores for chemical-looping combustion (CLC) and chemical-looping with oxygen uncoupling (CLOU). *Int. J. Greenhouse Gas Control* 43: 179–188.

109 Sundqvist, S. and Mattisson, T. (2016). Manganese ore screening and solid fuel testing. 4th International Conference on Chemical Looping, Nanjing, China.

110 Arjmand, M., Keller, M., Leion, H. et al. (2012). Oxygen release and oxidation rates of $MgAl_2O_4$-supported CuO oxygen carrier for chemical-looping with oxygen uncoupling (CLOU). *Energy Fuels* 26: 6528–6539.

111 Eyring, E., Konya, G., Lighty, J. et al. (2011). Chemical looping with copper oxide as carrier and coal as fuel. *Oil Gas Sci. Technol. – Rev. IFP* 66: 209–221.

112 Peterson, S., Konya, G., Clayton, C. et al. (2013). Characteristics and CLOU performance of a novel SiO_2-supported oxygen carrier prepared from CuO and B-SiC. *Energy Fuels* 27: 6040–6047.

113 Wang, K., Yu, Q., and Qin, Q. (2013). Reduction kinetics of Cu-based oxygen carriers for chemical-looping air separation. *Energy Fuels* 27: 5466–5474.

114 Sahir, A., Sohn, H., Leion, H., and Lighty, J. (2012). Rate analysis of chemical-looping with oxygen uncoupling (CLOU) for solid fuels. *Energy Fuels* 26: 4395–4404.

115 Arjmand, M., Hedayati, A., Azad, A.-M. et al. (2012). $Ca_xLa_{1-x}Mn_{1-y}M_yO_3$ (M = Fe, Ti, Mg, Cu) as oxygen carriers for chemical-looping with oxygen uncoupling (CLOU). *Energy Fuels* 27: 4097–4107.

8

Mixed Metal Oxide-Based Oxygen Carriers for Chemical Looping Applications

Fanxing Li[1], *Nathan Galinsky*[2], *and Arya Shafiefarhood*[1]

[1] *North Carolina State University, Department of Chemical Engineering, 911 Partners Way, Raleigh, NC, USA*
[2] *Oak Ridge Institute Science and Engineering and National Energy Technology Laboratory, 3610 Collins Ferry Road, Morgantown, WV, USA*

8.1 Overview

Early studies on oxygen carriers generally focus on monometallic oxides of nickel, iron, and copper. While oxides of cobalt and manganese were also considered, they do not possess suitable redox properties for chemical looping combustion (CLC) in their monometallic oxide forms [1–3]. It is also noted that although Ni, Cu, and Fe oxides are suitable chemical looping materials, they face challenges in terms of toxicity (Ni), high cost (Ni, Cu), defluidization (Cu), or relatively low activity (Fe). Although strategic selection of support and synthesis strategies can significantly enhance the performance of these oxygen carriers, the limited number of monometallic oxides suitable for chemical looping greatly restricts the material design space and their potential applications. The "re-discovery" and investigation of mixed metal oxides as chemical looping oxygen carriers opened up an exciting area for chemical looping research by significantly expanding the design space for primary (oxygen carrying) oxide selection and applications. For instance, mixed oxides can offer high tunability in terms of their bulk thermodynamic, structural, and surface properties. Such properties can be valuable towards rationalization of the design of oxygen carriers for various chemical looping applications. Figure 8.1 summarizes year-over-year publication counts on oxygen carrier-related publications. Besides the upward trend in the overall publication counts, it is worth noting that the number of studies on mixed metal oxides has steadily increased from virtually nonexistent prior to 2008. With this in mind, the current chapter focuses on (first-row) transition metal-containing mixed metal oxides and their applications in both CLC and selective oxidation applications. It should be noted that mixed metal oxides are covered under specific applications in Chapters 6 and 7. Chapter 6 discusses iron and copper oxide systems for CLC whereas Chapter 7 covers some Mn-containing mixed metal oxide oxygen carriers for chemical looping with oxygen uncoupling (CLOU). This chapter aims to provide a general overview of various exciting opportunities enabled by mixed metal oxides. It will also

Handbook of Chemical Looping Technology, First Edition. Edited by Ronald W. Breault.

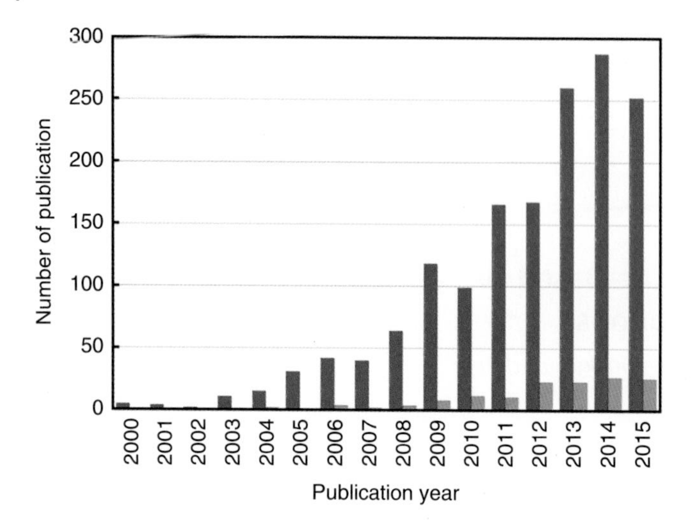

Figure 8.1 Year-over-year chemical looping publication counts on the general chemical looping topic (blue bars) and mixed oxide-based oxygen carriers (yellow bars). (Source: SciFinder database with keywords "chemical looping" and "chemical looping" + "mixed oxide," respectively.)

discuss in extensive detail with respect to the development, status, and the potential of mixed oxides as tunable oxygen carriers or redox catalysts for various chemical looping applications including CLOU, in situ gasification chemical looping combustion (iG-CLC), and chemical looping partial oxidation (CLPOx) such as chemical looping reforming (CLR). Furthermore, this chapter will also discuss the importance of bulk and surface properties for oxygen carrier design in selective oxidation applications. Future research directions of mixed oxide-based oxygen carriers will also be discussed.

Chemical looping processes rely on the redox properties of transition metal oxides for the oxidation of carbon-containing fuels. Although the rates of carbonaceous fuel oxidation reactions and product distributions are largely affected by kinetic parameters such as surface catalytic activity and conductivity of bulk lattice oxygen (O^{2-}) and electrons, thermodynamic analyses can be very effective in providing general guidance on oxygen carrier selection. From a thermodynamic standpoint, chemical looping processes involve indirect and cyclic exchange of oxygen between a fuel and one or more oxidants such as air and/or steam. Although such exchange is facilitated by the oxide-based oxygen carrier, a simple thermodynamic analysis can be carried out by investigating the independent reactions below, using methane and air as an example.

Oxygen carrier reduction step	$2CH_4 + O_2 \leftrightarrow 2CO + 4H_2$
	$2CO \leftrightarrow C + CO_2$
	$2CO + O_2 \leftrightarrow 2CO_2$
	$2H_2 + O_2 \leftrightarrow 2H_2O$
	$2MeO_x \leftrightarrow 2MeO_{x-1} + O_2$
Oxygen carrier regeneration step	$2MeO_{x-1} + O_2 \leftrightarrow MeO_x$

Here, the metal oxide is merely treated as an oxygen source and sink from a thermodynamic standpoint. As such, the equilibrium oxygen chemical potential or partial pressure (P_{O_2}) for the MeO_x/MeO_{x-1} redox pair directly affects the equilibrium product distributions from the fuel oxidation reactions. In addition, Equilibrium P_{O_2} also determines the regenerability of the reduced metal oxide. That is, the MeO_x/MeO_{x-1} pair with high equilibrium P_{O_2} (e.g. >0.05 atm) will be difficult to be reoxidized by air without significant amount of unconverted oxygen exiting the oxidizer or air reactor. The relationship between redox pair equilibrium P_{O_2} and the corresponding chemical looping applications can be illustrated on an Ellingham diagram (Figure 8.2). As can be seen, redox pairs with relatively high P_{O_2} can be suitable for CLOU applications due to the capacity for spontaneous O_2 release under chemical looping conditions. The upper bound P_{O_2} for CLOU material is set at 0.05 atm as required for reasonable conversion in the reoxidation step. A sample monometallic redox pair residing within this region is CuO/Cu_2O. The green region below the CLOU zone defines redox pairs with suitable properties for the iG-CLC applications. With 95% CO_2 yield defining

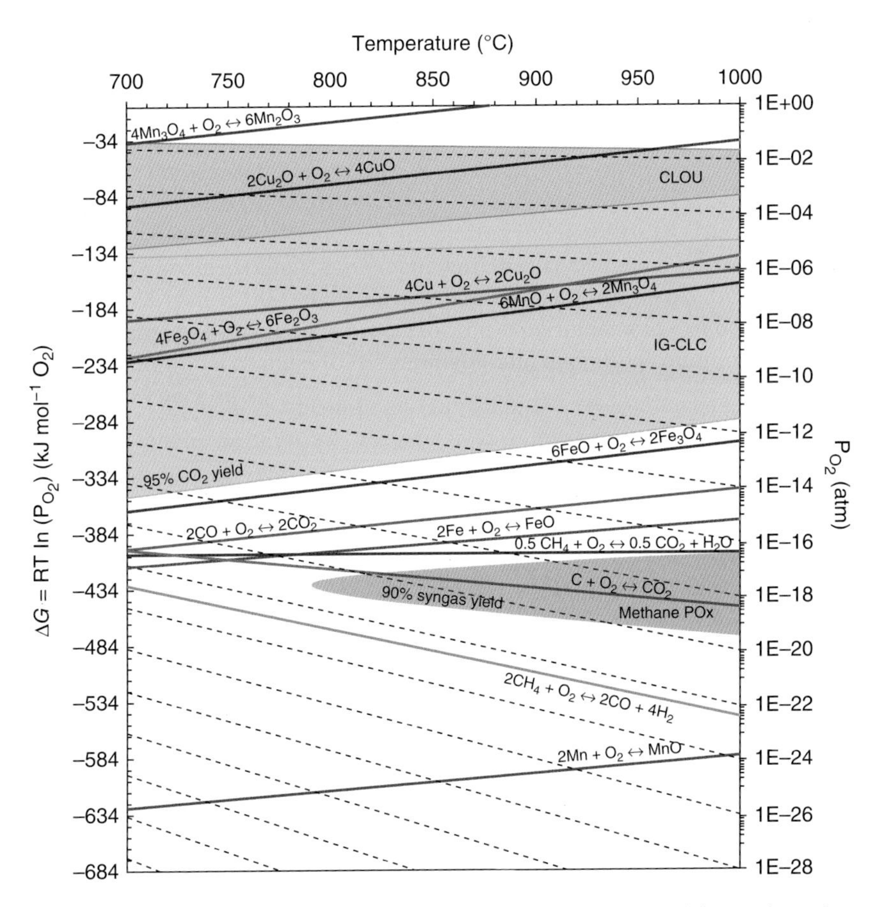

Figure 8.2 Oxygen carrier selection criteria based on equilibrium P_{O_2} and thermodynamic analysis.

the lower bound of the region, metal oxides with properties in this region can generate concentrated CO_2 from carbonaceous fuels. Sample redox pairs in this region include NiO/Ni and Fe_2O_3/Fe_3O_4. Compared to the CLOU and iG-CLC regions that generate combustion products, the purple region highlights an area where high syngas (CO and H_2) selectivity can be anticipated since the carbonaceous fuels will not be over- or under-oxidized provided that an oxygen carrier redox pair possesses equilibrium properties residing in such a region. It is worth noting that none of the known first row transition metal oxides possess redox properties defined by this partial oxidation (POx) zone. As will be illustrated in the following sections, redox properties of a number of mixed oxides can be tailored to this POx zone. This is in a stark contrast to the limited options for monometallic oxides. We also noted that the analysis above is made purely from a thermodynamic standpoint. One can certainly envision a metal oxide with high P_{O_2} exhibiting high syngas selectivity, provided that the oxide surface is tailored, kinetically or catalytically, to favor POx reactions. In the following sections, mixed oxides with properties suitable for CLOU, iG-CLC, and CLPOx are discussed, respectively.

8.2 Mixed Oxides for Chemical Looping with Oxygen Uncoupling (CLOU)

CLOU differs from traditional CLC primarily in terms of the way oxygen is delivered from the oxygen carrier to the fuel. In CLOU, the oxygen carrier spontaneously releases molecular oxygen under typical fuel reactor operating temperature and P_{O_2}. As such, CLOU involves both gaseous and lattice oxygen, leading to both homogeneous gas-phase reactions and heterogeneous gas–solids reactions. In comparison, molecular oxygen is not present in the fuel combustion and metal oxide reduction reactions in conventional CLC. CLOU is particularly advantageous for solid fuel combustion as gaseous oxygen can significantly enhance fuel combustion kinetics. In addition to the improved kinetics, the fuel reactor's heat management in CLOU can be simpler than that in conventional CLC since a high equilibrium P_{O_2} for the oxygen carrier leads to less endothermic or even exothermic metal oxide–fuel reactions. Despite these advantages, selection of CLOU oxygen carriers can be challenging due to the delicate requirements for oxygen carrier redox properties, as evidenced by the narrower CLOU region in Figure 8.2 compared to the (conventional) iG-CLC region. Figure 8.3 compares the redox properties of several commonly encountered monometallic redox pairs. With the exception of CuO/Cu_2O and Co_3O_4/CoO, none of the monometallic oxides possess desirable CLOU properties. Co_3O_4 is typically considered to be unsuitable for chemical looping due to toxicity concerns. Moreover, Co_3O_4 is challenging to reoxidize because of its high equilibrium P_{O_2}.

Although CuO has been extensively investigated for CLOU, it faces challenges such as sintering and defluidization, when reduced to metallic copper, and high cost. To address defluidization concerns, inert support materials with high sintering resistance are often mixed with (or to "support") CuO to improve

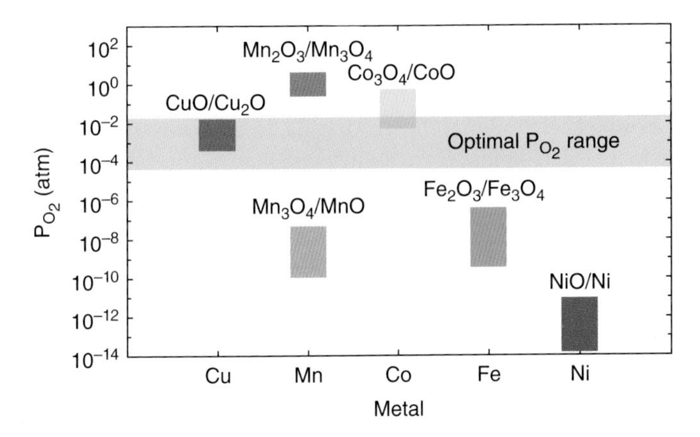

Figure 8.3 Thermodynamic properties of various monometallic redox pairs relative to the desired P_{O_2} range for CLOU (between 800 and 950 °C).

the mechanical stability of the oxygen carriers under CLOU operations. These supports include TiO_2, Al_2O_3, $MgAl_2O_4$, SiO_2, MgO, and ZrO_2 [4–7]. This limits the oxygen-carrying capacity of copper oxide-based oxygen carriers. Although CuO has a theoretical CLOU capacity of 10 wt%, the use of inert support, which represents at least 40 wt% of CuO-based oxygen carriers, significantly limits their oxygen-carrying capacity [8–11]. This can be improved upon through synthesis techniques, such as co-precipitation, that allow for high dispersion of the support and copper oxide phases in which up to 80 wt% CuO can be achieved [12]. However, scale-up is a concern in these types of synthesis procedures. Additionally, CuO tends to form solid solutions or mixed oxides with a variety of metals, leading to varying CLOU properties compared to its pure oxide forms. As such, mixed oxides containing CuO must balance the oxygen decomposition kinetics while improving the mechanical stability that pure CuO lacks. A few studies of mixed oxides utilizing copper involve the Cu—Mn system: Azad et al. investigated a Cu—Mn oxygen carrier containing a 1 : 1 molar ratio of Cu—Mn [13]. The authors showed that at 800 °C in pure N_2 the oxygen concentration was nearly 0.8%, which is significant at this relatively low temperature. $Cu_3Mn_3O_8$ was determined to be the primary phase of the oxygen carrier with $Cu_{0.1}Mn_{2.9}O_4$ as a minor component. Over the short term, no agglomeration was observed, and the oxidation reaction was fast, which is a concern with Mn-containing oxides. A separate study by Pour et al. investigated the effect of Cu:Mn ratio on the mixed oxide CLOU properties [14]. The amount of CuO ranged from 5 to 60 wt%. The authors only observed a spinel phase of $(Cu,Mn)_3O_4$ above 10 wt% CuO. The 5, 10, and 61 CuO wt% oxygen carriers performed the best, showing up to 1.4 wt% oxygen capacity. Additionally, complete conversion of the char for these oxygen carriers required a minimum ratio of oxygen carrier (OC)/char of approximately 30, which was half the ratio needed for the oxygen carriers containing 20 and 31 CuO wt%. The authors determined that, in general, oxygen carriers containing more Cu were less prone to attrition. Additionally, copper-containing ores have been investigated for their CLOU properties. More details on these systems were covered in Chapter 7.

Manganese oxides offer the advantage of being able to release their oxygen at lower temperatures than CuO. However, the oxidation rate from Mn_3O_4 to Mn_2O_3 is kinetically slow. This is due to the low reaction temperature required for Mn_3O_4 oxidation from a thermodynamic standpoint. In addition to the slow reoxidation kinetics, Mn_2O_3 also suffers from similar mechanical stability issues as copper oxides, such as agglomeration. Manganese forms solid solutions with several metals that can modify the redox thermodynamics, improve the oxidation kinetics, and/or improve the mechanical stability. Some of these systems include Mn—Fe—O [15–21], Mn—Si—O [16, 22–25], Mn—Mg—O [26–28], and Mn—Ni—O [1]. Besides synthetic mixed Mn oxides, manganese ores offer potential for CLOU. Sundqvist et al. investigated 11 different manganese ores for CLOU and conversion of syngas and methane [29]. The primary Mn-containing phase for these ores was Mn_3O_4, but also included Mn_2O_3 and bixbyite phases. Oxygen concentrations in the reactor varied between 0.2% and 0.85% for all the ores between 900 and 1000 °C. The ores achieved varying yields of CO_2 when methane was used as the fuel, with 90% being the maximum. When CO was used as a fuel, CO_2 yield varied between 50% and 90%. It should be noted that attrition, mechanical stability, and sulfidation issues are of concern for manganese ores. It was concluded that Slovakian and Metmin ores, which contain higher contents of Fe and/or Ca than other ores, have the highest O_2 release. Attrition, however, is a concern for these ores.

Perovskites represent a family of mixed oxides that offer highly tunable redox properties. Typical perovskites take the form of $ABO_{3-\delta}$ where A is a larger cation usually of an alkali or rare earth metal and B is typically a transition metal. With regard to Mn-based perovskites, $CaMnO_3$ offers potential as an oxygen carrier for CLOU. $CaMnO_3$ has lower oxygen uncoupling properties than copper oxides but offers lower costs as it can be synthesized using cheap manganese ores and calcium precursors. $CaMnO_3$ has been studied extensively as an oxygen carrier for both CLOU and CLC applications [30–42]. However, $CaMnO_3$ suffers from potential degradation with redox cycling. The result of the degradation is due to the formation of undesirable spinel ($CaMn_2O_4$) and Ruddlesden–Popper (Ca_2MnO_4) phases [43]. High sulfur-containing solid fuels create additional challenges for $CaMnO_3$ due to the potential formation of $CaSO_4$, which could also affect its long-term stability [32, 41]. To overcome this potential challenge of degradation during redox cycles, many researchers have investigated doping of A- and/or B-site metals into the perovskite structure to improve the oxygen carrier stability and performance. A-site metals added to the structure include La, Ba, and Sr while B-site dopants include Ti, Mg, and Fe. Most studies involving dopants use a trial-and-error approach in attempting to achieve long-term stability with gaseous and solid fuels.

Recent studies by Galinsky et al. offered an interesting strategy to tune the oxygen carrier's oxygen release properties through doping of the $CaMnO_3$ perovskite A- and B-sites. With regard to the A-site, the authors used Sr- and Ba as potential dopants to replace a portion of the Ca in the oxygen carrier [31]. The authors determined that through the addition of Ba, oxygen uncoupling properties were significantly reduced. In addition to the reduced oxygen uncoupling properties, Ba-doped $CaMnO_3$ additionally did not prevent decomposition into undesired phases at high temperature. In contrast, Sr is able to incorporate well into the $CaMnO_3$. Additionally, Sr-doped $CaMnO_3$ does not alter significantly the total amount of oxygen released; however, the oxygen release temperature is lowered by approximately 200 °C. Additionally, noticeable amounts of potential α-oxygen (chemisorbed) were observed. Sr-doped oxygen carriers showed long-term (100 cycles) and high-temperature resistance toward decomposition into undesired phases. As a follow-up study, the same authors investigated how different B-site metals affected the oxygen release properties and overall chemical stability of the $CaMnO_3$ oxygen carrier [35]. B-site dopants included Fe, Ni, Co, V, and Al. Fe exhibited the best incorporation into the perovskite phase where the other metals produced undesired secondary or ternary oxide phases. Doping of 5% of Fe into the B-site improved the oxygen carrier's high-temperature stability while improving the overall oxygen carrier CLOU activity.

Besides using perovskites as a primary CLOU oxygen carrier, a study by Shafiefarhood et al. used perovskites as supports with mixed oxide CLOU oxygen carriers based on Fe—Mn and Fe—Co oxides [44]. The authors used perovskite-based mixed conductive supports to enhance the oxygen anion/electron mobility through the catalyst. They investigated the addition of $La_{0.8}Sr_{0.2}M_{1-x}Fe_xO_{3-\delta}$ (M = Mn and Co) support in a similar B-site cation ratio found in the primary mixed oxide phase between Fe—Mn and Fe—Co. The addition of perovskite support improved the overall CLOU properties of the oxygen carrier by decreasing the initial decomposition temperature by up to 9%. Oxygen-carrying capacity for Fe—Co oxides more than doubled for the supported oxygen carrier. While the Fe—Mn supported oxides did not exhibit significant oxygen release under isothermal cyclic redox conditions, the activity for H_2 and CH_4 conversion was significantly improved compared to the unsupported oxygen carriers. The authors attribute the poor CLOU properties of the supported Fe—Mn to a redistribution of Mn and Fe ions in the perovskite phase whereas the Fe—Co supported oxygen carrier exhibited stability in terms of its composition. A summary of key CLOU mixed oxide oxygen carriers is shown in Table 8.1.

8.3 Mixed Oxides for iG-CLC

iG-CLC is an alternative chemical looping scheme for solid fuels conversion. In iG-CLC, solid fuels are gasified into syngas using steam or CO_2 as a gasifying agent. These reactions are shown in Reactions (8.1) and (8.2):

$$C + CO_2 \rightarrow 2CO \tag{8.1}$$

$$C + H_2O \rightarrow H_2 + CO \tag{8.2}$$

Subsequently, the syngas is further oxidized by the oxygen carriers' lattice oxygen into CO_2 and H_2O via Reactions (8.3) and (8.4).

$$CO + Me_xO_y \rightarrow CO_2 + Me_xO_{y-1} \tag{8.3}$$

$$H_2 + Me_xO_y \rightarrow H_2O + Me_xO_{y-1} \tag{8.4}$$

The net production of gasifying agents, i.e. CO_2 and H_2O, from Reactions (8.1)–(8.4) increases as the reactions proceed. Therefore, enhanced conversion

Table 8.1 Summary of key mixed oxide oxygen carriers for CLOU.

Oxygen carrier system	Main phases	Synthesis	Tested scale	Pros	Cons	References
Cu—Mn—O	$(Cu,Mn)_3O_4$	Slurry	Laboratory-scale FB	–Fast oxidation kinetics –Low temperature oxygen release –High solid fuel conversion	–No long term study –Mechanical stability over long term	[13, 14]
Mn—Fe—O	$(Mn_yFe_{1-y})O_x$	Spray drying	Laboratory-scale FB	–Better O_2 release than Fe-containing Mn ores	–Lack of detailed phase analysis –Mechanical stability	[17, 21]
Mn—Fe—Ti—O	$(Mn,Fe)_3O_4$, Fe_2TiO_4, Fe_2O_3, $MnTiO_3$, Mn_2O_3, Mn_3O_4	Solid-state reaction	TGA	–Detailed phase and composition analysis on oxygen carrier properties –Ferromagnetic materials would allow easy separation from coal ash –Good mechanical stability	–Oxidation reaction in some materials a concern	[15]
Mn—Si—O	Mn_7SiO_{12}, Mn_2O_3, SiO_2	Spray drying	Laboratory-scale FB	–High sulfur resistance –High temperature capability	–Mechanical stability –Long-term operation is a concern	[22–25]
Mn—Mg—O	Mg_2MnO_4	Freeze granulation	Laboratory-scale FB	–High fuel conversion	–Mechanical stability –Lower oxygen uncoupling properties	[26]
Mn Ores	Mn_3O_4, bixbyite, Mn_2O_3	N/A	Laboratory-scale FB	–Cheap, abundant –High gaseous fuel reactivity	–Sulfidation –Long-term mechanical stability –Deactivation during redox cycles	[29]
Ca—Mn—O	$CaMnO_3$	Spray drying, sol–gel	TGA, lab scale FB, 10 kW_{th} FB	–Low-temperature oxygen uncoupling –Easily tunable with dopants –Good fuel conversion	–Sulfidation –Deactivation through phase decomposition	[30–35, 37, 38, 41, 42]

FB, fluidized bed; FBR, fixed bed reactor; and TGA, thermogravimetric analysis.

rates for both the fuel and oxygen carrier particles can be anticipated when a relatively small amount of gasification agents are used to initiate the reactions. In addition to the reactions above, several other reactions including the water gas shift (WGS) reaction, devolatilization reactions, and reactions between volatile components of the solid fuels and oxygen carriers can take place. The main challenge of iG-CLC is that the reverse Boudouard reaction (Reaction (8.1)) and steam carbon reaction (Reaction (8.2)) are relatively slow. This poses a problem as unconverted carbon will be transported to the air reactor or carried out through the cyclone, leading to decreased overall carbon capture efficiency [45, 46]. Additionally, chemical looping applications involving solid fuels can suffer from oxygen carrier deactivation caused by ash and sulfur in coal, and oxygen carrier loss in ash removal and separation can be high. Therefore, utilization of cheaper oxygen carriers is desired. As a result, mixed oxides investigated in the iG-CLC space are often composed of oxide ores. We note that specifically synthesized Cu—Fe based oxygen carrier materials have also been extensively investigated for iG-CLC. This topic is extensively covered in Chapter 7.

Ilmenite ($FeTiO_3$) and hematite (Fe_2O_3) ores are more commonly used in iG-CLC [47–61]. Iron ores are potentially attractive oxygen carrier candidates due to their low cost and abundance. Additionally, they tend to possess good mechanical stability. Cuadrat et al. investigated the effects of temperature, gasifying agents, activity with fuels, and fluidization gases on ilmenite oxygen carriers for iG-CLC of South African bituminous coal [54]. Ilmenite is known to increase its activity during redox cycles due to increased pore formation and surface structures [62]. Cuadrat et al. found that the ilmenite oxygen carrier would be fully activated in the fuel reactor only after 7 cycles, but in the oxidation reaction full activation was not observed even after 16 cycles. The authors attributed this to the low variation in ilmenite conversion between each redox cycle. This result was further confirmed by Cuadrat et al. in a paper in which they tested ilmenite oxygen carriers in a 500-W_{th} CLC unit [52]. In determining the effects of gasifying agents, the coal gasification rate increased when steam/CO_2 was added to the reactor in the presence of the ilmenite oxygen carrier. Additionally, it was determined that steam led to the production of more H_2 whereas CO_2 gasifying agent led to more CO. It was concluded that steam is a preferred gasifying agent because the ilmenite oxygen carrier has higher activity for H_2 combustion than CO. The authors' final analysis showed that the solids inventory needed to be approximately 1000–1500 kg/MW_{th} when achieving at least 90% conversion of the solid fuels. Cuadrat et al. also investigated the effect of coal rank on the overall performance of ilmenite in iG-CLC applications [53]. The authors used lignite, high and medium volatile bituminous coals, and anthracite as the solid fuels. Carbon capture efficiencies of over 90% without a carbon separation system was obtained for the lignite coal at 920 °C due to its fast gasification rates. The authors determined that oxygen demand, defined as the fraction of additional oxygen needed to fully combust the fuel reactor exhaust, was less than 10% for all fuel types. One potential issue was found for coals with higher volatile contents as certain volatiles were difficult to convert using the ilmenite oxygen carrier. No change was observed in experiments using the lignite coal depending on the gasifying agent, but with anthracite coals that have slower gasification rates a significant performance drop was observed. Berguerand and Lyngfelt investigated ilmenite oxygen carriers conversion with

Mexican petroleum coke and South African bituminous coal in a 10-kW_{th} CLC unit [48–50]. The authors operated the reactor between 10 and 140 hours for the testing with varying results. The oxygen demand reported in these papers averaged between 25% and 30% while the solid conversions ranged from 50% to 80% depending on the fuel. The low conversion was mainly attributed to the poor efficiency of the cyclone that they used in their reactor set-up. Attrition rates of the ilmenite oxygen carriers were found to be between 0.01 and 0.03 wt% h^{-1}.

Besides ilmenite, other low-cost iron oxide-based ores and industrial wastes such as hematite ore and bauxite waste (Fe_2O_3/Al_2O_3) have been investigated as potential oxygen carriers in iG-CLC applications. Mendiara et al. compared different low-cost oxygen carriers including hematite ore, ilmenite ore, bauxite waste, a copper oxide-enriched iron oxide ore (1.0 wt% CuO), and a synthetic iron oxide [56]. Of the oxygen carriers investigated, the hematite ore and synthetic iron oxide resulted in the highest combustion efficiency followed by the bauxite waste. The copper oxide enriched iron oxide ore did not show any CLOU effects and cycling produced no noticeable changes in activity. In addition to its high combustion efficiency, the hematite ore, similarly to ilmenite, exhibited an increase in its activity with cycle number. As a follow-up study, Mendiara et al. investigated a Spanish hematite ore mined in Tierga (Zaragoza, Spain) as an oxygen carrier for various solid fuels [55]. The authors used anthracite, bituminous, and lignite coals in a 500-W_{th} continuous CLC unit at temperatures ranging from 875 to 930 °C. The authors reported less than 4% oxygen demand for all the fuels investigated. In addition, similar carbon capture efficiencies (>95%) were obtained compared to bauxite and ilmenite oxygen carriers. These ores also contained a high amount of CaO (~5 wt%) that could potentially capture SO_2 in the fuel reactor. However, the authors noted that make-up rates would likely be too high to maintain the retention capacity. Additional improvement to gasification rates using hematite ores can be done through addition of potassium dopants to the ore.

Bauxite waste was studied in several other studies [56, 63–65]. Mendiara et al. investigated the potential to use bauxite waste, which is a byproduct during the production of alumina in the digestion step of bauxite in the Bayer process. The sand fraction that is obtained from the suspension when aluminum is extracted through dissolving of bauxite contains Fe_2O_3 (70 wt%) as a primary phase and β-Al_2O_3 as a secondary phase. The authors tested the oxygen carrier for fuel conversion with char produced from bituminous coal under various temperatures and gasifying mixtures [64, 65]. The bauxite waste oxygen carriers demonstrated higher activity toward the gasification products compared to ilmenite. A 70/30 molar mixture of steam/CO_2 gasifying agents was adequate to achieve a combustion efficiency of 0.95 at both 900 and 980 °C. Solids inventory for the bauxite waste was determined to be 812 kg/MW_{th} to achieve a combustion efficiency of 0.99. Mendiara et al.'s follow-up study included testing the bauxite waste carrier with different coal ranks. Similar combustion efficiencies were found for each rank of coal. The same authors also investigated the oxygen carrier in a continuous 500-W_{th} CLC unit with bituminous coal [63]. Combustion efficiencies of up to 90% were obtained, and oxygen demand was lower than 5%. Continuous operations exhibited no defluidization or agglomeration issues.

While many manganese ores were investigated in the context of CLOU, some Mn-containing ores can also be applied for IG-CLC. Brazilian Mn ores were shown to exhibit catalytic properties to improve the gasification rate. However, the ores suffer from mechanical stability issues. These ores are comprised primarily of MnO (70 wt%) and contain significant Al_2O_3 (~7.5 wt%), SiO_2 (~9 wt%), and Fe_2O_3 (~8 wt%) while, in addition, trace amounts of alkali and alkali earth metal oxides such as potassium, sodium, barium, and calcium can be found. Arjmand et al. examined the steam gasification rate of petcoke in the presence of Brazilian manganese ore compared to ilmenite [66]. The authors found that the steam gasification of Mexican petroleum coke in the presence of the manganese ore was five times faster than with the ilmenite. Since the manganese ores often produce fines, which is undesirable, Linderholm et al. suggested combining the manganese ore with ilmenite to produce an oxygen carrier with high reactivity and mechanical stability inside the reactor [47]. The authors indicated that adding even a small amount of manganese ore (about 8% of the total inventory) to ilmenite can eliminate nearly half of the unconverted gas produced from the gasification reaction of bituminous coal combustion.

It is expected that the presence of oxygen carriers would lead to higher coal gasification rates than if the bed is filled with inert particles [46, 67, 68]. However, the underlying reasons behind the increased rates could be one of the following two possibilities, in addition to the propagation of gasification agents via Reactions (8.1)–(8.4): (i) release of gas-phase oxygen by the oxygen carrier and (ii) catalytic effect that enhances carbon gasification rate (e.g. in the presence of alkali metals). Mechanism 1 occurs in oxygen carriers with high partial pressure of oxygen as a part of CLOU, as discussed earlier. It should be noted that while many of the manganese ores do exhibit minor CLOU properties, their oxygen release is much smaller or even negligible when compared with synthetic Mn—Fe oxygen carriers. The mechanisms behind the increased gasification rates of solid fuels in the presence of manganese ores were studied by Keller et al. [69]. To simulate the performance of CLOU on the effect of gasification rates, the authors introduced a small, fixed amount of oxygen into the stream containing steam. The authors found no additive relationship between the overall char conversion and the steam gasification when the gaseous O_2 was introduced. Moreover, the authors found that adding more O_2 to the reactor caused the gasification reaction to be slightly suppressed. Elemental analysis of the typical Brazilian manganese coals reveals contents of barium, potassium, and iron, which can contribute to catalytically enhanced gasification reactions. Through the use of electron dispersive spectroscopy (EDX), the authors analyzed partially gasified petroleum coke mixed in a bed of sand and manganese ore. Little change was determined between the fresh and partially gasified petroleum coke when sand was used in the bed. However, noticeable amounts of potassium could be observed when the Mn ore was used. Furthermore, EDX mapping showed that the potassium was evenly distributed throughout the surface of the partially gasified coke particles, leading to a likely effective gas-phase transport of the potassium from the ores to the carbon surface. These results were confirmed in a later study by Arjmand et al. using different manganese ores with varying levels of K and Na contents [66]. Despite the improved activity and gasification rates when manganese ores are used, mechanical stability of the ores is of major concern. A summary of commonly used mixed oxides for iG-CLC is shown in Table 8.2.

Table 8.2 Summary of key mixed oxide oxygen carriers for iG-CLC.

Oxygen carrier system	Main phases	Synthesis	Tested scale	Pros	Cons	References
Ilmenite ore	Ilmenite ($FeTiO_3$)	Natural ore	Laboratory-scale FB, 10–50 kW_{th} CLC unit	–Cheap, abundant ores –Good mechanical properties –Good conversion of gasification products	–Poor solid conversion and gasification rates –Poor volatile conversion rate	[48, 49, 52–54, 70, 71]
Hematite ores	Fe_2O_3, SiO_2, CaO, MgO	Natural ore	TGA, laboratory-scale batch FB, 1–100 kW_{th} CLC unit	–Better fuel conversion than ilmenite based ores –Mechanical stability –Cheap and abundant	–High circulation rates needed –Poor solid conversion rates	[56, 57, 59–61]
Bauxite waster	Fe_2O_3, Al_2O_3	Produced from the Bayer process	TGA, batch FB, 500 W_{th} FB reactor	–Higher conversion rates/low inventory needed compared to ilmenite –Cheap and easily obtained –Mechanically stable	–Poor solid conversion rates –Sulfur effects not studied	[56, 57, 63–65]
Mn ores	MnO, Fe_2O_3, SiO, Al_2O_3	Natural ore	TGA, laboratory-scale FB, 10–100 kW_{th} CLC unit	–Improves gasification rates –High solid and gas conversions –Cheap and abundant	–Mechanical stability –Long-term operation is a concern	[47, 66, 69, 72, 73]

FB, fluidized bed; FBR, fixed bed reactor; and TGA, thermogravimetric analysis.

8.4 Mixed Oxides for Chemical Looping Reforming (CLR)

8.4.1 Chemical Looping Reforming

Although both CLC and CLOU research focuses on carbon capture from coal conversion, the abundance of natural gas compared with petroleum and its notably lower emission have led to renewed interest in converting methane to value-added products [74–77]. Currently, most methane conversion processes adopt an indirect route due to its demonstrated reliability and economic feasibility. In these processes, methane is converted to syngas (a mixture of carbon monoxide and hydrogen) in a reforming step. The syngas is further processed to value-added products such as hydrogen, liquid fuels, or chemical feedstocks through one or more conversion and purification steps. Being the common operation among these methane conversion processes, methane reforming has been a topic of interest for decades. At present, all commercial reforming processes are performed in the presence of a heterogeneous catalyst and using a gaseous oxidant such as oxygen (partial oxidation), steam (steam reforming), or carbon dioxide (dry reforming) [78–80]. Dry reforming faces severe catalyst deactivation problem as a result of coke formation and is not commercially practiced [81, 82]. While steam methane reforming (SMR) and POx processes are currently commercially implemented, their efficiencies are limited due to their high energy demand for steam generation, reforming reactions, or cryogenic air separation [83, 84].

Compared to conventional reforming technologies, CLR offers a potentially more efficient and environmentally friendly option for methane reforming. Similar to CLC, CLR process is performed in two interconnected reactors in the presence of a solid oxygen carrier, also known as redox catalyst following a definition used in catalysis literature. The redox catalyst donates its lattice oxygen (O^{2-}) to partially oxidize methane to syngas in the reducer (fuel reactor); the oxygen-depleted redox catalysts are then transferred to the oxidizer to regenerate with a gaseous oxidant [2, 85–87]. As illustrated in Figure 8.4, gaseous oxidants

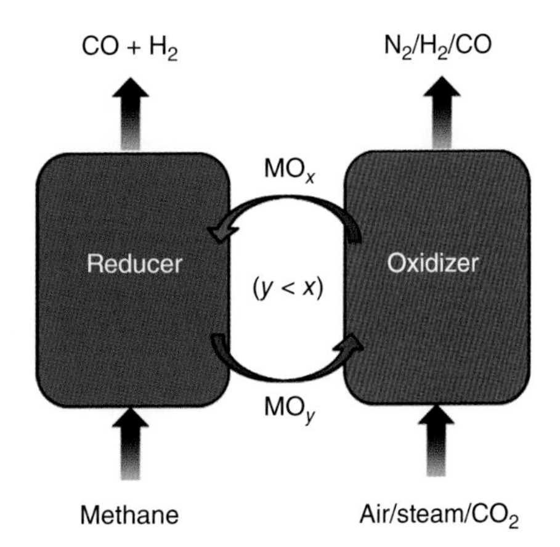

Figure 8.4 Schematic of chemical looping reforming; various regeneration options are shown.

that are being investigated in CLR are air, steam, and carbon dioxide. Use of air as an oxidant provides in situ air separation and eliminates the need for the energy-intensive air separation unit (ASU) in POx. The air regeneration also leads to significant heat release, allowing an autothermal operation for the CLR process. Compared to methane POx, avoidance of direct contact between methane and gaseous oxidant can also enhance process safety. Use of steam and/or CO_2 for redox catalyst regeneration, on the other hand, leads to production of hydrogen and/or CO respectively. The added value of such products can enhance the overall efficiency and economics of the process. In such cases, however, the CLR process needs external heat inputs. The heat required can be satisfied by combustion of additional fuels or concentrated solar energy. Despite its potentials, CLR is still in the early developmental stages and faces many challenges in both synthesis of more active, selective, and stable redox catalyst, and more effective reactor design to improve the gas–solid interaction and heat integration. It should be noted that the term CLR is also referred to the process of combusting pressure swing adsorption (PSA) offgas in CLC scheme and using the generated steam and heat for SMR. However, the oxygen carrier design principles for this process are similar to the CLC process and hence are not covered in this section.

8.4.2 Monometallic Redox Catalysts for CLR

Compared to typical CLC oxygen carriers whose primary function is to oxidize the carbon and hydrogen atoms in fuels to their most stable gaseous states, i.e. CO_2 and water, the requirements on redox catalyst is more stringent since a high selectivity toward syngas, a partial oxidation product, is desired. As such, the efficiency and product yield of the CLR process are highly dependent on the performance of the redox catalyst. To be effective for CLR applications, a redox catalyst must possess several characteristics, namely good reactivity with methane, high syngas selectivity, long-term recyclability and stability, and high oxygen capacity. To date, extensive research activities have been devoted to the design and synthesis of such redox catalysts.

Typically, redox catalysts contain a primary (transition metal) oxide phase for oxygen storage and a support phase for activity/stability enhancement. The support is commonly considered as an "inert" phase and does not have major contributions to the redox reactions. Among all the CLR redox catalysts, nickel-based and iron-based oxygen carriers attracted the most attention during the early stages of CLR research [2, 85, 88–90]. Nickel seemed to be a logical option due to its high oxygen-carrying capacity and well-established catalytic activity toward partial oxidation [91–94]. Initial tests on Ni-based oxygen carriers showed very promising activity, which led to pilot-scale tests [95, 96]. Pilot-scale CLR tests are performed at a 140-kW plant at Vienna University of Technology using $NiAl_2O_4$-supported NiO as the redox catalyst. They reported close to complete methane conversion and up to 60% CO selectivity at 747–903 °C operating temperatures with decreasing CO selectivity and methane conversion at higher temperatures and higher global air/fuel ratios. Despite good activity, research interests in nickel-based redox catalysts saw significant decrease over the years due to their low CO selectivity in oxidized forms, sulfur poisoning, deactivation, cost, toxicity, and environmental concerns [97, 98]. Iron was also considered because of its abundance, high oxygen content, and environmental benignity, but research on developing iron-based redox catalysts also declined due to its low activity with methane, low selectivity, and agglomeration issues [2, 89, 90]. Other monometallic oxides such as oxides of manganese, copper, cobalt, and cerium were also tested. However, they all faced challenges such as poor stability/recyclability, agglomeration, low selectivity, slow kinetics, and toxicity [2, 89, 90]. It was shown, however, that addition of inert supports such as Al_2O_3, TiO_2, SiO_2, YSZ, and $MgAl_2O_4$ help alleviate some of these problems [88, 99–111].

While the inert support phase does not contribute to the redox reactions, their presence can be crucial for the activity and long-term performance of the redox catalysts, especially when using monometallic oxides such as Fe_2O_3, CuO, Mn_2O_3, and NiO as the oxygen carriers. The presence of a support phase not only provides a stable matrix for metal oxide particles and prevents them from sintering and agglomeration in reduced states, but also facilitates the oxygen anion conduction through the solid particles by reducing the energy barrier for oxygen migration [112]. Adánez et al. studied the effect of different inert supports and sintering temperatures on the performance of the abovementioned metal oxides and reported the best inert supports on the basis of reactivity with methane and crushing strength. The Cu-based oxygen carriers performed the best when supported on SiO_2 or TiO_2. Among the Fe-based oxygen carriers, Al_2O_3 and ZrO_2 showed better performance, and TiO_2 and ZrO_2 are shown to be effective for Ni-based and Mn-based oxygen carriers respectively. The underlying reasons for the enhancement in redox performance and support–primary oxide interactions, however, were not reported.

Regardless of the support, most monometallic oxides exhibit low syngas selectivity when reacting with methane, especially during the early stages of the reaction. This low selectivity is the result of both the thermodynamic limitations of redox pairs and the inherent low selectivity of methane oxidation on metal oxide surfaces. As shown in Figure 8.2, the equilibrium P_{O_2} for most common redox pairs is not suitable for methane partial oxidation. Hu and Ruckenstein extensively studied the kinetics and reaction mechanisms of methane oxidation on Ni oxide surfaces both with and without the presence of gaseous oxidants. They reported methane oxidation proceeding through Eley–Rideal reaction mechanism on NiO surface, which gives rise to methane combustion reaction. However, on reduced Ni, methane is converted through the Langmuir–Hinshelwood mechanism, which is more selective toward partial oxidation. Although the mentioned reaction mechanisms are aimed at explaining the reaction mechanisms for heterogeneous catalytic reactions, they corroborate well with the pilot-scale testing results on NiO-based oxygen carriers, which showed low syngas selectivity in CLR.

These limitations of supported monometallic oxides call for improvements in surface and/or thermodynamic properties of the redox catalysts in order to achieve higher syngas yield. Use of mixed metal oxides as either a support or a standalone redox catalyst is recently introduced as an approach that can potentially address both of the abovementioned limitations.

8.5 Redox Catalyst Improvement Strategies

As discussed earlier, both thermodynamic and kinetic limitations contribute to low selectivity of methane partial oxidation in redox conditions. Most metal oxide redox pairs do not provide suitable oxygen partial pressure for partial oxidation reactions. Moreover, oxide surfaces are often not selective toward syngas. With the promising possibilities mixed oxides offer, two potential approaches can be adopted to enhance the properties of mixed oxides for CLR: (i) design and synthesis of single-phase mixed oxides with suitable bulk properties to thermodynamically limit the over-oxidation of syngas products and (ii) modification of redox catalysts' surface catalytic properties to enhance the syngas formation. The latter can be achieved by development of core–shell structured redox catalysts with a metal oxide with high oxygen capacity at the core and a selective mixed oxide shell to regulate the oxygen donation and provide a catalytic surface, or by promoting the surface of the redox catalysts with small amounts of catalytic sites. Both strategies have been put into practice in recent years and very promising results have been reported.

Mixed oxides often possess thermodynamic and surface properties that are different from those of their parent metal oxides. This can be used to our advantage to design and synthesize endless number of metal oxide combinations for various purposes. Many different combinations of first row transition metals have been tested for CLR [2, 89, 90]. Without the presence of a support phase, however, most of these mixed oxides experienced irreversible phase segregation and phase changes through cyclic redox reaction, leading to deactivation or slow kinetics. Additionally, the majority of these mixed oxides showed poor syngas yields due to thermodynamic limitations, which is predictable from the Ellingham diagram (Figure 8.2).

Recent studies showed that mixed oxides with stable structures such as spinel and perovskite can be effective as support materials [113–115]. In fact, some of these mixed oxides such as $La_xSr_{1-x}Fe_yCo_{1-y}O_{3-\delta}$ possess mixed ionic-electronic conductivity (MIEC), which allows countercurrent diffusion of electrons and lattice oxygens, which further facilitates the oxygen migration [113–115]. These advantages, however, come at a cost. Use of mixed oxide as a support for monometallic oxides introduces a new level of complexity in the process conditions; cations can redistribute to form more stable mixed oxide with less favorable redox properties. As a result, long-term phase compatibility investigations are essential when designing a mixed oxide-supported oxygen carrier. It was also found that some of these oxides can withstand some degree of oxygen non-stoichiometry and donate small amounts of lattice oxygens before going through any phase change [44, 90, 116]. These characteristics, along with providing more active catalytic sites on the surface, make mixed oxides a very appealing option not only as a support material, but also as a stand-alone redox catalyst.

Introduction of mixed oxides as stand-alone oxygen carriers started a new era in redox catalyst synthesis. With practically endless number of metallic combinations with different crystal structures, redox properties, and mechanical characteristics, numerous mixed oxides can be designed, synthesized, and fine-tuned for partial oxidation applications [2, 88]. Because of its low cost and environmental impact, and high oxygen capacity, iron is the most common metal in mixed oxides [2, 89, 90]. Other transition metals such as copper, manganese, cobalt, and nickel were also tested extensively due to their higher equilibrium partial pressure and catalytic activities [117–120]. Despite some reported improvements in terms of selectivity, activity, and tunability, most of the tested mixed oxides suffered from coke formation and agglomeration issues.

Among all mixed oxides for CLR applications, perovskite-structured materials showed very promising results, making them the most widely studied redox catalysts [36, 38, 41, 42, 121–126]. These oxides have the general formula of $ABO_{3-\delta}$ with the A-site representing a large cation, usually from alkali earth or rare earth metals, and the B-site representing a smaller transition metal. Each of these sites can also be filled with multiple cations, extending the number of possibilities. Figure 8.5 illustrates the possible combinations of metals in perovskite structures. Various combination of cations can lead to perovskites with a wide range of structures from cubic to more complex orthorhombic and rhombohedral variations. These classes of material have characteristics that are very attractive for CLR. They can withstand moderate deviation from their perfect oxidation states and donate some of their lattice oxygen without collapsing their structures. They also demonstrated vastly reversible phase changes in reductive and oxidative environments, which make them highly stable redox catalysts. Different combinations of perovskite materials have different stabilities. A semi-empirical parameter named "Goldschmidt tolerance factor" is commonly used to compare their relative stability [127]. This factor is defined as

$$t = \frac{r_A + r_O}{\sqrt{2}(r_B + r_O)}$$

with r_A, r_B, and r_O being radii of A-site and B-site cations and oxygen anions respectively. The value of the Goldschmidt factor is about 0.9–1 for the ideal cubic structures and deviates from it for other structure variations.

Lanthanum ferrite and calcium manganate-based materials are among the perovskites that are studied the most for CLR applications. Various dopants have been added to the A-site (Ca, Ba, Sr, and La) and B-site (Mn, Fe, Co, and Ni) of these materials to tune their mixed conductivity, oxygen-carrying capacity, equilibrium partial pressure, coke resistance, catalytic activity, etc. [31–33, 36, 38, 42, 128–130]. The presence of cations with lower oxidation states in the A-site contributes to formation of oxygen vacancies, which facilitates the anion migration while cations with higher oxidation states stabilize the structure. B-site dopants on the other hand reduce faster and provide catalytic surface sites.

Among all the perovskite structures investigated for chemical looping applications, $La_xSr_{1-x}Fe_yCo_{1-y}O_{3-\delta}$ (LSCF) [121, 123], $La_xSr_{1-x}FeO_3$ (LSF) [131], and $CaMnO_3$-based (CM) [30, 32–34, 36, 38, 40–42, 125, 126] showed superior performance for CLR, CLC, and CLOU respectively. Not only did LSF (with x ranging from 0.6 to 0.8) show close to 100% selectivity toward syngas at reasonable oxygen donation capacity and reaction kinetics [131] but it also significantly enhanced the syngas selectivity of single metal oxides when used as a support phase. Recent research attributes the higher selectivity of LSF to both controlled release of oxygen, which regulates the partial oxidation reaction and

Atomic # → | Atomic symbol → | English element name → | Ions commonly formed ← | Atomic mass (rounded) ←

Example: 29 +2,1 Cu copper 63.55

A-site atom | B-site atom

Period	1 I A	2 II A	3 III B	4 IV B	5 V B	6 VI B	7 VII B	8 VIII B	9 VIII B	10 VIII B	11 I B	12 II B	13 III A	14 IV A	15 V A	16 VI A	17 VII A	18 VIII A
1 1s	1 ±1 H hydrogen 1.008																	2 He helium 4.003
2 2s / 2p	3 +1 Li lithium 6.968	4 +2 Be beryllium 9.012											5 +3 B boron 10.81	6 −4 C carbon 12.01	7 −3 N nitrogen 14.01	8 −2 O oxygen 16.00	9 −1 F fluorine 19.00	10 Ne neon 20.18
3 3s / 3p	11 +1 Na sodium 22.99	12 +2 Mg magnesium 24.31											13 +3 Al aluminum 26.98	14 −4 Si silicon 28.09	15 −3 P phosphorus 30.97	16 −2 S sulfur 32.07	17 −1 Cl chlorine 35.45	18 Ar argon 39.95
4 4s / 3d / 4p	19 +1 K potassium 39.10	20 +2 Ca calcium 40.08	21 +3 Sc scandium 44.96	22 +4,3,2 Ti titanium 47.87	23 +5,2,3,4 V vanadium 50.94	24 +3,2,6 Cr chromium 52.00	25 -2,3,4,6,7 Mn manganese 54.94	26 +3,2 Fe iron 55.85	27 +2,3 Co cobalt 58.93	28 +2,3 Ni nickel 58.69	29 +2,1 Cu copper 63.55	30 +2 Zn zinc 65.38	31 +3 Ga gallium 69.72	32 +4,2 Ge germanium 72.63	33 −3 As arsenic 74.92	34 −2 Se selenium 78.97	35 −1 Br bromine 79.90	36 Kr krypton 83.90
5 5s / 4d / 5p	37 +1 Rb rubidium 85.47	38 +2 Sr strontium 87.62	39 +3 Y yttrium 88.91	40 +4 Zr zirconium 91.22	41 +5,3 Nb niobium 92.91	42 +6,3,5 Mo molybdenum 95.95	43 +7,4,6 Tc technetium 98	44 +4,3,6,8 Ru ruthenium 101.1	45 +3,4,6 Rh rhodium 102.9	46 +2,4 Pd palladium 106.4	47 +1 Ag silver 107.9	48 +2 Cd cadmium 112.4	49 +3 In indium 114.8	50 +4,2 Sn tin 118.7	51 +3,5 Sb antimony 121.8	52 −2 Te tellurium 127.6	53 −1 I iodine 126.9	54 Xe xenon 131.3
6 6s / † 5d / 6p	55 +1 Cs cesium 132.9	56 +2 Ba barium 137.3	71 +3 Lu lutetium 175.0	72 +4 Hf hafnium 178.5	73 +5 Ta tantalum 180.9	74 +6,4 W tungsten 183.8	75 +7,4,6 Re rhenium 186.2	76 +4,6,8 Os osmium 190.2	77 +4,3,6 Ir iridium 192.2	78 +4,2 Pt platinum 195.1	79 +3,1 Au gold 197.0	80 +2,1 Hg mercury 200.6	81 +1,3 Tl thallium 204.4	82 +2,4 Pb lead 207.2	83 +3,5 Bi bismuth 209.0	84 +4,2 Po polonium 209	85 At astatine 210	86 Rn radon 222
7 7s / ‡ 6d / 7p	87 +1 Fr francium 223	88 +2 Ra radium 226	103 +3 Lr lawrencium 262	104 Rf rutherfordium 267	105 Db dubnium 268	106 Sg seaborgium 271	107 Bh bohrium 272	108 Hs hassium 270	109 Mt meitnerium 276	110 Ds darmstadtium 281	111 Rg Roentgenium 280	112 Cn copernicium 285	113 Nh nihonium 284	114 Fl flerovium 289	115 Mc moscovium 288	116 Lv livermorium 293	117 Ts tennessine 292	118 Og oganesson 294

Lanthanides (rare earth metals) † 4f	57 +3 La lanthanum 138.9	58 +3,4 Ce cerium 140.1	59 +3,4 Pr Praseodymium 140.9	60 +3 Nd neodymium 144.2	61 +3 Pm promethium 145	62 +3,2 Sm samarium 150.4	63 +3,2 Eu europium 152.0	64 +3 Gd gadolinium 157.3	65 +3,4 Tb terbium 158.9	66 +3 Dy dysprosium 162.5	67 +3 Ho holmium 164.9	68 +3 Er erbium 167.3	69 +3,2 Tm thulium 168.9	70 +3,2 Yb ytterbium 173.1
Actinides ‡ 5f	89 +3 Ac actinium 227	90 +4 Th thorium 232.0	91 +5,4 Pa protactinium 231.0	92 +6,3,4,5 U uranium 238.0	93 +5,3,4,6 Np neptunium 237	94 +4,3,5,6 Pu plutonium 244	95 +3,4,5,6 Am americium 243	96 +3 Cm curium 247	97 +3,4 Bk berkelium 247	98 +3 Cf californium 251	99 +3 Es einsteinium 252	100 +3 Fm fermium 257	101 +3,2 Md mendelevium 258	102 +2,3 No nobelium 259

Figure 8.5 Possible elements for occupying the A-site or the B-site of the perovskite structures mixed oxides.

prevents over-oxidation, and presence of metallic sites with catalytic activity toward the partial oxidation route. The most prominent drawbacks of perovskite materials are their low oxygen capacity relative to pure oxides and high cost due to the presence of rare earth metals. The latter can be addressed by using less expensive alkali earth metals such as calcium and barium.

To address the cost issue, the perovskites can also be mixed with cheaper metal oxides with higher oxygen capacity. Recent findings showed that use of high selectivity perovskites as supports can significantly enhance the selectivity and activity of iron oxide-based oxygen carriers [44, 113, 114, 132]. This enhancement can be further improved by forming core–shell structures and minimizing the exposure of the less selective oxides to the surface [116]. This significantly reduces the cost of redox catalyst production by minimizing the use of more expensive phases and concentrating them on the surface to fully exploit their selective surface and regulate oxygen release. Despite very promising results, the presence of more than one phase introduces more variables such as phase compatibility and diffusion, and significantly increases the complexity of the system. This is demonstrated in recent studies on methane activation and partial oxidation mechanism on FeO_x@LSF core–shell redox catalyst [116, 133, 134]. The core–shell redox catalyst goes through multiple reduction stages/regions with markedly different mechanism and consequently different product selectivities [133, 135]. This exacerbates the dynamic nature of the CLR process, which makes overall process design more challenging.

Catalytic surface promotion is another alternative that has the potential to further increase the selectivity and activity of the mixed oxide redox catalyst. Addition of very small amounts of catalytically active metals such as alkali metals (to change the acidity of the surface) or platinum group metals (which are known to have good C—H bond activation and/or partial oxidation activity) can potentially lead to higher CO selectivity and/or higher activity at lower temperatures. The promoting effect of Rh_2O_3 has been demonstrated by Nakayama et al. who used Rh_2O_3 to promote Fe_2O_3/YSZ for high syngas selectivity through methane partial oxidation with air [136]. More recent studies by Shafiefarhood et al. confirmed the promoting effect of Rh on the kinetic and thermodynamic behavior of $CaMnO_3$ and $LaCeO_{3.5}$. They reported up to 10 times faster reaction kinetics and up to 300 °C reduction in the onset temperature of methane partial oxidation. This option has been proposed very recently and not been tested extensively yet.

The discussed bulk (thermodynamic) and surface (kinetic) improvement strategies provide us practically endless possibilities to develop more effective redox catalysts for more efficient methane partial oxidation using the CLR scheme.

8.6 Mixed Oxides for Other Selective Oxidation Applications

As discussed earlier, chemical looping scheme offers quite a few advantages over methane combustion and reforming by preventing direct contact between the fuel and gaseous oxygen. However, these advantages are not limited to methane

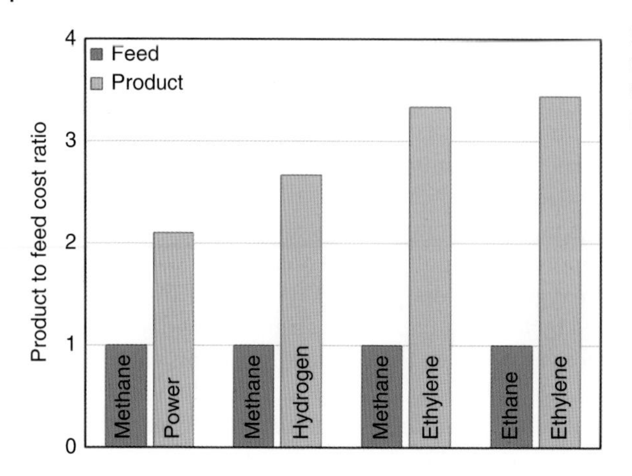

Figure 8.6 Value creation from light hydrocarbon feedstocks to potential products.

oxidation to syngas. In principle, any oxidation (or reduction) reaction can be carried out via the chemical looping scheme. The modification of chemical looping scheme in order to adapt for different applications, however, can be challenging. The main challenges lie in the design of effective redox catalysts.

Light olefin production, for instance, is one of the most important industrial processes that can be enhanced using the chemical looping scheme. Olefins in general, and ethylene and propylene in particular, are one of the most important classes of materials in chemical, petrochemical, and polymer industries. From 2010 to 2015, the global production of ethylene has been increased by about 15% (from 145 to 167 million metric tons per year) and the global capacity is projected to increase to over 200 million metric tons per year by 2020. As shown in Figure 8.6, converting light hydrocarbons to ethylene results in higher value creation compared to power or hydrogen/syngas production. This has motivated the development of novel and more efficient technologies for olefin production from light hydrocarbon feedstocks. Currently, steam cracking of ethane is one of the most commonly used commercial technologies for ethylene production, representing nearly 45% of the global ethylene production capacity, with the remaining provided by steam cracking of heavier feeds such as naphtha. Typical steam cracking facilities have an ethylene production capacity in the order of 1 million tons per year due to economy of scale. The primary challenges to scale down steam cracking reside in its high operating temperature (up to 1200 °C), high endothermicity and energy requirements (16 GJ per ton of ethylene), and equilibrium limitations. In Sections 8.6.1 and 8.6.2, we will discuss two novel applications of chemical looping process for ethylene production.

8.6.1 Oxidative Coupling of Methane

Oxidative coupling of methane (OCM) is an alternate route for ethylene production from methane feedstocks that was initially proposed by Keller and Bhasin in 1982 [137]. OCM has attracted more attention in recent years due to advances in fracking technologies that enable cheaper production of natural gas from shale gas reserves. This process is based on formation of methyl radicals by

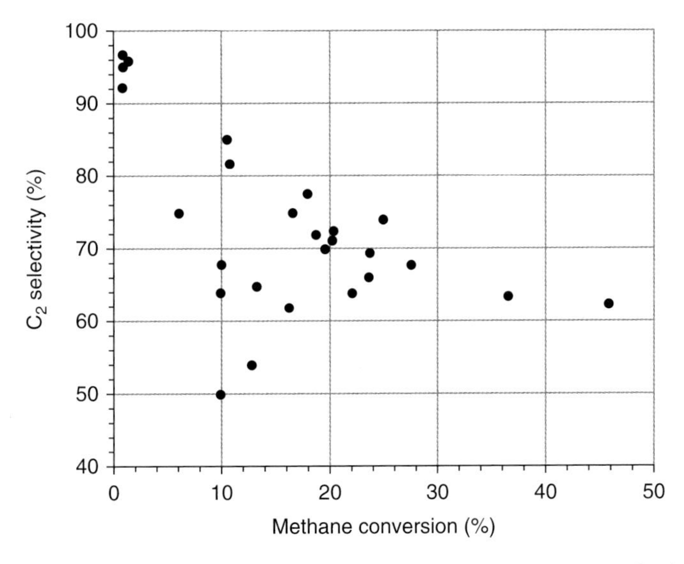

Figure 8.7 Comparison of C_2 selectivity and methane conversion of redox OCM experiments reported by 2016.

catalytic activation of methane and C—H bond cleavage, and recombination of methyl radicals in the gas phase to form ethane [138–140]. The main challenges of the OCM process are the parallel activation of coupled products along with the methane and their deep oxidation to combustion products as a result of the high temperature needed for methane activation. This results in low selectivity and yield of C_2 products [140].

The chemical looping oxidative coupling of methane (CL-OCM) process, which was proposed more recently, has the potential to enhance the OCM process by replacing O_2 with lattice oxygen and minimizing the gas-phase combustion route [141]. Like before, in CL-OCM the redox catalyst functions as both oxygen reservoir and catalyst. Studies conducted on the use of $Na_2WO_4/Mn/SiO_2$ as redox catalyst confirmed higher selectivities at lower methane conversions [140, 141]. Typically, CL-OCM results in ~10% C_{2+} yield increase compared to OCM with O_2 co-feed. Figure 8.7 compares the C_2 selectivity and methane conversions of redox OCM experiments reported by 2016.

8.6.2 Oxidative Dehydrogenation (ODH) of Ethane

Ethane oxidative dehydrogenation (ODH) in the presence of heterogeneous catalysts, which has been extensively studied, has the potential to address the challenges associated with ethane steam cracking. It can potentially provide higher olefin yields, is autothermal (H_2 oxidation provides most of the required energy), and can effectively remove the equilibrium constraints of ethane conversion. Despite all these benefits, the need for a cryogenic air separation plant, complexity of reactor design, and the safety concerns over O_2–ethane mixing make conventional ODH unattractive when compared to steam cracking.

The chemical looping-oxidative dehydrogenation process (CL-ODH) addresses the limitations in both cracking and conventional ODH. The chemical looping scheme and use of a redox catalyst, which acts both as a catalyst and an air separation agent, avoid O_2–ethane mixing, provide in situ oxygen separation, balance the process energy requirements, and enhance ethylene selectivity. Recent process analysis indicates that nearly an order of magnitude reduction in energy consumption can be achieved with CL-ODH when compared to state-of-the-art cracking processes. The recent discovery of "low-temperature" redox catalysts, which is enabled through surface catalytic promotion (e.g. lithium), allows efficient CL-ODH processes at low to intermediate temperatures (650–850 °C) and offers additional advantage for modular ethylene production at small scales. CL-ODH is also self-sufficient from a process heat requirement standpoint, making it a highly efficient and low emission technology ideal for distributed systems.

In the CL-ODH process, ethane is partially oxidized with lattice oxygen from a mixed metal oxide-based redox catalyst such as a promoted Ruddlesden–Popper phase (e.g. $LiO_x/La_2Sr_{2-x}FeO_{4-\delta}$) to produce water and ethylene (Reaction 8.5). The removal of hydrogen by converting it to water significantly improves equilibrium conversion of the ethane to ethylene at lower temperatures. In a separate step, air is introduced into the packed bed to regenerate the redox catalyst's lattice oxygen (Reaction 8.6). The general process scheme is identical to CLC and CLR. The exothermic regeneration provides heat to both the catalyst bed and the endothermic reactions in parallel channels. This cyclic operation can be achieved through alternation of gas feed in reactor channels, similar to the industrially proven Houdry process, or by using moving bed reactors to circulate the redox catalyst between the two reactors. In a recent study, Neal et al. reported up to 66% ethane conversion, 61% ethylene yield, and over 80% steam selectivity (percentage of H_2 converted to steam) using a sodium tungstate-promoted Mg_6MnO_8 redox catalyst. Gao et al. also reported promising results (61% ethane conversion, 55% ethylene yield, and close to 100% steam selectivity) using lithium-promoted $La_xSr_{2-x}FeO_4$ redox catalyst. They attributed the higher ethylene selectivity to regulating the rate of O^{2-} conduction and surface evolution caused by surface enrichment of Li cations on the surface of the redox catalyst. The same scheme can also potentially be used for ODH of propane to propylene. While many studies have been conducted on propane ODH on heterogeneous catalysts, not much effort has been devoted to CL-ODH of propane as of now.

$$C_2H_6 + MO_x \rightarrow C_2H_4 + H_2O + MO_{x-1} \quad (8.5)$$

$$MO_{x-1} + 1/2O_2 \rightarrow MO_x \quad (8.6)$$

8.7 Toward Rationalizing the Design of Mixed Metal Oxides

With emerging applications for chemical looping-based processes, the need for rational design of mixed oxide-based redox catalysts for different purposes is felt more than ever. Deeper understanding of the surface reaction mechanisms and bulk properties of different phases are crucial for a truly rational material selection. However, the dynamic nature of redox reactions makes detailed investigation of the redox system rather challenging. An example of this dynamic behavior is demonstrated in the FeO_x@LSF core–shell redox catalyst system, which undergoes several reduction regions with markedly different reaction mechanisms and selectivities during a reduction half-cycle.

As effective as the experimental methods are for investigating the bulk thermodynamic and kinetic properties of the mixed oxide, the repetitive nature of the experimental techniques and virtually endless number of possible mixed oxide combinations make trial and error very inefficient and time consuming. The use of available resources such as thermodynamic and phase diagram databases along with softwares such as HSC chemistry, FactSage, and ASPEN can help with more educated selection when choosing materials for different applications. The more recent ab initio calculations are shown to have the potential to efficiently predict the bulk and surface properties of complex oxides. For instance, density functional theory (DFT) can be used to determine vacancy formation energies, anion/cation migration energies, and electronic properties such as band gap. These methods, however, can be computationally expensive. Moreover, the magnetic properties for the mixed oxides and inaccuracies caused by electron self-interaction with d-Electrons of many transition metal oxides can impose additional challenges and complexity for DFT calculations.

On top of bulk thermodynamics and kinetics, surface properties are very important to consider when designing oxygen carriers and redox catalysts, especially the ones with core–shell structures and/or promoters present on the surface. The aforementioned limitations and complexities are often intensified when surface properties of the redox catalyst are taken into account. This is due to the uncertainties about the surface terminations of different phases. As confirmed by experimental evidence, the topmost layers of the heterogeneous catalyst can be entirely different from the bulk of the material. This is a very important factor to consider when designing a redox catalyst as the surface catalytic properties may alter the overall selectivity of the reactions. Experimental methods are often limited to investigating the apparent activation or adsorption energies or predicting the rate-determining steps and fall short from determining the actual adsorption, dissociation, and desorption routes/energies and reaction mechanisms. Computational methods, when supplemented with experimental data, can be very useful for studying site-specific properties and predicting the reaction mechanism.

8.8 Future Directions

Considering the promising outlook of mixed oxides in chemical looping applications in general and partial oxidation applications in particular, the future success of these fields is tied with developing active and cost-effective mixed oxides and exploiting their full potentials. This mandates continued investigations on rational design of mixed oxides, exploration of natural and readily available ores, and design of oxygen carriers or redox catalysts with tailored bulk and surface properties. Moreover, chemical looping research may find applications well beyond power generation and CO_2 capture. Development of redox catalysts that are suitable for chemical production via the chemical looping scheme can create significant opportunities for value creation while reducing the energy intensity and carbon emissions in conventional industrial processes.

References

1 Mattisson, T. (2013). Materials for chemical-looping with oxygen uncoupling. *ISRN Chem. Eng.* 2013: 1–19.

2 Adanez, J., Abad, A., Garcia-Labiano, F. et al. (2012). Progress in chemical-looping combustion and reforming technologies. *Prog. Energy Combust. Sci.* 38 (2): 215–282.

3 Fan, L.-S. (2010). *Chemical Looping Systems for Fossil Energy Conversions.* Hoboken, NJ: Wiley.

4 Chuang, S., Dennis, J., Hayhurst, A., and Scott, S. (2008). Development and performance of Cu-based oxygen carriers for chemical-looping combustion. *Combust. Flame* 154 (1–2): 109–121.

5 Chuang, S.Y., Dennis, J.S., Hayhurst, A.N., and Scott, S.A. (2009). Kinetics of the chemical looping oxidation of CO by a co-precipitated mixture of CuO and Al_2O_3. *Proc. Combust. Inst.* 32 (2): 2633–2640.

6 Gayán, P., Forero, C.R., Abad, A. et al. (2011). Effect of support on the behavior of Cu-based oxygen carriers during long-term CLC operation at temperatures above 1073 K. *Energy Fuels* 25 (3): 1316–1326.

7 Imtiaz, Q., Hosseini, D., and Müller, C.R. (2013). Review of oxygen carriers for chemical looping with oxygen uncoupling (CLOU): thermodynamics, material development, and synthesis. *Energy Technol.* 1 (11): 633–647.

8 Abad, A., Adánez-Rubio, I., Gayán, P. et al. (2012). Demonstration of chemical-looping with oxygen uncoupling (CLOU) process in a 1.5 kW_{th} continuously operating unit using a Cu-based oxygen-carrier. *Int. J. Greenhouse Gas Control* 6: 189–200.

9 Adánez-Rubio, I., Gayán, P., Abad, A. et al. (2012). Evaluation of a spray-dried $CuO/MgAl_2O_4$ oxygen carrier for the chemical looping with oxygen uncoupling process. *Energy Fuels* 26 (5): 3069–3081.

10 Adánez-Rubio, I., Abad, A., Gayán, P. et al. (2012). Identification of operational regions in the chemical-looping with oxygen uncoupling (CLOU) process with a Cu-based oxygen carrier. *Fuel* 102: 634–645.

11 He, F., Linak, W.P., Deng, S., and Li, F. (2017). Particulate formation from a copper oxide-based oxygen carrier in chemical looping combustion for CO_2 capture. *Environ. Sci. Technol.* 51 (4): 2482–2490.

12 Imtiaz, Q., Broda, M., and Müller, C.R. (2014). Structure–property relationship of co-precipitated Cu-rich, Al_2O_3- or $MgAl_2O_4$-stabilized oxygen carriers for chemical looping with oxygen uncoupling (CLOU). *Appl. Energy* 119: 557–565.

13 Azad, A.-M., Hedayati, A., Rydén, M. et al. (2013). Examining the Cu–Mn–O spinel system as an oxygen carrier in chemical looping combustion. *Energy Technol.* 1 (1): 59–69.

14 Mohammad Pour, N., Leion, H., Rydén, M., and Mattisson, T. (2013). Combined Cu/Mn oxides as an oxygen carrier in chemical looping with oxygen uncoupling (CLOU). *Energy Fuels* 27 (10): 6031–6039.

15 Abián, M., Abad, A., Izquierdo, M.T. et al. (2017). Titanium substituted manganese-ferrite as an oxygen carrier with permanent magnetic properties for chemical looping combustion of solid fuels. *Fuel* 195: 38–48.

16 Arjmand, M., Frick, V., Rydén, M. et al. (2015). Screening of combined Mn–Fe–Si oxygen carriers for chemical looping with oxygen uncoupling (CLOU). *Energy Fuels* 29 (3): 1868–1880.

17 Azimi, G., Leion, H., Mattisson, T., and Lyngfelt, A. (2011). Chemical-looping with oxygen uncoupling using combined Mn–Fe oxides, testing in batch fluidized bed. *Energy Procedia* 4: 370–377.

18 Bhavsar, S., Tackett, B., and Veser, G. (2014). Evaluation of iron- and manganese-based mono- and mixed-metallic oxygen carriers for chemical looping combustion. *Fuel* 136: 268–279.

19 Mattisson, T., Jing, D., Lyngfelt, A., and Rydén, M. (2016). Experimental investigation of binary and ternary combined manganese oxides for chemical-looping with oxygen uncoupling (CLOU). *Fuel* 164: 228–236.

20 Rydén, M., Leion, H., Mattisson, T., and Lyngfelt, A. (2014). Combined oxides as oxygen-carrier material for chemical-looping with oxygen uncoupling. *Appl. Energy* 113: 1924–1932.

21 Rydén, M., Lyngfelt, A., and Mattisson, T. (2011). Combined manganese/iron oxides as oxygen carrier for chemical looping combustion with oxygen uncoupling (CLOU) in a circulating fluidized bed reactor system. *Energy Procedia* 4: 341–348.

22 Arjmand, M., Rydén, M., Leion, H. et al. (2014). Sulfur tolerance and rate of oxygen release of combined Mn–Si oxygen carriers in chemical-looping with oxygen uncoupling (CLOU). *Ind. Eng. Chem. Res.* 141208064420002.

23 Frick, V., Rydén, M., Leion, H. et al. (2015). Screening of supported and unsupported Mn–Si oxygen carriers for CLOU (chemical-looping with oxygen uncoupling). *Energy* 93 (1): 544–554.

24 Hanning, M., Frick, V., Mattisson, T. et al. (2016). Performance of combined manganese–silicon oxygen carriers and effects of including titanium. *Energy Fuels.*

25 Jing, D., Arjmand, M., Mattisson, T. et al. (2014). Examination of oxygen uncoupling behaviour and reactivity towards methane for manganese silicate oxygen carriers in chemical-looping combustion. *Int. J. Greenhouse Gas Control* 29: 70–81.

26 Shulman, A., Cleverstam, E., Mattisson, T., and Lyngfelt, A. (2011). Chemical-looping with oxygen uncoupling using Mn/Mg-based oxygen carriers – oxygen release and reactivity with methane. *Fuel* 90 (3): 941–950.

27 Johansson, M., Mattisson, T., and Lyngfelt, A. (2006). Investigation of Mn_3O_4 with stabilized ZrO_2 for chemical-looping combustion. *Chem. Eng. Res. Des.* 84 (9): 807–818.

28 Zafar, Q., Abad, A., Mattisson, T. et al. (2007). Reduction and oxidation kinetics of Mn_3O_4/Mg–ZrO_2 oxygen carrier particles for chemical-looping combustion. *Chem. Eng. Sci.* 62 (23): 6556–6567.

29 Sundqvist, S., Arjmand, M., Mattisson, T. et al. (2015). Screening of different manganese ores for chemical-looping combustion (CLC) and chemical-looping with oxygen uncoupling (CLOU). *Int. J. Greenhouse Gas Control* 43: 179–188.

30 Pishahang, M., Larring, Y., McCann, M., and Bredesen, R. (2014). $Ca_{0.9}Mn_{0.5}Ti_{0.5}O_{3-\delta}$: a suitable oxygen carrier material for fixed-bed chemical looping combustion under syngas conditions. *Ind. Eng. Chem. Res.* 53 (26): 10549–10556.

31 Galinsky, N., Mishra, A., Zhang, J., and Li, F. (2015). $Ca_{1-x}A_xMnO_3$ (A = Sr and Ba) perovskite based oxygen carriers for chemical looping with oxygen uncoupling (CLOU). *Appl. Energy* 157: 358–367.

32 Sundqvist, S., Leion, H., Rydén, M. et al. (2013). $CaMn_{0.875}Ti_{0.125}O_{3-\delta}$ as an oxygen carrier for chemical-looping with oxygen uncoupling (CLOU)-solid-fuel testing and sulfur interaction. *Energy Technol.* 1 (5–6): 338–344.

33 Källén, M., Rydén, M., Dueso, C. et al. (2013). $CaMn_{0.9}Mg_{0.1}O_{3-\delta}$ as oxygen carrier in a gas-fired 10 kW_{th} chemical-looping combustion unit. *Ind. Eng. Chem. Res.* 52 (21): 6923–6932.

34 Rydén, M., Lyngfelt, A., and Mattisson, T. (2011). $CaMn_{0.875}Ti_{0.125}O_3$ as oxygen carrier for chemical-looping combustion with oxygen uncoupling (CLOU)—Experiments in a continuously operating fluidized-bed reactor system. *Int. J. Greenhouse Gas Control* 5 (2): 356–366.

35 Galinsky, N., Sendi, M., Bowers, L., and Li, F. (2016). $CaMn_{1-x}B_xO_{3-\delta}$ (B = Al, V, Fe, Co, and Ni) perovskite based oxygen carriers for chemical looping with oxygen uncoupling (CLOU). *Appl. Energy* 174: 80–87.

36 Hallberg, P., Rydén, M., Mattisson, T., and Lyngfelt, A. (2014). $CaMnO_{3-\delta}$ made from low cost material examined as oxygen carrier in chemical-looping combustion. *Energy Procedia* 63: 80–86.

37 Arjmand, M., Hedayati, A., Azad, A.-M. et al. (2013). $Ca_xLa_{1-x}Mn_{1-y}M_yO$ (M = Mg, Ti, Fe, or cu) as oxygen carriers for chemical-looping with oxygen uncoupling (CLOU). *Energy Fuels.*

38 Hallberg, P., Jing, D., Rydén, M. et al. (2013). Chemical looping combustion and chemical looping with oxygen uncoupling experiments in a batch reactor using spray-dried $CaMn_{1-x}M_xO_{3-\delta}$ (M = Ti, Fe, Mg) particles as oxygen carriers. *Energy Fuels* 27 (3): 1473–1481.

39 Schmitz, M., Linderholm, C., and Lyngfelt, A. (2014). Chemical looping combustion of sulphurous solid fuels using spray-dried calcium manganate particles as oxygen carrier. *Energy Procedia* 63: 140–152.

40 Pour, N.M., Azimi, G., Leion, H. et al. (2014). Production and examination of oxygen-carrier materials based on manganese ores and $Ca(OH)_2$ in chemical looping with oxygen uncoupling. *AIChE J.* 60 (2): 645–656.

41 Arjmand, M., Kooiman, R.F., Rydén, M. et al. (2014). Sulfur tolerance of $Ca_xMn_{1-y}M_yO_{3-\delta}$ (M = Mg, Ti) perovskite-type oxygen carriers in chemical-looping with oxygen uncoupling (CLOU). *Energy Fuels* 28 (2): 1312–1324.

42 Leion, H., Larring, Y., Bakken, E. et al. (2009). Use of $CaMn_{0.875}Ti_{0.125}O_3$ as oxygen carrier in chemical-looping with oxygen uncoupling. *Energy Fuels* 23 (10): 5276–5283.

43 Bakken, E., Norby, T., and Stolen, S. (2005). Nonstoichiometry and reductive decomposition of CaMnO. *Solid State Ionics* 176 (1–2): 217–223.

44 Shafiefarhood, A., Stewart, A., and Li, F. (2015). Iron-containing mixed-oxide composites as oxygen carriers for chemical looping with oxygen uncoupling (CLOU). *Fuel* 139: 1–10.

45 Cuadrat, A., Abad, A., Gayán, P. et al. (2012). Theoretical approach on the CLC performance with solid fuels: optimizing the solids inventory. *Fuel* 97: 536–551.

46 Leion, H., Mattisson, T., and Lyngfelt, A. (2008). Solid fuels in chemical-looping combustion. *Int. J. Greenhouse Gas Control* 2 (2): 180–193.

47 Linderholm, C., Schmitz, M., Knutsson, P., and Lyngfelt, A. (2016). Chemical-looping combustion in a 100-kW unit using a mixture of ilmenite and manganese ore as oxygen carrier. *Fuel* 166: 533–542.

48 Berguerand, N. and Lyngfelt, A. (2008). Design and operation of a 10 kW_{th} chemical-looping combustor for solid fuels – testing with South African coal. *Fuel* 87 (12): 2713–2726.

49 Berguerand, N. and Lyngfelt, A. (2009). Operation in a 10 kW_{th} chemical-looping combustor for solid fuel—Testing with a Mexican petroleum coke. *Energy Procedia* 1 (1): 407–414.

50 Berguerand, N. and Lyngfelt, A. (2009). Chemical-looping combustion of petroleum coke using ilmenite in a 10 kW_{th} unit–high-temperature operation. *Energy Fuels* 23 (10): 5257–5268.

51 Berguerand, N., Lyngfelt, A., Mattisson, T., and Markström, P. (2011). Chemical looping combustion of solid fuels in a 10 kW_{th} unit. *Oil Gas Sci. Technol. – Rev. d'IFP Energies Nouv.* 66 (2): 181–191.

52 Cuadrat, A., Abad, A., García-Labiano, F. et al. (2011). The use of ilmenite as oxygen-carrier in a 500 Wth chemical-looping coal combustion unit. *Int. J. Greenhouse Gas Control* 5 (6): 1630–1642.

53 Cuadrat, A., Abad, A., García-Labiano, F. et al. (2012). Relevance of the coal rank on the performance of the in situ gasification chemical-looping combustion. *Chem. Eng. J.* 195–196: 91–102.

54 Cuadrat, A., Abad, A., de Diego, L.F. et al. (2012). Prompt considerations on the design of chemical-looping combustion of coal from experimental tests. *Fuel* 97: 219–232.

55 Mendiara, T., de Diego, L.F., García-Labiano, F. et al. (2014). On the use of a highly reactive iron ore in chemical looping combustion of different coals. *Fuel* 126: 239–249.

56 Mendiara, T., Pérez, R., Abad, A. et al. (2012). Low-cost Fe-based oxygen carrier materials for the iG-CLC process with coal. 1. *Ind. Eng. Chem. Res.* 51 (50): 16216–16229.

57 Abad, A., Cuadrat, A., Mendiara, T. et al. (2012). Low-cost Fe-based oxygen carrier materials for the iG-CLC process with coal. 2. *Ind. Eng. Chem. Res.* 51 (50): 16230–16241.

58 Ge, H., Guo, W., Shen, L. et al. (2016). Biomass gasification using chemical looping in a 25 kW_{th} reactor with natural hematite as oxygen carrier. *Chem. Eng. J.* 286: 174–183.

59 Gu, H., Shen, L., Zhong, Z. et al. (2014). Potassium-modified iron ore as oxygen carrier for coal chemical looping combustion: continuous test in 1 kW reactor. *Ind. Eng. Chem. Res.* 53 (33): 13006–13015.

60 Linderholm, C. and Schmitz, M. (2016). Chemical-looping combustion of solid fuels in a 100 kW dual circulating fluidized bed system using iron ore as oxygen carrier. *J. Environ. Chem. Eng.* 4 (1): 1029–1039.

61 Yang, W., Zhao, H., Ma, J. et al. (2014). Copper-decorated hematite as an oxygen carrier for in situ gasification chemical looping combustion of coal. *Energy Fuels* 28 (6): 3970–3981.

62 Adánez, J., Cuadrat, A., Abad, A. et al. (2010). Ilmenite activation during consecutive redox cycles in chemical-looping combustion. *Energy Fuels* 24 (2): 1402–1413.

63 Mendiara, T., de Diego, L.F., García-Labiano, F. et al. (2013). Behaviour of a bauxite waste material as oxygen carrier in a 500 Wth CLC unit with coal. *Int. J. Greenhouse Gas Control* 17: 170–182.

64 Mendiara, T., García-Labiano, F., Gayán, P. et al. (2013). Evaluation of the use of different coals in chemical looping combustion using a bauxite waste as oxygen carrier. *Fuel* 106: 814–826.

65 Mendiara, T., Gayán, P., Abad, A. et al. (2013). Performance of a bauxite waste as oxygen-carrier for chemical-looping combustion using coal as fuel. *Fuel Process. Technol.* 109: 57–69.

66 Arjmand, M., Leion, H., Mattisson, T., and Lyngfelt, A. (2014). Investigation of different manganese ores as oxygen carriers in chemical-looping combustion (CLC) for solid fuels. *Appl. Energy* 113: 1883–1894.

67 Keller, M., Leion, H., Mattisson, T., and Lyngfelt, A. (2011). Gasification inhibition in chemical-looping combustion with solid fuels. *Combust. Flame* 158 (3): 393–400.

68 Azimi, G., Keller, M., Mehdipoor, A., and Leion, H. (2012). Experimental evaluation and modeling of steam gasification and hydrogen inhibition in chemical-looping combustion with solid fuel. *Int. J. Greenhouse Gas Control* 11: 1–10.

69 Keller, M., Leion, H., and Mattisson, T. (2013). Mechanisms of solid fuel conversion by chemical-looping combustion (CLC) using manganese ore: catalytic gasification by potassium compounds. *Energy Technol.* 1 (4): 273–282.

70 Berguerand, N. and Lyngfelt, A. (2008). The use of petroleum coke as fuel in a 10 kW_{th} chemical-looping combustor. *Int. J. Greenhouse Gas Control* 2 (2): 169–179.

71 Abad, A., Pérez-Vega, R., de Diego, L.F. et al. (2015). Design and operation of a 50kW_{th} chemical looping combustion (CLC) unit for solid fuels. *Appl. Energy* 157: 295–303.

72 Arjmand, M., Leion, H., Lyngfelt, A., and Mattisson, T. (2012). Use of manganese ore in chemical-looping combustion (CLC) – effect on steam gasification. *Int. J. Greenhouse Gas Control* 8: 56–60.

73 Frohn, P., Arjmand, M., Azimi, G. et al. (2013). On the high-gasification rate of Brazilian manganese ore in chemical-looping combustion (CLC) for solid fuels. *AIChE J.* 59 (11): 4346–4354.

74 Rahimpour, M.R., Mirvakili, A., and Paymooni, K. (2011). A novel water perm-selective membrane dual-type reactor concept for Fischer–Tropsch synthesis of GTL (gas to liquid) technology. *Energy* 36 (2): 1223–1235.

75 Elia, J.A., Baliban, R.C., Xiao, X., and Floudas, C.A. (2011). Optimal energy supply network determination and life cycle analysis for hybrid coal, biomass, and natural gas to liquid (CBGTL) plants using carbon-based hydrogen production. *Comput. Chem. Eng.* 35 (8): 1399–1430.

76 Bao, B., El-Halwagi, M.M., and Elbashir, N.O. (2010). Simulation, integration, and economic analysis of gas-to-liquid processes. *Fuel Process. Technol.* 91 (7): 703–713.

77 Krishna, R., Ellenberger, J., and Sie, S.T. (1996). Reactor development for conversion of natural gas to liquid fuels: a scale-up strategy relying on hydrodynamic analogies. *Chem. Eng. Sci.* 51 (10): 2041–2050.

78 Rostrup-Nielsen, J.R., Sehested, J., Nørskov, J.K. et al. (2002). Hydrogen and synthesis gas by steam- and CO_2 reforming. In: *Advances in Catalysis*, vol. 47, 65–139. Academic Press.

79 Dissanayake, D., Rosynek, M.P., Kharas, K.C.C., and Lunsford, J.H. (1991). Partial oxidation of methane to carbon monoxide and hydrogen over a Ni/Al_2O_3 catalyst. *J. Catal.* 132 (1): 117–127.

80 Rostrupnielsen, J.R. and Hansen, J.H.B. (1993). CO_2-reforming of methane over transition metals. *J. Catal.* 144 (1): 38–49.

81 Tomishige, K., Chen, Y., and Fujimoto, K. (1999). Studies on carbon deposition in CO_2 reforming of CH_4 over nickel–magnesia solid solution catalysts. *J. Catal.* 181 (1): 91–103.

82 Wang, S., Lu (Max), G.Q., and Millar, G.J. (1996). Carbon dioxide reforming of methane to produce synthesis gas over metal-supported catalysts: state of the art. *Energy Fuels* 10 (4): 896–904.

83 Simpson, A.P. and Lutz, A.E. (2007). Exergy analysis of hydrogen production via steam methane reforming. *Int. J. Hydrogen Energy* 32 (18): 4811–4820.

84 Wilhelm, D.J., Simbeck, D.R., Karp, A.D., and Dickenson, R.L. (2001). Syngas production for gas-to-liquids applications: technologies, issues and outlook. *Fuel Process. Technol.* 71 (1–3): 139–148.

85 Protasova, L. and Snijkers, F. (2016). Recent developments in oxygen carrier materials for hydrogen production via chemical looping processes. *Fuel* 181: 75–93.

86 Fan, L.-S. (2011). *Chemical Looping Systems for Fossil Energy Conversions.* Wiley.

87 Zafar, Q., Mattisson, T., and Gevert, B. (2005). Integrated hydrogen and power production with CO_2 capture using chemical-looping reforming redox reactivity of particles of CuO, Mn_2O_3, NiO, and Fe_2O_3 using SiO_2 as a support. *Ind. Eng. Chem. Res.* 44 (10): 3485–3496.

88 Adánez, J., de Diego, L.F., García-Labiano, F. et al. (2004). Selection of oxygen carriers for chemical-looping combustion. *Energy Fuels* 18 (2): 371–377.

89 Tang, M., Xu, L., and Fan, M. (2015). Progress in oxygen carrier development of methane-based chemical-looping reforming: a review. *Appl. Energy* 151: 143–156.

90 (2015). Chemical looping technology: oxygen carrier characteristics. *Annu. Rev. Chem. Biomol. Eng.* 6 (1): 53–75.

91 Vernon, P.D.F., Green, M.L.H., Cheetham, A.K., and Ashcroft, A.T. (1990). Partial oxidation of methane to synthesis gas. *Catal. Lett.* 6 (2): 181–186.

92 Christian Enger, B., Lødeng, R., and Holmen, A. (2008). A review of catalytic partial oxidation of methane to synthesis gas with emphasis on reaction mechanisms over transition metal catalysts. *Appl. Catal., A* 346 (1–2): 1–27.

93 Hu, Y.H. and Ruckenstein, E. (2003). Multiple transient response methods to identify mechanisms of heterogeneous catalytic reactions. *Acc. Chem. Res.* 36 (10): 791–797.

94 Ruckenstein, E. and Hu, Y.H. (1995). Near 100% CO selectivity in CH_4 direct catalytic oxidation at low temperatures (<700°C) under unsteady state conditions. *Catal. Lett.* 35 (3–4): 265–269.

95 Kolbitsch, P., Pröll, T., Bolhar-Nordenkampf, J., and Hofbauer, H. (2009). Operating experience with chemical looping combustion in a 120 kW dual circulating fluidized bed (DCFB) unit. *Energy Procedia* 1 (1): 1465–1472.

96 Pröll, T., Kolbitsch, P., Bolhàr-Nordenkampf, J., and Hofbauer, H. (2011). Chemical looping pilot plant results using a nickel-based oxygen carrier. *Oil Gas Sci. Technol. – Rev. d'IFP Energies Nouv.* 66 (2): 173–180.

97 García-Labiano, F., de Diego, L.F., Gayán, P. et al. (2009). Effect of fuel gas composition in chemical-looping combustion with Ni-based oxygen carriers. 1. Fate of sulfur. *Ind. Eng. Chem. Res.* 48 (5): 2499–2508.

98 Tian, H., Simonyi, T., Poston, J., and Siriwardane, R. (2009). Effect of hydrogen sulfide on chemical looping combustion of coal-derived synthesis gas over bentonite-supported metal–oxide oxygen carriers. *Ind. Eng. Chem. Res.* 48 (18): 8418–8430.

99 Ryu, H.-J., Lim, N.-Y., Bae, D.-H., and Jin, G.-T. (2003). Carbon deposition characteristics and regenerative ability of oxygen carrier particles for chemical-looping combustion. *Korean J. Chem. Eng.* 20 (1): 157–162.

100 Villa, R. (2003). Ni based mixed oxide materials for CH_4 oxidation under redox cycle conditions. *J. Mol. Catal. A: Chem.* 204–205: 637–646.

101 Shulman, A., Linderholm, C., Mattisson, T., and Lyngfelt, A. (2009). High reactivity and mechanical durability of $NiO/NiAl_2O_4$ and $NiO/NiAl_2O_4/MgAl_2O_4$ oxygen carrier particles used for more than 1000 h in a 10 kW CLC reactor. *Ind. Eng. Chem. Res.* 48 (15): 7400–7405.

102 Erri, P. and Varma, A. (2009). Diffusional effects in nickel oxide reduction kinetics. *Ind. Eng. Chem. Res.* 48 (1): 4–6.

103 Blas, L., Dorge, S., Michelin, L. et al. (2015). Influence of the regeneration conditions on the performances and the microstructure modifications of $NiO/NiAl_2O_4$ for chemical looping combustion. *Fuel* 153: 284–293.

104 Baek, J.-I., Yang, S.-R., Eom, T.H. et al. (2015). Effect of MgO addition on the physical properties and reactivity of the spray-dried oxygen carriers prepared with a high content of NiO and Al_2O_3. *Fuel* 144: 317–326.

105 Hamers, H.P., Gallucci, F., Williams, G., and van Sint Annaland, M. (2015). Experimental demonstration of CLC and the pressure effect in packed bed reactors using $NiO/CaAl_2O_4$ as oxygen carrier. *Fuel* 159: 828–836.

106 Kooiman, R.F., Hamers, H.P., Gallucci, F., and van Sint Annaland, M. (2015). Experimental demonstration of two-stage packed bed chemical-looping combustion using syngas with CuO/Al_2O_3 and $NiO/CaAl_2O_4$ as oxygen carriers. *Ind. Eng. Chem. Res.* 54 (7): 2001–2011.

107 Hossain, M.M., Lopez, D., Herrera, J., and de Lasa, H.I. (2009). Nickel on lanthanum-modified γ-Al_2O_3 oxygen carrier for CLC: reactivity and stability. *Catal. Today* 143 (1–2): 179–186.

108 Mattisson, T., Johansson, M., and Lyngfelt, A. (2006). The use of NiO as an oxygen carrier in chemical-looping combustion. *Fuel* 85 (5–6): 736–747.

109 Corbella, B.M., de Diego, L.F., García-Labiano, F. et al. (2006). Performance in a fixed-bed reactor of titania-supported nickel oxide as oxygen carriers for the chemical-looping combustion of methane in multicycle tests. *Ind. Eng. Chem. Res.* 45 (1): 157–165.

110 Siriwardane, R., Poston, J., Chaudhari, K. et al. (2007). Chemical-looping combustion of simulated synthesis gas using nickel oxide oxygen carrier supported on bentonite. *Energy Fuels* 21 (3): 1582–1591.

111 Zafar, Q., Mattisson, T., and Gevert, B. (2006). Redox investigation of some oxides of transition-state metals Ni, Cu, Fe, and Mn supported on SiO_2 and $MgAl_2O_4$. *Energy Fuels* 20 (1): 34–44.

112 Li, F., Sun, Z., Luo, S., and Fan, L.-S. (2011). Ionic diffusion in the oxidation of iron – effect of support and its implications to chemical looping applications. *Energy Environ. Sci.* 4 (3): 876–880.

113 Galinsky, N.L., Huang, Y., Shafiefarhood, A., and Li, F. (2013). Iron oxide with facilitated O_2 – transport for facile fuel oxidation and CO_2 capture in a chemical looping scheme. *ACS Sustainable Chem. Eng.* 1 (3): 364–373.

114 Galinsky, N.L., Shafiefarhood, A., Chen, Y. et al. (2015). Effect of support on redox stability of iron oxide for chemical looping conversion of methane. *Appl. Catal., B* 164: 371–379.

115 Chen, Y., Galinsky, N., Wang, Z., and Li, F. (2014). Investigation of perovskite supported composite oxides for chemical looping conversion of syngas. *Fuel* 134: 521–530.

116 Shafiefarhood, A., Galinsky, N., Huang, Y. et al. (2014). Fe_2O_3@$La_xSr_{1-x}FeO_3$ core–shell redox catalyst for methane partial oxidation. *ChemCatChem* 6 (3): 790–799.

117 He, F., Galinsky, N., and Li, F. (2013). Chemical looping gasification of solid fuels using bimetallic oxygen carrier particles – feasibility assessment and process simulations. *Int. J. Hydrogen Energy* 38 (19): 7839–7854.

118 Adánez-Rubio, I., Arjmand, M., Leion, H. et al. (2013). Investigation of combined supports for Cu-based oxygen carriers for chemical-looping with oxygen uncoupling (CLOU). *Energy Fuels* 27 (7): 3918–3927.

119 Moghtaderi, B. and Song, H. (2010). Reduction properties of physically mixed metallic oxide oxygen carriers in chemical looping combustion. *Energy Fuels* 24 (10): 5359–5368.

120 Johansson, M., Mattisson, T., and Lyngfelt, A. (2006). Creating a synergy effect by using mixed oxides of iron- and nickel oxides in the combustion of methane in a chemical-looping combustion reactor. *Energy Fuels* 20 (6): 2399–2407.

121 Readman, J.E., Olafsen, A., Larring, Y., and Blom, R. (2005). $La_{0.8}Sr_{0.2}Co_{0.2}Fe_{0.8}O_{3-\delta}$ as a potential oxygen carrier in a chemical looping type reactor, an in-situ powder X-ray diffraction study. *J. Mater. Chem.* 15 (19): 1931.

122 Dai, X.P., Li, J., Fan, J.T. et al. (2012). Synthesis gas generation by chemical-looping reforming in a circulating fluidized bed reactor using perovskite $LaFeO_3$-based oxygen carriers. *Ind. Eng. Chem. Res.* 51 (34): 11072–11082.

123 Rydén, M., Lyngfelt, A., Mattisson, T. et al. (2008). Novel oxygen-carrier materials for chemical-looping combustion and chemical-looping reforming; $La_xSr_{1-x}Fe_yCo_{1-y}O_{3-\delta}$ perovskites and mixed-metal oxides of NiO, Fe_2O_3 and Mn_3O_4. *Int. J. Greenhouse Gas Control* 2 (1): 21–36.

124 Zhao, K., He, F., Huang, Z. et al. (2014). Three-dimensionally ordered macroporous $LaFeO_3$ perovskites for chemical-looping steam reforming of methane. *Int. J. Hydrogen Energy* 39 (7): 3243–3252.

125 Arjmand, M., Hedayati, A., Azad, A.-M. et al. (2013). $Ca_xLa_{1-x}Mn_{1-y}M_yO_{3-\delta}$ (M = Mg, Ti, Fe, or Cu) as oxygen carriers for chemical-looping with oxygen uncoupling (CLOU). *Energy Fuels* 27 (8): 4097–4107. doi: 10.1021/ef3020102.

126 Cabello, A., Abad, A., Gayán, P. et al. (2014). Effect of operating conditions and H_2S presence on the performance of $CaMg_{0.1}Mn_{0.9}O_{3-\delta}$ perovskite material in chemical looping combustion (CLC). *Energy Fuels* 28 (2): 1262–1274.

127 Kronmüller, H. and Parkin, S. (2007). *Handbook of Magnetism and Advanced Magnetic Materials: Spintronics and Magnetoelectronics*. Wiley.

128 Mihai, O., Chen, D., and Holmen, A. (2012). Chemical looping methane partial oxidation: the effect of the crystal size and O content of $LaFeO_3$. *J. Catal.* 293: 175–185.

129 Gao, P., Li, N., Wang, X. et al. (2013). Perovskite $La_{1-x}Sr_xFe_{1-y}Mn_yO_3$ hollow nanospheres: synthesis and application in catalytic hydrogen peroxide decomposition. *Mater. Lett.* 111: 89–92.

130 Mishra, A., Galinsky, N., He, F. et al. (2016). Perovskite-structured $AMn_xB_{1-x}O_3$ (A = Ca or Ba; B = Fe or Ni) redox catalysts for partial oxidation of methane. *Catal. Sci. Technol.* 6 (12): 4535–4544.

131 He, F., Li, X., Zhao, K. et al. (2013). The use of $La_{1-x}Sr_xFeO_3$ perovskite-type oxides as oxygen carriers in chemical-looping reforming of methane. *Fuel* 108: 465–473.

132 He, F. and Li, F. (2015). Perovskite promoted iron oxide for hybrid water-splitting and syngas generation with exceptional conversion. *Energy Environ. Sci.* 8 (2): 535–539.

133 Neal, L.M., Shafiefarhood, A., and Li, F. (2014). Dynamic methane partial oxidation using a Fe_2O_3@$La_{0.8}Sr_{0.2}FeO_{3-\delta}$ core–shell redox catalyst in the absence of gaseous oxygen. *ACS Catal.* 4 (10): 3560–3569.

134 Neal, L., Shafiefarhood, A., and Li, F. Effect of core and shell compositions on $MeO_x@La_ySr_{1-y}FeO_3$ core–shell redox catalysts for chemical looping reforming of methane. *Appl. Energy*.

135 Shafiefarhood, A., Hamill, J.C., Neal, L.M., and Li, F. (2015). Methane partial oxidation using $FeO_x@La_{0.8}Sr_{0.2}FeO_{3-\delta}$ core–shell catalyst – transient pulse studies. *Phys. Chem. Chem. Phys.* 17 (46): 31297–31307.

136 Nakayama, O., Ikenaga, N., Miyake, T. et al. (2008). Partial oxidation of CH_4 with air to produce pure hydrogen and syngas. *Catal. Today* 138 (3–4): 141–146.

137 Keller, G.E. and Bhasin, M.M. (1982). Synthesis of ethylene via oxidative coupling of methane. *J. Catal.* 73 (1): 9–19.

138 Shahri, S.M.K. and Alavi, S.M. (2009). Kinetic studies of the oxidative coupling of methane over the $Mn/Na_2WO_4/SiO_2$ catalyst. *J. Nat. Gas Chem.* 18 (1): 25–34.

139 Taheri, Z., Seyed-Matin, N., Safekordi, A.A. et al. (2009). A comparative kinetic study on the oxidative coupling of methane over LSCF perovskite-type catalyst. *Appl. Catal., A* 354 (1–2, 152): 143.

140 Fleischer, V., Littlewood, P., Parishan, S., and Schomäcker, R. (2016). Chemical looping as reactor concept for the oxidative coupling of methane over a $Na_2WO_4/Mn/SiO_2$ catalyst. *Chem. Eng. J.* 306: 646–654.

141 Chung, E.Y., Wang, W.K., Nadgouda, S.G. et al. (2016). Catalytic oxygen carriers and process systems for oxidative coupling of methane using the chemical looping technology. *Ind. Eng. Chem. Res.* 55 (50): 12750–12764.

9

Oxygen Carrier Structure and Attrition

Nathan Galinsky[1,2], Samuel Bayham[2], Esmail Monazam[2,3], and Ronald W. Breault[2]

[1] *Oak Ridge Institution of Science and Education, 3610 Collins Ferry Road, Morgantown, WV, USA*
[2] *National Energy Technology Laboratory, 3610 Collins Ferry Road, Morgantown, WV 26507, USA*
[3] *REM Engineering Services, 3537 Collins Ferry Road, Morgantown, WV 26507, USA*

9.1 Introduction

The oxygen carrier (OC) material will be subject to severe conditions of temperature and pressure in the chemical looping process. The material must cope with mechanical forces due to thermal and chemical expansion as well as hydrodynamically caused impacts and shearing. It can be expected that these conditions will introduce changes to the physical and chemical properties of the oxygen carrier structure. From a mechanical point of view, the material must be strong enough to support its own weight initially and to cope with the hydrodynamically caused impacts and shearing after several thousand cycles. Additionally, other problems can be expected, such as variations in heat flux, downstream fouling, and so on. Mechanical failure is due to the brittle fracture arising from a sudden catastrophic growth of a critical flaw under stress induced within the bulk particle. The oxygen carrier can be porous and full of defects, crystal edges, dislocations, and non-identical materials. Any discontinuation that appears in the oxygen carrier bulk may be treated as a flaw and hence an origin of stress concentration. Variations in size, shape, and orientation of these flaws result in the wide scatter of the strength data of the oxygen carrier pellets.

Particle attrition is defined as the breakdown of solids in a system whose main design intent is not comminution [1]. Among various reasons why attrition is undesirable in a chemical looping combustion (CLC) system, the greatest is the cost required to replace the attrited oxygen carrier. For novel, emerging processes such as chemical looping, this has the potential to offset the economic benefits that the emerging technologies are trying to attain. The CLC concept has the advantage that the oxygen needed for fuel combustion is supplied by means of a cyclically reacted oxygen carrier instead of being taken directly from air. The oxygen carrier is usually circulated between two fluidized beds, one in which the fuel is oxidized by the carrier, and the other in which the carrier is reoxidized by air. As a result, the particles are subjected to high mechanical stresses caused

Handbook of Chemical Looping Technology, First Edition. Edited by Ronald W. Breault.

by interparticle collision and bed to wall impacts, thereby resulting in particle attrition. It is also well known that both thermal stresses and chemical reaction can influence the propensity of a particle to undergo attrition. Therefore, particle attrition is a major issue that can cause many problems such as inventory losses that impact significantly the operating costs. Thus, the choice of oxygen carrier particles is a key parameter for the process development of CLC to allow for minimization of cost.

9.2 Oxygen Carrier Structure

This section will discuss in depth the oxygen carrier structure as affected by thermal and chemical stresses encountered in chemical looping systems. The primary focus will be on hematite and hematite-based oxygen carriers, understanding how thermal and chemical stresses affect the structure of the oxygen carriers, and its importance to activity and mechanical stability in chemical looping processes.

9.2.1 Unsupported Oxygen Carriers

Unsupported oxygen carriers are not typically tested in chemical looping systems because of mechanical issues such as agglomeration. However, understanding the changes in surface and bulk structures of unsupported oxides can improve understanding of the degradation of performance of the oxygen carrier while undergoing thermal and chemical stresses encountered by typical chemical looping systems. These structural changes can additionally cause mechanical issues such as agglomeration, which can lead to defluidization (in the case with copper oxides). Additionally, the particles can degrade and form cracks and pores in the bulk structure that can lead to higher attrition rates.

9.2.1.1 Surface

The thermal and chemical stresses that oxygen carriers undergo during chemical looping can lead to significant changes in the surface of the oxygen carrier. This fundamental change to the surface structure correlates to significant changes in the observed activity of the oxygen carrier with gaseous fuels. Hematite (Fe_2O_3)-based oxygen carriers are especially susceptible to surface changes. Tang et al. [2] used scanning tunneling microscopy (STM) results to introduce the theory of a triphase depicting the various topmost Fe atomic layers at 850 °C and above, under reducing conditions. The study defines a third structure as a γ-Fe_2O_3 domain, which prompts instability on the Fe_2O_3 oxidizing surface [2]. These areas of instability appear to be the surface layers missing from the oxidizing surface. The lath structure (a dense platelet or crystal plane of magnetite) could result from the instability, opening an artery line into the Fe_2O_3/Fe_3O_4 interface. Breaking down the Fe_2O_3 surface into Fe_3O_4 layers is through reduction, and it is difficult to regenerate the Fe_3O_4 layers during oxidation. During reoxidation, it is proposed by Kim et al. [3] that once an Fe_2O_3 layer is formed on the sample surface, it may act as a barrier that prevents oxygen from penetrating into the lattice underneath. This has been confirmed through a separate study by Breault et al. [4].

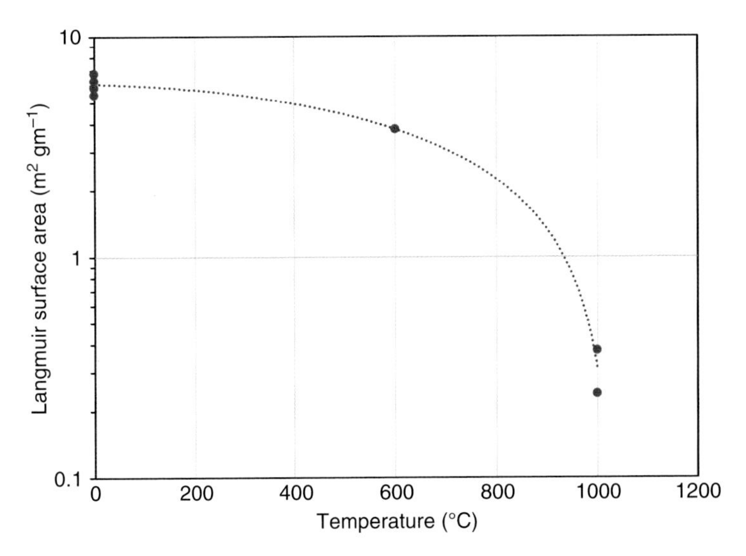

Figure 9.1 Surface area as a function of pretreatment temperature of unreacted hematite [5].

Oxygen carriers are often thermally pretreated to reduce the initial thermal stress during operation. Breault et al. investigated the effects of pretreatment temperature on hematite ore surface area [5]. Figure 9.1 shows the effects of various pretreatment temperatures on the Langmuir surface area of hematite before the reduction process begins. The surface area is significantly reduced as the pretreatment temperatures rise above 600 °C and decreases by more than an order of magnitude by 1000 °C. As can be seen with the line fitted, the surface area drops from ~6.0 $m^2 g^{-1}$ for the fresh carrier to 0.3 $m^2 g^{-1}$ when pretreated at 1000 °C. With the surface area of a spherical hematite particle being on the order of 0.001 $m^2 g^{-1}$, most of the surface area can be attributed to an internal area associated with pores and grains. The change in surface area by a factor of 20 implies that significant internal changes occur as the grains grow [6].

Furthermore, as an oxygen carrier is cycled through reduction and oxidation reactions, the surface area can be significantly altered if no pretreatment is conducted before the redox process begins. Figure 9.2 shows the changes in the surface area as a function of the number of reduction/oxidation cycles for both untreated and treated hematite ore-based oxygen carrier. The surface area of the oxygen carrier that does not undergo any thermal pretreatment is reduced by more than 50 times in just 10 cycles. Additionally, the oxygen carrier's reaction rate is also observed to continually decrease as a result of the surface area decrease [7]. This reduction in surface area for the untreated material is compared to the surface area for the treated material, also shown in Figure 9.2. There is a slight degradation in the surface area from 0.301 to 0.226 $m^2 g^{-1}$ (25% change). As reduction cycles are continued from the 25th to the 75th cycle, the surface area increases from 0.226 to 0.557 $m^2 g^{-1}$, a 250% increase. This new surface area is from the formation of gas pores and lath structure penetrating the particle, which helps to increase the surface area by about 175% over that for the fresh hematite.

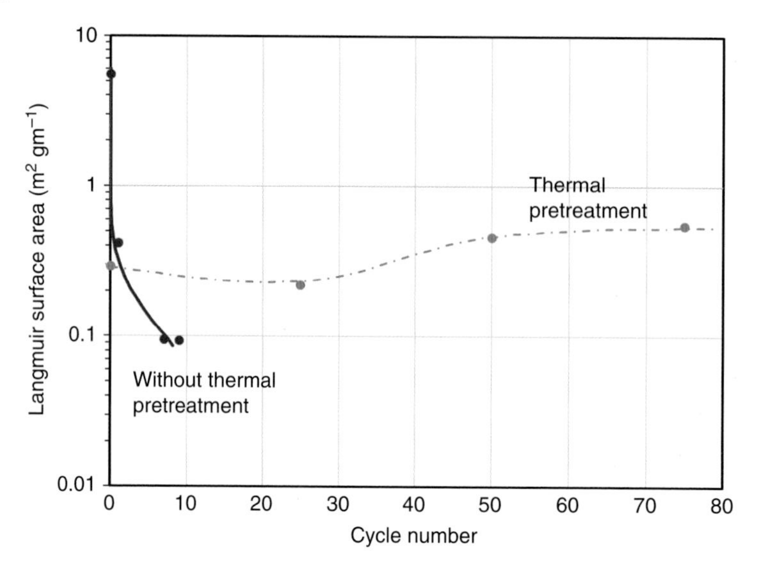

Figure 9.2 Surface area of thermally pretreated and untreated hematite oxygen carrier [5].

For the untreated oxygen carrier, the surface structural components change as the reduction process proceeds. A probable explanation of this phenomenon is that the hematite particle is made up of grains varying in size. Owing to the thermal grain growth [6], large grains consume smaller grains. This process opens lath structures between large and growing grains as the small grains are consumed. As the surface phases change and volume expansion occurs on the oxide surface, the growth of the lath structure is slowed [8–10]. Hayes and coworkers conducted an experiment confirming the phase of the surface and change in structure formations with temperature and bulk gas content [11].

These lath structure paths form according to the crystal structures defining the inner grains of the pellet. Additionally, thermal expansion and stresses may also cause a shift in these crystal structures and cut off the lath structures that form within the grain. The material goes through a biphase change of Fe_2O_3 and Fe_3O_4 between the temperatures of 650 and 810 °C until it stabilizes at 850 °C [10]. This may create a lag in formation of a lath structure into the hematite grain. The effects of thermal exposure are illustrated in Figure 9.3.

9.2.1.2 Structure

Besides the surface, bulk structures can undergo very large chemical stresses during the redox process. For example, hematite particles experience volume changes during the reduction process in the following trend: Fe_2O_3 (100%) → Fe_3O_4 (124%) → FeO (131%) → Fe (126%) [12–14]. Yi et al. [8] investigated the phase composition of iron containing particles between 800 and 1000 °C in an atmosphere of 1 : 1 H_2 and CO by analysis of X-ray diffraction (XRD). The authors observed the magnetite, wüstite, and iron phases that coexisted in the swollen pellet at 800 °C. As the temperature increases, magnetite

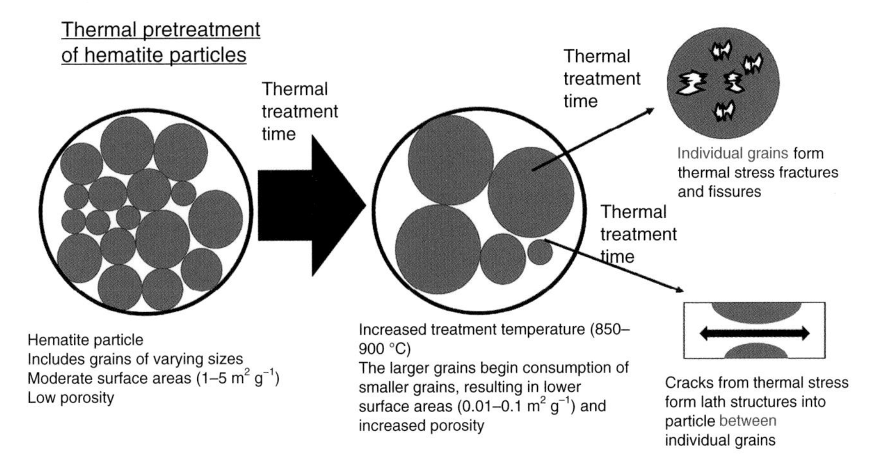

Figure 9.3 Thermal exposure effects on hematite.

and iron phases dissipate as wüstite increases. At 1000 °C, the magnetite phase is completely reduced and wüstite becomes 87.5% of the particle. Meanwhile, the proportion of wüstite in particles at 800 and 900 °C was only 55.5% and 70.3%, respectively. This volume change leads to mechanical property changes. Particle strength has shown strong correlation with volume change in hematite. The strength sharply decreases to 20%, from ~2500 N per pellet to the lowest ~500 N per pellet at 10–20 minutes reduction times [8]. At reduction times of 10–20 minutes, particle formations consisted primarily of the wüstite phase. The conclusion is that most structural damage occurs in the site stage where the particle showed the maximum swelling and lowest strength. Thereafter, bonding between grains enhanced with the iron phase formation and particles experienced volume contraction and strength improvement.

Two distinct microstructures have been identified during the reduction of hematite to magnetite. It has been reported [15] that at low reduction temperatures a porous magnetite phase is formed, which advances equiaxially into the solid from nucleation sites at the oxide/gas interface. Figure 9.4a shows the porous growth morphology in a partially reduced hematite crystal and Figure 9.4b a fully reduced sample. For reaction temperatures greater than 700 °C, the magnetite grows in the form of dense plates or laths, which traverse the width of the particles while other areas of the oxide remain completely untransformed (Figure 9.4c) but partially reduced; Figure 9.4d shows the transformation across the width of the crystal. The experimental method used by Brill-Edwards et al. for the preparation of the reduced oxide is significant to the reduction morphology. $H_2/H_2O/Ar$ gas mixtures were adjusted so that the reduction of the pellets was achieved after an arbitrary time of six hours; no specific gas mixtures are quoted for the reduction temperatures.

A reaction mechanism between methane and hematite was observed based upon the earlier work of Breault and Monazam [7]. The reaction of methane with hematite takes place according to the process defined below. Methane migrates

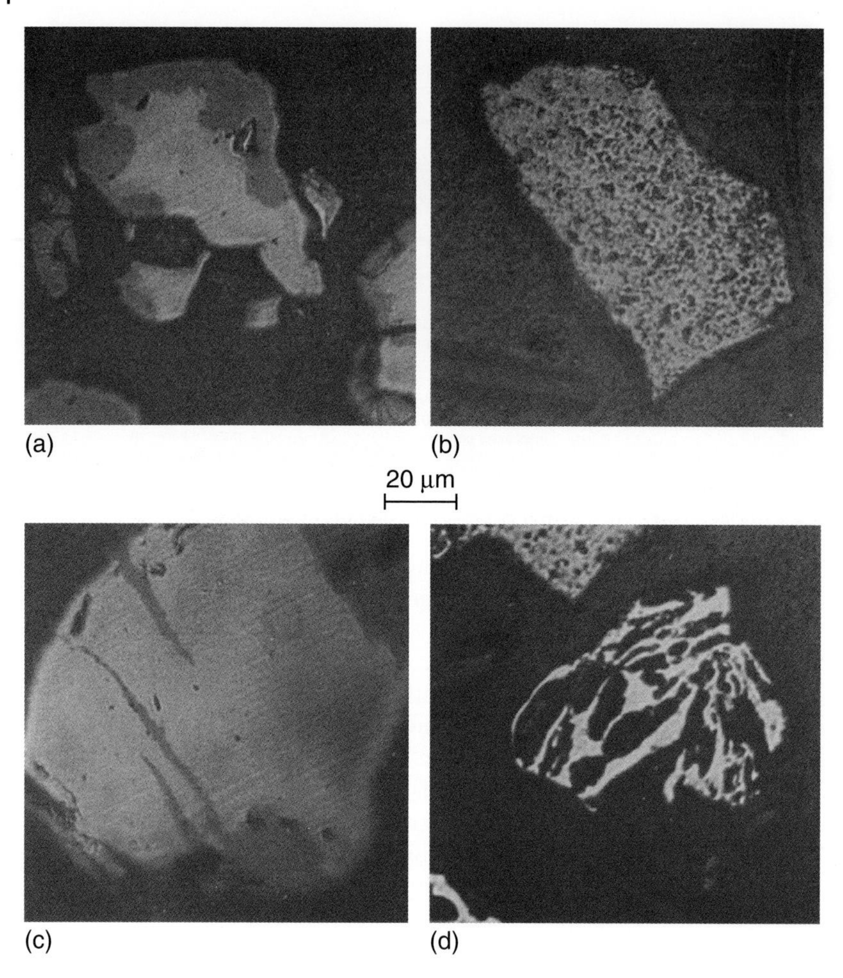

Figure 9.4 Microstructures observed during the $Fe_2O_3 \rightarrow Fe_3O_4$ transformation: (a, b) porous magnetite; (c, d) lath magnetite (Fe_2O_3 has the lighter appearance). Source: Hayes and Grieveson 1981 [15]. Reprinted with permission from Springer Nature.

to the particle surface region (outer shell of particles/grains) and reacts within a shallow layer on the order of 70 Å associated with the shell to produce CO and H_2 species and likely the solid species FeO. This is shown in the sketch presented in Figure 9.5. The carbon monoxide and hydrogen can either react very rapidly with hematite in this shell or will escape from the particle through the large pores to react downstream. Hematite and hydrogen reaction kinetics [17–19] are so fast that the observed free hydrogen is usually at the detection limit of most laboratory gas analyzers except when the methane undergoes cracking late into the particle reduction stage. Once a reasonable large share of the surface-available oxygen has been consumed, oxygen diffuses from the core (particle/grain) to the surface region where it reacts with the migrating methane in the same chemical reactions, although the rate is controlled by the speed at which oxygen can diffuse from the core hematite lattice to the attached methane.

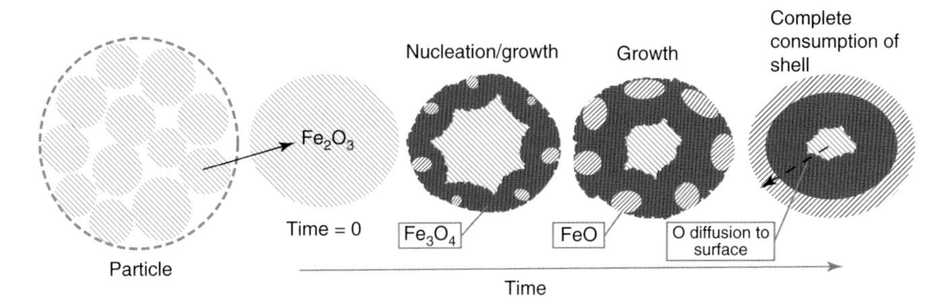

Figure 9.5 Particle model concept for hematite. Source: Breault and Monazam 2015 [16]. Reprinted with permission from Elsevier.

9.2.1.3 Gas Pores and Diffusion

As discussed in the above sections, the actual surface of the oxygen carrier represents minimal Langmuir surface area. This means that an intricate pore structure is obtained for oxygen carriers. As the oxygen carrier's surface area decreases with exposure to high temperatures (>800 °C), generation of a new surface area during the reduction process through the diffusion of gas pockets in the core of the pellet to the outer surface is needed. Fick's law of diffusion suggests that the rates should decrease with time because the diffusion distance increases as the reaction progresses [20]. However, through a number of cycles, there is a progressive movement of gas pores from the sample surface toward the dense Fe_2O_3 material as pores and larger lath structures are formed, which maintains a high rate of diffusion [21]. Instabilities on the surface form central gas pores. Gas micropores (<0.1 μm diameter) form 45° below the central gas pore [11]. Fracturing of the product oxide occurs at right angles from the pore as the phase changes from magnetite to wüstite. As the phase change occurs, volume expansion/shrinkage can occur, which prompts stresses at the growing pore tip, and the growth of these porous structures is influenced by the preexisting crystal structure [21]. Controlling and maintaining pore penetration with stable lath structures deep within and between grains are essential for stabilizing the particles' ability to oxidize the fuel. Figure 9.6 illustrates this phenomenon. The reaction front continues to grow until it intersects with the micropores, forming a dendritic structure. Between these

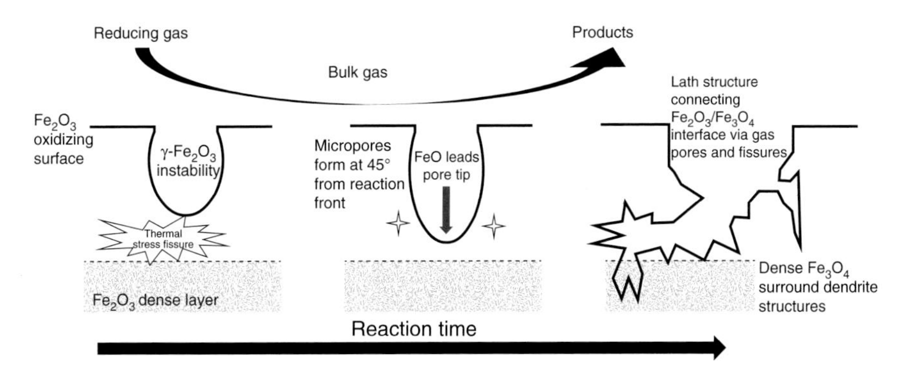

Figure 9.6 Development of lath structure into hematite material.

dendritic, treelike structures are volumes of dense magnetite reaction product. The lath structure penetrates deeper, growing slowly toward inner gas pores and fissions until a network between the inner grains of the pellet is completed after several cycles.

Hayes and coworkers showed that with higher thermodynamic driving forces, the formation of stress cracks, or fissures, was found within the porous product structures planar to the dense Fe_2O_3 material [8, 21]. These fissures can serve as interconnecting pathways as the gas pore penetrates the material. The grain growth and structure of the material are stabilized through thermal pretreatment. Within the 850–950 °C range, the surface of the pellet is stabilized and gas pores grow without impedance of shifting crystal structure.

9.2.1.4 Ilmenite

Ilmenite is a naturally obtained iron-based ore that contains titanium. Ilmenite undergoes significant structural changes during the redox process involved with chemical looping. The major change with respect to the initial ilmenite is that porosity increases substantially with the redox cycles as the surface area decreased. For a 100 cycle test using syngas as reducing agent [22], the initial porosity of calcined ilmenite was measured to be 1.2%, after 8 cycles it increased to 12.5%, after 20 cycles it was 27.5%, and after 100 cycles it reached the value of 38% [22]. Final Brunauer Emmett Teller (BET) surface area measurements for activated ilmenite were as follows: with CH_4, after 23 cycles, 0.4 $m^2\ g^{-1}$; with CO, after 20 cycles, 0.6 $m^2\ g^{-1}$; and with $CO + H_2$ mixture, 0.4 $m^2\ g^{-1}$. The initial calcined ilmenite had a BET surface of 0.8 $m^2\ g^{-1}$.

Figure 9.7 shows the morphological characterization of the ilmenite oxygen carrier during 100 redox cycles. Scanning electron micrography (SEM)

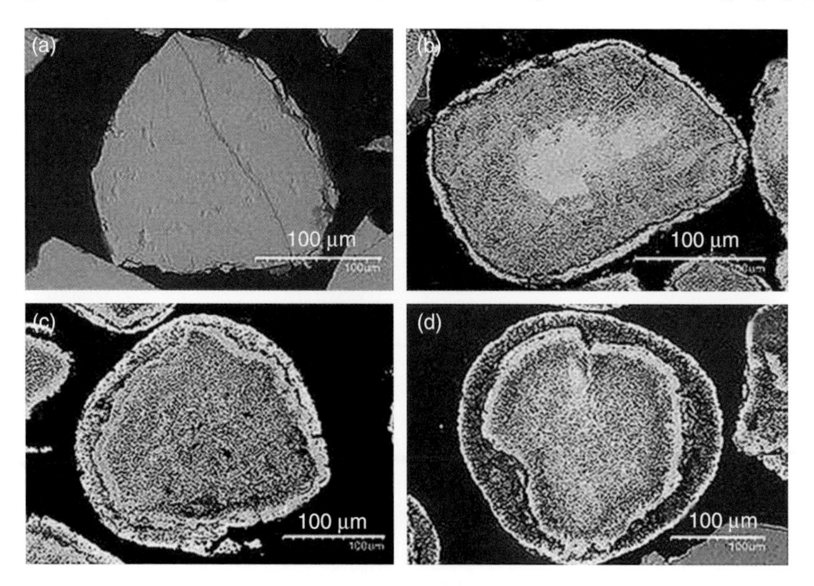

Figure 9.7 SEM–EDX images of cross-cut ilmenite particles (a) calcined and after (b) 16, (c) 50, and (d) 100 redox cycles. Source: Cuadrat et al. 2012 [22]. Reprinted with permission from Elsevier.

microphotographs confirm the low pore development for calcined ilmenite. Throughout the reduction–oxidation cycles, there is continuous appearance of cracks, which generate the abovementioned rise in porosity. Moreover, a gradual generation of an external layer slightly separated from the rest of the particle, which grows with the number of cycles, could be clearly observed. This space between the layer and the core also enhances the porosity measured for the particle.

Electron dispersive spectroscopy (EDX) analyses were done to determine Fe and Ti distributions throughout the particles. In calcined ilmenite both distributions were uniform, which agrees with the XRD analysis finding that Fe_2TiO_5 is the main component. With increasing number of cycles, Fe_2TiO_5 has been seen to undergo a physical segregation and the particle core gets titanium enriched, whereas the external part gets iron enriched. XRD analyses of the external layer found that this region is composed only of iron oxide; XRD of the internal core revealed the existence of TiO_2 and Fe_2TiO_5. After each cycle, there is an increase in the amount of TiO_2 in the core and iron oxide in the external shell, together with a decrease of iron titanates. This means that there is migration of iron oxide toward the external part of the particle, where there is no TiO_2 to form iron titanates. Since the active phase for a CLC application is the iron oxide, the fact that iron oxide is at the outer part of the solid facilitates the reaction.

9.2.2 Supported Oxygen Carriers

9.2.2.1 Copper Oxides

Copper oxides are highly active oxygen carriers owing to their ability to release gaseous oxygen as a potential chemical looping with oxygen uncoupling carrier and are talked more about in detail in Chapters 7 and 8. While pure CuO has mechanical stability issues such as agglomeration due to the relatively low melting point of pure Cu, supported Cu oxides such as that with alumina are commonly used in chemical looping applications. Forero et al. used a Cu–Al oxygen carrier extracted from the air reactor (AR) at different operation times to determine changes produced in the structure of the solid materials during operation [23]. Moreover, some particles were embedded in resin, cut, polished, and analyzed by SEM and EDX. Images of the cross section of fresh and used particles at different tests and operation times are shown in Figure 9.8. Fresh Cu–Al particles exhibited an outer shell of CuO, which disappeared during the first hours of operation by the process of attrition. This qualitatively agrees with the CuO loss observed in the particles during operation. The CuO content of the particles decreased from 14.2% in the fresh particles to about 11.8 wt% after 63 hours of operation at high temperature.

Samples of used Cu–Al particles presented important changes both in the integrity and CuO grain dispersion depending on the temperatures used. At the lower temperatures, $T_{FR} = 1073$ K and $T_{AR} = 1173$ K, the small CuO grains are dispersed throughout the entire matrix, which was corroborated with an EDX profile. An increase in AR temperature produces an increase in the CuO grain size by sintering of copper. At the highest temperatures, $T_{FR} = 1173$ K and $T_{AR} = 1223$ K, oxygen carrier particles start to break apart, forming fines.

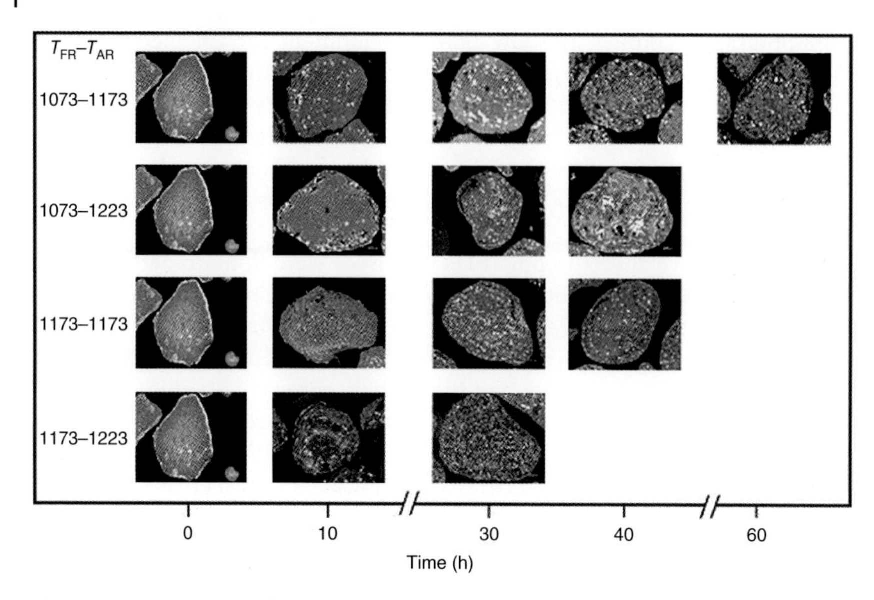

Figure 9.8 SEM pictures of a cross section of fresh and used Cu–Al particles at different pair FR–AR temperatures and operation times. Source: Forero et al. 2011 [23]. Reprinted with permission from Elsevier.

Although the exact mechanism of formation of these structural defects is unidentified, the sintering of copper and the formation of the metastable $CuAlO_2$ compound could help to break the particles.

9.2.2.2 Iron Oxides

Cho et al. looked at zirconia-supported iron-based oxygen carriers as they undergo morphological changes during repeated redox cycles with respect to the calcination temperature, degree of solid conversion, and gas used in the oxidation reaction [24]. Four types of activated particles with different reactivity could be obtained with respect to solid conversion (i.e. FeO or Fe) and the gas used for the oxidation reaction (i.e. steam or air) shown in Figure 9.9. These conditions determined mainly the distribution of iron species and the formation of cracks or pores during the repeated cycling. Cho et al. reported that reduction to FeO was favorable for the bottom product of the fuel reactor (FR) when CH_4 was used as a feedstock due to the large endothermic reaction between CH_4 and Fe_2O_3 in the FR [25, 26]. This reaction means that a reactor design and process operation based on the data from the Fe–Fe_2O_3 particle would not be correct if the bottom product of the FR was determined to be FeO or vice versa.

Cho et al. observed more cracks or pores formed after air oxidation for the iron-based oxygen carriers than after steam oxidation [24]. Crack or pore formation was attributed to the atomic volume increase of the iron oxide because such increases in gas–solid reactions could increase the internal stresses in the product layer, causing its possible fracture [27]. However, this explanation is not sufficient to explain the rare formation of pores or cracks as well as the aggregated surface of the activated Fe–Fe_3O_4 particle in which the molar volume also

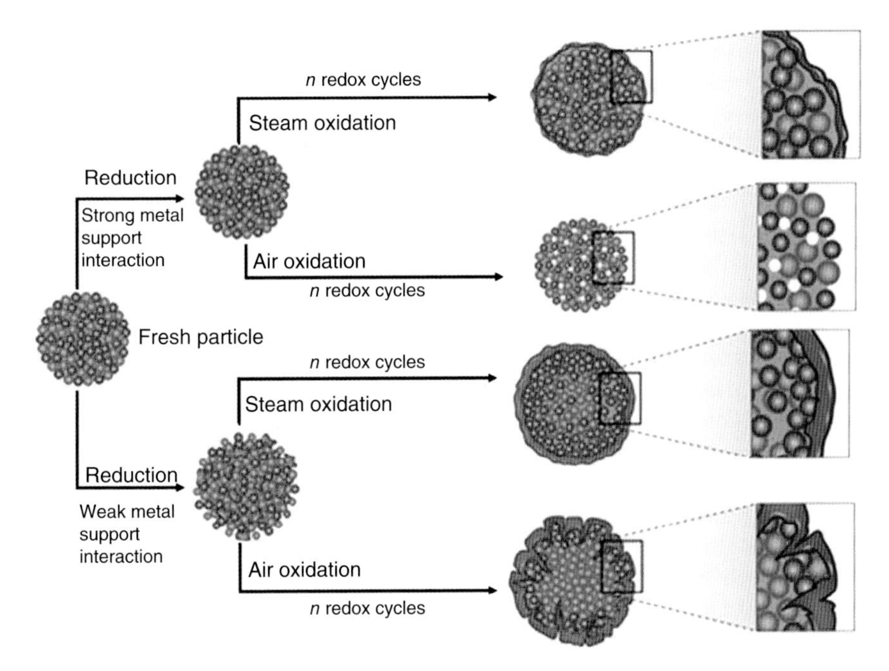

Figure 9.9 Activation of ZrO_2 supported iron oxides under various oxidation conditions. Source: Cho et al. 2015 [24]. Reprinted with permission from American Chemical Society.

increases. Mitchell and coworkers [28] observed spalling and decohesion of the outer hematite layer during the high-temperature oxidation of pure iron, and analysis of the resulting scrolled oxide indicated a strong compressive stress gradient. The observed growth stress at the Fe_2O_3–Fe_3O_4 interface was explained by a compressive stress in the Fe_2O_3 layer and a tensile stress in the Fe_3O_4. Additionally, the heat release in the particle in the oxidation step should be considered. During the oxidation step with O_2, there is no gaseous product that converts the heat away from the reaction site. All the exothermic energy is released within the particle. This results in very high local temperatures and rapid grain growth, which might result in local stress. When steam is used as the oxygen regeneration source, the heat of reaction is significantly less. Additionally, there is H_2, which is produced via the steam–Fe reaction, to convert a part of the heat away and might lead to less growth and fewer pores and cracks. Therefore, the formation of many pores or cracks found in activated FeO–Fe_2O_3 and activated Fe–Fe_2O_3 particles could be explained by a localized stress at the interface between the lower oxides and Fe_2O_3 and/or rapid changes in local temperature.

Besides the porosity, the distribution of iron species had a notable effect on the reactivity of the zirconia-supported iron oxide oxygen carrier. The diffusion of iron species during the activation, mainly due to the sintering of iron species, contributed to the formation of the activated particles covered by the aggregated iron oxide. This particle (i.e. Fe–Fe_2O_3) exhibited a relatively low yield of H_2 and the possible formation of unreacted sintered iron. This result implied that an oxygen carrier with iron oxide sintering potential is at risk of lowering the amount of H_2 from the steam reaction as well as loss of reactivity in the application to chemical looping. In addition, the low reactivity of the activated particles oxidized by only steam were mainly attributed to the agglomeration of iron oxide due to the growth of the atomic volume or sintering of the iron oxide. Finding the optimal activation procedure that fully invigorates the oxygen carrier will affect the start-up operation, reactor design, and overall system efficiency. Air oxidation after the reduction of iron oxide without oxidation by steam was found to be the optimal activation procedure with the minimum number of cycles.

Galinsky et al. investigated supported iron oxides for stability and activity in methane [29]. The authors investigated supports with various chemical and physical properties. These included perovskite ($Ca_{1-x}Sr_xTi_{1-y}Ni_y$), fluorite (CeO_2), and spinel ($MgAl_2O_4$) supports. The perovskite and fluorite supports are known mixed-ionic and electronic conductors (MIECs) while the spinel support is an inert ceramic. Mixed conductors can transfer electrons and oxygen anions at a high rate. For the $Ca_{1-x}Sr_xTi_{1-y}Ni_y$ supported oxygen carrier, the oxygen carrier deactivates slightly in the first few redox cycles, but exhibits high redox activity and stability over multiple redox cycles. The authors observe a surface area decrease by nearly a factor of 4 over the first 10 cycles. The corresponding deactivation is less than 15%. Additionally, no volume expansion is observed for this oxygen carrier over 50 redox cycles. CeO_2-supported iron oxide deactivates by iron oxide aggregation on the surface, through Fe^{x+} migration (similar to that seen with ilmenite oxygen carriers [22]). Aggregation of iron oxide causes deactivation through increased gaseous and solid-state diffusion resistances [30, 31]. Decrease in sample surface area, as observed in 800 °C sintered sample, is a consequence of iron migration/agglomeration as opposed to the primary

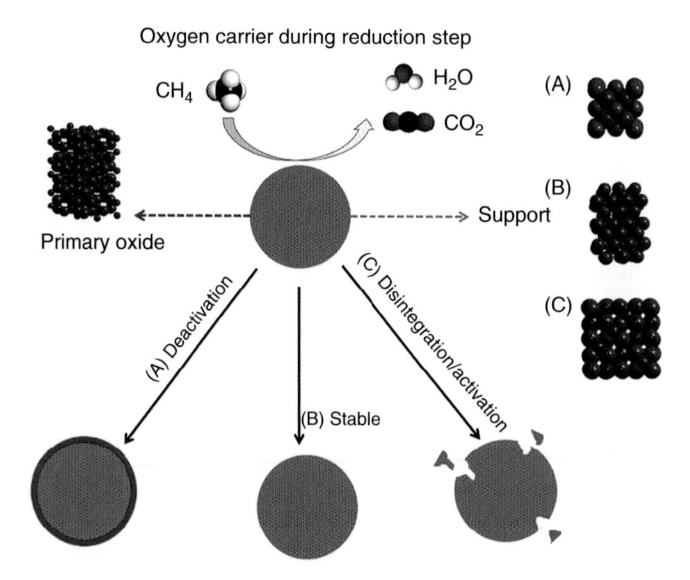

Figure 9.10 Summary of structural changes observed during redox reactions with CH_4/O_2 for variously supported iron oxides. Source: Galinsky et al. 2015 [29]. Reprinted with permission from Elsevier.

cause for deactivation. In contrast to the mixed conductors that both deactivated to some degree, spinel $MgAl_2O_4$ was observed to activate in redox cycles with methane and oxygen as the fuels. Increase in surface area was observed with repeating cycles. The increase of surface area is determined to be a result of deformation of the oxygen carrier particles. The deformation of the particle is due to growth of the filamentous carbon, causing interstitial gaps within the oxygen carrier. Visual inspection of the sample also indicates significant expansion of the oxygen carrier sample over multiple redox cycles. The same oxygen carrier deactivated over time when hydrogen was used as a fuel. This further confirms that the carbon filaments, because of the methane decomposition reaction, helped activate the oxygen carrier. A summary of the structural and activity stability of the oxygen carriers is shown in Figure 9.10.

9.3 Attrition

Because of the great variety of forces and energy sources in a circulating system, attrition in a chemical looping system has the potential to be multifaceted and ubiquitous. Various authors have defined the term attrition slightly differently. For the purposes of this chapter, attrition will be defined as the unintentional breakdown of the oxygen carrier. Figure 9.11 shows that attrition can be divided into two main types: abrasion and fragmentation [32]. Abrasion is the main mechanism of attrition when particles are subjected to lower velocities. The wearing of surface asperities leads to the production of very fine particles, which are easily elutriated. Given that the mother particles follow a single normal particle size distribution, abrasion results in a bimodal particle size distribution. When particles are subjected to higher velocities, attrition is controlled by particle fragmentation. Fragmentation causes the particles to break into two or

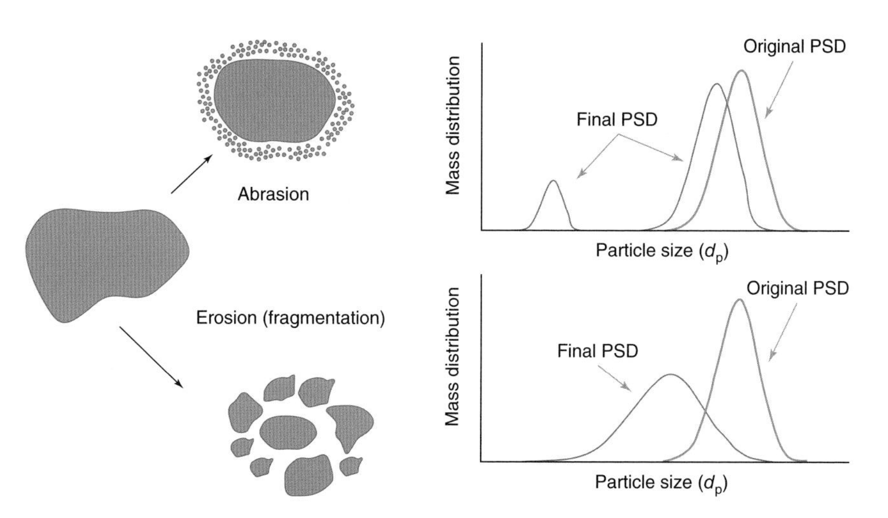

Figure 9.11 Particle size distributions of abrasion and erosion attrition mechanisms. Source: Bayham et al. 2016 [1]. Reprinted with permissions from Elsevier.

more fragments that are non-elutriable and can be subject to further attrition. Fragmentation is divided into primary and secondary fragmentation. Primary fragmentation is the result of feeding fresh particles into a hot furnace whereas secondary fragmentation occurs in a fluid bed or process [33]. In summary, abrasion is generally more responsible for fines production and economic losses from material elutriation and replacement of the lost sorbent, while fragmentation causes greater changes in the mean bed particle size, resulting in decreased performance and increased control issues [34]. The production of particle fragments, however, allows for the possibility of a greater rate of surface abrasion, since the daughter particles from fracture are non-spherical, and the protuberances are more easily worn compared to spherical particles.

9.3.1 Sources of Attrition

Attrition of particles can be expected when the material undergoes stress in any of the three categories: (i) thermal, (ii) chemical, and (iii) mechanical (or hydrodynamic) [35, 36]. These are described in detail as follows. Thermal stress is produced as a result of the uneven expansion or contraction of the particle as it is heated or cooled, causing decrepitation [37]. During the feeding of cold particles to a high temperature region, such as a hot fluidized bed, the material can spall due to thermal shock [35]. Examples of this include the addition of cold solid fuel to a hot fluidized bed combustor or the addition of cold makeup catalyst or sorbent to the system. Additionally, thermal stress is exacerbated when the particles are spatially confined, e.g. at the bottom of a standpipe or bed, and when the particles undergo cyclic heating and cooling, e.g. in a chemical looping system, where the particles undergo reduction at one temperature and oxidation at a higher temperature.

Chemical stress is produced as a result of a chemical reaction on the surface of the particle (e.g. fluidized catalytic cracking or CO_2 adsorption catalysts) or within the particle (e.g. coal or metal oxide oxygen carrier). Chemical reaction produces internal stresses that change the lattice structure. This weakens the particles and produces surface features that can be easily abraded upon application of small external stresses. For oxygen carriers that undergo both reducing and oxidizing reactions, the phase change of the parent material produces different physical properties because each possible phase in the particle exhibits different properties. For example, for the reduction of hematite by carbon monoxide or hydrogen, such as in a chemical looping system, the hardness of the material was shown to decrease due to the reduction in hardness of the corresponding phases [37]. Additionally, if the gas–solid reaction is highly exothermic, chemical stress can be coupled with thermal stress, causing temperature gradients to occur. This is common in CLC reactors as the oxidizing reaction is exothermic.

Mechanical stresses are those that are a result of compression, tensile, or shear stresses on a particulate material, which can be the result of the simple support of static loads or more complex due to the dynamic motion of particles. Mechanical stresses are divided into two: static and kinetic [38]. Static stress can be the result of particles undergoing a compressive load with no motion or a very low velocity of solids, such as in the lower portion of a standpipe, at the bottom of a bed,

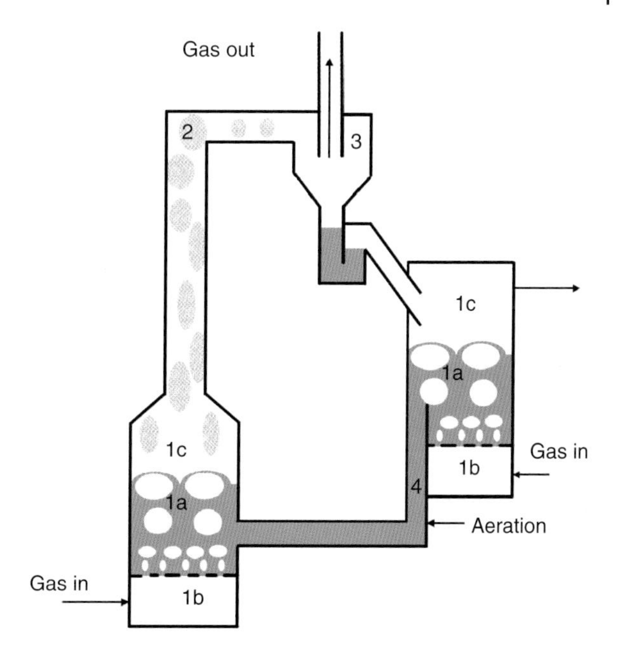

Figure 9.12 Locations of attrition in a chemical looping reactor system.

or in a silo. Kinetic mechanical stresses involve the motion of solids, whether at high velocity such as impacts with the wall in a riser, or at lower velocity where the particles rub together when bubbles pass through a fluidized bed. The particles can undergo both compressive and tensile stresses. When particles undergo impact, tensile stresses are exhibited as a portion of the material wedges the particle apart; after a low- or high-velocity impact, the particle can crack under the release of residual tensile stresses [39, 40].

A summary of the locations in chemical looping circulating fluidized beds (CFBs) where attrition due to mechanical (static and kinetic) stress is likely to occur is shown in Figure 9.12. The locations for attrition to occur consist of the (i) fluidized bed, (ii) conveying lines and risers, (iii) cyclones or other gas–particle separation devices, and (iv) standpipes.

The sources of attrition in a fluidized bed are threefold. The first area is bubble attrition (1a in Figure 9.12), which is the result of particles rubbing against each other due to bubbles passing through the bed [41]. Thus, the main mechanism of this type of attrition is particle–particle abrasion. The attrition rate is proportional to the "excess" velocity in the bed, namely the superficial gas velocity minus the minimum fluidization velocity. In beds that do not use porous distributor plates, or in beds where the introduction of fluidizing gas is highly localized, grid jet attrition can occur (1b in Figure 9.12). The jets have the propensity to provide a high amount of kinetic energy to the solids by entraining solids locally through the bottom of the jet and accelerating particles toward other particles. As a result, gas jets provide a location for both abrasion and erosion, since the impact can produce fragmentation or chipping due to the high impact velocity. The third area is the freeboard region (1c in Figure 9.12), where breakage of bubbles causes whole particles to become ejected into the freeboard region to impact other particles or the wall. This area is more difficult to quantify because it consists of a small region of low-velocity impacts. The attrition in the freeboard region can be preliminarily assumed to be negligible, except in the case of fluidized beds in the turbulent fluidization regime. Depending on the conditions, the freeboard attrition can also be assumed to be negligible if the particle–particle impact velocities are lower than the threshold velocities required to initiate chipping or fracturing.

For risers and conveying lines, the gas velocity required to convey particles has the potential to create high-velocity impacts for particles near sharp turns or bends [42]. Additionally, in some CFB designs, the particles may impact U-beams that intend to separate flue gas from the bed material [43] and anti-wear beams in the riser that intend to reduce the downward particle velocity in the annular region [44, 45]. Impact in these locations can cause significant particle fragmentation if the impact velocity is above the threshold velocity.

Because the principal of operation of a cyclone requires high velocities of gas and solid flow to separate gas and solids, the rate of attrition has the potential to be high. Zenz has stated that attrition in cyclones for a fluid catalytic cracker (FCC) process has been shown to be several orders of magnitude greater than in gas grids in the fluidized bed [46]. For a traditional reverse-flow cyclone, there are two main regions where particle attrition occurs: (i) high-velocity impacts of the particles at the tangential inlet section and (ii) particle abrasion on the walls of the cyclone as the system reaches steady state. The particles from the inlet are thrust toward the wall due to the vortex pressure gradient [47]. This is mainly prevalent for fresh materials, where the weaker fraction of the manufactured material is easily fractured as a result of impact. The produced daughter particles from the impacts are more easily worn down. When the system attrition reaches a steady state, the mode mainly shifts to particle abrasion, which is the result of particles sliding against the walls of the cyclone as they travel down toward the collection hopper or standpipe [48, 49]. For steady-state abrasion, the rate of attrition has been shown to be a function of the inlet gas velocity to the power of 2.5 [47]. Surprisingly, the rate of steady-state attrition in cyclone experiments has been shown to be independent of cyclone geometry [47].

Other sources of mechanical attrition in chemical looping CFB systems include standpipes. The main mode of attrition is by abrasion of particles with the wall of the standpipe and additionally by particle–particle abrasion due to radial solid velocity gradients. Additionally, for standpipes that have gas injected locally, such as non-mechanical valves, the attrition can be a result of fragmentation, similar to that for jets as described for fluidized beds. However, in population balance models used for determining particle size distribution as a function of time and attrition in CFBs, the abrasion in the standpipe has been assumed to be negligible [49, 50].

9.3.2 Solids Properties Relevant to Attrition

As described previously, the attrition of particulate solids can be divided into abrasion (wear) or erosion (fragmentation). This section will delve deeper into these topics to show how to quantify the rate of each process based on known material properties. In theory, with knowledge of the phenomenological rate laws for a material, the rate of attrition can be determined using information about the stresses and strains exhibited in various sections of a CFB system. Many of these parameters described in the following sections are summarized in Figure 9.13.

9.3.2.1 Hardness

Hardness is defined as the resistance to local plastic deformation on a material's surface. Historically, hardness has been categorized into three different types: (i) scratch, (ii) static indentation, and (iii) dynamic or rebound hardness [51]. Scratch hardness describes the ability of standard materials to scratch a sample, and its use is limited to simple characterization in mineralogy. The static indentation hardness is quantified by applying a load to an indenter of a certain shape onto a material and measuring the resulting area of indentation. Several examples include tests by Brinell, Vickers, Knoop, and Rockwell [6]. The hardness value mainly used for attrition research is the indentation hardness, and this parameter shows up in numerous models for abrasion and erosion, both for particulate and non-particulate (e.g. agglomerate) materials [52]. For a Vickers pyramid, the hardness, with a unit of pressure, is obtained by dividing the load applied by the area of indentation formed, based on the length of the indents from the pyramid. The Vickers hardness (HV), denoted as H, is then determined from a relationship between the load and the depth of penetration as shown in Eq. (9.1) [53]:

$$H = 1.8544\,\frac{F}{l^2} \tag{9.1}$$

where F is the contact force and l is indentation length. As described in later sections, the hardness of a material is an important parameter for determining the rate of wear and the loss of material upon impact. For applications of wear, assuming that the hardness of the asperities is the same as that of the entire material, asperities are broken off when the load applied is close to the flow pressure, which is related to the hardness [54].

Numerous factors affect the hardness of a material and the rate of attrition (both abrasion and erosion). The surface treatment of particles, such as in a fluidized bed or in a ball mill [55], causes the particles to become work hardened [56]. Work (or strain, or cold) hardening is the resulting increase of surface hardness due to the application of loads that cause plastic deformation. Surface hardening is a result of the increase in dislocations in the material [6]. Dislocations exhibit localized fields of compression and tension, and since the number of dislocations increases with strain, the motion of the dislocations is constrained due to repelling of the compression or tension fields [6]. This is usually reversed by applying heat to the material, which releases the stress around the dislocations. Kimura et al. studied the effects of ball milling on the microhardness of iron powder (−100 mesh), and the Vickers hardness increased from around 50 HV (~0.5 GPa) up to 950 HV (~9.5 GPa) due to the work hardening effect [55].

Another important factor for hardness is the temperature, which can change the hardness in two ways. The first is the presence of different phases inside a material at varying temperatures, and the second is the effect of temperature on each individual phase. The existence of different phases is also attributed to

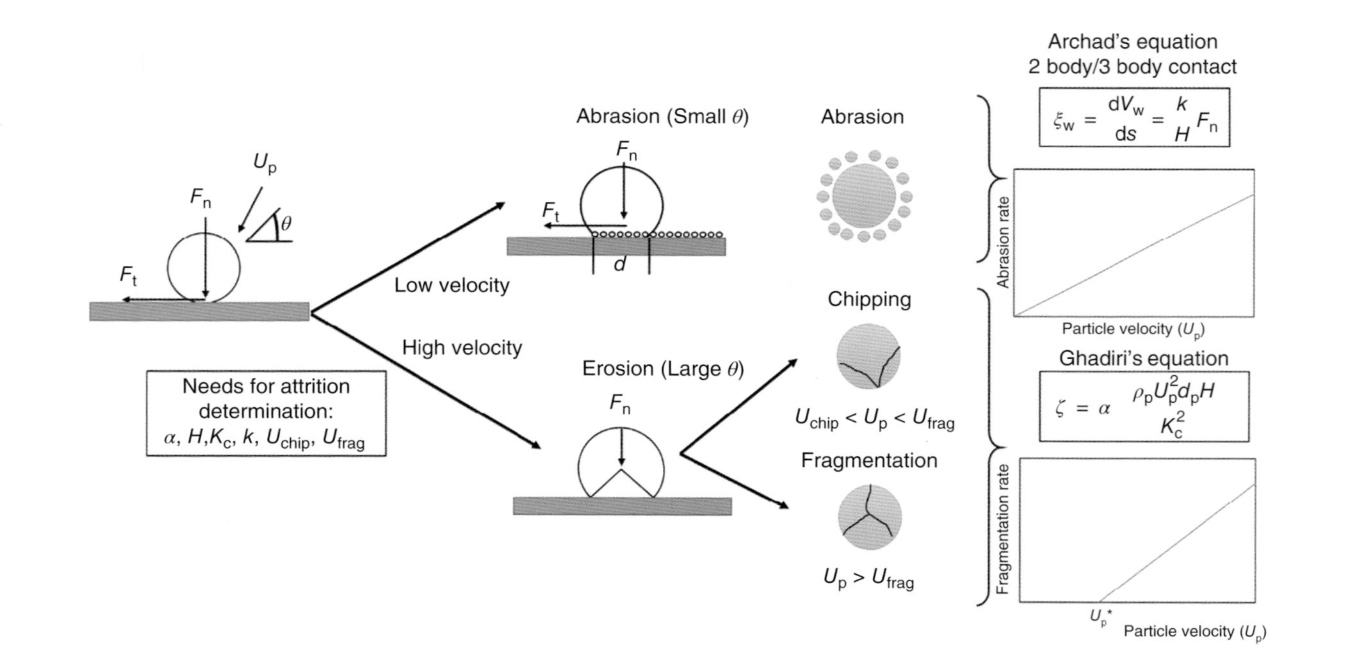

Figure 9.13 Summary of abrasion and erosion expressions. Source: Bayham et al. 2016 [1]. Modified with permissions from Elsevier.

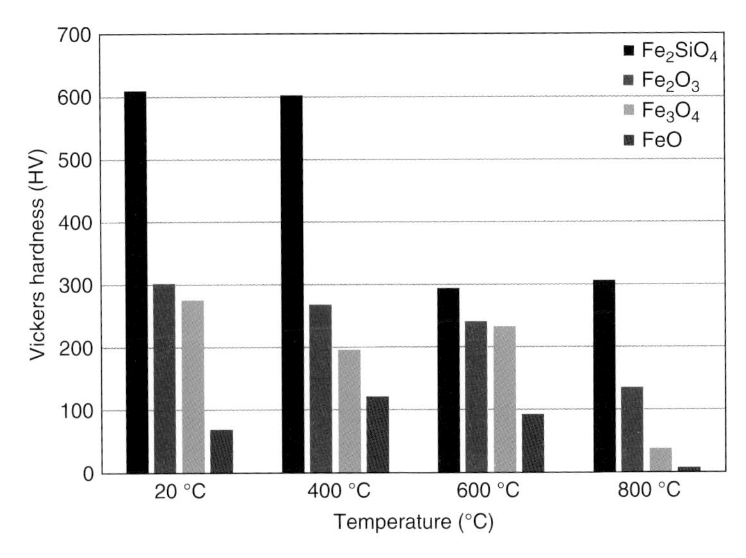

Figure 9.14 Hardness of iron oxides and an iron silicate as a function of temperature [57].

the partial pressure of the oxidizing or reducing gases in the particle environment. Generally, the hardness of a material decreases with temperature as dislocation strains are overcome by the added thermal energy. Takeda et al. showed the effects of temperature, from room temperature to 1000 °C, on iron oxides and an iron silicate. Figure 9.14 illustrates the Vickers hardness (HV) for the different iron oxide and iron silicate phases [57]. In each case, the hardness of oxides decreases as temperature increases, and the magnitude of decrease follows the order: Fe_2SiO_4, Fe_2O_3, Fe_3O_4, and FeO. Furthermore, the hardness of Fe_2SiO_4 and Fe_2O_3 is especially high at room temperature, but above 400 °C, the hardness values of the two are similar to the lower oxidation states. The lowest oxidation state (FeO) represents the softest material. From this analysis, it should be noted that iron oxide, as indicated, is a proposed oxygen carrier for CLC, so that as the hardness decreases with increase in temperature, the rate of wear will increase due to the softening of the material, ignoring other factors such as work hardening. Additionally, ignoring the chemical stress introduced by chemical reaction as the oxygen carrier becomes reduced in the FR, it drops to a lower oxidation state, which is also a softer material. Thus, these parameters need to be taken into account in the attrition models for particulate solids.

9.3.2.2 Fracture Toughness

A material may seem to be smooth and without imperfection to the naked eye. However, the material usually contains microscopic flaws such as gaps, cracks, or fissures. When an external load is applied to the material, the gaps in the material act as "stress raisers," or "stress concentrators," causing the stress at the tips of these cracks to go beyond the yield stress – or even the ultimate stress – of the material, even if the external stress applied is not above the yield stress. Depending on the curvature of the crack at the tip, the effective stress at the tip can be as high as 60 times the stress applied externally to the material [58].

For brittle materials, the theoretical stress derived from knowledge of cohesive forces and bond strength in a perfect material are much higher because most materials exhibit these microscopic gaps. The stress of these materials in terms of order of magnitude is usually one tenth of the Young's modulus [6].

In general, the state of stress around a crack in a material can be determined with knowledge of the crack size and load applied. The critical stress intensity factor (often termed the fracture toughness) of a material is the combination of applied external stress and the crack size [59]. Fracture toughness can be thought of as the resistance to brittle fracture, whereby a high value of the fracture toughness will indicate that failure is more likely to occur through ductile fracture. It is also an indication of the propensity for crack propagation, given a crack size and applied load.

The critical stress intensity factor for a material is a function of several parameters. Temperature plays a role in the degree of ductility the material exhibits: the higher the ductility, the more likely the material will undergo ductile fracture instead of brittle fracture. This is shown qualitatively in Figure 9.15. Hidaka et al. performed tensile tests on iron oxides under various temperatures and oxidation states [60]. For α-Fe_2O_3, the fracture of the specimen is via brittle fracture at 800 °C, via intergranular fracture at 1000 °C, and via ductile fracture indicated by necking at 1200 °C [60]. Fracture toughness is a strong function of the temperature and microstructural properties (e.g. grain size) of the material [6].

Composition has a strong influence on crack toughness, as it describes the propensity to produce plastic deformations around the crack tip. Rocha-Rangel studied the effect of metallic additions to aluminum oxide that formed composites on the indentation fracture toughness, the results of which are graphically shown in Figure 9.16 [62]. Furthermore, adding other ductile metals such as nickel and titanium to the aluminum matrix caused an increase in fracture toughness. The added metals are able to bridge the cracks, causing closing stresses [62]. Additionally, certain elements also help maintain a homogeneous

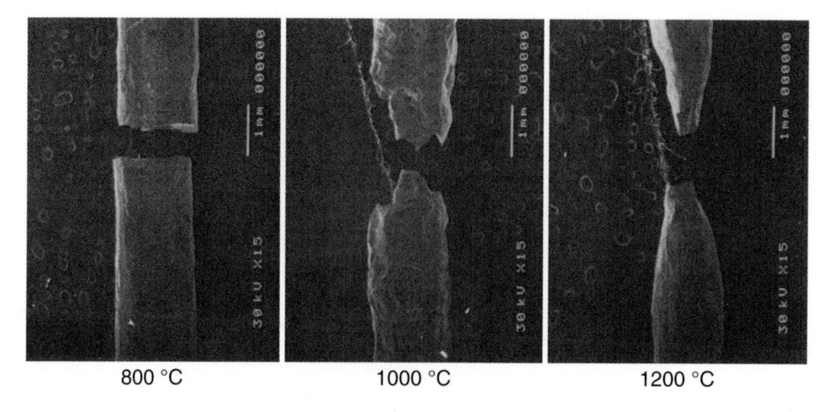

Figure 9.15 Fracture properties of α-iron oxide from tensile stress testing. Source: Hidaka et al. 2002 [60]. Reprinted with permissions from Springer Nature.

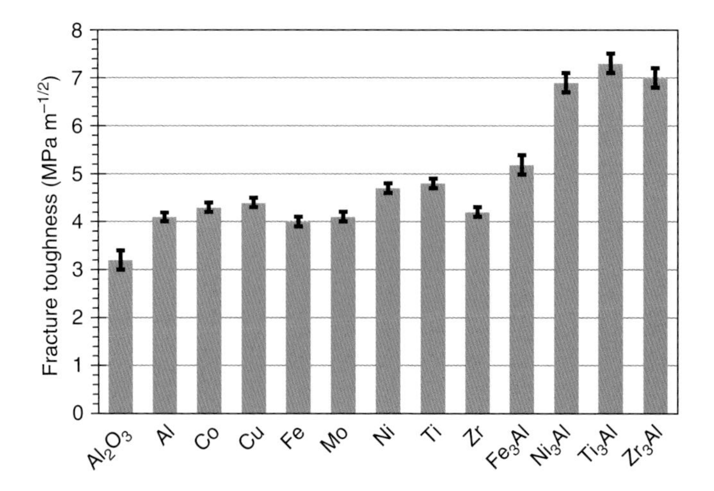

Figure 9.16 Fracture toughness of various metallic and oxide composites of alumina. Source: Viktorov et al. 2014 [61]. Reproduced with permission of Springer Nature.

microstructure in the material. In the study of the mineral grains in rocks, measurements by Viktorov et al. show that magnetite (Fe_3O_4) has greater fracture toughness than hematite (Fe_2O_3), which is due to the lower oxidation state being more ductile [61].

9.3.3 Mechanistic Modeling of Attrition

9.3.3.1 Attrition due to Wear (Abrasion)

The rate of wear (abrasion) can be deduced from laws in the area of tribology, which is the study of rubbing or sliding of solid surfaces [52]. While the study of wear is not new, a universal model for abrasion between two surfaces based on material properties, loads, and sliding distances is not available. On the other hand, there exist hundreds of models that describe the rate of wear based on parameters such as applied load, material properties, and sliding distance [63]. The wear models in the literature can be categorized as (i) empirical, (ii) contact-mechanics based, and (iii) material failure mechanism [63]. A contact-mechanics-based model that has found moderate success is derived from Archard [64]. Archard's equation argues that the distance rate of wear (the volume worn per distance traveled) of surface asperities is independent of the actual contact surface area and linearly dependent upon the normal load applied (P) between the two surfaces and inversely proportional to the material hardness (H) [64]:

$$\xi_w = \frac{dV_w}{ds} = \frac{k}{H}P \tag{9.2}$$

where k is a wear constant, which is added to account for the fact that not all asperity contacts result in breakage. Typically for two-body cases, k is in the range

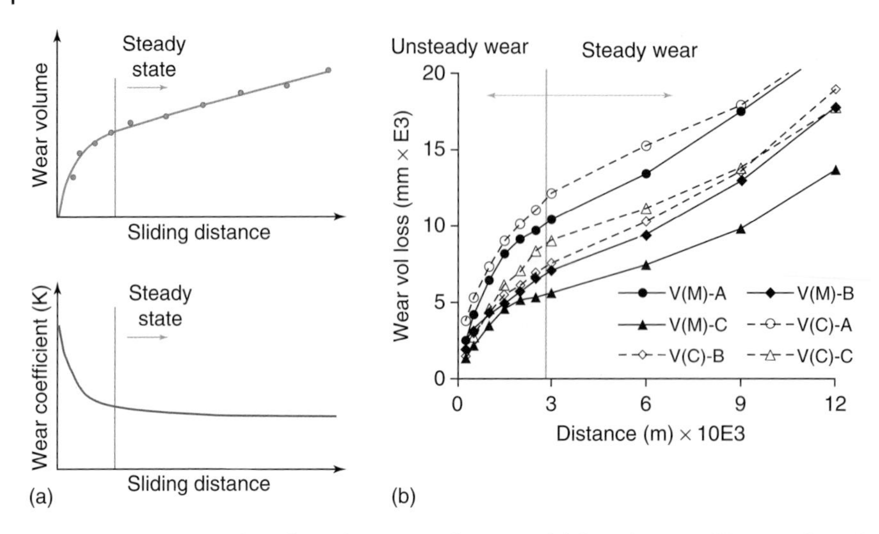

Figure 9.17 (a) Unsteady and steady wear as a function of sliding distance. (b)Wear volume for aluminum samples. Source: Yang 2005 [65]. Reprinted with permissions from Elsevier.

of $0.006 < k < 0.06$ and for three-body cases, it is in the range $0.0003 < k < 0.003$ [54]. The abrasion of particulate solids could be considered a multi-body problem, depending on the conditions. Thus, for a material with known hardness and wear characteristics, if the normal load is known as well as the distance traveled, the wear volume can be calculated.

An important factor to consider in wear experiments that utilize Archard's law is the effect of the initial unsteady-state "break-in" wearing period that is exhibited for fresh materials. For non-particulate samples in tribology experiments, it has been shown that this unsteady period at the start of the experiment is due to wearing down of very rough asperities, as shown in Figure 9.17 as a plot of the wear volume or the standard wear coefficient versus sliding distance. After the initial startup period, the wear coefficient approaches an asymptote and can describe the degree of wear in a system at steady state [65]. This is different than the "cushioning effect" found in comminution experiments in linear time-variant population balance models [66].

Wear has also been considered to be due to the formation of subsurface lateral cracks due to a chipping mechanism. Ning and Ghadiri propose the following relation for wear based on a material failure mechanism due to a particle sliding a curved distance L over another particle [67]:

$$V_{\mathrm{w}} = \frac{0.07P^{\frac{5}{4}}}{\pi^{\frac{5}{4}}K_{\mathrm{c}}H^{\frac{1}{4}}}\left(2L + \frac{\pi^{\frac{1}{4}}H^{\frac{1}{4}}P^{\frac{3}{4}}}{17.1K_{\mathrm{c}}}\right) \tag{9.3}$$

where P is the load applied, K_{c} is the fracture toughness, and H is the material hardness (assumed the same for both particles). The equation for wear is based on a material relation between the crack length and the indentation radius [67]; the contact radius can be related to the hardness and load applied (similar to Archard's law), and both surfaces undergo wear (which is an overestimate, according to Ning and Ghadiri).

9.3.3.2 Impact Attrition

A mechanistic model to describe the fragmentation or chipping process was also developed by Ghadiri and Zhang [40]. The formulation of the model is based on the indentation fracture mechanics of lateral cracks. A fractional loss per impact, ζ, is defined as the ratio of the volume removed from a particle to the volume of the original particle. According to the model, the fractional loss per impact is given by

$$\zeta = \alpha \frac{\rho_p U_p^2 d_p H}{K_c^2} \tag{9.4}$$

where α is a proportionality factor, ρ_p is the particle density, U_p is the impact velocity, d_p is a characteristic particle size, H is the material hardness, and K_c is the critical stress intensity factor [40]. As mentioned by Meier et al., Gahn proposes a slightly different fractional volume removal model upon impact, which is of the following form [68, 69]:

$$\zeta = \alpha \frac{\rho_p U_p^{\frac{8}{3}} d_p H^{\frac{2}{3}}}{G\beta_{max}} \tag{9.5}$$

where G is the shear modulus of the material and β_{max} is the maximum fracture energy of the material. Vogel and Peukert developed a different correlation based on Weibull statistics and parameters from the dimensional analysis of Rumpf [39] on comminution [70]. It relates the breakage probability to two "material" parameters, f_{mat}, which is a propensity of breakage parameter, and $W_{m,min}$, which is the minimum energy per unit mass required to cause breakage.

$$S = 1 - \exp(-f_{mat} d_p f_{coll}(W_{m,kin} - W_{m,min})) \tag{9.6}$$

where d_p is the particle size, f_{coll} is the frequency of impacts, and $W_{m,kin}$ is equal to the mass-specific kinetic energy of impact ($W_{m,kin} = \frac{1}{2} U_{impact}^2$). The parameter f_{mat} is based on the crack extension energy and parameters related to the deformation (Young's modulus and Poisson ratio), but the material parameters are usually derived from data fitting of post-mortem particle size in controlled impact experiments. The equation is fit from data with the particle velocity upon impact as the abscissa, which is known, and the fraction of particles broken after impact as the ordinate.

9.4 Attrition Modeling

An important aspect of attrition research is developing quantitative models for attrition rate prediction to help foresee the cost of solids makeup. This is especially important in CLC, where the makeup cost of the oxygen carrier may potentially outweigh the cost benefits of the technology. Thus, the following section discusses some of the quantitative or semi-quantitative attrition rate models derived from studies on previously developed technologies, such as FCC and CFB combustion. There are two groups of attrition rate models developed based on observations from these more established technologies: unsteady state and steady state. The unsteady-state phase is characterized by a higher attrition rate due to abrasion of the rough edges of the particle and the breakage of very weak particles. Once the protuberances and rough edges are smoothed down, the attrition rate decreases and settles to a nearly constant, steady state if the process conditions are constant. This can be described as having the characteristics of a Markov process [48, 71], occurring when around 3–5% of the initial bed mass has attrited [71]. For the latter period, models for specific unit operation have been developed. The two groups of models are discussed in Sections 9.4.1 and 9.4.2.

9.4.1 Unsteady-State Models

When fresh solids are added to a fluidized system and gas and solid flows are initiated, the attrition rate is initially high and changes with time, usually decreasing with time exponentially. This has caused researchers to attempt to fit the data according to a chemical kinetics analogy, with the attrition rate being proportional to the bed weight and/or time:

$$\frac{dm}{dt} = f(t, m) \tag{9.7}$$

The authors have used various functions $f(t, m)$ to represent the transient rate of attrition measured in their processes, and some of these functions are shown in Table 9.1. The simplest function is first order in the bed weight, which Paramanathan and Bridgwater have reported to describe the disappearance of particles in a given size interval in an annular shear cell with limited success [72]. Similarly, Cook et al. used a second-order model to fit data from attrition tests in a CFB [73].

The most successful and versatile unsteady-state model for describing attrition is the empirical formulation of Gwyn [74], in the original form of which the mass of fines produced is proportional to the time elapsed to some power, usually a negative value. Several authors have found this form effective in fitting attrition data, such as Neil and Bridgwater's experiments in annular shear cells, fluidized beds, and a screw pugmill [75–77]. Mathematically, an issue with Gwyn's formulation exists in that the model implies that the attrition rate is infinite at the initiation of attrition (time equals zero). Other authors have found limitations in

Table 9.1 List of expressions for estimating the unsteady-state rate of particulate attrition.

	$f(t, m)$	Used by authors
1	$-km$	Paramanathan and Bridgwater [72]
2	$-k(m^2 - m_{\text{min}}^2)$	Cook et al. [73]
3	$m_{\text{total}} k_g b t^{b-1}$ (original form: $\frac{m}{m_0} = k_g t^b$) where $k_g = a d_p^{-\frac{2}{3}}$	Gwyn [74]

using Gwyn's formulation as well. Under high normal stress in the annular shear cell unit, Ghadiri et al. [78] found that the data deviated from Gwyn's formulation, indicative of a more complex attrition mechanism. A weakness of this type of attrition modeling is that the exponents and the model form are unknown a priori, so detailed experimental studies are required to determine the form of this model and the corresponding constants for new materials and novel experimental apparatuses.

9.4.2 Steady-State Models

The source of attrition in a fluidized bed is due to the gas jet impingement at the bottom of the bed and the passing of bubbles through the bed [48]. The attrition mechanism for jet impingement in a bubbling fluidized bed is due to the impact of solids at the top of the jet and abrasion of the solids due to circulation of the solid material around the jet [79]. Werther and Xi developed a steady-state grid jet attrition model based on the efficiency of kinetic energy from the gas jet to produce new surface area of the bed material. The resulting expression, as shown in Table 9.2, is a function of the gas density, the orifice diameter squared, and the gas orifice velocity to the third power [82]. A bubbling attrition rate expression can be similarly derived assuming that the energy balance is between the excess gas velocity (i.e. the superficial gas velocity minus the minimum fluidization velocity) and the energy required to create new surface area of the bed material [35, 38]. The derivation involves an assumed bubble frequency, the energy required to create new surface, and the velocity of the particles upon collision. The main parameters in the derived model are thus the excess gas velocity, the specific surface energy, the particle density, and dimensional parameters related to the mass of the bed (e.g. bed height). With important parameters lumped together, the attrition rate is then proportional to the Bond number, namely, the ratio of inertial forces to surface forces. Thus, the greater the excess gas velocity, the lower the specific surface energy for the material, and the greater the mass of the bed of solids, the higher the rate of attrition due to bubbling. It is assumed in these

Table 9.2 List of expressions for estimating steady-state rate of particulate attrition.

	Rate of attrition expression	Used by authors
Jet attrition	$R_{\mathrm{j,abr}} = \left(\frac{\pi}{8}\frac{\eta_{\mathrm{j,abr}}}{\sigma_{\mathrm{p}} S_{\mathrm{mi}}}\right)\rho d_{\mathrm{or}}^2 U_{or} = C_{\mathrm{jet}}\rho d_{\mathrm{or}}^2 U_{\mathrm{or}}^3$	Tajmid et al. [80]
Bubble-induced attrition	$\frac{R_{\mathrm{b,abr}} h}{(U-U_{\mathrm{mf}})} = C\left(\frac{g\rho_{\mathrm{p}} h^2}{\sigma_{\mathrm{p}}}\right)$	Ciliberti et al. [35] and Vaux et al. [38]
Cyclone	$\dot{m}_{\mathrm{c,abr}} = C_{\mathrm{cyc}}\dot{m}_{\mathrm{solids,in}}\mu^{-0.5}U_{\mathrm{gas,in}}^2$	Reppenhagen and Werther [47]
	$\dot{m}_{\mathrm{attr}} = K_{\mathrm{cyc}}\mu^n \dot{m}_{\mathrm{c,in}} u_{\mathrm{c,in}}^2$ $K_{\mathrm{cyc}} = 2\frac{k\rho_{\mathrm{s}}}{H}\cdot K_{\mathrm{geo}}\cdot k_\mu$ $K_{\mathrm{geo}} =$ $\frac{\overline{r_i}^2}{D_{\mathrm{c}}^3}\frac{\pi(D_{\mathrm{c}}^2-D_{\mathrm{e}}^2)S}{A}\sqrt{4\left(\frac{\overline{r_i}}{D_{\mathrm{c}}}\right)^2+\left(\frac{4A}{\pi(D_{\mathrm{c}}^2-D_{\mathrm{e}}^2)}\right)^2}$	Bayham et al. [81]

analyses that the bubbling and jetting attritions follow a principle of superposition and that their effects are additive.

A model for particulate wear in a cyclone has been developed by Reppenhagen and Werther [47] and assumes that the transfer of kinetic energy of the particles (i.e. the square of the cyclone inlet gas velocity, $U_{gas,in}$) is related to the energy of new surface creation, as shown in Table 9.2. From experiments on cyclone attrition, the attrition rate was also shown to be inversely proportional to the solids loading in the solids laden gas. The constant C_{cyc} is assumed to be proportional to the surface free energy of the material and independent of cyclone geometry [47]. At high velocities, the experimental data deviated from the derived expression owing to the particle impacting the wall, causing fragmentation of the particle [47]. Nevertheless, authors such as Tamjid et al. use this model to predict attrition from computational fluid dynamics (CFD) simulations based on the velocity of particles at each point in the cyclone [80]. Bayham et al. [81] developed a model similar to Reppenhagen and Werther but based on a wear-path mechanism. The model resulted in a similar functionality with respect to inlet gas velocity. Additionally, it also incorporated a geometric term that was a function of the dimensions of the cyclone.

9.4.3 System Modeling

The models previously mentioned can be combined to estimate the rate of attrition in a system, and a population balance can be used to track the fates of particle sizes. This is particularly important for systems with multi-modal particle size distributions such as CLC, where solid fuels are mixed with attrited oxygen carrier [83, 84]. A population balance consists of mass balances within size classes. The effect of attrition on a certain size of particle causes it to move to a smaller size class. The balance for a size class consists of a breakage rate function and a particle selection function. The breakage rate function acts as a kinetic constant that describes how quickly a particle undergoes attrition from a certain size class, while the particle selection function describes what fraction of the original particle size goes into which smaller size class. Restrictions exist on the breakage rate and the breakage distribution functions. First, particles in the finest size class cannot disappear. The analysis can also assume no agglomeration, which prevents the finer particle sizes from growing into a larger particle size. Finally, the conservation of mass causes the sum of the mass fractions to add to unity [66]. Variations of the model include constants that are based on time, such as Gwyn's equation, and include changes in density and chemical reaction [83, 85].

Kramp et al. simulated a steady-state 50-kW_{th} gaseous fuel process based on the laboratory scale and a 100-MW_{th} gaseous fuel CLC process based on a potential commercial-scale unit [84]. Three locations for attrition were used in this work: cyclone, grid jets, and bubbling bed. Since the attrition constants were not known for the oxygen carriers in question, representative constants were chosen from vanadium phosphorous oxide (VPO) and FCC catalysts from small-scale laboratory studies. The locations of loss of oxygen carrier from the vessel differed between the scales, where most of the fines elutriated out the FR at the small 50 kW_{th} scale, while at the large scale, most fines were lost from the AR cyclone.

The main source of the attrition in both the 50 kW_{th} and the 100 MW_{th} processes was due to the cyclones [84]. Unfortunately, the model does not consider attrition due to reaction.

9.5 Experimental Testing

To complement the attrition modeling approaches discussed in Section 9.4, experimental techniques to quantitatively and qualitatively compare materials are needed. Traditionally, attrition research has been performed using two approaches as described by Werther and Reppenhagen [32]. One is a test facility method, which uses a smaller version of the desired unit that consists of the types of stresses that the large-scale unit would exhibit. Small-scale tests facility can often give a rapid, inexpensive indication of material behavior, and results from several sources can be compared. For example, a small-scale fluidized bed attempts to mimic a larger-scale bed. The resulting model is usually a result of a few key variables, e.g. gas velocity and bed weight, times a bulk system constant, which is system and material dependent. Furthermore, a determination of reduction of bias and precision errors as well as data verification and identification process needs to be used to increase the accuracy. Since significant impact on this process may have been generated by bias error, its reduction and determination is the key issue. This approach works under the assumption that the small laboratory unit will maintain dynamic similarity to the large-scale unit; however, this may not necessarily be the case. Furthermore, the bulk constant is unknown without performing experiments.

The second alternative approach attempts to consider both the types of stresses and the range of magnitude of stresses exhibited on the particles [32]. While the testing apparatus does not necessarily involve simulating the actual components found in the large-scale system, the resulting force-attrition data can be coupled with multiphase fluid mechanical models to obtain the attrition of a large-scale unit. A number of the parameters described in Sections 9.3 and 9.4 can be determined from representative experiments that provide the same types of forces as would be seen in the actual CFB system. Most experimental test rigs for attrition are intended to isolate either abrasion- or impact-based attrition mechanisms.

9.5.1 Nanoindentation

The hardness of a material is historically considered the most important parameter in measuring the attrition of materials. Determination of the hardness of non-particulate materials is straightforward by measuring the projected area of the permanent indentation for a given load on a sample of the material [51]. Similarly, the fracture toughness can be determined by performing tensile tests on a material with a known crack size [6]. However, properties of a material in particulate form may deviate from its equivalent properties in larger non-particulate form or particles bound together in an agglomerate. Thus, a technique is needed that can provide a small enough indenter load and size to fit onto the surface of a particle. Because the creation of tensile samples of the same material as the catalyst powder and the same physical properties is not always possible, an analogous technique is required for fracture toughness. Additionally, since CFB processes are designed to run at high temperatures, the material properties should be tested at the respective temperature, with low partial pressure of oxygen in some cases, since some of these environments are reducing.

Nanoindentation is a powerful technique that can determine physical properties such as Young's modulus, hardness, and fracture toughness at a small scale [86]. During the indentation test, a diamond indenter tip penetrated the sample under a normal load that can be increased continuously up to a designated level. The load can then gradually decrease back to zero, resulting in a load–displacement (depth) curve. The maximum penetration depth (d), related to the hardness, is the displacement under the maximum load. The technique uses a small diamond indenter, such as a Vickers, Berkovich, and cubed corner, and sensitive transducers to apply a known loading rate to the sample while continuously measuring the corresponding displacement of the indenter in the sample. From the load–displacement curve, the Young's modulus and hardness can be calculated. The testing is usually programmed to stop at a maximum load, the indenter is held for a short period of time, and then, the indenter begins the unloading process. Owing to the locally plastic nature of the material, a hysteresis effect is observed upon unloading of the sample. The transducers for a nanoindentation instrument are very sensitive, with typical values of resolution for load around 1 μN and for depth on the order of angstroms [87]. The technique requires knowledge of the functional relationship between the indenter depth and projected area of the indenter. The projected area can be estimated from experiments that fit data from knowledge of the projected area and the indentation depth [88].

The nanoindentation technique can also be used to find the fracture characteristics if the load applied to the indenter can be greatly increased to cause the material to crack. For such small scales, the load required to crack the material using a Vickers probe is impractical, so a cube-corner probe is often used instead, which lowers the cracking load threshold [89]. While hardness is traditionally assumed to be independent of the load applied, care should be taken to indent the material sufficiently enough to prevent the phenomenon of the indentation size effect from hampering accurate measurements of the particle [87].

The versatility of the technique can be due to the very small nature of the indenter, allowing for measurement of individual particles and grains. The motivation behind nanoindentation was to determine hardness of thin films where traditional hardness indentation techniques would indent at a depth longer than the thickness of the material [90], so the technique naturally applies well to small particles. The technique has been used at room temperature to deduce the properties of pharmaceutical crystals [91], steam methane reforming catalyst [92], silica [86], and yttria stabilized zirconia [86]. Most importantly, the technique can also measure the material properties in inert environments at high temperatures, which is important for processes such as combustion and oxidation/reduction in CLC [93].

9.5.2 Fluidized Beds

The advantage of fluid bed methods is that samples can be prepared in the fluid bed itself, and the stresses experienced by the particles are representative of the actual process. Furthermore, fluid bed method has good mixing abilities, excellent heat transfer, and efficient gas–solid contacting. However, the main causes of particle attrition in fluid bed are fluid-dynamics-induced stresses such as high and low velocity impacts, stresses due to chemical reactions, and thermal stresses resulting in change in internal pressure. Depending on the types of solids used, a significant size reduction of solid pellets may occur due to attrition during fluidization. The size reduction resulting from attrition may affect the reactor performance as well as the fluidizing properties, or cause a significant loss of bed materials by elutriation during fluidization. Attrition in fluidized beds can be caused by either abrasion or fragmentation mechanisms.

The laboratory fluidized bed can take several forms. For studying jet attrition, the gas distribution will either consist of a single orifice [94] or three orifices, such as in the ASTM D5757 test [95]. Forsythe and Hertwig designed a fluidized bed attrition testing unit with a gas distributor consisting of a single orifice [94]. Additionally, to only study bubbling attrition, the plate is replaced with a perforated plate, such as by Ray et al. [71]. The obvious use of the fluidized bed attrition unit is to monitor the attrition due to bubbling or grid jets, but fluidized bed tests have also been used in an attempt to predict the susceptibility of attrition in pneumatic conveying lines [32]. Bemrose and Bridgwater provide a more detailed summary of prior work using fluidized beds [96].

Ray et al. studied limestone attrition (300–1000 μm feed size) in a small fluidized bed consisting of a porous plate distributor [71]. They developed a multicomponent attrition model to determine the relative rate of attrition between different particulate species, comparing the attrition of different sizes of limestone (high degree of attrition) and silicate sand (low degree of attrition) to validate their model. Ray et al. [71] also proposed a "surface reaction" model to simulate the distribution of attrition rates for a multicomponent mixture. Ray et al. [97] developed a population model for particle mixtures undergoing both attrition and reaction.

For CLC applications, Brown et al. studied the effects of bubbling fluidization on pure iron oxide (hematite) and alumina-supported copper oxide under three different hot conditions at 850 °C: (i) fluidization with air, (ii) single reduction stage with carbon monoxide/carbon dioxide, and (iii) oxidation–reduction over 10 cycles, as well as an 11th reduction after sieving and reheating the particles [98]. Attrition was measured by the elutriation rate, using a pair of filters at the gas outlet of the fluid bed, switching between new and used filters periodically throughout the run. The first set of experiments showed the typical exponential decay of the time dependence of attrition by abrasion. The second set showed the rate of attrition at multiple points during reduction of the oxygen carriers. For the unsupported iron oxide, attrition increased gradually through the reduction phase, then decreased with time after the reaction stopped. For the alumina-supported copper oxide, however, the attrition rate dropped as a function of time, even during chemical reaction. Multi-cycle testing showed that the unsupported iron oxide attrition rate increased dramatically with increased oxidation–reduction cycles, while the same for the supported copper oxide decreased, for the most part. The magnitudes of the attrition rates for the different materials were also vastly different, with unsupported iron oxide having a peak attrition rate of around 110 mg min^{-1} during the 10th oxidation cycle, with the alumina–copper oxide having a peak of around 2.7 mg min^{-1} during the first reduction cycle [98]. This is confirmed through work conducted by He et al. [99]; however, they observed a steady-state attrition rate for the alumina-supported copper oxide after approximately 1000 redox cycles.

Reactions that cause the loss of species such as CO_2 during calcination in fluidized bed combustion and lattice oxygen during reduction in CLC tend to cause higher attrition rates, while reactions that add to the lattice structure of the particles such as sulfation cause the attrition rate to be hindered. Finally, grid jets in fluidized beds cause a great degree of localized attrition, so the attrition rate results will depend on the size of the bed compared to the size of the jet.

9.5.3 Impact Testing

Particulate impact tests are used for determining the resistance to impact, the size of the daughter products, and the threshold velocity at different velocities. The particle will not experience significant breakage typically when the velocity is below the threshold velocity. Traditionally, these tests have been performed using a drop tube, whereby the particle free-falls to a hard target below, and the particle size distribution is measured post mortem [48]. However, researchers such as Yuregir et al. have developed an apparatus that utilizes an air-blown eductor to convey particles, usually one at a time, at a target [100]. This apparatus allows for a much greater range of impact velocities. The apparatus is shown in Figure 9.18 and consists of an eductor that allows solids to be fed into the air stream, a filter for the fines collection, and a camera or velocimetry device to capture the particle velocity upon impact.

Zhang and Ghadiri studied the impact attrition of cut crystals of MgO, NaCl, and KCl over 20 impacts at velocities of 4.3–8.4 m s^{-1}. They were able to extract the proportionality factor (α) for impact attrition, which was around $1.2–1.7 \times 10^{-4}$ for MgO, $4.3–8.2 \times 10^{-4}$ for NaCl, and $1.0–4.6 \times 10^{-4}$ for KCl and show that the impact attrition follows this model [101].

For CLC, Brown et al. studied impact attrition of iron oxide in the form of unsupported hematite and alumina-supported copper oxide, both after 1 and 10 reduction–oxidation cycles described previously in the fluidized bed experiments [98]. Hematite and magnetite particles (reduced form) after 1 reaction cycle showed a reasonable impact attrition resistance, with 4.6% and 11.7% of the original particle mass being fragmented at 45 m s^{-1}, respectively, but after the 10th oxidation and 11th reduction for hematite and magnetite, such as at 38 m s^{-1}, a > 97% mass fragmentation occurred [98]. Meanwhile, the alumina-supported copper oxide showed a much greater resistance to impact fragmentation, whereby the breakage after the 10th oxidation and 11th reduction cycles is around 18% of the original mass [98]. This indicates the importance of using a support for oxygen carriers.

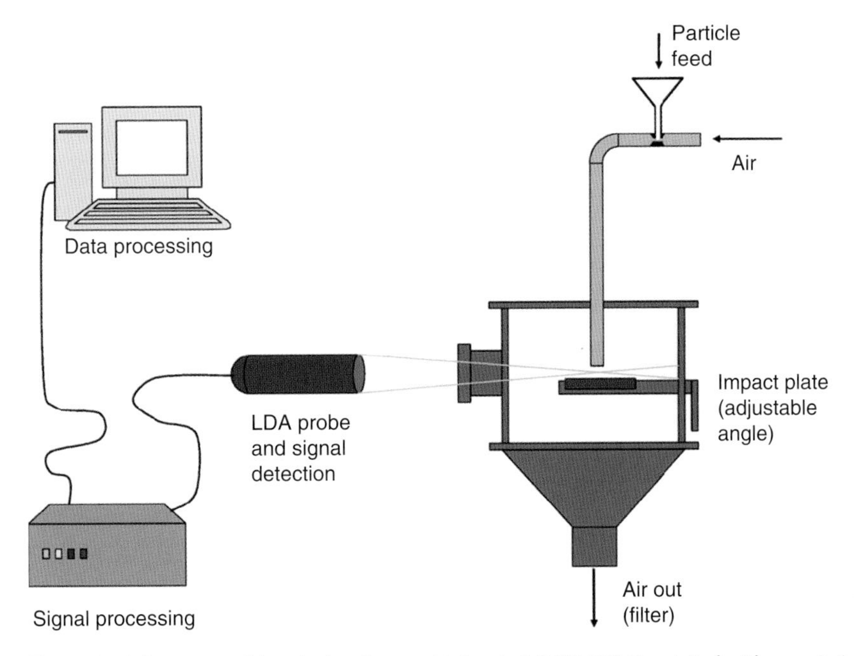

Figure 9.18 Impact attrition device. Source: Meier et al. 2009 [69]. Reprinted with permissions from Elsevier.

The impact velocity for solids depends on the properties of the solid and the conditions present, such as temperature and partial pressure of oxygen. Ghadiri and coworker found that the mother salt crystals tend to begin to chip around $4\,m\,s^{-1}$ and undergo fragmentation around $13\,m\,s^{-1}$ [101]. Chen et al. found that limestone particles begin to fragment at velocities around $8.5–13.5\,m\,s^{-1}$ from 25 to 580 °C [102], and Reppenhagen and Werther found that the abrasion model for attrition in a cyclone for FCC powder begins to deviate due to fragmentation of the particles around $20\,m\,s^{-1}$ [47]. Thus, the threshold velocity could be considered a material parameter (this is qualitatively shown in Figure 9.13).

9.5.4 Jet Cup

The Grace-Davidson jet cup is a common device used to compare the attrition characteristics of fine catalysts, usually in small quantities [32, 103]. As shown in Figure 9.19, it consists of a cup that holds the catalyst sample, with a tangential gas inlet that impinges a small but high-velocity gas jet, causing the particles to impact other particles and the walls of the cup. The gas and solids exit the cup through a disengagement section, allowing for gravity to separate most of the particles, while the remaining particles are captured with a filter, either internally or external to the device. This provides a good screening tool for catalyst development, since only a small amount is required for testing [32]; however, smaller sample sizes can result in larger measurement error [33]. Additionally, there is no standard for the dimensions of a jet cup nor is there a standard gas velocity [33]. The usual value that comes out of this test is the attrition index,

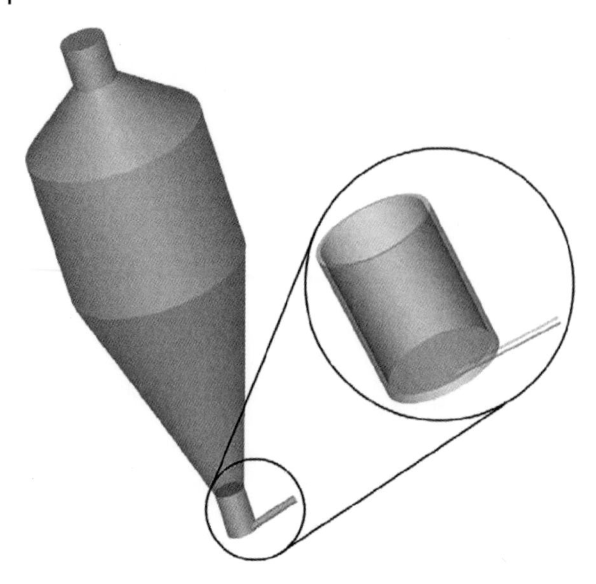

Figure 9.19 Grace-Davidson jet cup attrition device. Source: Cocco et al. 2010 [104]. Reprinted with permissions from Elsevier.

which is the mass of fines generated below either 44 or 20 μm divided by the mass of the unattrited sample. This is mainly prevalent for Group A particles. The Grace-Davidson jet cup usually is considered to represent the attrition in a cyclone in a CFB. Additionally, the incoming air is usually humidified to prevent the particles from sticking to the walls due to static electricity [33].

Amblard et al. studied the attrition characteristics of a Geldart Group A FCC catalyst and a Group B CLC mineral oxygen carrier, and their work attempted to develop a new attrition index for Group B particles, which are prevalent in CLC, since breakage by fragmentation of Group B particles is not below 44 μm [105]. Their new attrition index involved measuring the difference in particle size distribution before and after attrition, named the total particles generated index. This parameter tended to increase dramatically after a certain jet velocity, indicative of the transition from abrasion to fragmentation. CFD analysis using Barracuda showed that in order to have the same mechanical flow patterns of the particles, the volume of solids in the cup needed to be constant, as opposed to equal weight [96].

Rydén et al. [33] characterize the attrition in the jet cup test by measuring the weight of fines that are elutriated and recovered in a filter. However, one problem is that the number of particles elutriated at a constant gas velocity depends also on particle properties such as the density. Therefore, when comparing two solids with different densities, the number of fines elutriated to the filter might be lower for the heaviest solids but it does not mean that the attrition rate is lower.

Rydén et al. tested 25 different oxygen carrier materials in a conical jet cup unit as part of an assessment for the selection of oxygen carriers for CLC, both before and after oxidation and reduction reactions [33]. They ran conical jet-cup testing using 5 g of material with a jet velocity of around 94 $m\,s^{-1}$, weighing the filter every 10 min. Many of the used oxygen carriers exhibited a linear attrition rate, while many of the fresh oxygen carriers, which consisted of perhaps rough surfaces, tended to show the time dependence of attrition. They conclude that iron- and nickel-based oxygen carriers supported on alumina-type supports ($NiAl_2O_4$, $MgAl_2O_4$) and perovskite materials had a higher resistance to attrition after reaction, while copper-based and ZrO_2 and Mg–ZrO_2 supported material showed lower resistance to attrition. They attribute the resistance of attrition to the lower porosity of the material. Finally, with exception of crushing strengths above 2 N, they indicate that there was no strong correlation between the particle crushing strength and the performance in the larger continuous chemical looping units, but the jet cup correlated better, despite the fact that chemical and thermal stresses were not applied to the particles in the room-temperature jet cup experiments [33].

The design philosophy of a jet cup should be to cause as much of the sample to undergo attrition conditions. Therefore, Cocco et al. looked at various design changes to the original Grace-Davidson jet cup to determine which one minimizes the degree of particle stagnation, and they found that a conical jet cup with a single tangential inlet produced the most active material [104]. Tests were run with both an FCC catalyst and a proprietary catalyst, and the gas velocities ranged from 76 to 274 $m\,s^{-1}$. The cups were made of Plexiglas to be able to monitor the solids motion externally. They also simulated the original Davidson jet cup using CFD to monitor where regions of particulate trauma were high and low, and they indicate that the particles tend to hit the wall rather than slide over it. Furthermore, the particulate velocity near the jet was only around 12% of the gas velocity. For the simulation of the conical jet cup, the particulate trauma was higher as well as the collisional frequency [104]. Similar to the work by Rydén et al., Cocco et al. compared the jet cup testing to attrition in an actual unit, namely a 12″ fluidized bed with three recycle cyclones, and the conclusion showed limited correlation between the cylindrical jet cup performance and that at a larger scale [104].

References

1 Bayham, S.C., Breault, R., and Monazam, E. (2016). Particulate solid attrition in CFB systems – an assessment for emerging technologies. *Powder Technol.* 302: 42–62.

2 Tang, Y.Y., Qin, H.J., Wu, K.H. et al. (2013). The reduction and oxidation of Fe_2O_3(0001) surface investigated by scanning tunneling microscopy. *Surf. Sci.* 609: 67–72.

3 Kim, C.Y., Escuadro, A.A., Bedzyk, M.J. et al. (2004). X-ray scattering study of the stoichiometric recovery of the α-Fe_2O_3(0001) surface. *Surf. Sci.* 572: 239–246.

4 Breault, R.W., Monazam, E.R., and Carpenter, J.T. (2015). Analysis of hematite re-oxidation in the chemical looping process. *Appl. Energy* 157: 174–182.

5 Breault, R.W., Yarrington, C.S., and Weber, J.M. (2016). The effect of thermal treatment of hematite ore for chemical looping combustion of methane. *J. Energy Res. Technol.: ASME* 138.

6 Callister, W. (2003). Materials science and engineering: an introduction. In: *Materials & Design*, 6e, vol. 12, 59. New York, NY: Wiley.

7 Breault, R.W. and Monazam, E.R. (2015). Fixed bed reduction of hematite under alternating reduction and oxidation cycles. *Appl. Energy* 145: 180–190.
8 Yi, L.Y., Huang, Z.C., Jiang, T. et al. (2015). Swelling behavior of iron ore pellet reduced by H_2–CO mixtures. *Powder Technol.* 269: 290–295.
9 Bora, D.K., Braun, A., Erat, S. et al. (2012). Evolution of structural properties of iron oxide nano particles during temperature treatment from 250 °C-900 °C: X-ray diffraction and Fe K-shell pre-edge X-ray absorption study. *Curr. Appl. Phys.* 12: 817–825.
10 Lanier, C.H., Chiaramonti, A.N., Marks, L.D., and Poeppelmeier, K.R. (2009). The Fe_3O_4 origin of the "Biphase" reconstruction on α-Fe_2O_3(0001). *Surf. Sci.* 603: 2574–2579.
11 Simmonds, T., Chen, J., Jak, E., and Hayes, P. (2014). *Phase Transformations and Microstructure Changes in Iron Oxide CLC Based Carriers*. Brisbane: University of Queensland.
12 Wang, X. (1991). *Ferrous Metallurgy (Ironmaking Part)*. Beijing, China: Metallurgical Industry Press.
13 Xiao, Q. (1991). *Theory and Practice of Pelletizing*, 93. Changsha: Central South University of Technology Press.
14 Li, J. (2007). *A Study on Mechanism and Process of Direct Reduction of Pellets Made from Concentrate and Composite Binder*. Changsha: Central South University Press.
15 Hayes, P.C. and Grieveson, P. (1981). Microstructural changes on the reduction of hematite to magnetite. *Metall. Trans. B* 12: 579–587.
16 Breault, R.W. and Monazam, E.R. (2015). Analysis of fixed bed data for the extraction of a rate mechanism for the reaction of hematite with methane. *J. Ind. Eng. Chem.* 29: 87–96.
17 Monazam, E.R., Breault, R.W., and Siriwardane, R. (2014). Kinetics of hematite to wustite by hydrogen for chemical looping combustion. *Energy Fuel* 28: 5406–5414.
18 Pineau, A., Kanari, N., and Gaballah, I. (2006). Kinetics of reduction of iron oxides by H_2 – Part I: Low temperature reduction of hematite. *Thermochim. Acta* 447: 89–100.
19 Pang, J.M., Guo, P.M., Zhao, P. et al. (2009). Influence of size of hematite powder on its reduction kinetics by H_2 at low temperature. *J. Iron Steel Res. Int.* 16: 7–11.
20 Fick, A. (1995). On liquid diffusion (Reprinted from the London, Edinburgh, and Dublin Philosophical Magazine and Journal of Science, Vol 10, Pg 30, 1855). *J. Membr. Sci.* 100: 33–38.
21 Radke, E., Chen, J., and Hayes, P. (2014). *Reduction of Mn–Fe Oxides in CO/CO_2 Gas Atmospheres Relevant to Chemical Looping Combustion Application*. Brisbane: University of Queensland.
22 Cuadrat, A., Abad, A., Adánez, J. et al. (2012). Behavior of ilmenite as oxygen carrier in chemical-looping combustion. *Fuel Process. Technol.* 94: 101–112.
23 Forero, C.R., Gayán, P., García-Labiano, F. et al. (2011). High temperature behaviour of a CuO/γAl_2O_3 oxygen carrier for chemical-looping combustion. *Int. J. Greenhouse Gas Control* 5: 659–667.

24 Cho, W.C., Kim, C.G., Jeong, S.U. et al. (2015). Activation and reactivity of iron oxides as oxygen carriers for hydrogen production by chemical looping. *Ind. Eng. Chem. Res.* 54: 3091–3100.

25 Cho, W.C., Lee, D.Y., Seo, M.W. et al. (2014). Continuous operation characteristics of chemical looping hydrogen production system. *Appl. Energy* 113: 1667–1674.

26 Cho, W.C., Seo, M.W., Kim, S.D. et al. (2012). Reactivity of iron oxide as an oxygen carrier for chemical-looping hydrogen production. *Int. J. Hydrogen Energy* 37: 16852–16863.

27 Rode, H., Orlicki, D., and Hlavacek, V. (1995). Noncatalytic gas–solid reactions and mechanical-stress generation. *AIChE J.* 41: 1235–1250.

28 Voss, D.A., Butler, E.P., and Mitchell, T.E. (1982). The growth of hematite blades during the high-temperature oxidation of iron. *Metall. Trans. A* 13: 929–935.

29 Galinsky, N.L., Shafiefarhood, A., Chen, Y.G. et al. (2015). Effect of support on redox stability of iron oxide for chemical looping conversion of methane. *Appl. Catal. B: Environ.* 164: 371–379.

30 Bleeker, M.F., Veringa, H.J., and Kersten, S.R.A. (2009). Deactivation of iron oxide used in the steam–iron process to produce hydrogen. *Appl. Catal. A: Gen.* 357: 5–17.

31 Turkdogan, E.T. and Vinters, J.V. (1971). Gaseous reduction of iron oxides. 1. Reduction of hematite in hydrogen. *Metall. Mater. Trans. B* 2: 3175.

32 Werther, J. and Reppenhagen, J. (1998). Attrition in fluidized beds and pneumatic conveying lines. In: *Fluidization, Solids Handling, and Processing*, 435–491. Westwood, NJ: William Andrew Publishing.

33 Rydén, M., Moldenhauer, P., Lindqvist, S. et al. (2014). Measuring attrition resistance of oxygen carrier particles for chemical looping combustion with a customized jet cup. *Powder Technol.* 256: 75–86.

34 Lupianez, C., Scala, F., Salatino, P. et al. (2011). Primary fragmentation of limestone under oxy-firing conditions in a bubbling fluidized bed. *Fuel Process. Technol.* 92: 1449–1456.

35 Ciliberti, D., Ulerich, N., Sun, C.C. et al. (1980). *Experimental/Engineering Support for EPA's FBC Program: Final Report.* US Environmental Protection Agency, Industrial Environmental Research Laboratory.

36 Shamlou, P.A., Liu, Z., and Yates, J.G. (1990). Hydrodynamic influences on particle breakage in fluidized-beds. *Chem. Eng. Sci.* 45: 809–817.

37 Huang, Z.C., Yi, L.Y., and Jiang, T. (2012). Mechanisms of strength decrease in the initial reduction of iron ore oxide pellets. *Powder Technol.* 221: 284–291.

38 Vaux, W.G. and Keairns, D.L. (1980). Particle attrition in fluid-bed processes. In: *Fluidization* (ed. J.R. Grace and J.M. Matsen), 437–444. Boston, MA: Springer.

39 Rumpf, H. (1973). Physical aspects of comminution and new formulation of a law of comminution. *Powder Technol.* 7: 145–159.

40 Ghadiri, M. and Zhang, Z. (2002). Impact attrition of particulate solids. Part 1: A theoretical model of chipping. *Chem. Eng. Sci.* 57: 3659–3669.

41 Werther, J. and Reppenhagen, J. (1999). Catalyst attrition in fluidized-bed systems. *AIChE J.* 45: 2001–2010.
42 Kalman, H. (1999). Attrition control by pneumatic conveying. *Powder Technol.* 104: 214–220.
43 Kitto, J. and Stultz, S. (2005). *Steam: It's Generation and Use*, 41e. The Babcock & Wilcox Company.
44 Xia, Y.F., Cheng, L.M., Huang, R. et al. (2016). Anti-wear beam effects on water wall wear in a CFB boiler. *Fuel* 181: 1179–1183.
45 Xia, Y.F., Cheng, L.M., Yu, C.J. et al. (2015). Anti-wear beam effects on gas–solid hydrodynamics in a circulating fluidized bed. *Particuology* 19: 173–184.
46 Zenz, F. (1974). Help from project EARL. *Hydrocarbon Process.* 53: 119–124.
47 Reppenhagen, J. and Werther, J. (2000). Catalyst attrition in cyclones. *Powder Technol.* 113: 55–69.
48 British Materials Handling Board (1987). *Particle Attrition: State of the Art Review*, Series on Bulk Materials Handling. Clausthal-Zellerfield, Germany: Trans Tech Publishing.
49 Ogawa, A. (1997). Mechanical separation process and flow patterns of cyclone dust collectors. *Appl. Mech. Rev.* 50: 97–130.
50 Werther, J. and Hartge, E.U. (2004). A population balance model of the particle inventory in a fluidized-bed reactor/regenerator system. *Powder Technol.* 148: 113–122.
51 Tabor, D. (1951). Hardness measurements with conical and pyramidal indenters. In: *The Hardness of Metals*. Oxford, UK: Oxford University Press.
52 Ludema, K.C. (1996). *Friction, Wear, Lubrication: A Textbook in Tribology*. CRC Press.
53 Biscans, B. (2004). Impact attrition in crystallization processes. Analysis of repeated impacts events of individual crystals. *Powder Technol.* 143: 264–272.
54 Archard, J. (1980). Wear theory and mechanisms. In: *Wear Control Handbook*, vol. 58. New York, NY: American Society of Mechanical Engineers (ASME).
55 Kimura, Y., Hidaka, H., and Takaki, S. (1999). Work-hardening mechanism during super-heavy plastic deformation in mechanically milled iron powder. *Mater. Trans. JIM* 40: 1149–1157.
56 Ghadiri, M., Yuregir, K.R., Pollock, H.M. et al. (1991). Influence of processing conditions on attrition of NaCl-crystals. *Powder Technol.* 65: 311–320.
57 Takeda, M., Onishi, T., Nakakubo, S., and Fujimoto, S. (2009). Physical properties of iron-oxide scales on Si-containing steels at high temperature. *Mater. Trans.* 50: 2242–2246.
58 Bartholomew, C.H. (2001). Mechanisms of catalyst deactivation. *Appl. Catal. A: Gen.* 212: 17–60.
59 Broek, D. (2012). *Elementary Engineering Fracture Mechanics*. Springer Science+Business Media.
60 Hidaka, Y., Anraku, T., and Otsuka, N. (2002). Tensile deformation of iron oxides at 600–1250 °C. *Oxid. Met.* 58: 469–485.

61 Viktorov, S.D., Golovin, Y.I., Kochanov, A.N. et al. (2014). Micro- and nano-indentation approach to strength and deformation characteristics of minerals. *J. Min. Sci.* 50: 652–659.
62 Rocha-Rangel, E. (2011). *Fracture Toughness Determinations by Means of Indentation Fracture*. INTECH Open Access Publisher.
63 Meng, H.C. and Ludema, K.C. (1995). Wear models and predictive equations – their form and content. *Wear* 181: 443–457.
64 Archard, J.F. (1953). Contact and rubbing of flat surfaces. *J. Appl. Phys.* 24: 981–988.
65 Yang, L.J. (2005). A test methodology for the determination of wear coefficient. *Wear* 259: 1453–1461.
66 Bilgili, E., Yepes, J., and Scarlett, B. (2006). Formulation of a non-linear framework for population balance modeling of batch grinding: beyond first-order kinetics. *Chem. Eng. Sci.* 61: 33–44.
67 Ning, Z.M. and Ghadiri, M. (2006). Distinct element analysis of attrition of granular solids under shear deformation. *Chem. Eng. Sci.* 61: 5991–6001.
68 Gahn, C. (1997). *Die Festigkeit von Kristallen und ihr Einfluss auf die Kinetik in Suspensionskristallisatoren*. Herbert Utz Verlag.
69 Meier, M., John, E., Wieckhusen, D. et al. (2009). Influence of mechanical properties on impact fracture: prediction of the milling behaviour of pharmaceutical powders by nanoindentation. *Powder Technol.* 188: 301–313.
70 Vogel, L. and Peukert, W. (2003). Breadage behaviour of different materials – construction of mastercurve for the breakage probability. *Powder Technol.* 129: 101–110.
71 Ray, Y.C., Jiang, T.S., and Wen, C.Y. (1987). Particle attrition phenomena in a fluidized-bed. *Powder Technol.* 49: 193–206.
72 Paramanathan, B.K. and Bridgwater, J. (1983). Attrition of solids. 2. Material behavior and kinetics of attrition. *Chem. Eng. Sci.* 38: 207–224.
73 Cook, J.L., Khang, S.J., Lee, S.K., and Keener, T.C. (1996). Attrition and changes in particle size distribution of lime sorbents in a circulating fluidized bed absorber. *Powder Technol.* 89: 1–8.
74 Gwyn, J.E. (1969). On the particle size distribution function and the attrition of cracking catalysts. *AIChE J.* 15: 35–39.
75 Neil, A.U. and Bridgwater, J. (1994). Attrition of particulate solids under shear. *Powder Technol.* 80: 207–219.
76 Neil, A.U. and Bridgwater, J. (1999). Towards a parameter characterising attrition. *Powder Technol.* 106: 37–44.
77 Bridgwater, J. (1987). Attrition of high-density polyethylenes. *Powder Technol.* 50: 243–252.
78 Ghadiri, M., Ning, Z., Kenter, S.J., and Puik, E. (2000). Attrition of granular solids in a shear cell. *Chem. Eng. Sci.* 55: 5445–5456.
79 Stein, M., Seville, J.P.K., and Parker, D.J. (1998). Attrition of porous glass particles in a fluidised bed. *Powder Technol.* 100: 242–250.
80 Tamjid, S., Hashemabadi, S.H., and Shirvani, M. (2010). Fluid catalytic cracking: prediction of catalyst attrition in a regeneration cyclone. *Filtr. Sep.* 47: 29–33.

81 Bayham, S.C., Breault, R., and Monazam, E. (2016). Applications of tribology to determine attrition by wear of particulate solids in CFB systems. *Powder Technol.*

82 Werther, J. and Xi, W. (1993). Jet attrition of catalyst particles in gas-fluidized beds. *Powder Technol.* 76: 39–46.

83 Wang, Q.H., Luo, Z.Y., Ni, M.J., and Cen, K.F. (2003). Particle population balance model for a circulating fluidized bed boiler. *Chem. Eng. J.* 93: 121–133.

84 Kramp, M., Thon, A., Hartge, E.U. et al. (2011). The role of attrition and solids recovery in a chemical looping combustion process. *Oil Gas Sci. Technol.* 66: 277–290.

85 Ouchiyama, N., Rough, S.L., and Bridgwater, J. (2005). A population balance approach to describing bulk attrition. *Chem. Eng. Sci.* 60: 1429–1440.

86 Zbib, M.B., Parab, N.D., Chen, W.N.W., and Bahr, D.F. (2015). New pulverization parameter derived from indentation and dynamic compression of brittle microspheres. *Powder Technol.* 283: 57–65.

87 Elmustafa, A.A. and Stone, D.S. (2003). Nanoindentation and the indentation size effect: kinetics of deformation and strain gradient plasticity. *J. Mech. Phys. Solids* 51: 357–381.

88 Oliver, W.C. and Pharr, G.M. (1992). An improved technique for determining hardness and elastic-modulus using load and displacement sensing indentation experiments. *J. Mater. Res.* 7: 1564–1583.

89 Morris, D.J., Vodnick, A.M., and Cook, R.F. (2005). Radial fracture during indentation by acute probes: II. Experimental observations of cube-corner and Vickers indentation. *Int. J. Fract.* 136: 265–284.

90 Doerner, M.F. and Nix, W.D. (1986). A method for interpreting the data from depth-sensing indentation instruments. *J. Mater. Res.* 1: 601–609.

91 Taylor, L.J., Papadopoulos, D.G., Dunn, P.J. et al. (2004). Mechanical characterisation of powders using nanoindentation. *Powder Technol.* 143: 179–185.

92 Couroyer, C., Ghadiri, M., Laval, P. et al. (2000). Methodology for investigating the mechanical strength of reforming catalyst beads. *Oil Gas Sci. Technol.* 55: 67–85.

93 Trenkle, J.C., Packard, C.E., and Schuh, C.A. (2010). Hot nanoindentation in inert environments. *Rev. Sci. Instrum.* 81.

94 Forsythe, W.L. and Hertwig, W.R. (1949). Attrition characteristics of fluid cracking catalysts – laboratory studies. *Ind. Eng. Chem.* 41: 1200–1206.

95 ASTM D5757-11 (2011). Standard test method for determination of attrition and abrasion of powdered catalysts by air jets. West Conshohocken, PA: ASTM.

96 Bemrose, C.R. and Bridgwater, J. (1987). A review of attrition and attrition test methods. *Powder Technol.* 49: 97–126.

97 Ray, Y.C., Jiang, T.S., and Jiang, T.L. (1987). Particle-population model for a fluidized-bed with attrition. *Powder Technol.* 52: 35–48.

98 Brown, T.A., Scala, F., Scott, S.A. et al. (2012). The attrition behaviour of oxygen-carriers under inert and reacting conditions. *Chem. Eng. Sci.* 71: 449–467.

99 He, F., Linak, W.P., Deng, S., and Li, F.X. (2017). Particulate formation from a copper oxide-based oxygen carrier in chemical looping combustion for CO_2 capture. *Environ. Sci. Technol.* 51: 2482–2490.

100 Yuregir, K.R., Ghadiri, M., and Clift, R. (1986). Observations on impact attrition of antigranulocytes solids. *Powder Technol.* 49: 53–57.

101 Zhang, Z. and Ghadiri, M. (2002). Impact attrition of particulate solids. Part 2: Experimental work. *Chem. Eng. Sci.* 57: 3671–3686.

102 Chen, Z.X., Lim, C.J., and Grace, J.R. (2007). Study of limestone particle impact attrition. *Chem. Eng. Sci.* 62: 867–877.

103 Weeks, S.A. and Dumbill, P. (1990). Method speeds FCC catalyst attrition resistance determinations. *Oil Gas J.* 88: 38–40.

104 Cocco, R., Arrington, Y., Hays, R. et al. (2010). Jet cup attrition testing. *Powder Technol.* 200: 224–233.

105 Amblard, B., Bertholin, S., Bobin, C., and Gauthier, T. (2015). Development of an attrition evaluation method using a jet cup rig. *Powder Technol.* 274: 455–465.

Section 3

Commercial Design Studies of CLC Systems

10

Computational Fluid Dynamics Modeling and Simulations of Fluidized Beds for Chemical Looping Combustion

Subhodeep Banerjee and Ramesh K. Agarwal

Washington University in St. Louis, Department of Mechanical Engineering and Materials Science, 1 Brookings Drive, St. Louis, MO 63130, USA

10.1 Introduction

CO_2 and other gases such as CH_4 produced from burning fossil fuels trap thermal radiation from the Sun in the Earth's atmosphere and lead to an increase in the Earth's surface temperature by the greenhouse effect. Following the industrial revolution at the turn of the twentieth century, the world began to consume fossil fuels for energy at an ever-increasing rate. As a result of the rapid combustion of fossil fuels, the level of CO_2 in the atmosphere has risen by almost 30% compared to the pre-industrial times and the International Panel on Climate Change reported that the "warming of the climate system is unequivocal" and "most of the observed increase in global average temperatures since the mid-twentieth century is very likely due to the observed increase in anthropogenic greenhouse gas concentrations" [1]. This global warming is projected to cause an increase in sea levels caused by melting of the polar ice caps as well as increase in the frequency and intensity of extreme weather events. While renewable energy is expected to account for an increasing portion of the energy supply in the future, fossil fuels will still remain the dominant energy source for at least the next 25 years as shown in Figure 10.1 [2]. Hence, addressing carbon emissions from power plants has become an active area of research.

In recent years, several technologies have been demonstrated to capture CO_2 emissions from fossil-fueled power plants and greatly reduce emissions into the atmosphere. These technologies can be broadly categorized as pre-combustion capture, such as the integrated gas combined cycle (IGCC); post-combustion capture, such as sorbent-based absorption; and oxy-fuel combustion. However, each of these technologies requires a separate process to isolate CO_2 from the other gases, which consumes much of the total energy produced by the plant and can lead to a significant increase in the cost of electricity [3, 4]. On the other hand, one technology that has shown great promise for high-efficiency low-cost carbon capture is chemical-looping combustion (CLC). The CLC process typically utilizes dual fluidized bed reactors and a metal oxide oxygen carrier that circulates between the two reactors, as shown in Figure 10.2a. Another setup for

Handbook of Chemical Looping Technology, First Edition. Edited by Ronald W. Breault.

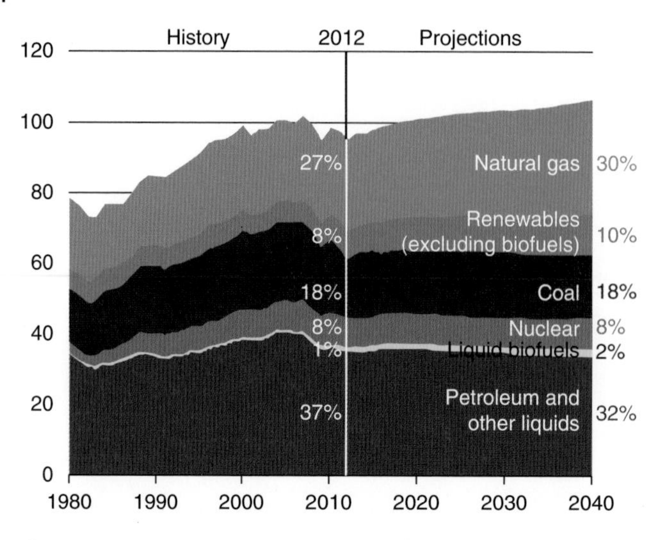

Figure 10.1 Primary energy consumption by fuel (quadrillion Btu) [2].

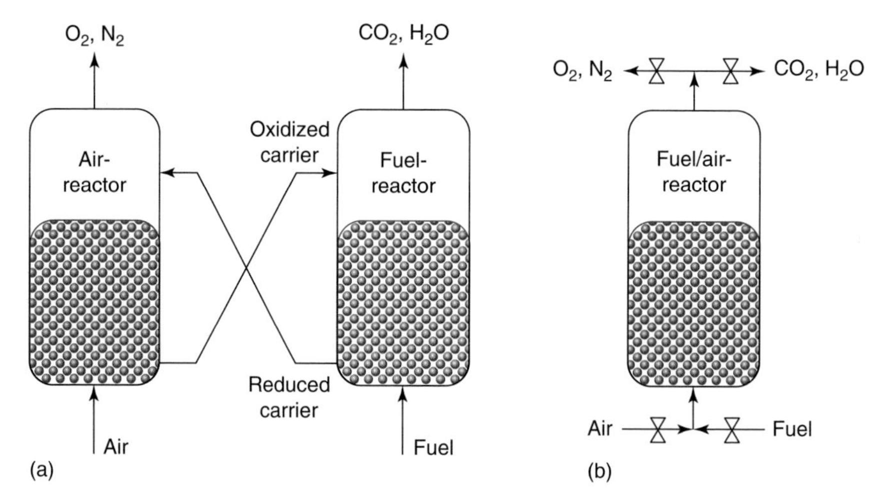

Figure 10.2 Schematic representation of a chemical-looping combustion system with (a) interconnected fluidized beds and (b) packed bed with alternating flow [5].

CLC that has been documented in the literature employs a single vessel with a packed bed of oxygen carrier that is alternatingly used as an air and fuel reactor via a high-temperature gas switching system, shown in Figure 10.2b.

The primary advantage of CLC is that the combustion of fuel in the fuel reactor takes place in the absence of air using oxygen provided by the oxygen carrier; the flue stream from the fuel reactor is not contaminated or diluted by other gases such as nitrogen. This provides a high-purity carbon dioxide stream available for capture at the fuel reactor outlet without the need for an energy-expensive gas separation process. The reduced oxygen carrier from the fuel reactor is pneumatically transported to the air reactor where it is reoxidized by oxygen from air and circulated back to the fuel reactor to complete the loop. The only energy cost of separation associated with CLC is the cost of solid recirculation; research has shown that this is only about 0.3% of the total energy released by the CLC process [6]. This is considerably lower than the benchmark for technologies such as oxy-fuel combustion where the oxygen separation process can consume about 15% of the total energy. Several studies on the energy and exergy of CLC systems in the literature suggest that power efficiencies greater than 50% can be achieved along with nearly complete CO_2 capture [7–11]. Recently, a technoeconomic study to assess the benefits of CLC has reported that the cost of electricity for a CLC plant using Fe_2O_3 oxygen carrier is $115.1 per MWh, which compares favorably against the cost of $137.3 per MWh for a conventional pulverized coal boiler when additional amine-based CO_2 absorption is considered [12].

A lot of the early research in the area of CLC focused primarily on the use of gaseous fuels such as natural gas and syngas. However, as shown in Figure 10.1, since coal is projected to remain one of the dominant fossil fuels in the near future, the use of coal for CLC has garnered significant interest in recent years. One way to utilize coal in a CLC process is to first gasify the coal into syngas in a stand-alone gasifier and then inject the freshly converted syngas into the fuel reactor. To ensure the absence of nitrogen and other gases in the syngas, the gasification must be carried out with oxygen instead of air, which requires an additional air separation unit. As such, this approach introduces the inefficiencies associated with oxy-fuel combustion and similar technologies. From the perspective of the CLC process, this scenario is identical to the one that uses gaseous fuel. An alternate approach is to inject pulverized coal directly into the fuel reactor, a process known as coal-direct chemical-looping combustion (CD-CLC). The CD-CLC concept eliminates the necessity of a separate gasifier and reduces the complexity of the power plant. Within CD-CLC, two alternatives have been proposed as to how the metal oxide will participate in the coal combustion since the solid–solid reaction rate of coal with the metal oxide is negligible [13]. One option, which is considered in this report, is to gasify the coal in the fuel reactor with CO_2 or H_2O as the fluidized agent and react the oxygen carrier with the products of gasification [14]; this process is known as CD-CLC with in situ gasification. The other option, based on a patent by Lewis and Gilliland [15] and discussed in the context of chemical looping by Mattisson et al. [16], is known as chemical-looping with oxygen uncoupling (CLOU), which utilizes special oxygen carriers that release gaseous oxygen under reactor conditions that can sustain the combustion of solid coal in the fuel reactor.

The work of Leion et al. [17] has identified that the rate of fuel conversion in the CD-CLC process is limited by the char gasification step. The agglomeration between oxygen carrier and coal ash is another concern as it has been reported to reduce the reactivity of the metal oxide particles [18]. These concerns are addressed by utilizing a spouted fluidized bed for both reactors with relatively large diameter particles, unlike in CLC using gaseous fuels that can use a bubbling or fast fluidized bed for the fuel reactor. The larger particles correspond to Group D or spoutable particles according to Geldart's powder classification [19]. In a spouted fluidized bed, a high velocity jet of pulverized coal and the fluidizing agent is injected at the center of the fluidized bed to induce strong circulation rates for the solid particles and enhance the solid–gas mixing. The increased friction from the mixing of solids can also serve to slough off the ash build-up on the metal oxide particles and restore reactivity [20].

The bulk of the literature on CLC so far involves laboratory-scale experiments of CLC systems. Setting up and executing such an experiment can be an expensive and laborious process. On the other hand, simulations at the system level using computational fluid dynamics (CFD) models provide an efficient means to analyze the performance of a CLC system in terms of the energy balance and chemical kinetics as well as the intricacies of the fluidization behavior. There are several excellent reviews on the chemical looping process from a chemical engineering perspective in the literature [21–23] but a review of the simulation aspect of CLC is lacking. Given the significant recent progress in this area, the authors feel it is timely to present a summary of recent developments in reactor-level simulations in chemical looping combustion. The work presented in this chapter provides valuable insights into the various process and design optimizations critical to the efficient operation and performance of the CLC process. The successful design and optimization of a CLC system requires the development of credible simulation models for multiphase flow, which include solid particle dynamics and chemical kinetics.

10.2 Reactor-Level Simulations of CLC Using CFD

CFD provides an efficient means to analyze the performance of a CLC system and characterize the fluidization and chemical kinetics in the system. In order to establish a credible simulation for an interconnected dual circulating fluidized bed configuration typical of a CD-CLC setup with particulate metal oxides and fuel present, it is critical to accurately capture the solid circulation and separation as a result of the solid–gas two-way coupling and the solid–solid interaction. Numerical modeling of multiphase flows involving a granular solid and a gas of the kind seen inside a CLC fuel reactor can be achieved with varying degrees of accuracy with varying computational cost depending on the Eulerian–Eulerian or Eulerian–Lagrangian modeling approach considered in the simulation.

The Eulerian–Eulerian approach or multi-fluid model circumvents the high computational demand of the particle-based models while retaining fidelity. In the multi-fluid models, the solid phase is also considered as a continuum fluid and particle variables such as mass, velocity, temperature, etc. are averaged over a region that is large compared to the particle size. As such, this approach only accounts for the bulk behavior of the solid phase. Constitutive equations for the solid-phase pressure and solid-phase viscosity, which are required to model the interactions between the solid and the gas phases, are provided by the kinetic theory of granular flow. The kinetic theory of granular flow is an extension of the classical kinetic gas theory that includes the inelastic particle–particle interactions [24, 25].

Initial CFD studies employing the Eulerian–Lagrangian approach focused on demonstrating the capability of computational methods to model a multiphase gas–solid system and were not based on any particular experiment [26, 27]. Later, the work of Mahalatkar et al. [28, 29] based on a single reactor setup similar to Figure 10.2b showed that the multi-fluid model is able to match the reaction mechanics inside a CLC fuel reactor with reasonable accuracy. However, the single reactor setup "cannot be operated with solid fuels and the design and operation of the hot gas switching system is problematic" [30]. To design a CLC system for operation using solid coal given its likelihood to remain a dominant fossil fuel in the near future, the use of the dual fluidized bed setup shown in Figure 10.2a is necessary. The single reactor simulations that exist in literature do not provide any information about the circulation of oxygen carrier inside a dual fluidized bed setup.

The drawback of the Eulerian–Eulerian simulation is that the exact particle dynamics in the system cannot be determined and optimized. Moreover, the work of Gryczka et al. [31] with 2-mm-diameter particles has suggested that accurate numerical representation of particle dynamics is not likely to be achieved for spouted beds using the multiphase granular solid-phase approximation due to "the inadequacies of the continuum model." The inaccuracy arises from the non-physical closure terms used in the Eulerian–Eulerian model such as the frictional solids viscosity or solids pressure based on the kinetic theory of granular flow. To address these shortcomings, the Eulerian–Lagrangian particle-based model known as discrete element method (DEM) has been proposed. The trajectory of each individual particle in DEM is resolved based on a force balance calculation. The particle tracking is coupled with the CFD solution for the fluid phase by considering the interaction between the particles and the fluid separately for each particle. The coupled CFD-DEM simulation can model the multiphase solids–gas flow with high accuracy, although the high computational cost associated with tracking the individual particles and their collisions is the reason behind the scarcity of particle-based models for CLC simulation in the literature to date. For the relatively large Geldart Group D particles, cold-flow simulations using the coupled CFD-DEM model have proved capable in accurately matching the particle dynamics of various laboratory-scale fluidized bed experiments [32–34]. The coupled CFD-DEM approach was recently employed by Parker [35] to develop a comprehensive model of the circulating reactor system at the National Energy Technology Laboratory with reacting flow for CD-CLC but the complexity of the work was such that it took 81 days to complete 50 seconds of simulation.

In a multiphase CFD simulation, the interaction between the solid phase and the gas flow manifests as a drag force on the particles and a corresponding drop in pressure in the gas phase. The Stokes' drag law for a single spherical particle in a viscous fluid significantly under-predicts the actual drag force when the particles are clustered in a fluidized bed; the drag is calculated using empirical relations instead. The choice of the drag model plays a crucial role in the accuracy of the multiphase simulation results. The Ergun equation [36] describes the drag for both low Reynolds number and high Reynolds number flows but is only accurate for packed beds. For bubbling or fast fluidized beds that include both dilute and dense regions, the Gidaspow drag law [37] is a good choice as it switches between the Ergun prediction and the Wen and Yu drag model [38] based on the local solids volume fraction. The Syamlal–O'Brien drag law [39] is well suited for spouted fluidized beds because it has a correlation based on the terminal velocity, the minimum fluid velocity that is large enough to lift the particle out of the bed, which is an important parameter for characterizing a spouted bed. Recently, a non-empirical model based on Lattice Boltzmann simulations of a particle-laden flow was derived by Beetstra et al. [40] that promises superior prediction of drag for binary particle systems but its usefulness is impeded by inconsistencies in the drag formulation presented and a scarcity of validation cases in the literature.

10.3 Governing Equations

All modeling and simulation work in this chapter have been performed using the Eulerian–Lagrangian approach using the commercial CFD solver ANSYS Fluent v14.5 [41, 42]. The Navier–Stokes equations of fluid motion are slightly modified to account for the presence of the solid particles by including the porosity or fluid volume fraction α_f in the computational cell where the equations are applied. The continuity equation and momentum equations for a cold flow simulation are given by

$$\frac{\partial}{\partial t}(\alpha_f \rho_f) + \nabla \cdot (\alpha_f \rho_f u_f) = 0 \tag{10.1}$$

$$\frac{\partial}{\partial t}(\alpha_f \rho_f u_f) + \nabla \cdot (\alpha_f \rho_f u_f u_f) = -\alpha_f \nabla p_f + \nabla \cdot \bar{\bar{\tau}}_f + \alpha_f \rho_f g - R_{sg} \tag{10.2}$$

The source term R_{sg} in Eq. (10.2) accounts for the gas–solid momentum exchange and is obtained from the average of the drag forces acting on all the discrete particles in the given cell.

For an incompressible Newtonian fluid, the shear stress tensor $\bar{\bar{\tau}}_f$ is simply the Cauchy stress tensor with zero bulk viscosity

$$\bar{\bar{\tau}}_f = \mu_f \left[(\nabla u_f + \nabla u_f^T) - \frac{2}{3} \nabla \cdot u_f \bar{\bar{I}} \right] \tag{10.3}$$

For reacting flow simulations, additional source terms are added to Eqs. (10.1) and (10.2) to account for the interphase transfer of mass and momentum due to the reaction.

$$\frac{\partial}{\partial t}(\alpha_f \rho_f) + \nabla \cdot (\alpha_f \rho_f u_f) = \sum (\dot{m}_{sg} - \dot{m}_{gs}) \tag{10.4}$$

$$\frac{\partial}{\partial t}(\alpha_f \rho_f u_f) + \nabla \cdot (\alpha_f \rho_f u_f u_f) = -\alpha_f \nabla p_f + \nabla \cdot \bar{\bar{\tau}}_f + \alpha_f \rho_f g - \sum (R_{sg} + \dot{m}_{sg} u_{sg} - \dot{m}_{gs} u_{gs}) \tag{10.5}$$

In addition, the chemical reactions in the flow require additional equations to characterize the local mass fractions of each species Y_j in the computational cell. The species conservation equations can be written as

$$\frac{\partial}{\partial t}(\rho Y_j) + \nabla \cdot (\rho u Y_j) = -\nabla \cdot J_j + R_j + S_j \tag{10.6}$$

where J_j is the diffusion flux of the species due to concentration gradients in the flow field, R_j is the net rate of production of the species due to chemical reactions, and S_j is the rate of creation of the species from devolatilization.

The equation for the conservation of energy for the fluid phase can be expressed in terms of the internal energy as

$$\frac{\partial}{\partial t}(\alpha_f \rho_f E_f)\nabla \cdot (\alpha_f u_f(\rho_f E_f + p_f)) = \nabla \cdot \left(k_f \nabla T_f - \sum h_j J_j + (\overline{\overline{\tau}}_f \cdot u_f)\right) + S_h \tag{10.7}$$

The trajectory of each particle is computed by integrating the force balance on the particle, which can be written in the Lagrangian frame per unit particle mass as

$$\frac{\partial u_p}{\partial t} = g\frac{(\rho_f - \rho_p)}{\rho_p} + F_D(u_f - u_p) + F_{coll} \tag{10.8}$$

where the subscript p denotes an individual particle. The terms on the right-hand side of Eq. (10.8) account for the gravitational and buoyant forces, the drag force, and an additional force due to particle–particle or particle–wall collisions. Forces such as the virtual mass force and pressure gradient force can be neglected for gas–solid flows given that ρ_p far exceeds ρ_f. The net drag coefficient F_D is given by

$$F_D = \frac{18\mu_f}{\rho_p d_p^2}\frac{C_D Re_p}{24} \tag{10.9}$$

where d_p is the particle diameter, C_D is the particle drag coefficient, and Re_p is the Reynolds number based on the particle diameter defined as

$$Re_p = \frac{\rho_f d_p |u_f - u_p|}{\mu_f} \tag{10.10}$$

The drag coefficient can be modeled using various empirical relations. The collision force in Eq. (10.8) is computed using the soft-sphere model, which decouples its normal and tangential components. The normal force on a particle involved in a collision is given by

$$F_{coll}^n = (K\delta + \gamma(u_{12}e))e \tag{10.11}$$

In Eq. (10.11), δ is the overlap between the particles pair involved in the collision as illustrated in Figure 10.3 and γ is the damping coefficient, a function of the particle coefficient of restitution η; e is the unit vector in the direction of u_{12}. Previous research by Link [43] has demonstrated that for large values of K, the results of the soft-sphere model are identical to those obtained using a hard-sphere model. The tangential collision force is a fraction μ of the normal force with μ as a function of the relative tangential velocity v_r given as

$$\mu(v_r) = \begin{cases} \mu_{stick} + (\mu_{stick} - \mu_{glide})(v_r/v_{glide} - 2)(v_r/v_{glide}) & \text{if } v_r < v_{glide} \\ \mu_{glide} & \text{if } v_r \geq v_{glide} \end{cases} \tag{10.12}$$

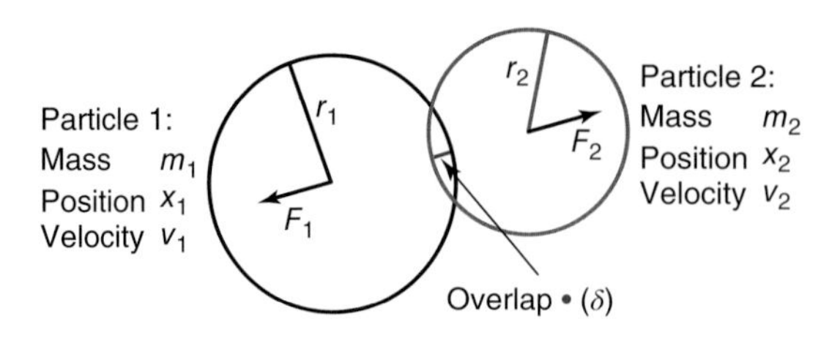

Figure 10.3 Schematic of particle collision model for DEM [42].

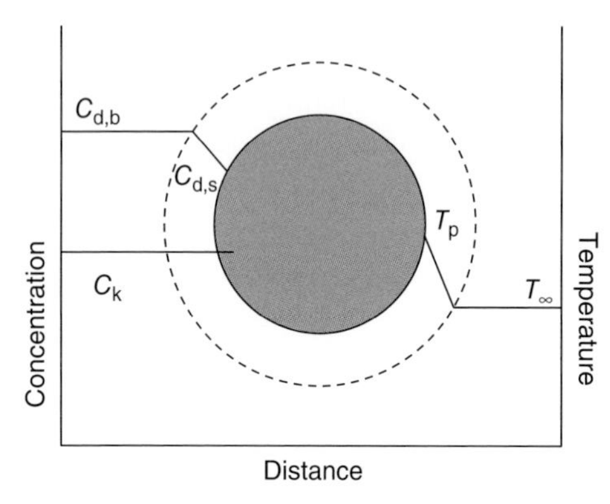

Figure 10.4 Reacting particle in the multiple surface reactions model [42].

For reacting flow simulations, the multiple surface reactions model is used for the reaction of the oxygen carrier particles with the injected gases. To understand this model, consider a reacting particle as shown in Figure 10.4. For a reaction of order 1, the depletion rate of particle species j in kg s^{-1} is given by

$$\overline{R}_j = A_p \eta_r Y_j p_n \frac{R_{kin} D_0}{D_0 + R_{kin}} \tag{10.13}$$

where A_p is the spherical surface area of the particle, η_r is the effectiveness factor, Y_j is the mass fraction of the species j, p_n is the bulk partial pressure of the reacting fluid species n, and D_0 is the diffusion rate coefficient for the reaction defined as

$$D_0 = C_1 \frac{[(T_p + T_\infty)/2]^{0.75}}{d_p} \tag{10.14}$$

where C_1 is the diffusion rate coefficient obtained empirically.

The remaining term in the Eq. (10.18) for depletion rate R_{kin} is the kinetic reaction rate obtained from literature for the particular reaction considered. The net depletion of particle mass dm_p/dt due to the reduction of the metal oxide Fe_2O_3 provides the source term $\dot{m}_{sg}$ used in the continuity equation. Heat transfer to the particle is governed by the equation for particle heat balance, which can be written as

$$m_p c_p \frac{dT_p}{d_t} = hA_p(T_\infty - T_p) - f_h \frac{dm_p}{dt} H_{reac} \tag{10.15}$$

where c_p is the particle heat capacity, h is the convective heat transfer coefficient, H_{reac} is the heat released by the reaction, and f_h is the fraction of the energy produced that is captured by the particle; the remaining portion $(1 - f_h)$ is applied as the heat source in the energy equation.

10.4 Eulerian–Lagrangian Simulation of a Spouted Fluidized Bed in a CLC Fuel Reactor with Chemical Reactions

The current focus in the area of CLC is to develop a CLC system that works well with solid coal fuel given its likelihood to remain the dominant energy source in the near future [2]. In the real world operation of CD-CLC, the implementation of a spouted fluidized bed fuel reactor offers several technical benefits. The spouted fluidized bed was proposed by Mathur and Gishler [44] to overcome the limitation of a typical bubbling or fast fluidized bed to handle particles larger than a few hundred micrometers in diameter. Relatively larger particles of the oxygen carrier are beneficial for CD-CLC operation for easier separation of the smaller coal and ash particles from the recirculating oxygen carrier; based on the diameter and density, these particles can be classified as Group D particles according to Geldart's powder classification [19]. The spouted fluidized bed utilizes a high velocity gas stream to create a local high velocity region at the center of the bed (known as the spout) where the particles and voids (bubbles) move in a structured manner with little radial displacement [45]. The high velocity jet injection can create strong solid–gas mixing in the spout and high circulation avoiding loss of reactivity due to the ash agglomeration with the oxygen carrier [18].

The fluidization performance of spouted beds in cold flow settings has been investigated in detail by Sutkar et al. [32, 45, 46]. However, there have been no reacting flow experiments employing a spouted bed configuration in the literature to date. Owing to the scarcity of experimental data, the transient cold flow simulations of Zhang et al. [34] of a spouted bed fuel reactor are extended to incorporate a chemical reaction between the metal oxide oxygen carrier and a gaseous fuel.

Most coupled CFD/DEM simulations of the spouted fluidized bed configuration for CLC in the literature have been limited to cold flow simulations using glass beads [32–34]. Previous work by the authors has shown that if pure Fe_2O_3 ($\rho_p = 5240\,kg\,m^{-3}$) is used as the bed material, the high density precludes successful fluidization [47]. In this work, the glass beads from the cold flow simulation of Zhang et al. [34] are replaced with an oxygen carrier consisting of 60% Fe_2O_3 by weight on $MgAl_2O_4$ support sintered at 1100 °C. This material, designated F60AM1100, was shown to provide excellent reactivity and sufficient hardness for CLC and its apparent density of $2225\,kg\,m^{-3}$ making it an ideal choice for fluidized bed operation [48].

The geometry and computational model of the CLC fuel reactor are shown in Figure 10.5. The geometry is derived from the pseudo-3D Plexiglas test rig used in the cold flow experiment conducted at Technische Universität Darmstadt with the cyclone, down-comer, and loop-seal [38]; a chute is added to the geometry based on the work of Zhang et al. [34] to improve the particle circulation. The mesh is generated such that the solution is stable when using second-order numerical schemes with minimal under-relaxation to achieve faster convergence at each time step. The total particle load in the bed is doubled compared to the work of Zhang et al. [34] to partly offset the tighter packing associated with the

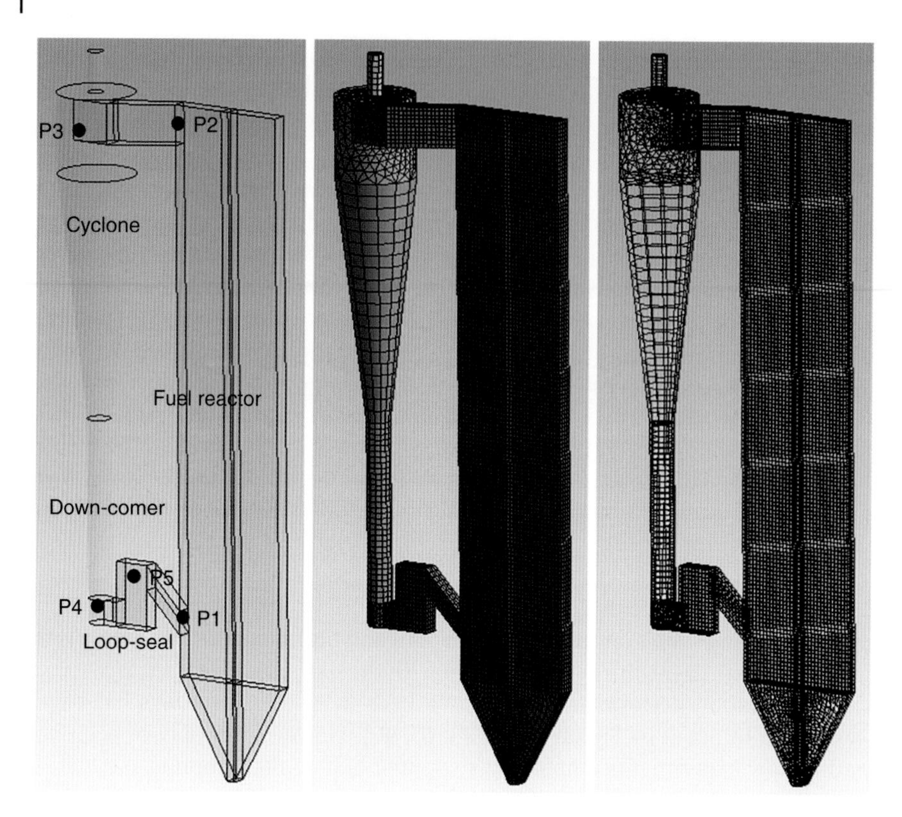

Figure 10.5 Geometry outline with pressure taps, mesh, and wireframe of the CLC fuel reactor.

heavier particles. Additional particles are also deposited in the down-comer and loop-seal to ensure adequacy of the number of particles for recirculation; a total of 87 320 particles are employed in the entire system in the present work.

The drag law proposed by Syamlal and O'Brien [49] is selected for the spouted fluidized beds simulations in this chapter. The Syamlal–O'Brien drag law is a good choice because it uses a correction based on the terminal velocity of the particle, which is the minimum velocity that is large enough to lift the particle out of the bed and is an important parameter for characterizing a spouted bed. The Syamlal–O'Brien drag law defines

$$C_D = \left(0.63 + \frac{4.8}{\sqrt{Re_p / v_{r,p}}} \right)^2 \tag{10.16}$$

In Eq. (10.16), $v_{r,p}$ is the terminal velocity correction for the particulate phase given by

$$v_{r,p} = 0.5 \left(A - 0.06\, Re_p + \sqrt{(0.06\, Re_p)^2 + 0.12\, Re_p (2B - A) + A^2} \right) \tag{10.17}$$

where

$$A = \alpha_f^{4.14} \text{ and } B = \begin{cases} 0.8\alpha_f^{1.28} & \text{if } \alpha_f \leq 0.85 \\ \alpha_f^{2.65} & \text{if } \alpha_f > 0.85 \end{cases}$$

The central jet velocity and the background velocity are slightly increased to compensate for the slightly higher minimum fluidization velocity of the reacting particles and maintain the same spouted fluidization regime as in the cold flow work of Zhang et al. [34]. The flow injection in the fuel reactor consists of 10% CH_4 and 10% H_2O by mass fraction. The remaining 80% of the flow injection is inert nitrogen. The absence of CO_2 in the gas injection is maintained so that the mass fraction of CO_2 generated by the reaction in the fuel reactor can be tracked without it being overshadowed by the injected mass fraction of CO_2. Similarly, the aeration gas in the down-comer and the loop-seal comprises solely of N_2 so that the recirculation of particles from the loop-seal to the fuel reactor can be easily identified; the particles that originate in the loop-seal have a smaller mass fraction of Fe_3O_4 since they were initially exposed to inert flow. Since the $MgAl_2O_4$ is non-reactive, the only reaction that takes place on the particle surface is the reduction of Fe_2O_3 to Fe_3O_4. The stoichiometric reaction is given as

$$12Fe_2O_{3(s)} + CH_4 \rightarrow 8Fe_3O_{4(s)} + CO_2 + 2H_2O \tag{10.18}$$

The kinetics for the reduction of Fe_2O_3 to Fe_3O_4 by CH_4, R_{kin} in Eq. (10.13), are based on the experimental work of Son and Kim [50]. All other parameters in the simulation are kept unchanged from the cold flow simulation of Zhang et al. [34]. Some of the key numerical parameters in the present simulation are summarized in Table 10.1.

Table 10.1 Key modeling parameters for reacting flow simulation in a CLC fuel reactor.

Primary phase	Gaseous mixture of CH_4, H_2O, CO_2, and N_2
Discrete phase	F60AM1100 particles
Particle diameter	0.0025 m
Particle density	2225 kg m^{-3}
Mass load of particle in bed	~0.7 kg
Inlet boundary condition	Velocity inlet with central jet velocity of 40 m s^{-1}, background flow velocity of 2 m s^{-1}
Outlet boundary condition	Pressure outlet with zero gage pressure
Drag law	Syamlal–O'Brien [39]
Particle collision model	Soft-sphere model
Spring constant	10 000 N m^{-1}
Coefficient of restitution	0.97
Friction coefficient	0.5
Time step size	Particle: 1×10^{-5} s, Fluid: 1×10^{-4} s

10.5 Spouted Fluidized Bed Simulation Results

The spouted bed simulation is run for 1600 ms to complete a few cycles of the bed spouting and its settling back down. Figure 10.6 shows the particle distributions and velocities at 80 ms intervals for the simulation duration.

From Figure 10.6, a prominent gas bubble can be observed to form immediately upon gas injection in the CLC fuel reactor. The leading front of the spout reaches the top of the fuel reactor around 400 ms and a large number of particles are deposited into the cyclone through the connecting duct. As the bed expands, the pressure buildup in the fuel reactor dissipates and the remaining particles in the fuel reactor fall back into the fluidized bed while the particles in the cyclone fall into the down-comer. Some recirculation of the particles from the loop-seal back into the fuel reactor is also evident. Once the particles start to settle back into a dense bed, aided by the recirculation of the particles from the loop-seal, the pressure buildup due to the jet injection is partially restored and subsequent gas bubbles are formed around 960 and 1440 ms. However, in these cases, the kinetic energy transferred to the particles is insufficient to carry them to the top of the reactor and the bubbles collapse prematurely. This can be explained by the bypass pathway formed at the same time by the high velocity jet in the absence of the initial packed bed, which allows the energy in the jet to bypass the dense bed region and prevents the critical pressure buildup in the fuel reactor. Since the gas bubble formation and the particle recirculation are both driven by the pressure at various locations in the system, the static pressure readings at pressure taps P1 through P5 are investigated to better understand the behavior observed from the particle tracks in Figure 10.6; the pressure tap data is presented in Figure 10.7.

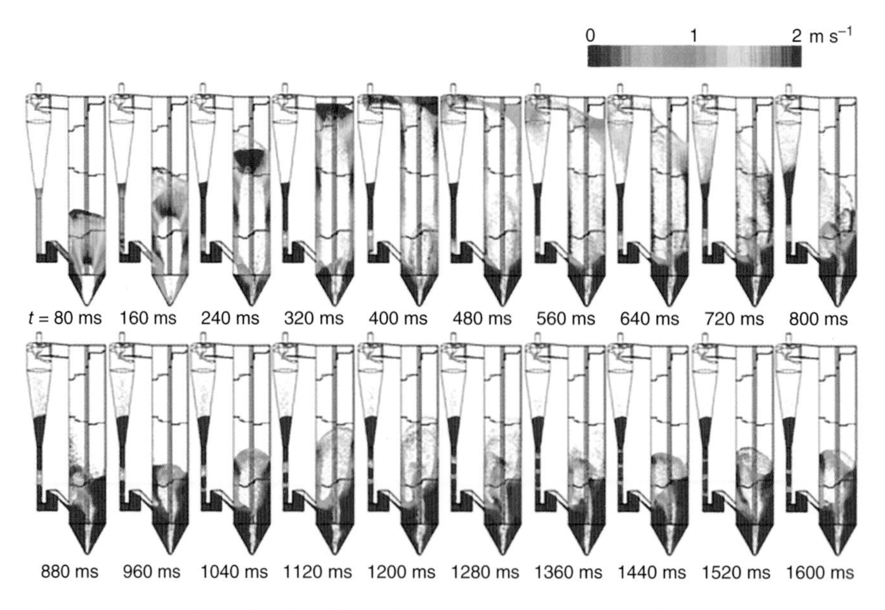

Figure 10.6 Particle tracks colored by velocity magnitude in reacting flow with F60AM1100 particles.

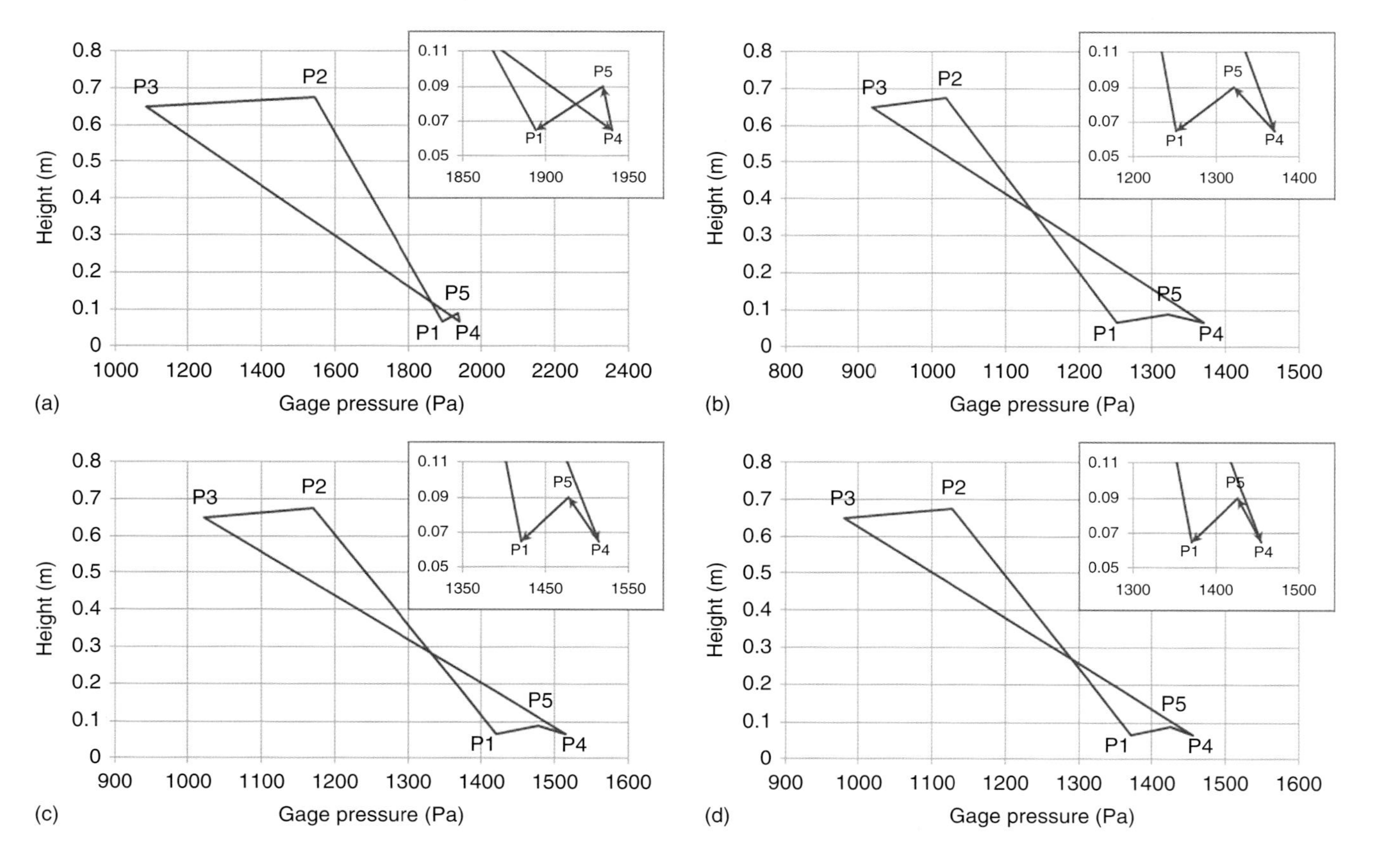

Figure 10.7 Static pressure at pressure taps P1–P5 in the CLC fuel reactor of Figure 10.5 at $t = 400$ ms (a), 800 ms (b), 1200 ms (c), and 1600 ms (d) in reacting flow.

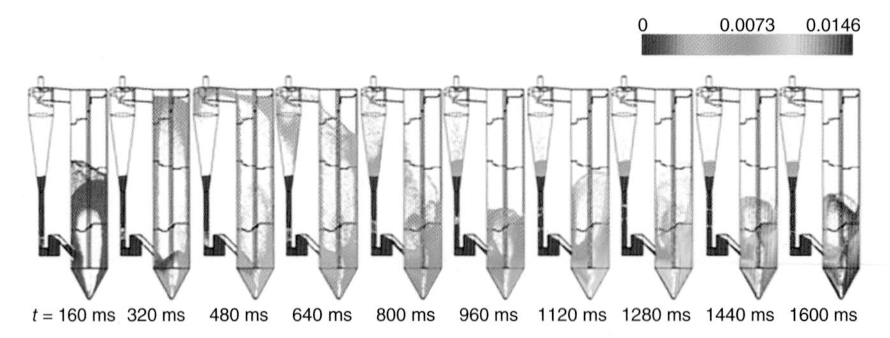

Figure 10.8 Particle tracks colored by mass fraction of Fe_3O_4 relative to original mass of Fe_2O_3.

The sub-plot at 400 ms in Figure 10.5 shows a large pressure buildup of approximately 1900 Pa at P1 (base of the reactor). At 800 ms, when the bubble has collapsed, the initial pressure buildup is lost as the pressure at P1 drops to nearly 1250 Pa. Subsequently, at 1200 and 1600 ms, the pressure at P1 increases slightly to nearly 1400 and 1350 Pa respectively. This buildup of pressure is in line with the observation of the second and third gas bubbles formations; the slight increase in pressure compared to the initial bubble also explains why the subsequent bubbles did not carry sufficient kinetic energy to reach the top of the reactor. From Figure 10.7, it can also be noted that there is a consistent positive pressure differential between taps P4 (base of the loop-seal) and P1 of around 100 Pa, which affirms the continuous recirculation of particles from the loop-seal back into the fuel reactor as can also be observed from the particle tracks.

In the absence of experimental results for reacting flow operation of a CLC fuel reactor in a spouted bed configuration, the successful incorporation of chemical reactions into the multiphase flow simulation is evaluated by inspecting the formation of Fe_3O_4 and CO_2 as a result of the oxygen carrier reduction reaction given in Eq. (10.15). These results are shown in Figures 10.8 and 10.9 respectively. From Figure 10.8, it can be seen that the mass fraction of Fe_3O_4 increases with time for the particles inside the fuel reactor as the simulation advances, as expected. The continuous particle recirculation is also evident from the consistent presence of a small number of particles in the fuel reactor with a lower mass fraction of Fe_3O_4 that originated in the loop-seal region where no reaction takes place. According to Figure 10.9, the mass fraction of CO_2 rises quickly in the first 640 ms of simulation, after which it drops slightly before spiking again around 1120 ms. These fluctuations are due to the inherently unsteady nature of the solid–gas mixing in the fuel reactor; similar results have also been reported in the literature for other fluidized bed reactors [51].

Comparing the current simulation results with the authors' previous work using pure Fe_2O_3 particles [47], it is clear that changing the oxygen carrier to a lighter material significantly improves the fluidization performance of the reactor by addressing the inadequacy of the central jet to provide sufficient momentum to the particles to reach the top of the reactor and by producing continuous particle recirculation. However, the formation of the bypass pathway through the bed after the first bubble collapses still remains a concern

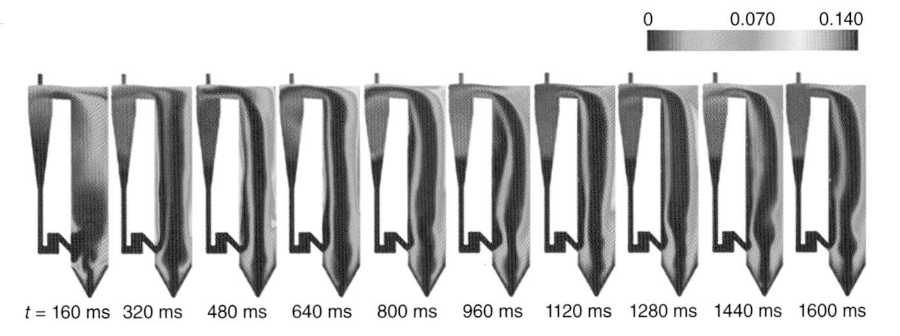

Figure 10.9 Contours of CO_2 mass fraction produced by reaction of Fe_2O_3 with CH_4.

and hinders the pressure buildup required for subsequent bubbles to reach the top of the reactor. Since F60AM1100 is already among the lightest Fe-based oxygen carriers studied by Johansson et al. [48] and lighter alternatives are likely to be expensive if developed, one alternate way to mitigate this issue is to use a cyclic flow injection whereby the jet is turned off intermittently to allow the bed particles to re-settle down into the dense bed configuration, which resets the fluidization behavior to the initial bubble formation stage once the jet is turned back on. Cyclic injections have already been used in laboratory-scale CLC experiments such as in the work of Son and Kim to switch between N_2 and CH_4 in lieu of separate fuel and air reactors [50] and their operational feasibility in an industrial setting can be readily studied in future work.

10.6 Eulerian–Lagrangian Simulation of a Binary Particle Bed in a Carbon Stripper

A typical CLC setup utilizes a cyclonic separator to isolate the oxygen carrier particles from the flue gases after the fuel reactor and the air reactor before transporting the solids between the reactors to continue to loop. Since the char gasification is a slow process [52], unburnt char particles often remain in the flue stream of the fuel reactor. If these are transported to the air reactor along with the oxygen carrier particles, the carbon capture efficiency of the CD-CLC process would be reduced. Several approaches have been proposed to prevent char particles from reaching the air reactor. One way is to provide sufficient residence time in the fuel reactor to ensure that the gasification reaction is complete. This can be achieved either by increasing the size of the reactor or by reducing the fluidizing gas velocity, but both options can impede the fluidization behavior of the bed, particularly in a spouted bed configuration as discussed in the previous section [53]. To avoid the poor fluidization while still maintaining an increased residence time, a multi-staged fuel reaction concept was recently proposed and investigated [54].

Figure 10.10 shows the differences in size between the particles of pulverized coal and a typical oxygen carrier (ilmenite) used in CD-CLC operation. One reason for considering the spouted fluidized bed configuration is that it overcomes

Figure 10.10 Size difference between particles of coal and ilmenite used in coal-direct CLC.

the limitation of a bubbling or fast fluidized bed to handle particles larger than a few hundred micrometers in diameter [44]. Thus, one way of preventing the leakage of unburnt char into the air reactor is to take advantage of the differences in size and density, and hence the terminal velocity, to separate the lighter char from the heavier oxygen carrier particles. Since char already has a lower density than the oxygen carrier, using pulverized coal particles smaller than or almost of the same size as the oxygen carrier particles should invariably lead to satisfactory separation results. The devolatilization and gasification processes that the coal undergoes further decrease the char particle size, enhancing the separation effect. The device that separates the char particles from the char and oxygen carrier mixture stream exiting the fuel reactor is known as a carbon stripper. The char particles from the carbon stripper can be returned to the fuel reactor to complete the gasification step while the oxygen carrier particles are transported to the air reactor to be regenerated.

By preventing the combustion of the unburnt char in atmospheric air, the carbon stripper also eliminates the formation of pollutants such as CO_2 and NO_X in the air reactor, as highlighted by Kramp et al. [55] and Mendiara et al. [56] and is deemed critical for CD-CLC operation despite the increased hydrodynamic complexity associated with implementing the carbon stripper compared to increasing the residence time in the fuel reactor. In recent years, carbon strippers operating with fluidizing velocity in the range of 0.15–0.40 $m\,s^{-1}$ have been incorporated into CD-CLC experiments by Markström et al. [57], Ströhle et al. [58], Abad et al. [59], and Sun et al. [60]. The results of these experiments indicated that the fluidization velocity should be increased further to increase the particle separation.

Later, Sun et al. [61] conducted cold-flow studies using a riser-based carbon stripper operating in the fast fluidized bed regime to investigate the effect of gas velocity on the separation ratio. The goal of Sun et al.'s design [61] was to achieve a high separation ratio to minimize the leakage of char particles into the air reactor with a low fluidizing gas velocity to keep operational costs low. However, the specific nature of the multiphase solid–gas flow inside the carbon stripper and how its geometry affects the design targets is not well understood from the experiment. In order to identify these relationships, a CFD-DEM coupled simulation is developed in this section for the carbon stripper consisting of a binary particle bed of coal and oxygen carrier particles and is validated against the experiment of Sun et al. [61].

The carbon stripper used in the cold-flow experiment by Sun et al. [61] consisted of a riser, 4 m tall with a diameter of 0.7 m. A schematic of the experimental setup is presented in Figure 10.11. The solids mixture contained 95% ilmenite

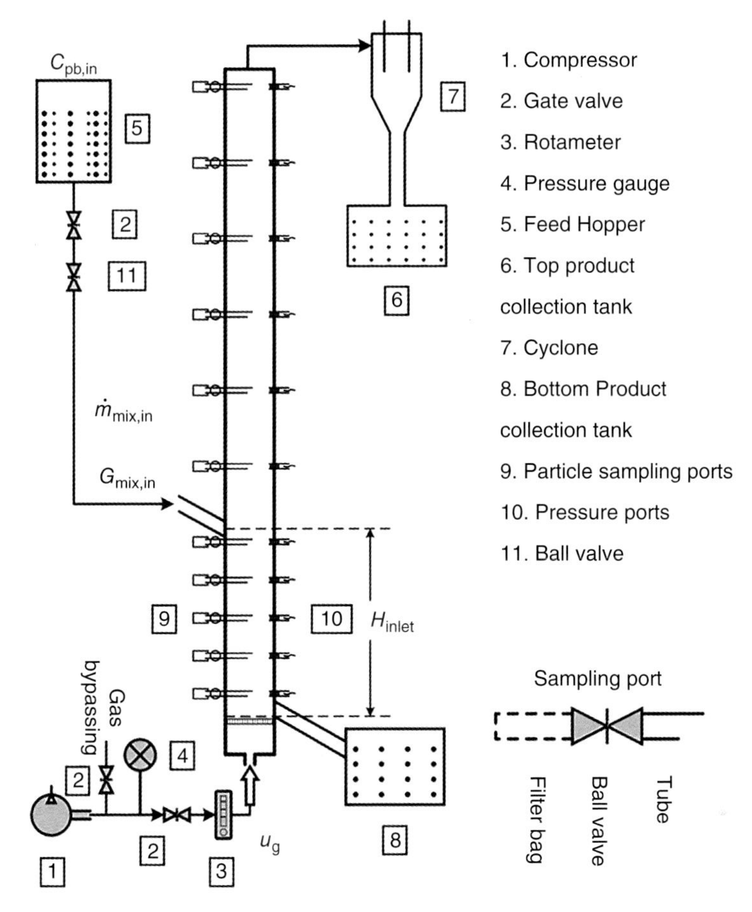

Figure 10.11 Schematic of riser-based carbon stripper used by Sun et al. 2016 [61]. Reproduced with permission of American Chemical Society.

Table 10.2 Properties of ilmenite particles and plastic beads used by Sun et al. 2016 [61].

Particle	d_p (0.5) (μm)	ρ_p (kg m^{-3})	u_t (m s^{-1})
Ilmenite	257	4260	5.65
Plastic beads	94	960	0.39

Reproduced with permission of American Chemical Society.

particles by mass and 5% plastic beads representing the unburnt char particles in the system. The physical properties of ilmenite and plastic beads are listed in Table 10.2. The riser was fluidized from the bottom by air with the fluidizing velocity u_g in the range of 1.50–2.75 m s^{-1} increasing at 0.25 m s^{-1} intervals. u_g was selected to fall between the terminal velocities u_t for the ilmenite and plastic beads such that the plastic beads will be carried out of the bed and exit the riser from the top into a tank while the ilmenite particles remain in the bed and collect in the bottom tank. The solids mixture is injected from the side of the riser at a height of 1 m above the bottom collection tank.

The separation ratio λ is defined as the mass of particles collected from the top tank to the mass of particles collected from the top and bottom tanks combined, as given by

$$\lambda = \frac{m_{\text{mix,top}}}{m_{\text{mix,top}} + m_{\text{mix,btm}}} \tag{10.19}$$

The concentration of plastic beads in each mixture sample in the experiment was determined by burning the mixture and measuring the change in weight. The plastic beads completely combusted to form CO_2 and H_2O while the weight loss of the oxygen carrier was approximately 1% due to the reduction of ilmenite. λ is calculated for each u_g based on the experimental results and is plotted in Figure 10.12 for a solids mixture feeding rate $G_{\text{mix, in}}$ of 12.2 kg m^{-2}-s; the same value of $G_{\text{mix, in}}$ is used in the simulations.

The geometry used in the CFD-DEM simulation of the carbon stripper used by Sun et al. [61] uses the exact dimensions of the riser presented in Figure 10.11. Since the solids flow from the feed hopper into the riser is of no consequence to the simulation, the solids inlet is simply modeled as a partial pipe. The top collection tank is also eliminated and the solids flow at the top is measured directly at the riser outlet. The bottom collection tank is modeled as a simple closed boundary in order to ensure the accurate pressure boundary condition at the bottom of the riser; the solids flow into the bottom tank is measured at the surface between the riser and the tank. A structured grid is generated for all elements of the geometry and is shown in Figure 10.13. The total number of cells is 51 884

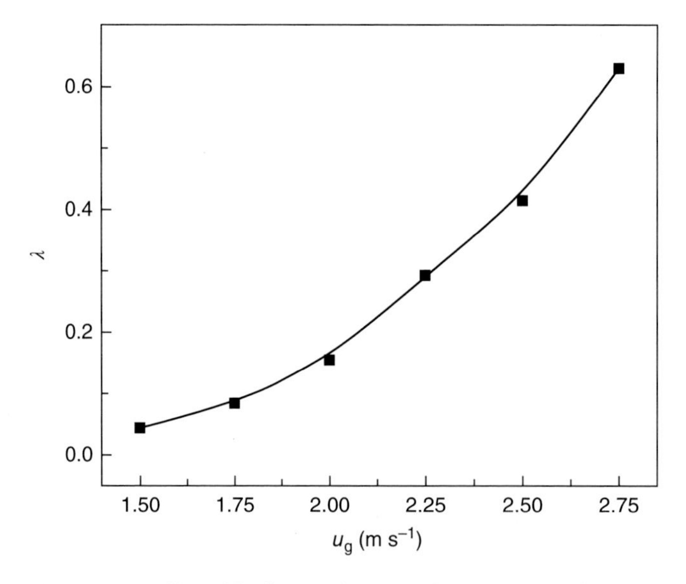

Figure 10.12 Effect of fluidizing velocity on the separation efficiency λ for a solids mass flux of 12.2 kg m^{-2}-s. Source: Sun et al. 2016 [61]. Reproduced with permission of American Chemical Society.

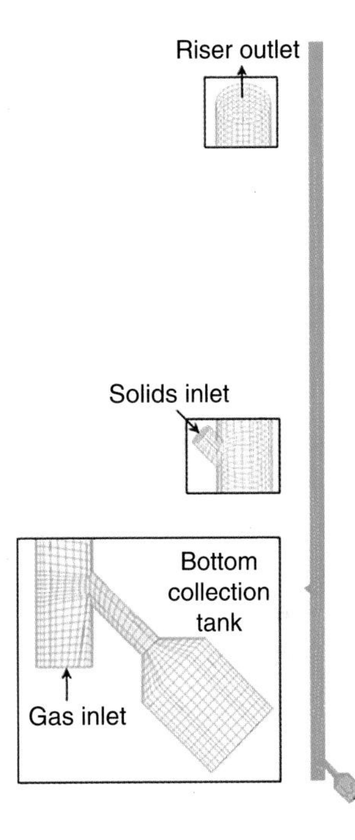

Figure 10.13 Geometry with detailed views used for CFD-DEM simulation of the experiment of Sun et al. 2016 [61]. Reproduced with permission of American Chemical Society.

in order to maintain a minimum cell volume greater than the particle (parcel) volume.

Given the small particle diameters of ilmenite and plastic beads used in the experiment (see Table 10.2), the number of particles in the system is very large. One simple approach available in ANSYS Fluent for reducing the computing load is to divide the particles into clusters called parcels. The motion of each parcel is determined as a whole by tracking a single representative particle [42]. Parcel collisions are evaluated in the same manner as shown in Figure 10.3 but the mass of the entire parcel is considered, not that of just a single representative particle. The parcel diameter is that of a sphere whose volume is the sum of the volumes of its constituent particles. Hence, specifying a parcel diameter equal to twice the particle diameter leads to a reduction in the number of objects tracked by the DEM solver by a factor of 8, with an even larger decrease in the number of collisions. The parcel approach is employed in the current simulation with a parcel diameter of 0.002 m.

Two solids injections are used corresponding to the ilmenite and the plastic beads; the injection mass flow rates are calculated based on $G_{\text{mix, in}}$, the riser cross section A_r ($= \pi D_r^2/4$), and the concentration of plastic beads in the solids flow c_{pb}, as outlined here.

$$\dot{m}_{\text{mix,in}} = G_{\text{mix,in}} A_r \tag{10.20}$$

$$\dot{m}_{\text{ilm,in}} = \frac{\dot{m}_{\text{mix,in}}}{1 + c_{\text{pb}}}; \dot{m}_{\text{pb,in}} = c_{\text{pb}} \frac{\dot{m}_{\text{mix,in}}}{1 + c_{\text{pb}}} \tag{10.21}$$

Table 10.3 Key modeling parameters for reacting flow simulation in the CLC fuel reactor.

Primary phase	Air
Discrete phase(s)	Ilmenite; plastic beads
Parcel diameter	0.2 m
Gas inlet fluidizing velocities	1.50, 1.75, 2.00, 2.25, 2.75 m s^{-1}
Solids injection velocity	$u_{\text{mix, in}} = 0.0034$ m s^{-1}
Solids injection flow rate	$\dot{m}_{\text{ilm,in}} = 0.044$ kg s^{-1}; $\dot{m}_{\text{pb,in}} = 0.0023$ kg s^{-1}
Outlet boundary condition	Pressure outlet at atmospheric pressure
Drag law	Gidaspow [37]
Particle collision model	Soft-sphere model
Spring constant	5000 N m^{-1}
Coefficient of restitution	0.97
Friction coefficient	0.5
Time step size	Particle: 5×10^{-5} s; fluid: 5×10^{-4} s

The volumetric flow rates can be determined given the respective densities of the two materials, and are used to determine the solids injection velocity.

$$q_{\text{ilm,in}} = \frac{\dot{m}_{\text{ilm,in}}}{\rho_{\text{ilm}}}; q_{\text{pb,in}} = \frac{\dot{m}_{\text{pb,in}}}{\rho_{\text{pb}}} \tag{10.22}$$

$$q_{\text{mix,in}} = q_{\text{ilm,in}} + q_{\text{pb,in}} \tag{10.23}$$

$$u_{\text{mix,in}} = \frac{q_{\text{mix,in}}}{A_r} \tag{10.24}$$

The soft sphere model shown in Figure 10.3 is used for all the particle–particle and particle–wall collisions. In order to keep the computing time low, the spring stiffness k_n is set at 5000 N m^{-1} to relax the minimum particle time step requirement. Bokkers [62] demonstrated that the results produced using this value of k_n are indistinguishable from those using larger values of k_n, which necessitate a smaller particle time step. The coefficient of restitution is set at 0.97. The numerical simulations are conducted using the phase-coupled SIMPLE scheme with second order discretization in space and first order in time. The simulation cases modeled and the key modeling parameters are summarized in Table 10.3.

10.7 Binary Particle Bed Simulation Results

Each CFD-DEM simulation of the binary particle bed in the riser-based carbon stripper is run for 20 seconds. The solids flow rate out of the riser outlet and into the bottom collection tank as well as the static pressure in the bed is recorded every 20 time steps (0.01 seconds). In the experiment of Sun et al. [61], after the initial development of fluidization caused by the solids injection, the pressure differences across sampling ports 1–3 and 8–11 shown in Figure 10.11 stabilized after approximately 10 seconds. The static pressure at 2 mm above

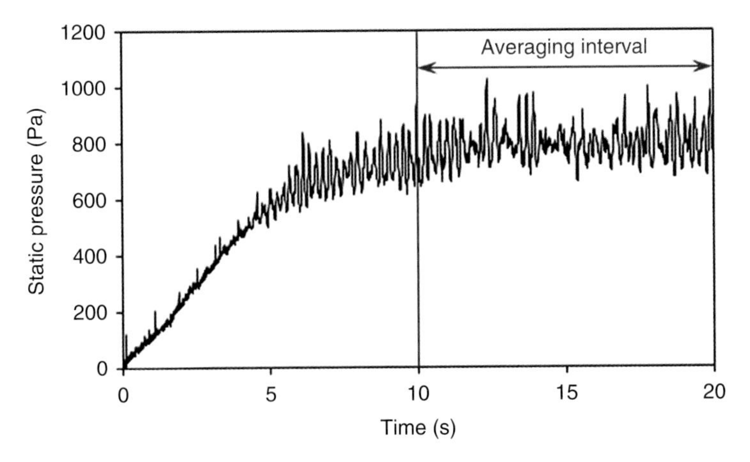

Figure 10.14 Static pressure at 2 mm for binary particle bed simulation with fluidizing velocity $u_g = 1.50\ \text{m s}^{-1}$.

the inlet is used to verify the stable bed in the numerical simulation; the results for the $u_g = 1.50\ \text{m s}^{-1}$ case are given in Figure 10.14. It can be seen that the static pressure in the bed stabilizes after approximately eight seconds. The final 10 seconds of simulation is used as the averaging interval for the solids flow rates in order to calculate the separation ratio λ, which is used to quantitatively validate the accuracy of the simulation against the experimental results [61]. To confirm that the averaging interval does not affect the value of λ, the simulation with $u_g = 2.00\ \text{m s}^{-1}$ was run for 30 seconds and the computed difference in λ was miniscule.

The development of solids flow into and out of the riser can be ascertained by examining the number of particles (parcels) of ilmenite and plastic beads held up in the riser after 20 seconds of simulation, as shown in Table 10.4. As u_g increases, the number of ilmenite parcels in the riser increases. This is because the increased gas velocity prevents the ilmenite from settling at the bottom of the riser and flowing into the bottom collection tank. However, this increase starts to diminish around $u_g = 2.50\ \text{m s}^{-1}$ and for $u_g = 2.75\ \text{m s}^{-1}$, the number of ilmenite particles in the riser decreases drastically. Although $2.50\ \text{m s}^{-1}$ is still lower than the terminal velocity of ilmenite, the decreased hold up suggests that at this velocity, the flow has sufficient energy to carry the particles out of the riser. On the other hand, the number of plastic beads in the riser steadily decreases as the fluidizing velocity increases. This is expected since the u_g/u_t ratio starts out at more than 1 at $u_g = 1.50\ \text{m s}^{-1}$ and as it gets larger, the flow is able to carry the plastic beads out with greater ease.

Table 10.4 Number of parcels in riser after 20 s of simulation for different fluidizing velocities.

Particle	$1.50\ \text{m s}^{-1}$	$1.75\ \text{m s}^{-1}$	$2.00\ \text{m s}^{-1}$	$2.25\ \text{m s}^{-1}$	$2.50\ \text{m s}^{-1}$	$2.75\ \text{m s}^{-1}$
Ilmenite	17 267	26 138	34 064	41 044	45 741	14 051
Plastic beads	4882	3570	2403	1383	1211	1059

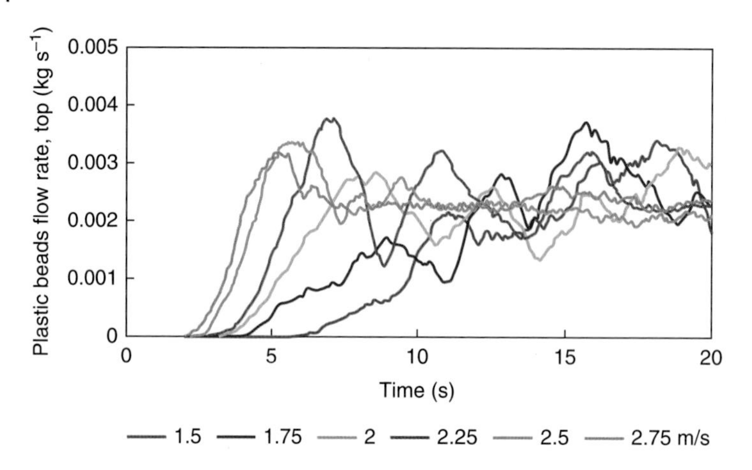

Figure 10.15 Plastic beads flow rate out of top of the riser for different fluidizing velocities.

For each case, the solids flow out of the riser outlet consists almost entirely of plastic beads with a few ilmenite particles, except at $u_g = 2.75\,\mathrm{m\,s^{-1}}$. On the other hand, the solids flow into the bottom collection tank is solely composed of ilmenite. This is expected given that the fluidizing velocities in each case lies between the terminal velocities of the plastic beads and ilmenite particles such that the fluid can carry the lighter plastic beads out of the bed but not the ilmenite particles. The flow rate of plastic beads out of the top riser outlet for different values of u_g is presented in Figure 10.15. As u_g increases, the plastic beads reach the outlet faster because of a higher induced particle velocity and the overall flow rate increases slightly until it stabilizes at a roughly constant value in each case equal to the injection flow rate of plastic beads; the plastic beads flow rate into the bottom collection tank is nil. The transient fluctuations in the flow rate are due to the highly unsteady flow in the fast fluidization regime associated with the riser.

Similar plots are generated for the flow rate of ilmenite out of the top of the riser and into the bottom collection tank and are shown in Figures 10.16 and 10.17 respectively. As mentioned above, the ilmenite flow rate out of the top outlet is

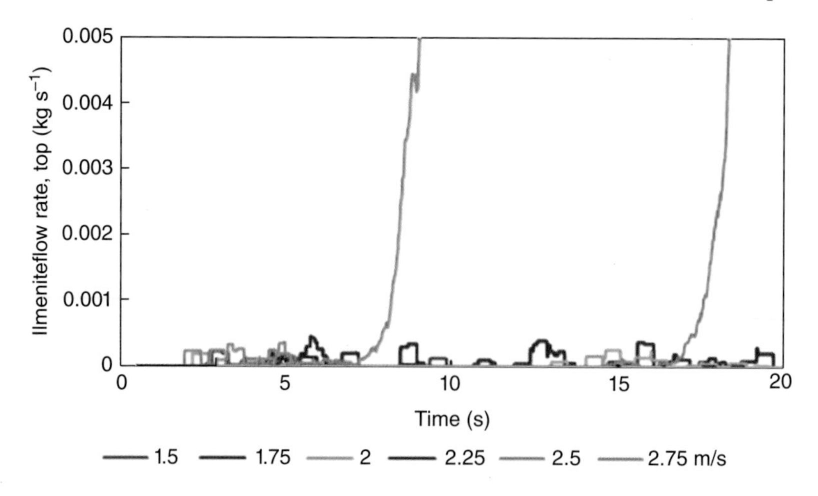

Figure 10.16 Ilmenite flow rate out of top of the riser for different fluidizing velocities.

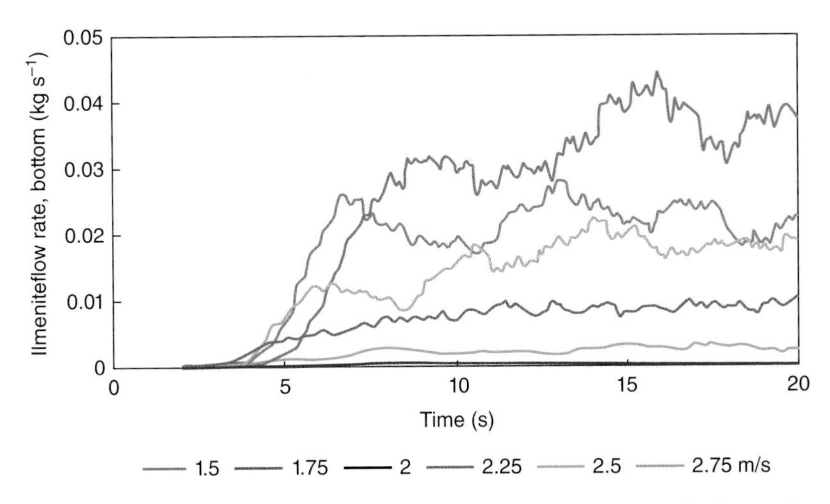

Figure 10.17 Ilmenite flow rate into bottom collection tank for different fluidizing velocities.

limited to isolated particles up to $u_g = 2.25\ \mathrm{m\ s^{-1}}$. The ilmenite flow rate into the bottom collection tank decreases as u_g increases. The flow rate plots confirm the solids flow behavior suggested by the parcel hold up numbers in Table 10.4.

The flow rates of the plastic beads out of the top of the riser and the ilmenite into the bottom collection tank shown in Figures 10.15 and 10.17 respectively are used to compute the separation ratio λ according to Eq. (10.19). The values of λ for different fluidization velocities are plotted in Figure 10.18. The values of λ in Figure 10.18 are in excellent agreement with the experimental values presented in Figure 10.12. Hence, the binary particle bed simulation shown in this chapter can be considered to be a credible model for the experiment and can be employed to examine additional changes to the geometry and operating conditions and investigate their effect on λ.

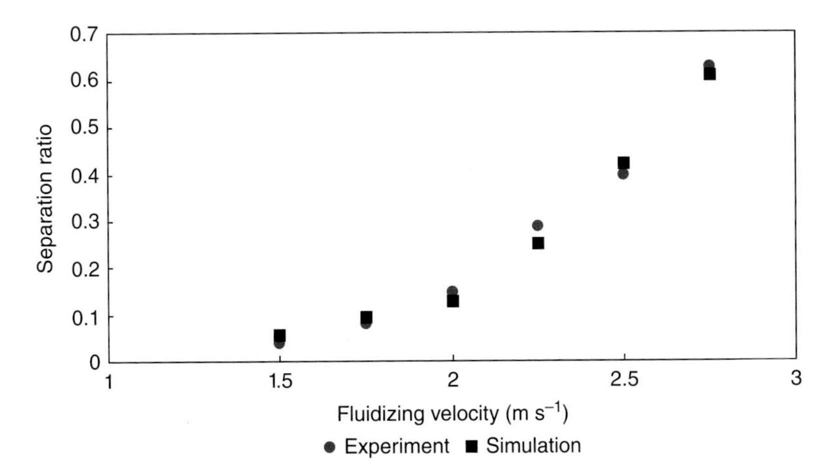

Figure 10.18 Effect of fluidizing velocity on separation ratio λ for a solids mass flux of $12.2\ \mathrm{kg\ m^{-2}\text{-}s}$ in the CFD-DEM simulation of the riser-based carbon stripper compared against the experiment of Sun et al. 2016 [61]. Reproduced with permission of American Chemical Society.

10.8 Summary and Conclusions

A CFD-DEM simulation of the multiphase flow inside a CLC fuel reactor has been performed. The initial simulation for cold flow validation in previous work showed excellent agreement with the experimental results obtained at TU-Darmstadt for a pseudo-3D CLC fuel reactor [39]. In this chapter, the fuel reactor model was expanded to a complete CD-CLC system including a cyclonic separator, down-comer, and loop-seal, and chemical reactions between Fe_2O_3 and CH_4 were added. Using 60% Fe_2O_3 supported on $MgAl_2O_4$ as the oxygen carrier for its favorable properties for CD-CLC operation [48], the fluidization behavior of the system was investigated. It was found that the initial bubble formation and continuous recirculation of particles from the loop-seal to the fuel reactor were adequate and it was determined that the chemical reactions were successfully incorporated.

CFD-DEM simulations were also conducted of the riser-based carbon stripper developed by Sun et al. [61] consisting of a binary particle bed of plastic beads (corresponding to coal) and ilmenite particles. The static pressure in the bed is used to assess the onset of a stable fluidization regime and the flow rates of the plastic beads and ilmenite out of the riser outlet and into the bottom collection tank respectively were averaged to determine the particle separation ratio. The results showed excellent agreement with the experimental data and established a credible model for a binary particle bed that can be used to optimize the design and operation of such systems in future. One point of disagreement between the simulation and the experiment is the absence of plastic beads in the bottom collection tank. In the experiment, a few plastic beads were collected in the bottom tank at lower values of u_g [61]. However, theoretically this behavior was unexpected since even the smallest value of $u_g = 1.50\ \mathrm{m\,s^{-1}}$ is well above the terminal velocity of the plastic beads and should carry the beads out of the riser.

References

1 IPCC (2007). *Climate Change 2007: Synthesis Report. Contribution of Working Groups I, II and III to the Fourth Assessment Report of the Intergovernmental Panel on Climate Change*, Core Writing Team (ed. R.K. Pachauri and A. Reisinger). Geneva: IPCC.

2 U.S. Energy Information Administration (2010). International Energy Outlook 2010. Report No. *DOE/EIA-0484*. Washington, DC: U.S. Energy Information Administration, U.S. Department of Energy.

3 Hong, J., Chaudhry, G., Brisson, J.G. et al. (2009). Performance of the pressurized oxy fuel combustion power cycle with increasing operating pressure. Proceedings 34th International Technical Conference on Clean Coal and Fuel Systems, Clearwater, FL.

4 Hong, J., Chaudhry, G., Brisson, J.G. et al. (2009). Analysis of oxy-fuel combustion power cycle utilizing a pressurized coal combustor. *Energy* 34 (9): 1332–1340.

5 Kruggel-Emden, H., Stepanek, F., and Munjiza, A. (2011). A study on the role of reaction modeling in multi-phase CFD-based simulations of chemical looping combustion. *Oil Gas Sci. Technol.* 66 (2): 313–331.

6 Lyngfelt, A., Leckner, B., and Mattisson, T. (2001). A fluidized-bed combustion process with inherent CO_2 separation; application of chemical-looping combustion. *Chem. Eng. Sci.* 56 (10): 3101–3113.

7 Ishida, M., Zheng, D., and Akehata, T. (1987). Evaluation of a chemical-looping-combustion power-generation system by graphic exergy analysis. *Energy* 12 (2): 147–154.

8 Ishida, M., Jin, H., and Okamoto, T. (1996). A fundamental study of a new kind of medium material for chemical-looping combustion. *Energy Fuels* 10 (4): 958–963.

9 Wolf, J., Anheden, M., and Yan, J. (2001). Performance analysis of combined cycles with chemical looping combustion for CO_2 capture. Proceedings 18th International Pittsburgh Coal Conference, Pittsburgh, PA.

10 Marion, J.L. (2006). Technology options for controlling CO_2 emissions from fossil fueled power plants. Proceedings 5th Annual Conference on Carbon Capture and Sequestration, Alexandria, VA.

11 ALSTOM Power Inc. (2008). Hybrid Combustion-Gasification Chemical Looping Coal Power Technology Development, Phase III – Final Report. Report No. *PPL-08-CT-25*. Windsor, CT: ALSTOM Power Inc.

12 National Energy Technology Laboratory (2010). Guidance for NETL's Oxy-Combustion R&D Program: Chemical Looping Combustion Reference Plant Designs and Sensitivity Studies. Report No. *DOE/NETL-2010/1643*. Pittsburgh, PA: U.S. Department of Energy.

13 Leion, H., Mattisson, T., and Lyngfelt, A. (2007). The use of petroleum coke as fuel in chemical-looping combustion. *Fuel* 86 (12–13): 1947–1958.

14 Cao, Y. and Pan, W. (2006). Investigation of chemical looping combustion by solid fuels: 1. Process analysis. *Energy Fuels* 20: 1836–1844.

15 Lewis, W. and Gilliland, E. (1954). Production of pure carbon dioxide. US Patent 2,665,972.

16 Mattisson, T., Lyngfelt, A., and Leion, H. (2009). Chemical-looping with oxygen uncoupling for combustion of solid fuels. *Int. J. Greenhouse Gas Control* 3 (1): 11–19.

17 Leion, H., Lyngfelt, A., and Mattisson, T. (2009). Solid fuels in chemical-looping combustion using a NiO-based oxygen carrier. *Chem. Eng. Res. Des.* 87 (11): 1543–1550.

18 Rubel, A., Zhang, Y., Liu, K., and Neathery, J. (2011). Effect of ash on oxygen carriers for the application of chemical looping combustion to a high carbon char. *Oil Gas Sci. Technol.* 66 (2): 291–300.

19 Geldart, D. (1973). Types of gas fluidization. *Powder Technol.* 7 (5): 285–292.

20 Shen, L., Wu, J., Gao, Z., and Xiao, J. (2009). Reactivity deterioration of NiO/Al_2O_3 oxygen carrier for chemical looping combustion of coal in a $10\,kW_{th}$ reactor. *Combust. Flame* 156 (7): 1377–1385.

21 Hossain, M.M. and de Lasa, H.I. (2008). Chemical-looping combustion (CLC) for inherent CO_2 separations—a review. *Chem. Eng. Sci.* 63: 4433–4451.

22 Adánez, J., Abad, A., García-Labiano, F. et al. (2012). Progress in chemical-looping combustion and reforming technologies. *Prog. Energy Combust. Sci.* 38 (2): 215–282.

23 Imtiaz, Q., Hosseini, D., and Müller, C.R. (2013). Review of oxygen carriers for chemical looping with oxygen uncoupling (CLOU): thermodynamics, material development, and synthesis. *Energy Technol.* 1: 633–647.

24 Patil, D.J., Annaland, M.V., and Kuipers, J.A.M. (2004). Critical comparison of hydro-dynamic models for gas–solid fluidized beds—Part I: bubbling gas–solid fluidized beds operated with a jet. *Chem. Eng. Sci.* 60 (1): 57–72.

25 Patil, D.J., Annaland, M.V., and Kuipers, J.A.M. (2004). Critical comparison of hydro-dynamic models for gas–solid fluidized beds—Part II: freely bubbling gas–solid fluidized beds. *Chem. Eng. Sci.* 60 (1): 73–84.

26 Jung, J. and Gamwo, I. (2008). Multiphase CFD-based models for chemical looping combustion process: fuel reactor modeling. *Powder Technol.* 183: 401–409.

27 Deng, Z.G., Xiao, R., Jin, B.S. et al. (2008). Multiphase CFD modeling for a chemical looping combustion process (fuel reactor). *Chem. Eng. Technol.* 31 (12): 1754–1766.

28 Mahalatkar, K., Kuhlman, J., Huckaby, E.D., and O'Brien, T. (2011). Computational fluid dynamic simulations of chemical looping fuel reactors utilizing gaseous fuels. *Chem. Eng. Sci.* 66 (3): 469–479.

29 Mahalatkar, K., Kuhlman, J., Huckaby, E.D., and O'Brien, T. (2011). CFD simulation of a chemical-looping fuel reactor utilizing solid fuel. *Chem. Eng. Sci.* 66 (16): 3617–3627.

30 Kruggel-Emden, H., Stepanek, F., and Munjiza, A. (2011). A study on the role of reaction modeling in multi-phase CFD-based simulations of chemical looping combustion. *Oil Gas Sci. Technol.* 66 (2): 313–331.

31 Gryczka, O., Heinrich, S., Deen, N.G. et al. (2009). Characterization and CFD modeling of the hydrodynamics of a prismatic spouted bed apparatus. *Chem. Eng. Sci.* 64: 3352–3375.

32 Sutkar, V.S., Deen, N.G., Mohan, B. et al. (2013). Numerical investigations of a pseudo-2D spout fluidized bed with draft plates using a scaled discrete particle model. *Chem. Eng. Sci.* 104: 790–807.

33 Alobaid, F., Ströhle, J., and Epple, B. (2013). Extended CFD/DEM model for the simulation of circulating fluidized bed. *Adv. Powder Technol.* 24 (1): 403–415.

34 Zhang, Z., Zhou, L., and Agarwal, R.K. (2014). Transient simulations of spouted fluidized bed for coal-direct chemical looping combustion. *Energy Fuels* 28 (2): 1548–1560.

35 Parker, J. (2014). CFD model for the simulation of chemical looping combustion. *Powder Technol.* 265: 47–53.

36 Ergun, S. (1952). Fluid flow through packed columns. *Chem. Eng. Prog.* 48: 89–94.

37 Gidaspow, D. (1992). *Multiphase Flow and Fluidization*. San Diego, CA: Academic Press.

38 Wen, C.Y. and Yu, H.Y. (1966). Mechanics of fluidization. *Chem. Eng. Prog. Symp. Ser.* 62: 100–111.
39 Syamlal, M. and O'Brien, T.J. (1989). Computer simulation of bubbles in a fluidized bed. *AIChE Symp. Ser.* 85: 22–31.
40 Beetstra, R., van der Hoef, M.A., and Kuipers, J.A.M. (2007). Drag force of intermediate Reynolds number flow past mono- and bi-disperse array of spheres. *AIChE J.* 53 (2): 489–501.
41 ANSYS (2012). *ANSYS Fluent User's Guide*. Canonsburg, PA: ANSYS, Inc.
42 ANSYS (2012). *ANSYS Fluent Theory Guide*. Canonsburg, PA: ANSYS, Inc.
43 Link, J.M. (1975). Development and validation of a discrete particle model of a spout-fluid bed granulator. PhD dissertation. Enschede: University of Twente.
44 Mathur, K.B. and Gishler, P.E. (1955). A technique for contacting gases with coarse solid particles. *AIChE J.* 157–164: 1.
45 Sutkar, V.S., Deen, N.G., and Kuipers, J.A.M. (2013). Spout fluidized beds: recent advances in experimental and numerical studies. *Chem. Eng. Sci.* 86: 124–136.
46 Sutkar, V.S., van Hunsel, T.J.K., Deen, N.G. et al. (2013). Experimental investigations of a pseudo-2D spout fluidized bed with draft plates. *Chem. Eng. Sci.* 102: 524–543.
47 Banerjee, S. and Agarwal, R.K. (2015). Transient reacting flow simulation of spouted fluidized bed for coal-direct chemical looping combustion with different Fe-based oxygen carriers. *Appl. Energy* 160: 552–560.
48 Johansson, M., Mattisson, T., and Lyngfelt, A. (2004). Investigation of Fe_2O_3 with $MgAl_2O_4$ for chemical-looping combustion. *Ind. Eng. Chem. Res.* 43: 6978–6987.
49 Syamlal, M. and O'Brien, T.J. (1989). Computer simulation of bubbles in a fluidized bed. *AIChE Symp. Ser.* 85: 22–31.
50 Son, S.R. and Kim, S.D. (2006). Chemical-looping combustion with NiO and Fe_2O_3 in a thermo-balance and circulating fluidized bed reactor with double loops. *Ind. Eng. Chem. Res.* 45: 2689–2696.
51 Berguerand, N. (2009). Design and operation of a 10 kW_{th} chemical-looping combustor for solid fuels. PhD dissertation. Göteburg: Chalmers University of Technology.
52 Leion, H., Mattisson, T., and Lyngfelt, A. (2007). The use of petroleum coke as fuel in chemical-looping combustion. *Fuel* 86 (12–13): 1947–1958.
53 Banerjee, S. and Agarwal, R.K. (2015). Transient reacting flow simulation of spouted fluidized bed for coal-direct chemical looping combustion. *J. Therm. Sci. Eng. Appl.* 7 (2): 021016.
54 Meng, W.X., Banerjee, S., Zhang, X., and Agarwal, R.K. (2015). Process simulation of multi-stage chemical-looping combustion using Aspen Plus. *Energy* 90 (2): 1869–1877.
55 Kramp, M., Thon, A., Hartge, E.U. et al. (2012). Carbon stripping – a critical process step in chemical looping combustion of solid fuels. *Chem. Eng. Technol.* 35: 497–507.

56 Mendiara, T., Izquierdo, M.T., Abad, A. et al. (2014). Release of pollutant components in CLC of lignite. *Int. J. Greenhouse Gas Control* 22: 15–24.

57 Markström, P., Linderholm, C., and Lyngfelt, A. (2013). Chemical-looping combustion of solid fuels – design and operation of a 100 kW unit with bituminous coal. *Int. J. Greenhouse Gas Control* 15: 150–162.

58 Ströhle, J., Orth, M., and Epple, B. (2014). Design and operation of a 1 MW_{th} chemical looping plant. *Appl. Energy* 113: 1490–1495.

59 Abad, A., Pérez-Vega, R., de Diego, L.F. et al. (2015). Design and operation of a 50 kW_{th} chemical looping combustion (CLC) unit for solid fuels. *Appl. Energy* 157: 295–303.

60 Sun, H., Cheng, M., Chen, D. et al. (2015). Experimental study of a carbon stripper in solid fuel chemical looping combustion. *Ind. Eng. Chem. Res.* 54: 8743–8753.

61 Sun, H., Cheng, M., Li, Z., and Cai, N. (2016). Riser-based carbon stripper for coal-fueled chemical looping combustion. *Ind. Eng. Chem. Res.* 55: 2381–2390.

62 Bokkers, G.A. (2005). Multi-level modelling of the hydrodynamics in gas phase polymerization reactors. PhD dissertation. Enschede: University of Twente.

11

Calcium- and Iron-Based Chemical Looping Combustion Processes

Robert W. Stevens Jr.[1], *Dale L. Keairns*[2], *Richard A. Newby*[3], *and Mark C. Woods*[3]

[1] *National Energy Technology Laboratory, U.S. Department of Energy, Systems Engineering & Analysis Directorate, 3610 Collins Ferry Road, Morgantown, WV 26507, USA*
[2] *Deloitte Consulting, LLP, National Energy Technology Laboratory, 626 Cochrans Mill Road, Pittsburgh, PA 15236, USA*
[3] *Keylogic Systems, Inc., National Energy Technology Laboratory, 626 Cochrans Mill Road, Pittsburgh, PA 15236, USA*

11.1 Introduction

One of the major areas of research activity within chemical looping combustion (CLC) has been on the development of an effective oxygen carrier, and a multitude of metal oxides have been proposed and tested under laboratory conditions. Carrier characteristics such as reactivity, oxygen capacity, durability, and price have been considered. To date, there is insufficient, long-term pilot testing and commercial assessment to select an optimum oxygen carrier. In this chapter, a limestone-based $CaSO_4$ carrier is evaluated and compared to a metal oxide-based, supported Fe_2O_3 carrier. These represent two significantly different types of oxygen carriers that have received much research interest, and have been speculated to have promising characteristics. The calcium-based carrier is low cost, low density, and possesses a high oxygen-carrying capacity, but has slow reaction kinetics. The reference iron-based carrier has better reactivity, but is also significantly higher in cost, higher density, and possesses lower oxygen-carrying capacity. It should be noted that while the calcium-based oxygen carrier is chemically similar to the oxygen carrier featured within GE/Alstom's chemical looping technology, the current chapter explores one embodiment of the calcium-based oxygen carrier based on a reactor design concept conceived by the US Department of Energy. Performance/cost estimates presented here should not be associated with expectations of the GE/Alstom CLC technology.

The application considered in this chapter is a base-load, utility-scale, coal-fueled CLC power plant with a 90% carbon capture requirement. Its estimated performance and cost (with both the calcium-based and iron-based oxygen carriers) are compared to that of a conventional pulverized coal (PC) power plant that uses amine-based, post-combustion CO_2 capture. Performance and cost

Handbook of Chemical Looping Technology, First Edition. Edited by Ronald W. Breault.

sensitivity analyses are also presented to provide further perspective on the CLC systems.

11.2 CLC Plant Design, Modeling, and Cost Estimation Bases

CLC reference power plant designs, with estimates of performance and cost, are developed in this chapter for a $CaSO_4$-based oxygen carrier, which is compared to an Fe_2O_3-based oxygen carrier using consistent assumptions. The general design basis for the power plant is identical to the design basis described in detail in the National Energy Technology Laboratory's (NETL) Bituminous Baseline report (BBR) for comparable, conventional, PC power plants [1]. The essential aspects of the design basis are presented in this chapter.

The CLC power plant contains two unique reactors, the Reducer and Oxidizer, that require significant modeling efforts to estimate their design features and performance. The general modeling approach applied is outlined in this section.

Most equipment components in the CLC power plant are conventional, and their performance and cost can be estimated by scaling from the BBR performance and cost estimates for conventional PC power plants. Other equipment components, the Reducer and Oxidizer and their associated subsystems, are developmental in nature, and their costs must be estimated by approximate sizing of the equipment and application of general cost correlations. The cost estimation approach applied is summarized in this section.

11.2.1 Design Basis

This is a base-load, utility-scale, power generation application. The reference CLC power plant is assumed to be located at a generic Midwestern U.S. plant site and the ambient conditions correspond to International Organization for Standardization (ISO) conditions listed in the Quality Guidelines for Energy System Studies (QGESS) process modeling parameters document [2]. The key features of the design basis are

- ambient conditions: ISO
- nominal net plant capacity: 550 MW
- design coal: Illinois No. 6
- steam conditions: supercritical {3500 psig (24 MPa), 1100 °F (593 °C), 1100 °F (593 °C)}
- carbon capture requirement: at least 90%
- CO_2 product purity requirement: at least 95 mol% CO_2
- CO_2 product delivery pressure: 2200 psig.

The Illinois No. 6 coal as described in the QGESS coal feedstock specification [2] was used in the present analysis.

Assumptions for the performance of major equipment components that are consistent with those used in the BBR are applied. Some major items are

- fan efficiency: 75% (polytropic)
- compressor efficiency: 86% (polytropic)

- electric motor efficiency: 97%
- generator efficiency: 98.5%.

Pressure drops and auxiliary loads representative of conventional equipment components in the CLC power plant are scaled directly from the comparable equipment components characterized in the BBR, Case 12. Such equipment components are

- coal handling
- coal pulverizers
- limestone handling
- ash handling
- forced and induced draft (ID) fans
- gas cleaning (baghouses, flue gas desulfurization (FGD), selective catalytic reduction (SCR))
- steam turbine
- cooling water system.

11.2.2 Cost Estimation Basis

Cost estimation for the CLC plants is consistent with the BBR, with the costs reported in 2011 dollars. Major costing premises are

- plant capacity factor: 85%
- plant financial classification: high-risk
- capital charge factor: 12.43% per year
- fixed and variable operating and maintenance (O&M) costs: Estimated analogously to BBR Case 12 [1].

Conventional equipment costs are scaled from the BBR Case 12 analogous equipment costs. Such equipment components are

- coal handling system
- coal preparation and feed systems
- feedwater and miscellaneous balance of plant (BOP) systems
- gas cleaning
- steam turbine generator system
- cooling water system
- accessory electric plant
- instrumentation and controls
- improvements to site
- buildings and structures.

For novel equipment or equipment unique to the CLC application, as listed below, approximate equipment sizing, and general cost correlations are applied:

- reducer reactor
- oxidizer reactor
- cyclones
- high-temperature piping
- solids cooling steam generator
- fuel recovery/CO_2 purification and compression

11.2.3 Reactor Modeling Basis

The Reducer and Oxidizer reactors are both circulating fluidized beds. The behavior and performance of each reactor are controlled by the characteristics of the oxygen carrier reaction kinetics, the coal gasification kinetics, and the fluidized bed hydrodynamics and mass transfer properties. The assumptions selected to represent each of these for the reference CLC plant are summarized in Table 11.1.

The calcium-based oxygen carrier and the reference iron-based carrier compared in this chapter have characteristics and properties extracted from the CLC literature. Table 11.1 summarizes the main characteristics of the oxygen carriers and addresses coal gasification assumptions of the Reducer reactor. Large literature exists on Fe_2O_3-based oxygen carrier CLC reaction behavior observed in laboratory testing, and a key reference has been applied that provides reaction kinetics correlations for a specific Fe_2O_3 oxygen carrier [3]. Alstom previously

Table 11.1 Reactor modeling basis.

Modeling factor	Fe_2O_3	$CaSO_4$
Oxygen carrier characteristics		
Oxygen carrier type	Supported metal oxide	Natural quarry material
Makeup carrier composition	45 wt% Fe_2O_3 on Al_2O_3 support	Limestone
Particle size (mm or in.)	0.28 (0.01)	0.5 (0.02)
Particle density (kg m^{-3} or lb ft^{-3})	3250 (203)	1571 (98)
Reduction and oxidation kinetics	Rapid; shrinking grain model behavior	Slow; shrinking grain model behavior
Reaction kinetic sources	Abad [3]	Song et al. [8], Tian and Guo [9], Song et al. [10]
Oxygen carrier reactivity with coal contaminants	Yes, with SO_2, H_2S, HCl	Yes, with SO_2, H_2S, HCl
Oxygen carrier price (delivered \$ ton^{-1})	High (2 000–10 000 \$ ton^{-1})	Low (100–300 \$ ton^{-1})
Coal gasification characteristics		
Coal devolatilization rate	Very fast (about 0.5 s gas residence time)	
Char gasification rate	Very slow compared to oxygen carrier reactor rates	
Char gasification kinetics sources	Johnson [11]	
Fluidized bed characteristics		
Oxygen carrier fluidization class [12] at reactor conditions	B	B
Type of Reducer fluidized bed regime	Turbulent at bottom and fast, circulating bed at top	Turbulent at bottom and fast, circulating bed at top
Type of Oxidizer fluidized bed regime	Fast, circulating bed	Fast, circulating bed
Fluid bed reactor model sources	Kunii and Levenspiel [13] and Abba et al. [12]	

conducted laboratory- and bench-scale studies for CLC using a $CaSO_4$ oxygen carrier [4], but did not report reaction kinetics correlations. Several literature resources on $CaSO_4$ laboratory reaction kinetics were applied, as listed in the exhibit, all testing only pure calcium anhydrite; however, none of these tested limestone-based $CaSO_4$.

The Reducer reactor is a coal gasifier with the oxygen carrier generating CO_2 and H_2O from reaction with the coal gasification products H_2 and CO. Because the rate of char gasification at the relatively low Reducer temperature is very small, the gasification rate controls the volume of the Reducer reactor.

With respect to the oxygen carrier delivered price, the Fe_2O_3 oxygen carrier is expected to be a relatively expensive fabricated material, in the range of $2000–10 000 per ton, while the $CaSO_4$ makeup material will be low-cost limestone, having a delivered price of $30–300 per ton. A cost sensitivity study considers the influence of the Fe_2O_3 oxygen carrier price on the CLC power plant cost of electricity (COE) if cheaper forms of Fe_2O_3-based oxygen carrier were used, such as a natural ore (e.g. hematite) or a waste material (e.g. red mud).

The circulating fluid bed reactor models used in this analysis represent the structure of the circulating fluid bed and the various mass transfer resistances within the fluid bed. These models have very uncertain behavior and are based primarily on small-scale test observations and contain several adjustable parameters. The oxygen carrier reaction kinetics are taken from laboratory testing reported in the open literature. Many aspects of the oxygen carrier behavior in the circulating fluid bed are uncertain, including the durability and reactivity degradation in this reactor system. The resulting reactor models can only be expected, at best, to represent approximate performance trends for the reducer and oxidizer reactors and to be an initial step in understanding the CLC system. Primary and secondary reactions involving the oxygen carriers and gaseous species present, gas-phase reactions, as well as coal gasification kinetics were modeled in a multi-regime fluidized bed reactor model. The modeling approach and resulting design equations applied for the circulating fluid bed CLC Reducer and Oxidizer reactors are described in detail elsewhere [5].

11.3 Chemical Looping Combustion Reference Plant Descriptions

The general configurations of the CLC plants using the two oxygen carrier types are very similar. The major subsystems and features of the CLC plants are

- Reducer circulating fluid bed reactor and associated cyclones, with char/oxygen carrier separation and char recycle to the Reducer to minimize CO_2 losses and to reduce the Reducer reactor volume;
- Oxidizer circulating fluid bed reactor and associated cyclones, with possible need for ash/oxygen carrier separation to minimize carrier losses;
- Reducer reactor off-gas heat recovery, particulate, Hg control, and FGD;
- Oxidizer off-gas heat recovery, particulate control, and SCR;
- circulating solids heat recovery unit;

- integrated CO_2 product stream compression and purification, with recovered fuel constituents fed to the Oxidizer;
- steam turbine power cycle (supercritical).

This section describes the CLC power plant configuration, equipment functions, and stream conditions. The operating conditions and reactor performances differ significantly between the Fe_2O_3 and $CaSO_4$ oxygen carrier cases.

11.3.1 General CLC Power Plant Configuration

Figure 11.1 illustrates a block flow diagram of the circulating fluidized bed chemical looping concept employed in the current study for both the $CaSO_4$- and Fe_2O_3-based oxygen carriers. The heart of the power plant is the Reducer and Oxidizer reactors coupled with circulating oxygen carrier. Coal is delivered to the base of the Reducer using conventional coal handling and coal feeding equipment. A stream of makeup oxygen carrier (Fe_2O_3 or $CaSO_4$) is also fed at the Reducer base. Steam and CO_2 recycled from the fuel recovery and compression system provide initial fluidization and coal devolatilization reactants for the Reducer. The CO_2 recycle is a slip-stream from the Reducer's raw-CO_2 stream that has passed through the first compression stage of the CO_2 compression system. It contains primarily CO_2, but also may contain some unconverted fuel constituents.

The Reducer is a circulating fluidized bed reactor, and a set of four parallel cyclones captures the entrained stream of oxygen carrier, coal char, and coal ash particles, and three recycle them to the base of the Reducer. One of the cyclones captures Reducer-entrained solids and transports them to the Oxidizer reactor. Because this is a circulating bed reactor, no separate solids transport system is needed.

The off-gas from the Reducer represents the raw CO_2 stream to be sequestered. It contains primarily CO_2 and H_2O, but also has portions of particulate, H_2, CO, and coal contaminants (SO_2, H_2S, HCl, Hg, etc.). The particulate and coal contaminants must be removed from the raw CO_2 stream using conventional cleaning equipment (baghouse, FGD, and mercury removal by activated carbon injected into the baghouse). A heat recovery steam generator (HRSG) precedes the gas cleanup equipment, producing a portion of the steam turbine steam.

The unconverted fuel constituents (primarily H_2 and CO) may need to be separated from the raw CO_2 stream and fed to the Oxidizer for utilization. This is a function of the fuel recovery and compression system, which is a near-term, low-temperature, phase separation technique for purification and compression of the CO_2 stream [6]. The Fe_2O_3 oxygen carrier is sufficiently reactive that the H_2 and CO content of the Reducer off-gas is low, and fuel recovery is not needed. In this case, conventional CO_2 stream dehydration and compression is applied. In contrast, the $CaSO_4$ oxygen carrier, having relatively low reactivity, results in a significant amount of H_2 and CO in the Reducer off-gas, and fuel recovery/CO_2 purification is applied to yield a purer CO_2 product and to utilize all the fuel constituents.

The Reducer reactor is the most complex of the two reactors, and conducts simultaneous coal gasification and the partial reduction of the oxygen carrier (Fe_2O_3 carrier to Fe_3O_4; or $CaSO_4$ to CaS) within a circulating fluidized bed

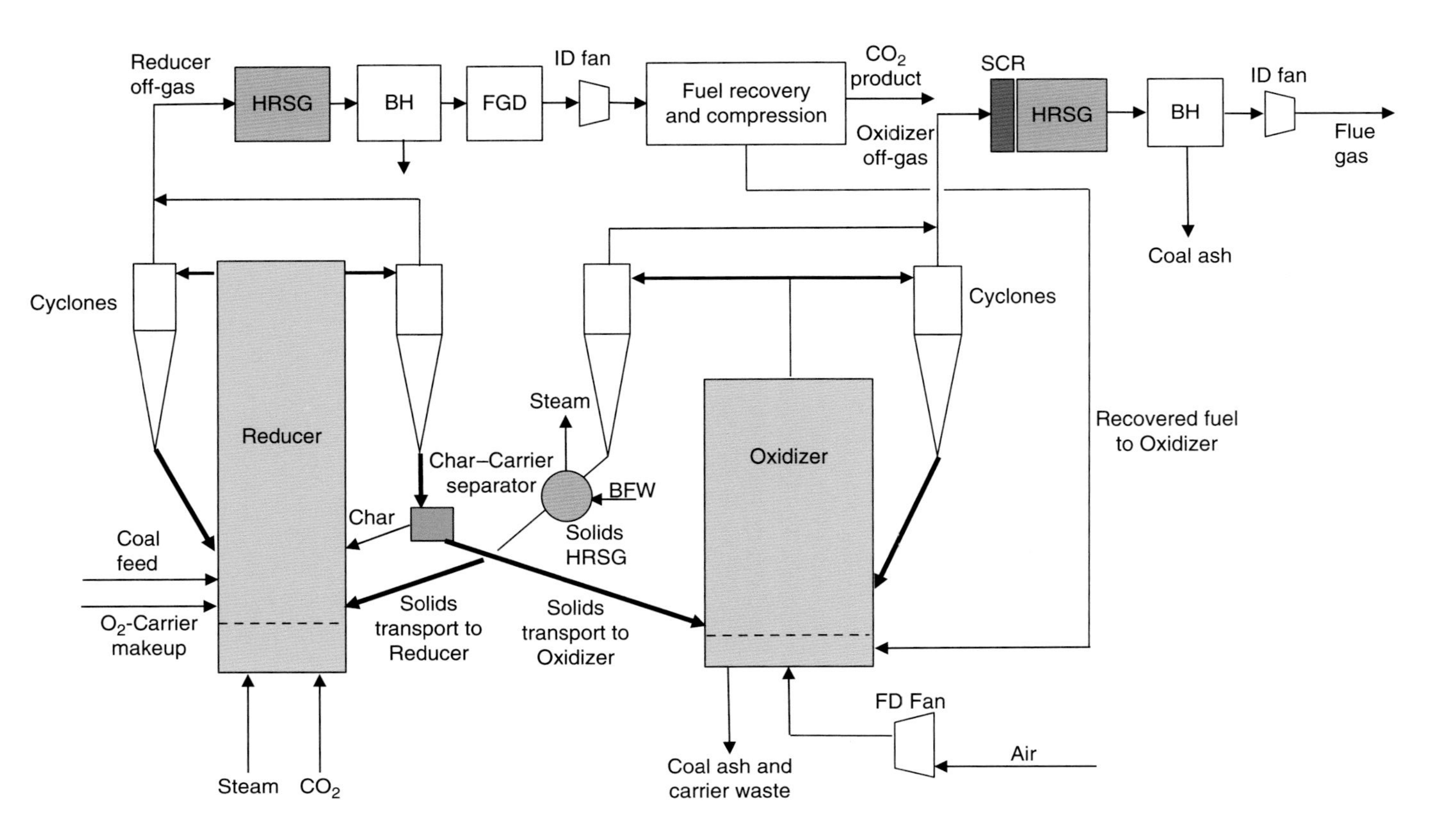

Figure 11.1 Reference CLC power plant block flow configuration.

environment. Because coal gasification is slow at the temperature of the Reducer, a significant amount of char may be unconverted and fed to the Oxidizer with the oxygen carrier solids circulation stream. It is expected that the carbon transferred to the Oxidizer would be significant enough to result in unacceptable power plant carbon capture. Thus, a device is inserted into the Reducer solids transport stream to separate char from the ash and oxygen carrier and recycle it to the Reducer. The separation will result in a more compact Reducer reactor by increasing the content and residence time of char particles in the Reducer reactor, and will yield acceptable carbon capture efficiency. The char–carrier–ash separator device is undeveloped and is a conceptual unit operation in the reference plant design. There are many char–carrier–ash separation mechanisms that can be attempted, such as

- particle segregation due to particle density differences
- particle segregation due to particle size differences
- particle separation based on differences in magnetic properties.

In the reference plant this device is treated as a separator block that separates out 80% of the unconverted char and recycles this char back to the Reducer.

The Oxidizer reactor is a much simpler reactor than the Reducer reactor. In it, air reacts with the partially reduced oxygen carrier particles from the Reducer, converting them back to a nearly fully oxidized form of the oxygen carrier. The oxygen carrier stream from the Reducer delivers the partially reduced oxygen carrier to the base of the Oxidizer vessel. This stream must be stripped of fuel gas constituents, and its flow is controlled by L-valve aeration. The Oxidizer is also a circulating bed reactor and uses eight parallel cyclones in the same way the Reducer reactor does: six for recycling of entrained solids back into the Oxidizer and the remaining two to transport a portion of the total entrained solids to the Reducer vessel, with this solids transport stream being stripped of air and controlled by aerated L-valves.

Because of the high particle velocities existing in the Oxidizer, and the limited volume of the Oxidizer reactor, no in-bed heat transfer surface is used. Instead, an external solids heat recovery unit (solids HRSG) is placed on the solids stream transported to the Reducer for the Fe_2O_3 oxygen carrier and on the solids transport stream leading to the Oxidizer for the $CaSO_4$ oxygen carrier. This is a moving bed heat exchanger, and solid heat exchangers of this type are currently commercially available at smaller capacities [7].

The off-gas from the Oxidizer primarily contains nitrogen with excess oxygen, H_2O, particulate, some CO_2, and small amounts of coal contaminants, as well as the possibility for some NO_x. It is expected that Oxidizer off-gas cleaning requirements will be limited to particulate control using a conventional baghouse and SCR (not shown). The baghouse is preceded by a HRSG. A conventional forced draft (FD) fan to supply the Oxidizer air flow and conventional ID fans are utilized.

The power plant must maintain a coal ash inventory balance within the reactors. This is normally performed in fluidized bed reactors by draining a portion of the Oxidizer reactor bed mass, which is a mixture of coal ash and oxygen carrier particles. In the reference plant configuration for Fe_2O_3 oxygen carrier it is assumed that the coal ash material balance is maintained by cyclone ash losses through the Reducer and Oxidizer cyclones. This is assumed to be accomplished by using moderately efficient cyclone designs that allow the lighter and smaller ash particles to penetrate the cyclones while capturing the larger, denser oxygen carrier particles. The reference plant configuration for Fe_2O_3 oxygen carrier requires no Oxidizer bed drain, with all the oxygen carrier losses also being through cyclone penetration, and the makeup oxygen carrier being based on these losses.

In contrast, with the $CaSO_4$ oxygen carrier it is assumed that high-efficiency cyclones are utilized that permit very little of the relatively low-density ash and oxygen carrier particles to penetrate. Oxidizer bed drainage is applied to maintain the plant coal ash material balance, and the $CaSO_4$ oxygen carrier makeup is fed to the Reducer to account for the oxygen carrier lost with the drained coal ash from the Oxidizer.

In cases where the plant ash material balance is maintained by Oxidizer bed drainage, large oxygen carrier losses may result. If this is the case, and if it results in an unacceptable operating cost for makeup oxygen carrier, an additional device will be needed to separate oxygen carrier from the bed ash to minimize oxygen carrier losses. Such a device is not required for the reference plants due to the assumptions applied.

11.3.2 Reference Plant Stream Conditions

The major reactor conditions and plant configuration features for the two oxygen carrier types are compared in Table 11.2. The Fe_2O_3 oxygen carrier reactors

Table 11.2 Comparison of reference plant reactor conditions and configuration features.

Oxygen carrier type	Fe_2O_3	$CaSO_4$
Reducer reactor type	Circulating fluid bed	Circulating fluid bed
Reducer outlet gas velocity (ft s^{-1})	30	29
Reducer temperature (°F)	1 745	1 800
Reducer pressure drop (psi)	21.4	2.9
Solids flow to Reducer (1000 lb h^{-1})	94 374	17 183
Oxidizer reactor type	Circulating fluid bed	Circulating fluid bed
Oxidizer outlet gas velocity (ft s^{-1})	31	26
Oxidizer temperature (°F)	1 800	2 000
Reducer off-gas H_2 and CO (mol%)	0.05	1.5
Oxidizer pressure drop (psi)	1.8	0.4
Oxygen carrier flow to Oxidizer (1000 lb h^{-1})	93 566	16 363
Oxidizer off-gas O_2 (mol%)	3.5	3.5
Location of solids cooler	Solids stream to Reducer	Solids stream to Oxidizer
Use of fuel recovery/CO_2 purification	None needed	Used for fuel recovery

operate at lower temperatures than the $CaSO_4$ oxygen carrier reactors. Much higher solids circulation rates and bed pressure drops result with the Fe_2O_3 oxygen carrier than with the $CaSO_4$ oxygen carrier. Better Reducer performance results with the Fe_2O_3 oxygen carrier than with the $CaSO_4$ oxygen carrier as indicated by the Reducer off-gas H_2 and CO content. As a result, there is no need for fuel recovery/CO_2 purification with the Fe_2O_3 oxygen carrier; with the $CaSO_4$ oxygen carrier it is assumed that purification is required so that the fuel constituents in the raw CO_2 stream can be utilized.

Table 11.3 lists stream conditions for the Fe_2O_3 oxygen carrier reference plant, referring to the selected streams numbered in the Figure 11.2 block flow diagram. Similarly, Table 11.4 lists stream conditions for the $CaSO_4$ oxygen carrier reference plant, referring to the selected streams numbered in the Figure 11.3 block flow diagram. Owing to the assumptions applied, stream 14 in Table 11.3 has negligible flow and is not included in the stream table.

11.4 Chemical Looping Combustion Reference Plant Performance

Table 11.5 lists the major Reducer reactor dimensions and some of its most important design characteristics. The Reducer operating velocity is high, but is within normal experience for circulating fluidized beds. There is a significant increase in the gas velocity across the Reducer reactor. Smaller oxygen carrier makeup rate is needed with Fe_2O_3 than with $CaSO_4$, but the Fe_2O_3 oxygen carrier has a much greater price. The Reducer vessel dimensions are similar for the two oxygen carrier types; this similarity results from the higher operating temperature assumed for the less reactive $CaSO_4$ oxygen carrier. Four very large cyclones are required to support the Reducer operation.

Some aspects of the circulating fluid bed average structure are included. These indicate that the Fe_2O_3 Reducer reactor solids have high oxygen carrier content relative to char and coal ash, while the $CaSO_4$ Reducer reactor solids have very high char and coal ash contents. Both Reducer reactors have low solids volumetric contents, with a large central core region containing small volume fractions of solids. These characteristics are representative of circulating fluidized beds.

Table 11.6 lists the Oxidizer reactor dimensions and some of its most important design characteristics. Like the Reducer reactor, the Oxidizer reactor operates with high gas velocity. The Fe_2O_3 Oxidizer is more compact than the $CaSO_4$ Oxidizer reactor, both having relatively low pressure drops. Again, many parallel cyclones are required. The Oxidizer circulating fluid bed structure has characteristics similar to those in the Reducer fluid bed.

Table 11.7 presents the CLC power plant performance and the breakdown of the auxiliary loads in the plants. Note that the Fe_2O_3 oxygen carrier process requires no fuel recovery, using conventional raw-CO_2 dehydration and compression, while the $CaSO_4$ oxygen carrier process requires fuel recovery. Comparison with a conventional PC power plant using amine absorption for carbon capture is included.

Table 11.3 Fe_2O_3-based chemical looping combustion power plant stream table.

	1	2	3	4	5	6	7	8	9	10	11	12	13	14	15
Gas composition (mole fraction)															
Ar	0	0	0	0	0	0	0	0	0	0.009	0	0	0.011	0	0.011
CH_4	0	0	0	0	0	0	0	0	0	0	0	0	0	0	0
CO	0	0	0	1.53E-05	0	0	7.93E-06	0	1.5516E-05	0	0	0	0	0	1.702 E.05
CO_2	0	0	0	0. 972	0	0	0.505	0	0.989	0.0003	0	0	0.0075	0	0.0075
H_2	0	0	0	9. 93E-06	0	0	5.16E-06	0	1.01E-05	0	0	0	0	0	0
H_2O	0	0	1	0.017	0	0	0.482	0	0	0.010	0	0	0.018	0	0.019
H_2S	0	0	0	0	0	0	0	0	0	0	0	0	0	0	0
N_2	0	0	0	0. 010	0	0	0.005	0	0.010	0. 773	0	0	0.928	0	0.927
NH_3	0	0	0	0	0	0	0.0004	0	0	0	0	0	0	0	0
O_2	0	0	0	0	0	0	0	0	0	0. 207	0	0	0.035	0	0.035
SO_2	0	0	0	0. 001	0	0	0.007	0	0.001	0	0	0	1.70E-05	0	0
Total	0	0	1	1	0	0	1	0	1	1	1	0	1	0	1
Gas Flow rate (kg mol h^{-1})	0	0	4 894	340	0	1 606	21 609	0	10 710	73 334	0	0	61 119	0	61 147
Gas Flow rate (kg h^{-1})	0	0	88 175	14 775	0	3 323	681 736	0	469 796	2 116 191	0	0	1 725 006	0	1 725 469
Solids Flow rate (kg h^{-1})	207 745	4 667	0	0	42 807 449	23 690	424	0	0	0	42 439 819	0	24 401	24 401	0

(continued)

Table 11.3 (Continued)

	1	2	3	4	5	6	7	8	9	10	11	12	13	14	15
Temperature (°C)	15	15	138	38	960	952	951	—	52	15	951	—	982	149	148
Pressure (MPa, abs)	0.101	0.101	0.276	0. 341	0.101	0.101	0.101	—	15.27	0. 101	0.101	—	0. 101	0.097	0.102
Enthalpy (kJ kg^{-1})	—	—	2 751.4	46.3	—	3 602.2	2 009.6	—	.158.6	30.2	.7 510.6	—	1 115.2	—	178.2
Density (kg m^{-3})	—	—	1.5	5.8	—	0	0.3	—	632.1	1.2	0	—	0.3	—	0.8
Gas molecular weight	—	—	18.0	43.4	—	0	31.5	—	43.9	28.9	0	—	28.2	—	28.2
Gas flow rate (lb mol h^{-1})	0	0	10 790	750	0	0	47 640	0	23 611	161 673	0	0	134 745	0	134 806
Gas flow rate (lb h^{-1})	0	0	194 392	32 572	0	0	1 502 971	0	1 035 722	4 665 403	0	0	3 802 988	0	3 804 007
Mass flow coal ash (lb h^{-1})	0	0	0		399 547	5 217	4			0	443 969	0	44 397	44 397	0
Mass flow O_2-carrier (lb h^{-1})	0	10 290	0		93 974 722	47 010	931			0	93 119 817	0	9 398	9 398	0
Solids Flow rate (lb h^{-1})	458 000	10 290	0	0	94 374 269	52 228	936	0	0	0	93 563 785	0	53 795	53 795	0

Temperature (°F)	59	59	280	100	1761	1745	1744	—	125	59	1744	—	1800	300	298
Pressure (psia)	14.7	14.7	40	49.5	14.7	14.7	14.7	—	2 214.5	14.7	14.7	—	14.7	14.1	14.8
Enthalpy (Btu lb^{-1})	—	—	1 182.9	19.9	—	1 548.7	864.0	—	.68.2	13.0	.3 229.0	—	479.5	—	76.6
Density (lb ft^{-3})	—	—	0.09	0 36	—	0	0.02	—	39.46	0 08	0	—	0 02	—	0. 05
O_2-carrier flow rate (lb mol h^{-1})															
Support Al_2O_3	0	71	0	0	647 760	0	6	0	0	0	647 824	0	65	65	0
Fe_2O_3	0	29	0	0	246 824	0	1	0	0	0	82 614	0	25	25	0
Fe_3O_4	0	0	0	0	12 164	0	1	0	0	0	121 655	0	1	1	0
Char carbon	0	0	0	0	10	3 889	0	0	0	0	972	0	0	0	0

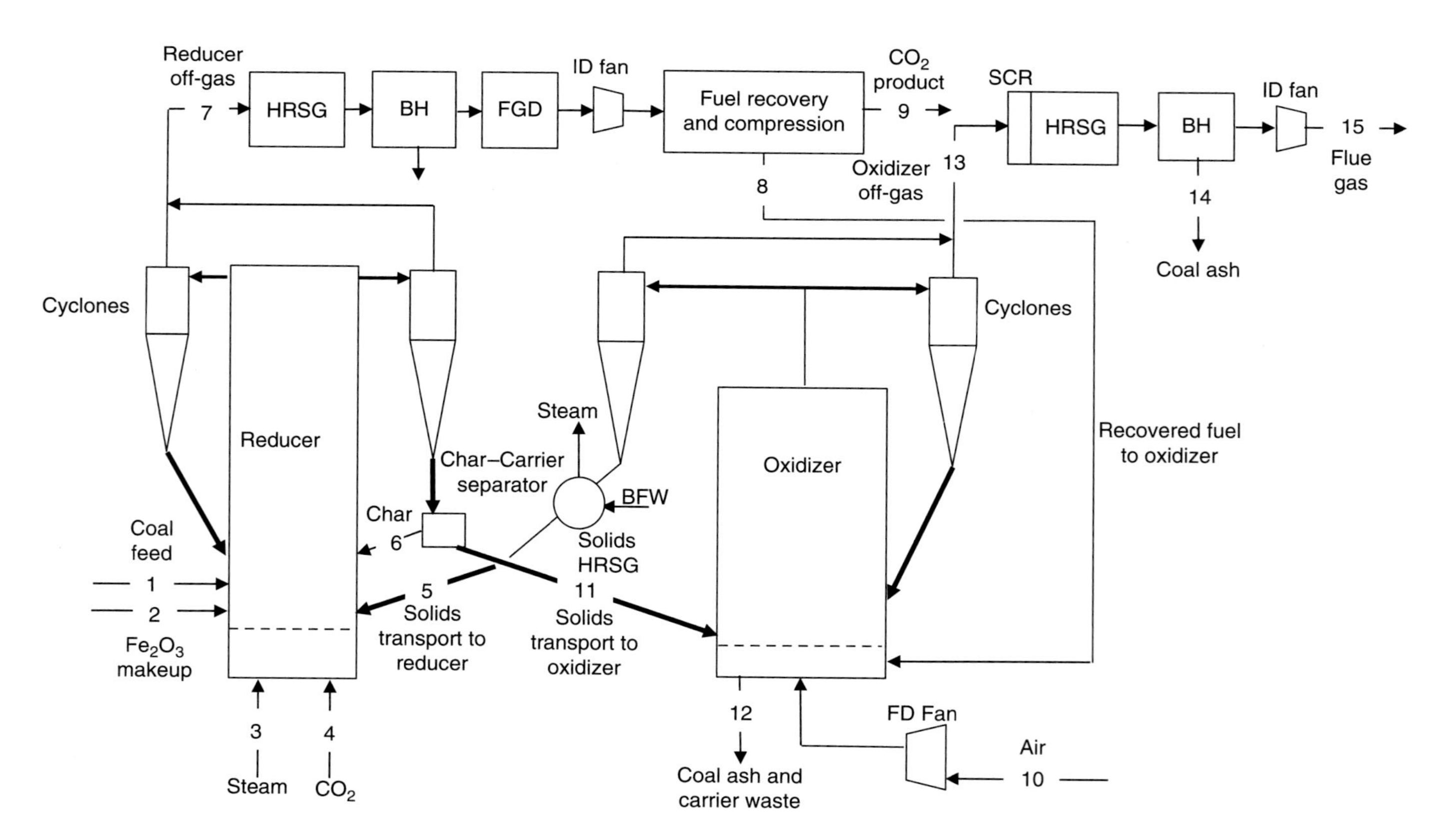

Figure 11.2 Fe_2O_3-based chemical looping combustion power plant stream flow diagram

Table 11.4 $CaSO_4$-based chemical looping combustion power plant stream table.

	1	2	3	4	5	6	7	8	9	10	11	12	13	15
Gas compostion (mole fraction)														
Ar	0	0	0	0	0	0	0	0	0	0.009	0	0	0.011	0.011
CH_4	0	0	0	0	0	0	0	0	0	0	0	0	0	0
CO	0	0	0	0.014	0	0	0.008	0.263	0	0	0	0	0	0
CO_2	0	0	0	0.944	0	0	0.503	0.335	0.997	0	0	0	0.016	0.016
H_2	0	0	0	0.014	0	0.998	0.007	0.254	0	0	0.998	0	0	0
H_2O	0	0	1	0.017	0	0	0.471	0	0	0.010	0	0	0.021	0.021
H_2S	0	0	0	0	0	0	0.005	0	0	0	0	0	0	0
N_2	0	0	0	0.008	0	0.002	0.004	0.148	0	0.773	0.002	0	0.918	0.918
NH_3	0	0	0	0	0	0	0	0	0	0	0	0	0	0
O_2	0	0	0	0	0	0	0	0	0	0.207	0	0	0.035	0.035
	0	0	0	0.003	0	0	0.002	0	0.003	0	0	0	0	0
Total	0	0	1	0	0	0	0	0	0	1	0	0	1	1
Gas Flow rate (kg mol/h)	0	0	5 279	485	0	1 729	23 257	644	11 055	78 139	432	0	65 932	65 949
Gas Flow rate (kg/h)	0	0	95 103	20 722	0	3 559	726 769	17 236	487 217	2 254 867	890	0	1 867 586	1 867 851
Solids flow rate (kg h^{-1})	223 689	16 069	0	0	7 794 191	25 508	0	0	0	0	7 422 116	33 331	0	0

(continued)

Table 11.4 (Continued)

	1	2	3	4	5	6	7	8	9	10	11	12	13	15
Temperature (°C)	15	15	127	38	1 093	15	982	129	52	15	980	1 093	1 093	158
Pressure (MPa, abs)	0.101	0.101	0.172	0.341	0.101	0.101	0.101	0.234	15.268	0.101	0.101	0.101	0.101	0.102
Enthalpy ($kJ\ kg^{-1}$)	—	—	2 733.6	47.0	—	314.1	2 053.3	155.6	163.8	30.2	1 439.3	—	1 266.2	192.4
Density ($kg\ m^{-3}$)	—	—	0.9	5.7	—	0.1	0.3	1.9	651.5	1.2	0.0	—	0.3	0.8
Gas molecular weight	—	—	18.0	42.7	—	2.1	31.3	26.8	44.1	28.9	2.1	—	28.3	28.3
Gas flow rate ($lb\ mol\ h^{-1}$)	0	0	11 638	1 070	0	3 811	51 272	1 419	24 373	172 268	953	0	145 356	145 393
Gas flow rate ($lb\ h^{-1}$)	0	0	209 667	45 685	0	7 847	1 602 250	37 998	1 074 130	4 971 131	1 962	0	4 117 322	4 117 906
Mass flow coal ash ($lb\ h^{-1}$)	0	0	0	—	11 141 616	5 618	0	—	—	0	11 189 452	47 647	—	—
Mass flow O_2-carrier ($lb\ h^{-1}$)	0	35 425	0	—	6 041 635	50 618	0	—	—	0	5 173 514	25 836	—	—
Solids flow rate ($lb\ h^{-1}$)	493 150	35 425	0	0	17 183 251	56 236	0	0	0	0	16 362 966	73 483	0	0

Temperature (°F)	59	59	260	100	1 999	59	1 799	264	125	59	1 796	2 000	2 000	317
Pressure (psia)	14.7	14.7	25.0	49.5	14.7	14.7	14.7	34.0	2 214.5	14.7	14.7	14.7	14.7	14.8
Enthalpy (Btu lb^{-1})	—	—	1 175.2	20.2	—	135.0	882.8	66.9	70.4	13.0	.618.8	—	544.4	82.7
Density (lb ft^{-3})	—	—	0.059	0.358	—	0.005	0.019	0 117	40.672	0 076	0.001	—	0.016	0.051
O_2-carrier flow rate (lb mol h^{-1})														
$CaSO_4$	0	0	0	—	14 328	0	0	—	—	0	0	61	—	—
CaS	0	0	0	—	0	0	0	—	—	0	14 391	0	—	—
CaO	0	0	0	—	65 458	0	0	—	—	0	65 395	280	—	—
$CaCO_3$	0	336	0	—	0	0	0	—	—	0	336	0	—	—
Char carbon	0	0	0	—	21	4 188	0	—	—	0	1 047	0	—	—

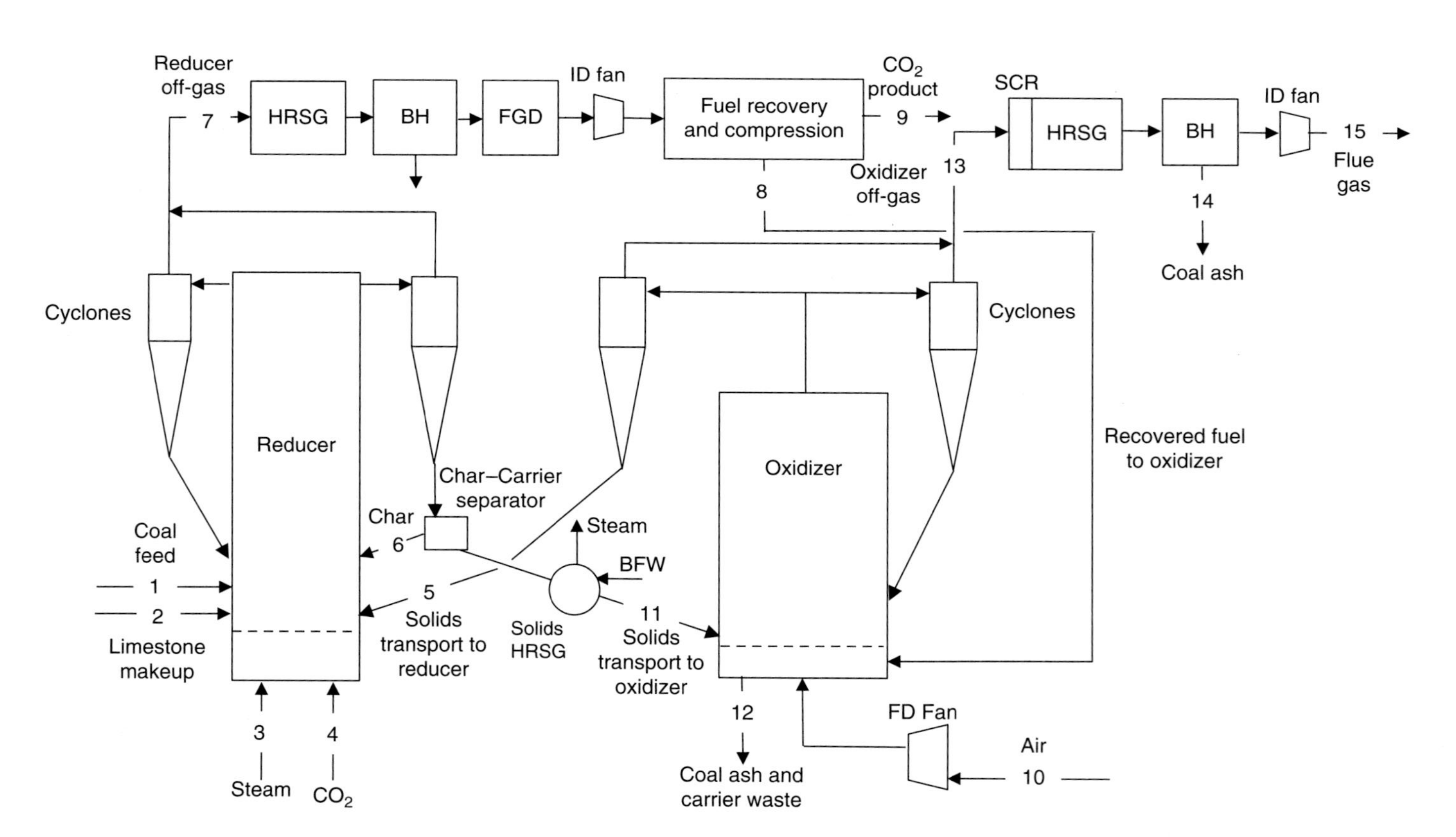

Figure 11.3 $CaSO_4$-based chemical looping combustion power plant stream flow diagram.

Table 11.5 Reducer reactor characteristics and vessel dimensions.

Oxygen carrier type	Fe_2O_3	$CaSO_4$
Inlet velocity (ft s^{-1})	20	20
Outlet velocity (ft s^{-1})	30	29
Oxygen carrier circulation rate (1000 lb h^{-1})	94 374	17 183
Oxygen carrier makeup rate (lb h^{-1})	10 290	35 425
Pressure drop (psi)	21.4	2.9
Carbon gasification efficiency (%)	96	96
Reducer off-gas H_2 and CO (mol%)	0.05	1.5
Reducer vessel shell diameter (ft)	39	41
Reducer vessel height (ft)	115	87
Reducer cyclone number	4	4
Reducer cyclone shell diameter (ft)	19	20
Reducer cyclone height (ft)	90	94
Solids content, % of total solids volume (oxygen carrier, char, ash)	92.8, 6.1, 1.1	25.3, 13.3, 61.4
Volume % of solids in bed core region	2.3	1.5

Table 11.6 Oxidizer reactor characteristics and vessel dimensions.

Oxygen carrier type	Fe_2O_3	$CaSO_4$
Inlet velocity (ft s^{-1})	32	30
Outlet velocity (ft s^{-1})	31	26
Pressure drop (psi)	1.8	0.4
Oxidizer vessel shell diameter (ft)	52	63
Oxidizer vessel height (ft)	39	54
Oxidizer cyclone number	8	8
Oxidizer cyclone shell diameter (ft)	23	25
Oxidizer cyclone height (ft)	108	117
Solids content, % of total solids volume (oxygen carrier, ash)	98.7, 1.3	26.1, 73.9
Volume % solids in bed core region	0.9	0.25

Higher plant thermal efficiency results with the CLC plants are primarily due to the much lower extracted steam usage in these plants compared to the conventional amine-based PC plant. The Fe_2O_3 oxygen carrier CLC plant has very high thermal efficiency resulting from the high Reducer performance, with no need for fuel recovery or CO_2 purification.

Table 11.7 Reference power plant performance comparison.

Plant performance factors	Fe_2O_3	$CaSO_4$	Conventional PC with CCS (BBR Case 12)
Plant output (kW)			
Steam turbine power	641 800	671 100	662 800
Auxiliary load (kW)			
Coal handling	460	480	510
Pulverizers	3 110	3 350	3 850
Sorbent and oxygen carrier handling	1 530	1 860	1 250
Ash and carrier waste handling	720	980	740
Forced draft fans	6 450	4 410	2 300
Induced draft fans	3 880	6 010	11 120
SCR	50	60	70
Baghouses	120	90	100
Wet FGD	6 440	4090	4 110
CO_2 removal	0	0	20 600
Fuel recovery and compression	55 920	86 170	44 890
Miscellaneous balance or plant	2 000	2 000	2 000
Steam turbine auxiliaries	400	400	400
Condensate pumps	870	890	560
Circulating water pumps	4 970	5 160	10 100
Ground water pumps	0	0	910
Cooling tower fans	2 570	2 670	5 230
Transformer losses	2 170	2 340	2 290
Plant performance			
Net auxiliary load (kW)	91 660	120 960	112 830
Net plant power (kW)	550 140	550 140	549 970
Net plant efficiency (% (HHV))	35.1	32.6	28.4
Coal feed flow rate (kg h^{-1} or lb h^{-1})	207 745 (458 000)	223 689 (493 150)	256 652 (565 820)
Thermal input (kW_{th})	1 565 887	1 686 064	1 923 519
Condenser duty (GJ h^{-1} or MMBtu h^{-1})	2 424 (2 298)	2 521 (2 389)	1 737 (1 646)
Carbon capture efficiency (%)	95.8	91.4	90.0

CCS, CO_2 capture and storage and HHV, high heating value.

11.5 Chemical Looping Combustion Reference Plant Cost

Table 11.8 shows the equipment cost breakdown for the two CLC power plants. Fuel recovery is not needed for the Fe_2O_3 oxygen carrier process, but is used with the $CaSO_4$ process.

Table 11.8 Total plant costbreakdown comparison.

Cost	Fe_2O_3 (\$ kW^{-1})	$CaSO_4$ (\$ kW^{-1})
Coal handling, prep and feed systems	88	92
Coal prep and feed systems	44	44
Feedwater and Miscellaneous BOP systems	181	185
Chemical looping combustion system	729	785
Reducer reactor	13	12
Reducer cyclones	13	14
Reducer high-temperature piping	5	5
Char–carrier separator	0	0
Oxidizer reactor	10	18
Oxidizer cyclones	37	44
Oxidizer high-temperature piping	9	9
Solids HRSG and convective HRSGs	326	351
CLC BOP (w/FD and ID fans)	315	331
Gas cleanup (FGD, Baghouses, SCR)	161	229
Fuel recovery and CO_2 compression	159	202
HRSG, ducting and stack	79	80
Steam turbine generator	292	301
Cooling water system	83	85
Ash and carrier waste handling system	133	144
Accessory electric plant	170	186
Instrumentation and control	58	61
Improvements to site	31	31
Buildings and structures	170	171
Total plant cost (TPC)	2 379	2 597
Total overnight cost (TOC)	2 975	3 204
Total as-spent cost (TASC)	3 392	3 653

These results indicate that the primary reactor vessels and cyclones represent a small cost contribution to the total CLC power plant cost. More significant costs are associated with the heat recovery units and the FD and ID fans.

Table 11.9 displays the initial and annual O&M expenses for the Fe_2O_3 CLC power plant. A relatively high cost is associated with the cost of makeup Fe_2O_3 oxygen carrier. Similarly, Table 11.10 displays the initial and annual O&M expenses for the $CaSO_4$ CLC power plant.

Table 11.11 shows the breakdown for the COE (first year, without transport and storage (T&S)) for the two CLC power plants and compares them against the conventional PC power plant with amine-based carbon capture. Even though the Fe_2O_3 oxygen carrier CLC power plant has higher thermal efficiency and lower capital cost than the $CaSO_4$ oxygen carrier CLC power plant, the COE is lower for the $CaSO_4$ CLC power plant due to the high cost of makeup oxygen carrier in the Fe_2O_3 CLC power plant.

Table 11.9 Fe_2O_3 CLC initial and annual O&M expenses.

Fe case – Fe_2O_3 chemical looping (1 x 550 MWnet) with CO_2 capture				**Cost base (Jun)**	**2011**
				Heat rate.net (Btu/kWh)	**9 712**
				MWe-net	**550**
				Capacity factor (%)	**85**
Operating and maintenance labor					
Operating labor					
Operating labor rate (base)	39.70	$ h^{-1}			
Operating labor burden	30.00	% of base			
Labor O.H charge rate	25.00	% of labor			
Operating labor requirements per shift			Total plant		
Skilled operator			2.0		
Operator			11.3		
Foreman			1.0		
Lab Tech's, etc.			2.0		
Total			16.3		
				Annual cost ($)	Annual unit cost ($/kW-net)
Annual operating labor cost				7 384 208	13.422
Maintenance labor cost				8 849 699	16.086
Administrative and support labor				4 058 477	7.377
Property taxes and insurance				26 173 671	47.576
Total fixed operating costs				**46 466 055**	84.462

Variable operating costs						
Maintenance material cost					**$13 274 549**	**$0.003 24**
Consumables	Initial fill	Consumption per day	Unit Cost	Initial fill Cost ($)		
Water(per 1000 gallons)	0	3 400	1.67	**0**	**1 766 069**	**0.000 43**
Chemicals						
MU and WT Chem. (lbs)	0	16 461	0.27	0	1 367 825	0.000 33
Limestone (ton)	0	293	33.48	0	3 040 413	0.000 74
Carbon (mercury removal (lb)	0	1 491	1.63	0	751 814	0.000 18
Fe_2O_3 oxygen transport (ton)	0	123	2 000	0	76 617 888	0.018 70
NaOH (tons)	0	0	671.16	0	0	0.000 00
H_2SO_4 (tons)	0	0	214.78	0	0	0.000 00
Corrosion inhibitor	0	0	0	115 072	5 480	0.000 00
Activated carbon (lb)	0	0	1.63	0	0	0.000 00
Ammonia (19% NH_3) ton	0	16	330.00	0	1 599 523	0.000 39
Subtotal chemicals				**115 072**	**83 382 943**	**0.020 36**
Other						
Supplemental fuel (MBtu)	0	0	0.00	0	0	0.000 00
SCR catalyst (m^3)	w/equip.	0.38	8 938.80	0	1 041 791	0.000 25
Emission penalties	0	0	0.00	0	0	0.000 00
Subtotal other				**0**	**1 041 791**	**0.000 25**

(continued)

Table 11.9 (Continued)

Fe case – Fe_2O_3 chemical looping (1 x 550 MWnet) with CO_2 capture				Cost base (Jun)		2011
				Heat rate.net (Btu/kWh)		9 712
				MWe-net		550
				Capacity factor (%)		85
Waste disposal						
Fly ash (ton)	0	657	25.11	0	5 116 465	0.001 25
Bottom ash (ton)	0	107	25.11	0	830 382	0.000 20
Subtotal waste disposal				**0**	**5 946 847**	**0.001 45**
By-products and emissions						
Gypsum (tons)	0	585	0.00	0	0	0.000 00
Sub total by-products				**0**	**0**	**0.000 00**
Total variable operating costs				**115 072**	**105 412 199**	**0.025 73**
Fuel (ton)	0	5 496	68.60	**0**	**116 972 192**	**0.028 56**

Table 11.10 $CaSO_4$ CLC initial and annual O&M expenses.

Ca case – $CaSO_4$ chemical looping (1 x 550 MW-net) with CO_2 capture				**Cost base (June)**	**2011**
				Heat rate.net (Btu/kWh)	**10 457**
				MWe-net	**550**
				Capacity factor (%)	**85**
Operating and maintenance labor					
Operating labor					
Operating labor rate(base)	39.70 $\$ h^{-1}$				
Operating labor burden	30.00%	Of base			
Labor O.H charge rate	25.00%	Of labor			
Operating labor requirements per shift			Total plant		
Skilled operator			2.0		
Operator			11.3		
Foreman			1.0		
Lab tech's, etc.			2.0		
Total			16.3		
				Annual cost ($)	Annual unit cost ($/kW-net)
Annual operating labor cost				7 384 208	13.422
Maintenance labor cost				9 629 603	17.504
Administrative and support labor				4 253 453	7.732
Property taxes and insurance				28 578 510	51.948
Total fixed operating costs				**49 845 774**	90.606

(continued)

Table 11.10 (Continued)

Ca case – $CaSO_4$ chemical looping (1 x 550 MW-net) with CO_2 capture				Cost base (June)		2011
				Heat rate.net (Btu/kWh)		10 457
				MWe-net		550
				Capacity factor (%)		85
Variable operating costs						
Maintenance material cost					**14 444 405**	**0.003 53**
Consumables	Initial fill	Consumption per day	Unit cost	Initial fill cost ($)		
Water(per 1000 gallons)	0	3 392	1.67	**0**	**1 761 443**	**0.000 43**
Chemicals						
MU & WT Chem.(lbs)	0	16 417	0.27	0	1 364 242	0.000 33
Limestone (ton)	0	66	33.48	0	688 590	0.000 17
Carbon (mercury removal) lb	0	1 546	1.63	0	779 694	0.000 19
Limestone/oxygen transport (ton)	0	425	33.48	0	4 415 606	0.001 08
NaOH (tons)	64	6	671.16	43 283	1 342 862	0.000 33
H_2SO_4 (tons)	0	0	214.78	0	0	0.000 00
Corrosion inhibitor	0	0	0	123 903	5 900	0.000 00
Activated Carbon (lb)	0	0	1.63	0	0	0.000 00
Ammonia (19% NH_3) ton	0	17	330.00	0	1 722 281	0.000 42
Subtotal chemicals				**167 186**	**10 319 175**	**0.002 52**

Other						
Supplemental fuel (MBtu)	0	0	0.00	0	0	0.000 00
SCR catalyst (m^3)	w/equip.	0.40	8 938.80	0	1 123 146	0.000 27
Emission penalties	0	0	0.00	0	0	0.000 00
Subtotal other				**0**	**1 123 146**	**0.000 27**
Waste disposal						
Fly ash (ton)	0	0	25.11	0	0	0.000 00
Bottom ash (ton)	0	882	25.11	0	6 869 485	0.001 68
Subtotal waste disposal				**0**	**6 869 485**	**0.001 68**
By-products and emissions						
Gypsum (tons)	0	67	0.00	0	0	0.000 00
Subtotal by-products				**0**	**0**	**0.000 00**
Total variable operating costs				**167 186**	**34 517 654**	**0.008 43**
Fuel (ton)	0	5 918	68.60	**0**	**125 949 425**	**0.030 75**

Table 11.11 Cost of electricity breakdown comparison.

Cost	Fe_2O_3 (\$ MWh^{-1})	$CaSO_4$ (\$ MWh^{-1})	Conventional PC BBR case 12(\$ MWh^{-1})
Capital	49.6	53.4	73.1
Fixed	11.3	12.2	15.7
Variable	25.7	8.4	13.2
Maintenance materials	3.2	3.5	4.7
Water	0.4	0.4	0.9
Carrier makeup[a)]	18.7	1.1	N/A
Other chemicals and catalyst	1.9	1.7	6.4
Waste disposal	1.5	1.7	1.3
Fuel	28.6	30.8	35.3
Total	115.2	104.7	137.3

a) Fe_2O_3 oxygen carrier makeup: 123 tons d^{-1} at \$2 000 per ton; limestone carrier makeup: 425 tons d^{-1} at \$33.5 per ton.

11.6 Chemical Looping Combustion Reference Plant Performance and Cost Sensitivities

The sensitivity parameters considered in this chapter relate to the reference circulating fluidized bed CLC concept. There are a host of parameters used in the design of the CLC power plant that can influence the CLC power plant performance and cost. There are also significant uncertainties associated with the design and performance parameters. Selected sensitivity evaluations illustrate the influence of the key parameters on the CLC plant performance and cost.

The key Reducer performance parameters are listed below, and their reference values are listed in Table 11.12.

- Steam and recycled-CO_2 feed rates
- Cyclone recycle ratio (solids rate recirculated to the Reducer/total solids collection rate) – the non-recycled material is transported to the Oxidizer
- Oxygen carrier Reducer outlet extent of conversion
- Reducer temperature
- Reducer gas velocity
- Reducer overall carbon conversion efficiency

The first two parameters are fixed at the reference plant values. The last four parameters were assessed in sensitivity evaluations. These factors influence the Reducer characteristics:

- Pressure drop
- Total vessel height
- Vessel shell diameter
- Outlet H_2 and CO gas content
- Overall carbon capture efficiency for the plant

Table 11.12 Reducer parameters.

Parameter	Fe_2O_3 case base values	$CaSO_4$ case base values
Base plant fixed parameters		
Steam feed rate (moles/mole carbon)	0.44	0.44
Recycled-CO_2 feed rate (moles/mole carbon)	0.031	0.044
Oxygen carrier feed rate (moles/mole carbon)	10.1	0.55
Oxygen carrier Reducer inlet extent of conversion (%)	6.9	0.0
Cyclone recycle ratio	4:1	4:1
Sensitivity parameters base plant values		
Oxygen carrier outlet extent of conversion (%)	68.7 Fe_3O_4	17.7 CaS
Reducer temperature (°F)	1 745	1 800
Reducer outlet gas velocity (ft s^{-1})	30	29
Reducer overall carbon conversion (%)	96	96
Plant performance sensitivity variables base values		
Reducer pressure drop (psi)	21.4	2.9
Reducer vessel height (ft)	115	87
Reducer vessel shell diameter (ft)	39	41
Reducer outlet H_2 and CO gas content (mol%)	0.05	1.5
Overall carbon capture efficiency for the plant (%)	95.8	91.4

Likewise, the Oxidizer performance parameters are listed below, and their reference plant values are listed in Table 11.13.

- Oxygen carrier feed rate and its inlet extent of conversion
- Outlet excess O_2
- H_2 and CO fed from the fuel recovery/CO_2 purification system, and the carbon sent from the Reducer
- Cyclone recycle ratio (solids rate recirculated to the Oxidizer/total solids collection rate) – the non-recycled material is transported to the Reducer
- Oxygen carrier outlet extent of conversion
- Oxidizer temperature
- Oxidizer gas velocity

The first five parameters are fixed at the reference plant values. The last two parameters were assessed in sensitivity evaluations. These factors influence the Oxidizer characteristics:

- Pressure drop
- Total vessel height
- Vessel shell diameter

Table 11.13 Oxidizer parameters.

Parameter	Fe_2O_3 case base values	$CaSO_4$ case base values
Base plant fixed parameters		
Oxygen carrier inlet extent of conversion (%)	31.3	0.0
Cyclone recycle ratio	3 : 1	3 : 1
Outlet excess O_2 (mol%)	3.5	3.5
Oxygen carrier outlet extent of conversion (%)	93.1 Fe_2O_3	100 $CaSO_4$
Sensitivity parameters base plant values		
Oxidizer temperature (°F)	1800	2000
Oxidizer outlet gas velocity (ft s^{-1})	32	30
Plant performance sensitivity variables base values		
Oxidizer pressure drop (psi)	1.8	0.4
Oxidizer vessel height (ft)	39	54
Oxidizer shell diameter (ft)	52	63
Oxidizer FD-Fan power (kW)	6 450	4 410

Two important plant cost sensitivities are as follows:

- Rate of oxygen carrier makeup and the makeup oxygen carrier delivered price
- Cost of char–carrier separation equipment

These two items influence the plant COE and are included in sensitivity studies.

11.6.1 Reactor Temperature Sensitivity

The Reducer temperature sensitivity results are displayed in Figure 11.4. The top chart shows results for the Fe_2O_3 Reducer and the bottom chart shows results for the $CaSO_4$ Reducer. The Reducer vessel height and off-gas H_2 and CO are shown as a function of the Reducer temperature. For the highly reactive Fe_2O_3 oxygen carrier the Reducer can operate at relatively low temperatures where char gasification is slow. Increasing the temperature greatly reduces the vessel height needed to gasify the coal char. This reduction in vessel height, however, results in an increase in the Reducer off-gas H_2 and CO content by a factor of 5. This increase will have minor impact on the plant performance and cost so long as the H_2 and CO content remain small at less than 0.1 mol%, so it is concluded that it would be beneficial to increase the Reducer temperature to the highest level that can be operated without secondary operating issues, such as oxygen carrier reactivity loss due to sintering, or fluid bed particle agglomeration.

In the $CaSO_4$ Reducer, the oxygen carrier has low reactivity and the Reducer must operate at higher temperatures where char gasification rates are also higher. Increasing the Reducer temperature results in a reduction in the Reducer vessel

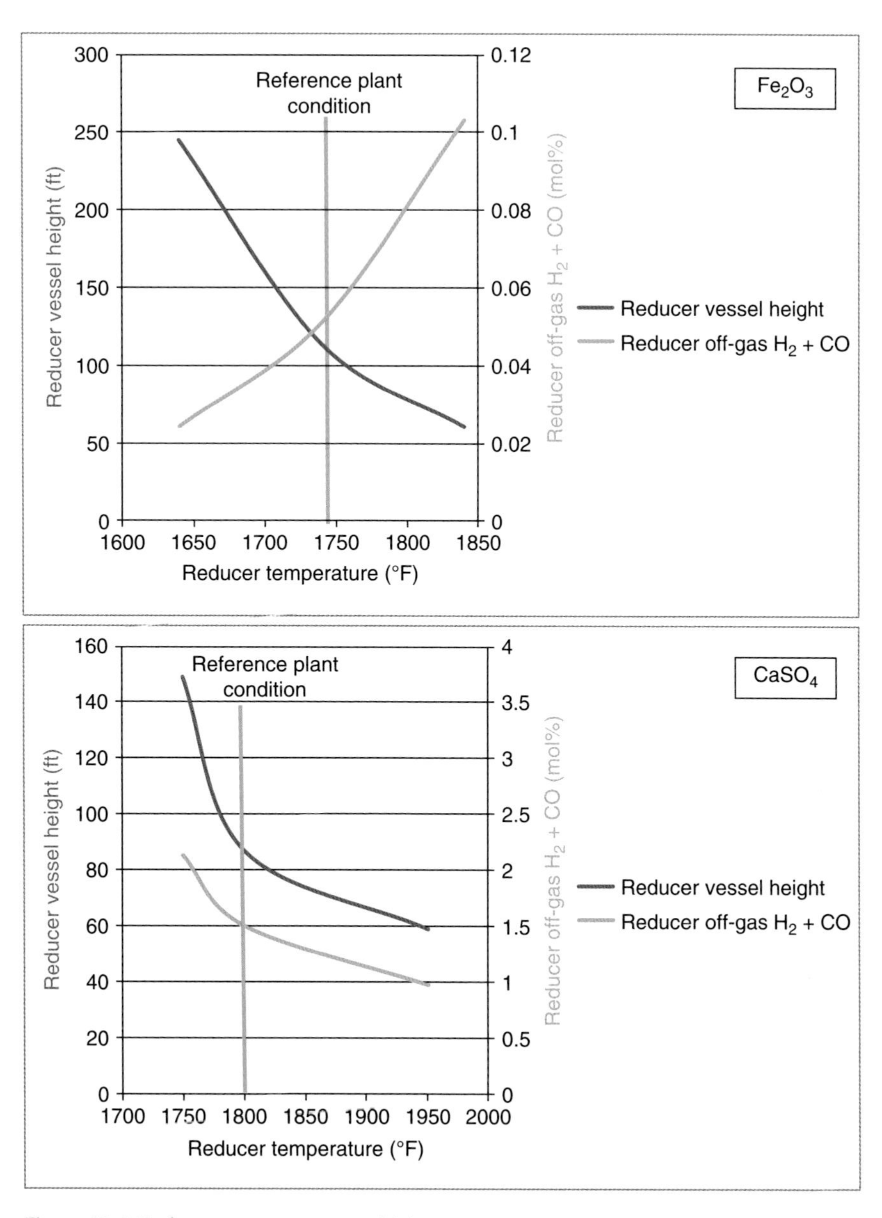

Figure 11.4 Reducer temperature sensitivity.

height and the off-gas H_2 and CO content, which ranges from 1 to 2 mol%. It is beneficial to increase the Reducer temperature to the highest level that can be operated without secondary operating issues, such as oxygen carrier reactivity loss due to sintering, or fluid bed particle agglomeration.

The Oxidizer temperature sensitivity results are displayed in Figure 11.5. Increase in the Reducer temperature will demand an increase in the Oxidizer temperature just due to the solid circulation heat balance. The exhibit shows

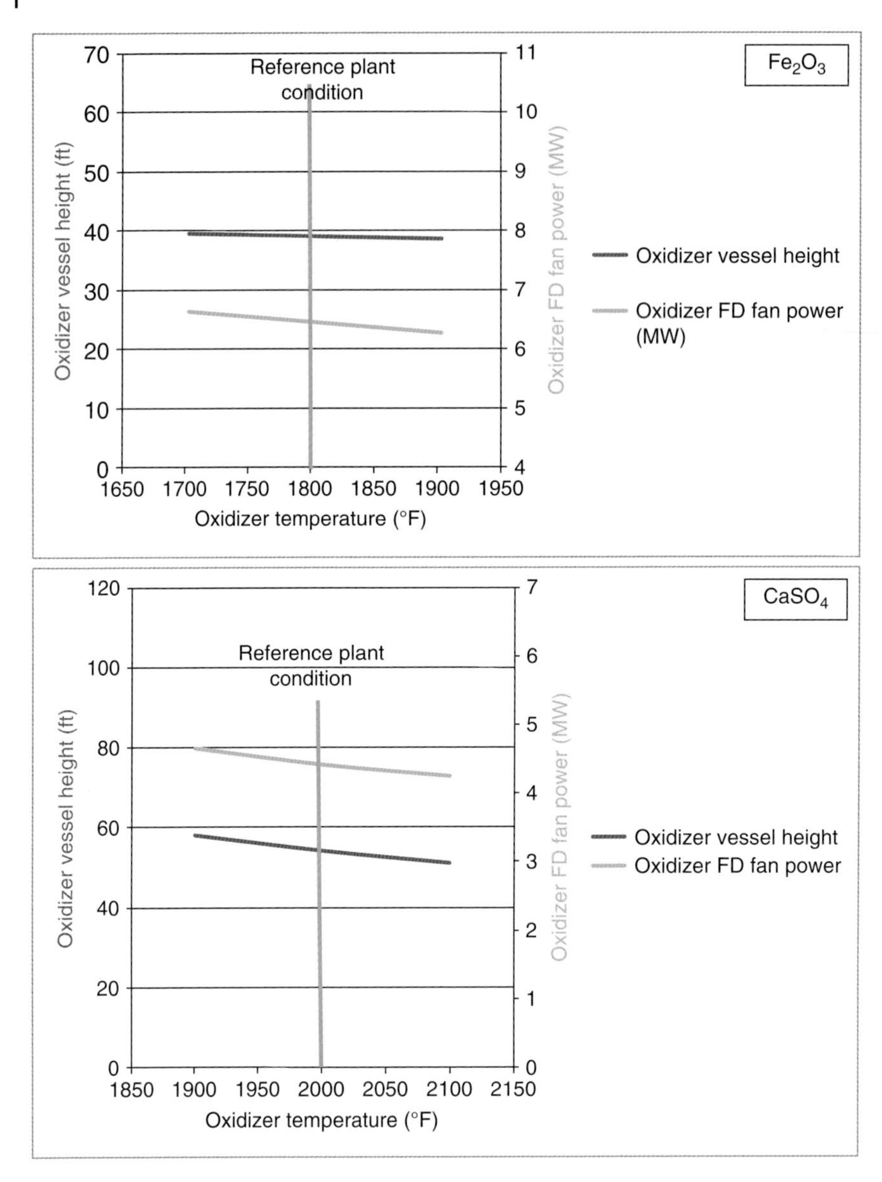

Figure 11.5 Oxidizer temperature sensitivity.

that for both the Fe_2O_3 and $CaSO_4$ oxygen carriers, increased temperature results in reduced Oxidizer vessel height and reducer FD fan auxiliary power consumption, although these improvements are very small for the high-reactivity Fe_2O_3 oxygen carrier. While these are helpful trends, the improvements will not result in significant improvements in the CLC plant performance or cost. It is concluded that the best operating temperature for the Reducer and Oxidizer vessels with respect to operational reliability and oxygen carrier durability needs to be identified experimentally, and the benefits of temperature increases above this temperature need to be considered relative to the detrimental impacts of these increases.

11.6.2 Reactor Velocity Sensitivity

The reducer velocity sensitivity results are displayed in Figure 11.6. Again, the top chart shows results for the Fe_2O_3 oxygen carrier, and the bottom chart shows results for the $CaSO_4$ oxygen carrier. Reducer vessel height, Reducer shell inner diameter, and off-gas H_2 and CO content are plotted against the Reducer outlet

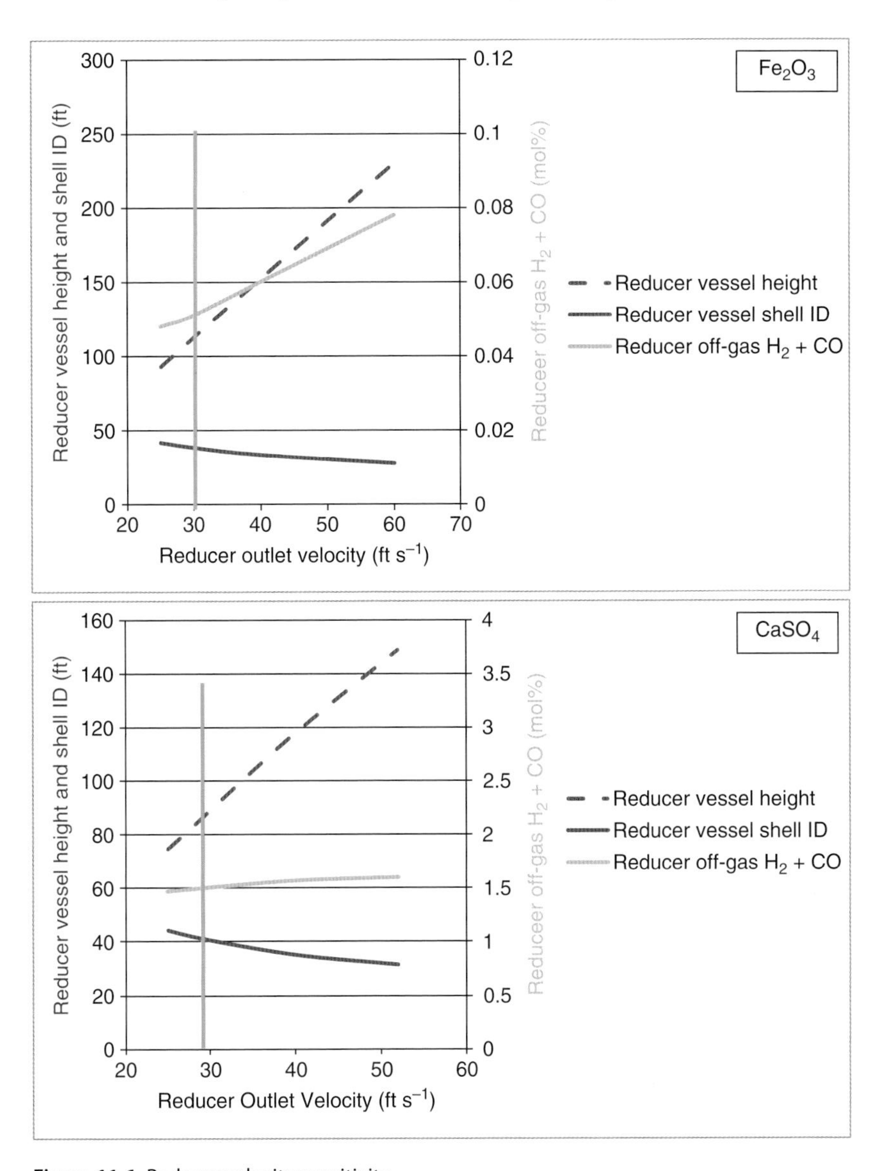

Figure 11.6 Reducer velocity sensitivity.

velocity. As the velocity increases, the vessel shell diameter decreases, but does not approach a vessel size that could be shop fabricated. Increasing velocity also results in greatly increased Reducer vessel height with a more moderate increase in the off-gas H_2 and CO content. There is no clear benefit resulting from increased Reducer velocity for either of the two oxygen carrier types. The impact of reactor footprint versus reactor vessel height needs to be assessed for given plant sites to provide further perspective and a basis for judging the sensitivity results.

The Oxidizer velocity sensitivity results are displayed in Figure 11.7. A similar trend is shown for the Oxidizer velocity sensitivity. Increasing the Oxidizer

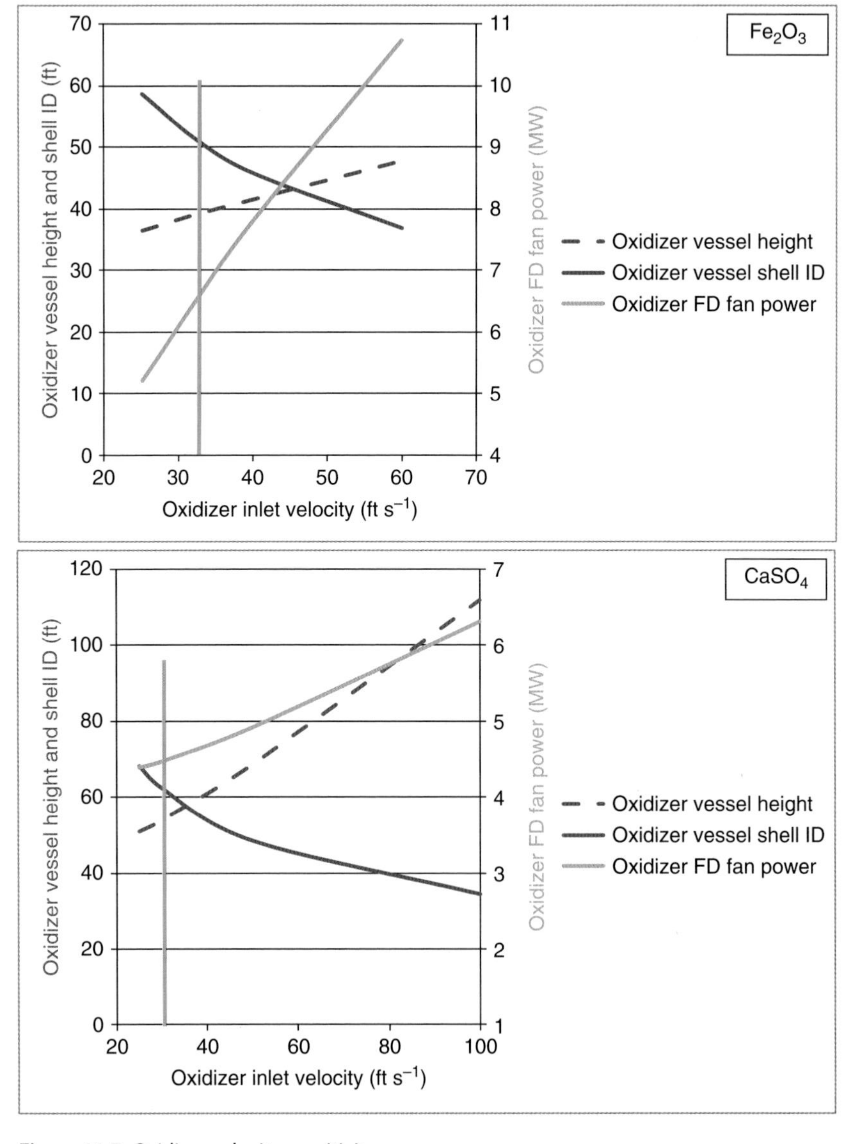

Figure 11.7 Oxidizer velocity sensitivity.

velocity for both oxygen carriers yields a reduction in the Oxidizer shell diameter, which is beneficial. This is accompanied by an increase in the Oxidizer vessel height and the Oxidizer FD fan power consumption due to higher Oxidizer vessel pressure drop. Again, it is concluded that there is no clear benefit to be shown for increasing the Oxidizer velocity. Operating velocities above 30 ft s^{-1} enter a region of limited commercial circulating fluid bed operational experience and would require significant development effort.

11.6.3 Carbon Gasification Efficiency Sensitivity

The Reducer carbon gasification efficiency sensitivity results are displayed in Figure 11.8. Plots are shown of the Reducer vessel height, cost, and off-gas H_2

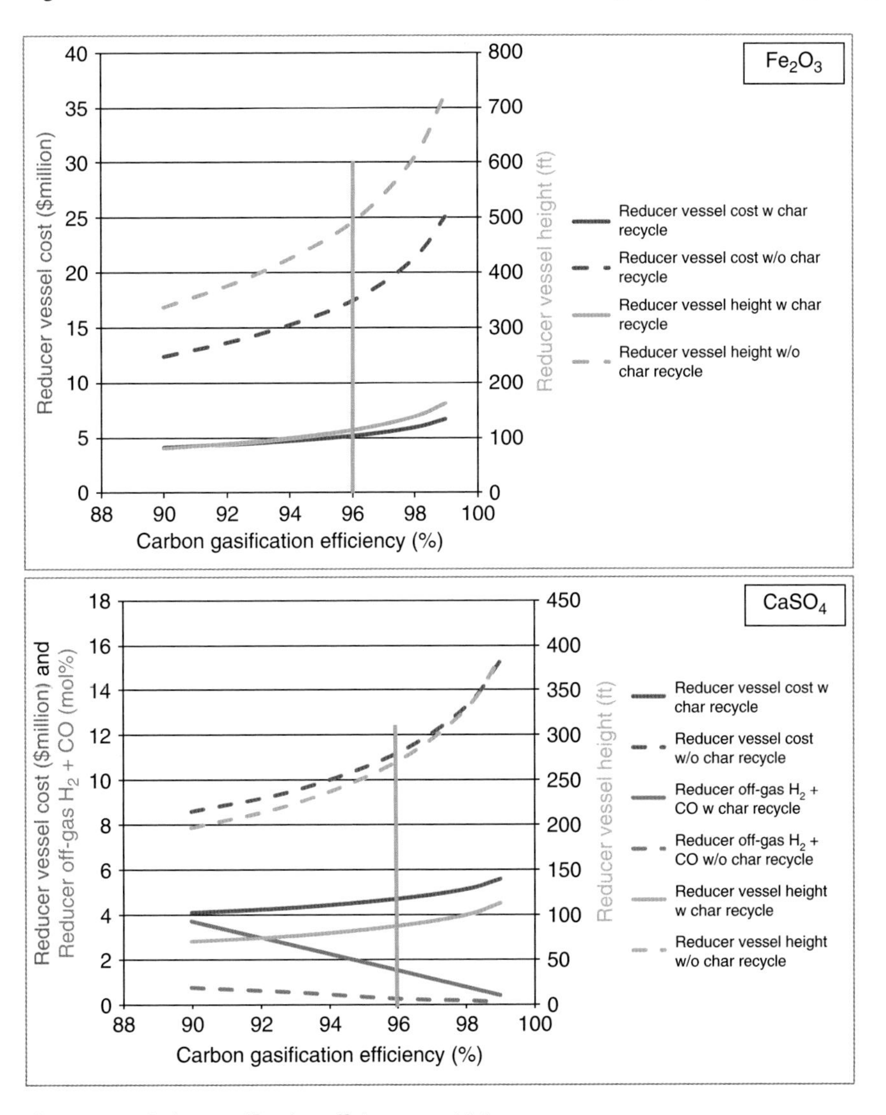

Figure 11.8 Carbon gasification efficiency sensitivity.

and CO content as a function of the carbon gasification efficiency for Reducers with and without char–oxygen carrier separators. For the Fe_2O_3 oxygen carrier, the off-gas H_2 and CO content curves are not shown, because the H_2 and CO content remain low with this highly reactive oxygen carrier. The carbon gasification efficiency is defined as the total coal carbon conversion rate to CO and CO_2 over the total coal carbon feed rate. The Reducer model provides estimates of how the carbon gasification efficiency might be increased by increasing the reducer temperature or by increasing the effectiveness of the char–oxygen carrier separation system.

The benefit of the char–oxygen carrier separation system appears to be clear from these plots. Without char–oxygen carrier separation the Reducer vessel height and vessel cost are dramatically increased to a point where the Reducer would not be a feasible reactor. Simultaneously, the Reducer off-gas H_2 and CO content would be reduced to very low levels due to the greatly increased gas residence time, but not to sufficient benefit to counter the greatly increased vessel height.

When using char–oxygen carrier separation, increased carbon gasification efficiency results in moderately greater Reducer vessel heights and vessel costs with slightly decreased off-gas H_2 and CO content. Increased carbon gasification efficiency will only serve to increase the plant CO_2 capture efficiency and will not impact the power plant thermal efficiency significantly since all the carbon not gasified in the Reducer will be burned in the Oxidizer reactor. There appears to be little need for Reducer carbon gasification efficiency greater than that needed to achieve 90% carbon capture efficiency.

The real need is to develop technically feasible and affordable char–oxygen carrier separation approaches for achieving even this level of carbon gasification efficiency. The challenge is the very high rate of solids flow, having very small content of char that is characteristic of these CLC processes.

11.6.4 Reducer Oxygen Carrier Conversion Sensitivity

The Reducer oxygen carrier conversion sensitivity results are displayed in Figure 11.9. The graphs plot the oxygen carrier circulation rate, the Reducer off-gas H_2 and CO content, and the Oxidizer FD fan power against the Reducer outlet oxygen carrier fractional conversion. Lower levels of oxygen carrier conversion will result in higher reactivity oxygen carrier and lower Reducer off-gas H_2 and CO contents. Lower levels of oxygen carrier conversion also result in moderately higher Oxidizer FD fan power consumption and a very large increase in the oxygen carrier circulation rate.

With the Fe_2O_3 oxygen carrier, the carrier is inherently of high reactivity, and higher conversions can be applied to avoid the huge oxygen carrier circulation rates that would result at low oxygen carrier conversion. A reference plant oxygen carrier conversion of approximately 69% is selected to minimize the circulation rate of the oxygen carrier and Oxidizer fan power while maintaining acceptable reactivity/kinetics and low H_2/CO emissions.

The $CaSO_4$ oxygen carrier is a low-reactivity material, with inherently low oxygen carrier circulation rates. Operating at lower levels of oxygen carrier

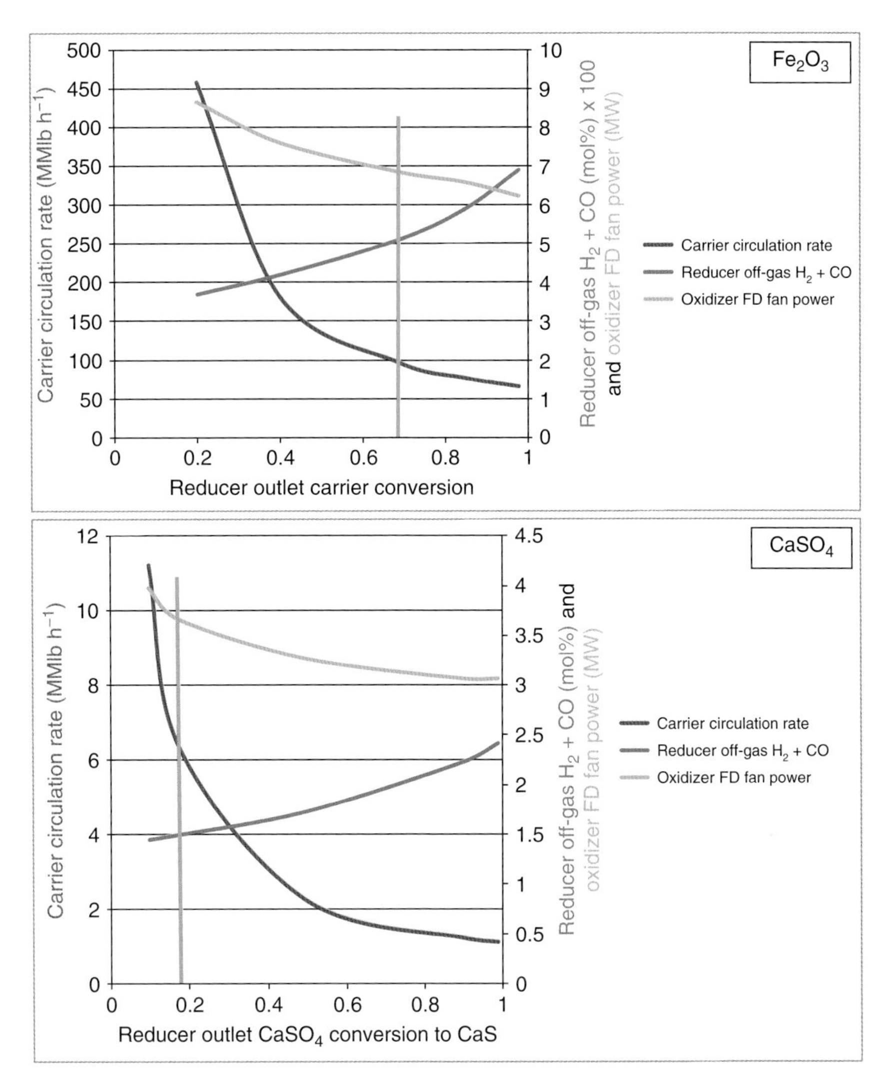

Figure 11.9 Reducer oxygen carrier conversion sensitivity.

conversion results in a relatively high reactivity in the Reducer, and also gives oxygen carrier circulation rates that are low compared to those found for the Fe_2O_3 oxygen carrier. A 19% $CaSO_4$ oxygen carrier reference plant conversion is chosen to keep the Reducer effluent H_2/CO composition to low levels.

Large oxygen carrier circulation rates are relatively easy to accommodate when using circulating fluidized bed reactors, because the Reducer and Oxidizer off-gases are the transport gases for the circulating solids and can generate high circulation rates if required. High rates are costlier and consume more auxiliary power with other reactor types, such as bubbling fluidized beds or moving beds. In these types of reactors, the oxygen carrier circulation system is a completely independent equipment system, and a separate transport gas system is needed.

11.6.5 COE Sensitivity to Oxygen Carrier Makeup Rate and Price

The oxygen carrier makeup rate and price sensitivity results are displayed in Figure 11.10. The COE for the CLC power plant is shown as a function of the oxygen carrier makeup rate and the price of the makeup material. The COE for the conventional PC power plant with amine-based CO_2 absorption is superimposed on the plots for reference.

The Fe_2O_3 oxygen carrier has an expected price of \$1 to \$5 lb^{-1}, and with high makeup rates the COE could exceed the COE of the conventional PC plant. For the Fe_2O_3 oxygen carrier CLC power plant to have lower COE than the $CaSO_4$ oxygen carrier CLC power plant the Fe_2O_3 makeup rate will need to be quite low. The dashed lines on the Fe_2O_3 oxygen carrier plot represent oxygen carrier prices

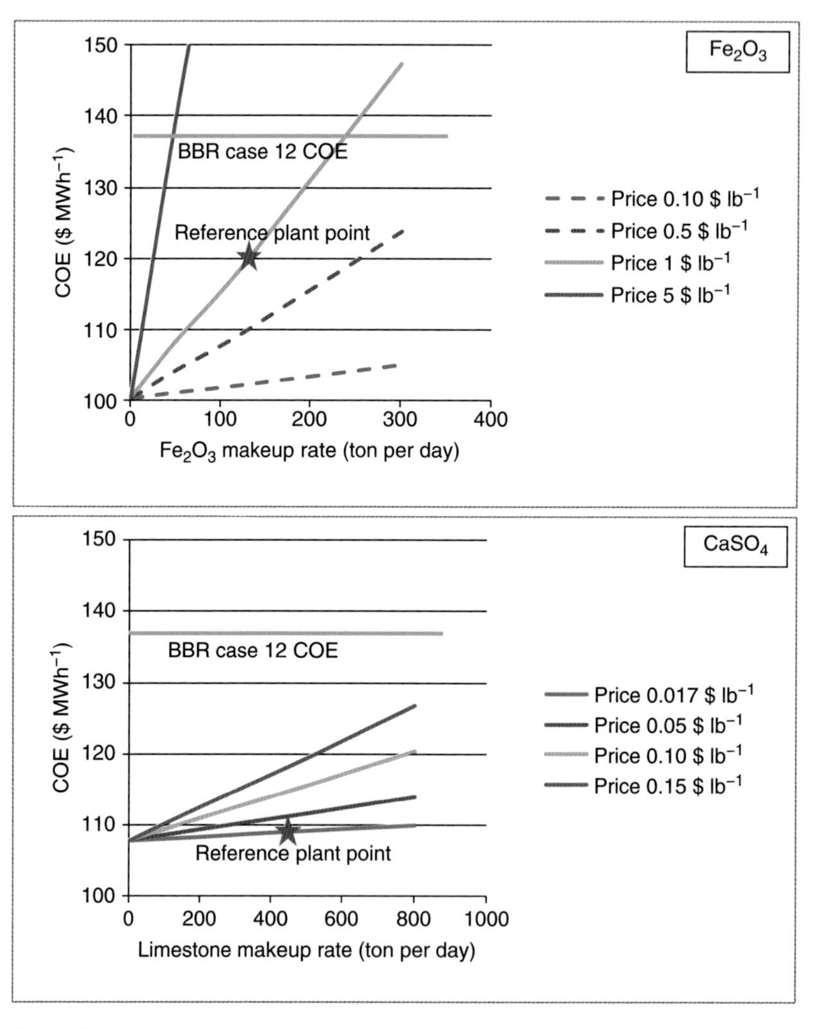

Figure 11.10 Oxygen carrier makeup and price sensitivity.

for cheaper Fe_2O_3 materials, for example raw ores or waste materials. While these materials may have lower reactivity than the reference plant supported Fe_2O_3 oxygen carrier, their lower price makes them candidates for development consideration. Minimizing Fe_2O_3 oxygen carrier losses is a priority for process development.

The $CaSO_4$ oxygen carrier, even if at a relatively high limestone price in the range of \$100–200 per ton, can accommodate high makeup rates and maintain a COE significantly lower than that in the conventional PC plant. At the lower price assumed in the reference plant design, \$33.5 per ton, the limestone makeup rate is not a significant consideration.

11.6.6 COE Sensitivity to Char–Oxygen Carrier Separator Cost

In the reference plant evaluations, the hypothetical char–oxygen carrier separation system was assumed to have zero cost. To understand how sensitive the CLC plant COE is to the potential capital cost of the char–carrier separation system, a range of char–carrier costs was applied that is equivalent to as much as 10 times the cost of the Reducer reactor. The characteristic duty of the reference plant char–carrier–ash separation device is also shown in the exhibit, with very large solids flow rates having very small char content. The COE sensitivity to the char–carrier separator cost results are displayed in Figure 11.11. Parallel lines for the two oxygen carriers result, with an increase in the CLC plant COE of as much as \$3 MWh^{-1} over the separator cost range considered. It is concluded that the capital cost of the char–carrier separation system is not likely to have a significant impact on the CLC power plant COE. The performance and reliability of the char–carrier separation system, however, will be critically important.

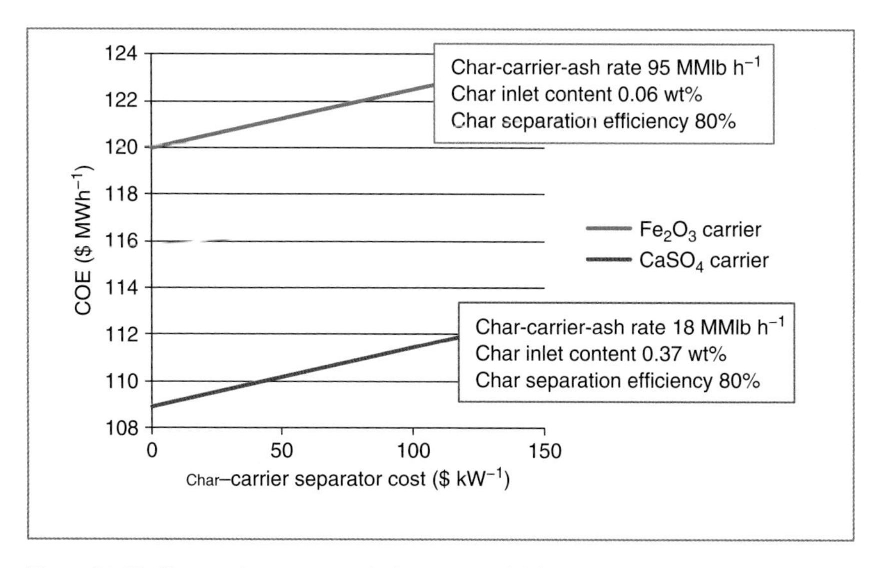

Figure 11.11 Char–carrier separator device cost sensitivity.

11.7 Summary and Conclusions

A calcium-based CLC process concept was considered in this chapter and compared to an iron-based reference plant design. The Reducer and Oxidizer reactors for both plant concepts are circulating fluidized beds operated with high gas velocities and with temperatures, pressure drops, solids circulation rates, and off-gas compositions characteristic of the oxygen carrier properties. The significance of these characteristics is described in this chapter.

Reference CLC power plant performance and cost have been estimated for these two oxygen carriers by means of process simulation to generate power plant energy and material balances. The reference CLC plant concept includes supercritical steam cycle and conventional carbon dioxide compression technology. Plant material and energy balance results are reported in Section 11.4. CLC reactor modeling has been integrated with the process simulation to estimate main reactor performance and dimensions.

The results indicate the following:

- Both CLC reference power plants show a sizable COE advantage over the comparable conventional PC power plant with amine-based post-combustion carbon capture technology.
- The Fe_2O_3 oxygen carrier CLC power plant has higher plant efficiency and lower plant capital cost, but the $CaSO_4$ oxygen carrier CLC power plant has lower COE.
- This lower COE is a direct result of the expected higher price of the makeup Fe_2O_3 oxygen carrier relative to the lower price for a $CaSO_4$ oxygen carrier makeup limestone.
- The carbon capture efficiency of the $CaSO_4$ oxygen carrier CLC plant, at 91.4%, is less than that of the Fe_2O_3 oxygen carrier CLC plant due to CO_2 losses in the CLC processing that ensures that fuel constituents (CO and H_2) are not lost and the CO_2 product stream is sufficiently pure.
- Alternative processing can be applied that will produce a lower purity CO_2 product stream while yielding lower CO_2 losses.

The CLC reference plant assessments have identified the status and potential issues associated with the CLC technology:

- The development status of CLC power generation is at a laboratory/bench scale; insufficient test data and data correlations are available to project plant performance and cost with any degree of certainty.
- The Reducer reactor is complex and is the major developmental component in the CLC process.
- The Reducer reactor is a simultaneous coal gasifier and oxygen carrier reducer, and it operates at temperatures where char gasification reaction rates are relatively slow.
- The Reducer reactor char gasification efficiency may limit the CLC power plant performance.
- To minimize the Reducer reactor size and meet the carbon capture requirement, a char–oxygen carrier separation process must also be developed as part of the Reducer system.

- The char–oxygen carrier separation process requires processing very large amounts of solids, an 18–100 million lb h^{-1} mixture of coal ash and oxygen carrier particles having a small amount of char particle content, to extract and recycle at least 80% of the char.
- The Reducer off-gas (the raw CO_2 stream) may contain substantial H_2 and CO, and purification with fuel recovery may be needed to maintain the plant efficiency and to meet CO_2 product purity specifications.
- CO_2 capture efficiency as high as 90% may be difficult to achieve, depending on how high the Reducer reactor carbon gasification efficiency can be maintained and how low the Reducer off-gas H_2 and CO content will be.

There is significant uncertainty in the CLC process performance and cost for the initial set of operating conditions and design parameters selected, and sensitivity studies have been performed to assess how sensitive the CLC power plant performance and cost is to the major operating conditions and design parameters. Sensitivity studies around the CLC reference plant designs have been completed, showing:

- Increased temperature results in reduced Oxidizer vessel height and reduced FD fan auxiliary power consumption. While these are helpful trends, these improvements will not result in significant improvements in the CLC plant performance or cost. It is concluded that the best operating temperature for the Reducer and Oxidizer vessels with respect to operational reliability and oxygen carrier durability needs to be identified experimentally. The benefits of temperature increase above this best temperature can then be considered relative to the detrimental impacts of these increases.
- As the Reducer velocity increases, the vessel shell diameter decreases, but does not approach a vessel size that could be shop fabricated. Increasing velocity also results in greatly increased Reducer vessel height with a moderate increase in the off-gas H_2 and CO content.
- There is certainly no clear benefit resulting from increased Reducer velocity for either of the two oxygen carrier types. The impact on reactor footprint versus reactor vessel height needs to be assessed for given plant sites to provide further perspective and a basis for judging these sensitivity results.
- A similar trend is shown for the Oxidizer velocity sensitivity. Increasing the Oxidizer velocity for both types of oxygen carriers yields a reduction in the Oxidizer shell diameter, which is beneficial. This is accompanied by an increase in the Oxidizer vessel height and the Oxidizer FD fan power consumption due to higher Oxidizer vessel pressure drop. Again, it is concluded that there is no clear benefit to be shown for increasing the Oxidizer velocity.
- The benefit of the char–oxygen carrier separator appears to be clear. Without char–oxygen carrier separation the Reducer vessel height and vessel cost are dramatically increased to a point where the Reducer would not be a feasible reactor to construct and install.
- When using char–oxygen carrier separation, increased carbon gasification efficiency results in moderately greater Reducer vessel heights and vessel costs, with slightly decreased off-gas H_2 and CO content. Increased carbon gasification efficiency will only serve to increase the plant CO_2 capture efficiency and

will not impact the power plant thermal efficiency significantly since all the carbon not gasified in the Reducer will be burned in the Oxidizer reactor.

- There appears to be little need for Reducer carbon gasification efficiency greater than that needed to achieve 90% carbon capture efficiency.
- The greatest need is to develop technically feasible and affordable char–oxygen carrier separation approaches for achieving even this limited level of carbon gasification efficiency. The technology challenge is the very high rate of solids flow having very small content of char that is characteristic of these CLC processes.
- Lower levels of oxygen carrier conversion will result in higher reactivity oxygen carrier and lower Reducer off-gas H_2 and CO contents. Lower levels of oxygen carrier conversion also result in moderately higher Oxidizer FD fan power consumption, and a very large increase in the oxygen carrier circulation rate.
- With the Fe_2O_3 oxygen carrier, the oxygen carrier is inherently of high reactivity, and higher conversions can be applied to avoid the huge oxygen carrier circulation rates that would result at low oxygen carrier conversion. The reference plant conversion level of about 69% appears to be a good design choice.
- The $CaSO_4$ oxygen carrier is a low-reactivity material, with inherently low oxygen carrier circulation rates. Operating at lower levels of oxygen carrier conversion results in a relatively high reactivity in the Reducer, and also gives oxygen carrier circulation rates that are low compared to those found for the Fe_2O_3 oxygen carrier. Again, the $CaSO_4$ oxygen carrier Reference plant conversion of about 19% appears to be a good design choice.
- Large oxygen carrier circulation rates are relatively easy to accommodate when using circulating fluidized bed reactors, because the Reducer and Oxidizer off-gases are the transport gases for the circulating solids and can generate high circulation rates if required. High rates are more costly and consume more auxiliary power with other types of reactors, such as bubbling fluidized beds or moving beds. In these types of reactors the oxygen carrier circulation system is a completely independent equipment system, and a separate transport gas system is needed.
- The Fe_2O_3 oxygen carrier has an expected price of $\$1–5\,lb^{-1}$, and with high makeup rates the COE could exceed that of the conventional PC plant. For the Fe_2O_3 oxygen carrier CLC power plant to have lower COE than the $CaSO_4$ oxygen carrier CLC power plant, the Fe_2O_3 makeup rate will need to be quite low. Minimizing Fe_2O_3 oxygen carrier losses is a priority for process development.
- The $CaSO_4$ oxygen carrier, even if at a relatively high limestone price, can accommodate high makeup rates and maintain a COE significantly lower than the conventional PC plant. At the lower price assumed in the reference plant design, the limestone makeup rate is not a significant consideration.
- In the reference plant evaluations, the hypothetical char–oxygen carrier separation system was assumed to have zero cost. It is found that the capital cost of the char–carrier separation system is not likely to have a significant impact on

the CLC power plant COE. The performance and reliability of the char–carrier separation system, however, will be critically important.

At this early stage of development of CLC technology the uncertainties in its performance and cost are great. The process simulations in this report have shown the possibility that CLC could provide sizable performance and cost advantages over conventional PC power plants using conventional, amine-based CO_2 capture technology. These findings may be optimistic given that the CLC plant operability and availability are assumed in this report to be the same as that of the conventional PC power plant. Operability and reliability issues are likely to represent the major challenges to be dealt with in continued, larger-scale development of CLC technology and may limit the technology's ultimate feasibility.

References

1 U.S. Department of Energy (DOE), National Energy Technology Laboratory (NETL) (2010). Cost and Performance Baseline for Fossil Energy Plants, Volume 1: Bituminous Coal and Natural Gas to Electricity. *DOE/NETL-2010/1397*. Pittsburgh, PA: DOE/NETL.
2 U.S. Department of Energy (DOE), National Energy Technology Laboratory (NETL) (2012). QGESS: Process Modeling Design Parameters. *DOE/NETL-341/081911*. Pittsburgh, PA: DOE/NETL.
3 Abad, A., Adánez, J., García-Labiano, F. et al. (2007). Mapping of the range of operational conditions for Cu-, Fe-, and Ni-based oxygen carriers in chemical looping combustion. *Chem. Eng. Sci.* 62: 533–549.
4 Andrus, H.E., Burns, G., Chiu, J.H., Liljedahl, G.N., et al. (2006). Hybrid Combustion-Gasification Chemical Looping Coal Power Technology Development, Phase II – Final Report. *DOE Report No. PPL-06-CT-27*.
5 U.S. Department of Energy (DOE), National Energy Technology Laboratory (NETL) (2014). Guidance for NETL's Oxycombustion R&D Program: Chemical Looping Combustion Reference Plant Designs and Sensitivity Studies. *DOE/NETL-2014/1643*. Pittsburgh, PA: DOE/NETL.
6 White, V. (2007). *Purification of Oxyfuel Derived CO_2 for Sequestration or EOR*. Windsor, CT: International Oxy-Combustion Research Network.
7 Moran, P.P. and Solex Thermal Science (2012). Energy recovery using indirect heat exchange: high temperature mineral applications. Enermin 2012, 2nd International Seminar on Energy Management in the Mining Industry.
8 Song, Q., Xiao, R., Deng, Z. et al. (2008). Effect of temperature on reduction of $CaSO_4$ oxygen carrier in chemical-looping combustion of simulated coal gas in a fluidized bed reactor. *Ind. Eng. Chem. Res.* 47: 8148–8159.
9 Tian, H. and Guo, Q. (2009). Investigation into the behavior of reductive decomposition of calcium sulfate by carbon monoxide in chemical-looping combustion. *Ind. Eng. Chem. Res.* 2009, 48: 5624–5632.

10 Song, Q., Xiao, R., Deng, Z. et al. (2008). Multicycle study on chemical-looping combustion of simulated coal gas with a $CaSO_4$ oxygen carrier in a fluidized bed reactor. *Energy Fuels* 22: 3661–3672.

11 Johnson, J.L. (1981). Fundamentals of coal asification. In: *Chemistry of Coal Utilization*, Chapter 23 (ed. M.A. Elliott). New York: Wiley.

12 Abba, A., Grace, J.R., Bi, H.T., and Thompson, M.L. (2003). Spanning the flow regimes: generic fluidized-bed reactor model. *AIChE J.* 49 (7): 1838–1848.

13 Kunii, D. and Levenspiel, O. (1997). Circulating fluidized bed reactors. *Chem. Eng. Sci.* 52 (15): 2471–2482.

12

Simulations for Scale-Up of Chemical Looping with Oxygen Uncoupling (CLOU) Systems

JoAnn S. Lighty[1], Zachary T. Reinking[2], and Matthew A. Hamilton[2]

[1] *Boise State University, Department of Mechanical and Biomedical Engineering, 1910 University Dr., Boise, ID 83725-2100, USA*
[2] *University of Utah, Department of Chemical Engineering, 50 S. Central Campus Dr. Room 3290 MEB, Salt Lake City, UT 84112-9203, USA*

12.1 Introduction

Process modeling and computational fluid dynamic (CFD) simulations are effective tools for the scale-up of chemical looping with oxygen uncoupling (CLOU) systems. Process modeling is defined as those simulations where material and energy balances are performed in the context of process units. Process modeling studies have been completed for both chemical looping combustion (CLC) and CLOU. CFD simulations take into account the complex fluid dynamics, reaction kinetics, and heat transfer in the systems, in addition to more global material and energy balances. In contrast to process modeling, there are few studies on CLOU specifically; since both CLC and CLOU are fluidized environments, both applications will be discussed in this chapter as the role of the simulations in scale up is similar.

12.2 Process Modeling

12.2.1 Background

In this discussion, the CLOU process, a variant of CLC, is defined as pictured in Figure 12.1. The oxygen carrier (OC) is assumed to be a manufactured material with a given content of metal oxide, which has been added on an inert. The reactions to the left of the schematic are the main reactions in the system. In the fuel reactor an endothermic reduction reaction for the OC takes place and releases oxygen, accompanied by an exothermic fuel reaction where the fuel is combusted with the oxygen [1]. The air reactor's OC oxidation reaction is exothermal. As such, heat may be recovered from the air reactor and, depending upon the conditions, the fuel reactor may supply heat or require heat. This is in contrast to CLC where the fuel reactor is endothermic due to the gasification reactions for the fuel as well as the reduction reaction.

Handbook of Chemical Looping Technology, First Edition. Edited by Ronald W. Breault.

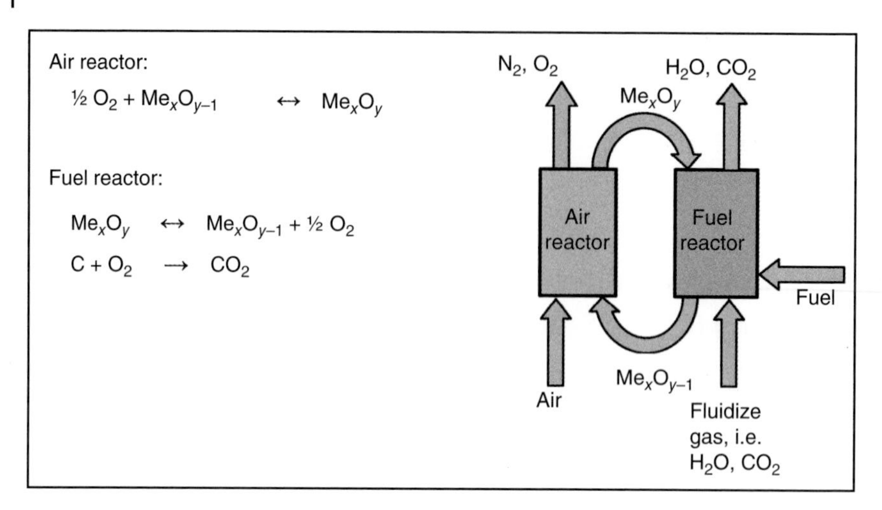

Figure 12.1 Schematic representation of CLOU system.

Process modeling, namely the determination of the mass and energy balances for a CLOU system, is useful to scale up situations for a variety of reasons. First, CLOU systems move a large quantity of material between the fuel and air reactors. The amount of material will be dependent upon the time for oxidation (air reactor) and reduction (fuel reactor) of the metal oxide carrier; the stoichiometric amount of oxygen required to combust the chosen fuel; and, the amount of metal oxide contained on the carrier. Heat release and consumption in each reactor is determined by the amount of OC, the concentration of metal oxide on the inert, and the heat of combustion for the chosen fuel, as well as the temperature difference between the two reactors. Process modeling can aid in the determination of the heat balance between the two reactors, ensuring that the fuel reactor remains exothermic, i.e. the temperature difference between the reactors is a central parameter to operation stability.

Commercial packages, such as Aspen Plus, have been used for CLC and CLOU process modeling. Challenges exist for commercial packages based on the limitations of the "blocks" provided, as discussed below. Other models, which do not utilize a commercial package, are also discussed, including a novel fuel reactor configuration that has been suggested.

12.2.2 Aspen Plus Modeling

A range of researchers studying CLC and, to a lesser extent, CLOU, have used Aspen Plus Modeling. Zhou et al. [2] used experimental CLC data to validate an Aspen Plus model they had created and found good agreement between the two. They modeled the fuel reactor using an RYIELD reactor in combination with an RGIBBS reactor, and the air reactor as an RSTOIC reactor. They compared their results with experiments and found excellent agreement. In addition, scaled-up simulations illustrated a nearly linear relationship where an increase in coal feed increased power output.

Meng et al. [3] modified the model developed by Zhou et al. [2] to be a CLC system that used two- and four-stage fuel reactors. This was done in an effort to increase the residence time and conversion of coal. Porrazzo et al. [4] modeled a CLC system where the air reactor was operated in the fast fluidization regime, and the fuel reactor modeled as a bubbling bed. The fuel reactor hydrodynamics were approximated as being based off of both a bubble and emulsion phase and were divided into multiple stages of parallel sub-reactors comprised of PFRs and CSTRs to create a more realistic model.

Sahir et al. [5] studied both CLC and CLOU for coal combustion. As discussed in Eyring et al. [6] and Sahir et al. [5], for investigations of CLC and CLOU combustion for a solid fuel, the first step in utilizing an Aspen Plus model is to create a structure to represent the fuel reactor and air reactor. For CLC, the study of Sahir et al. [5] utilized 60% Fe_2O_3 supported on Al_2O_3. The air reactor was modeled as a stoichiometric reactor, RSTOIC, with an 80% conversion of Fe_3O_4 to Fe_2O_3. Figure 12.2 illustrates this case. Two temperatures were considered, 935 and 1050 °C, with fuel reactor temperatures of 950 and 970 °C.

The fuel reactor is more complex. First, the gasification process was modeled with a combination of both RYIELD and RGIBBS, an equilibrium reactor, process units. The RYIELD decomposed the components of the coal into its elemental constituents, which were then gasified to form a syngas in the RGIBBS. The residence time was 10 minutes. The syngas was 32% CO, 5% CO_2, 50% hydrogen, and 13% H_2O.

The mass flow rates of the inlet metal oxide stream to the fuel reactor and the outlet metal oxide stream from the air reactor were equated to accomplish the recycle of the material. No OC was considered lost and the coal was completely converted. In Figure 12.2 the OC heat exchangers were used only to assess the overall heat balance to determine autothermal operation as discussed in Section 12.2.4.

Table 12.1 shows the exhaust gas from the fuel reactor under these conditions. As seen in the table, the major constituents are water and carbon dioxide. In terms

Table 12.1 Flow rate and composition of the exhaust gas from the fuel reactor for CLC case.

Constituent	Flow rate (Kg h^{-1})	Mass fraction
Water	1770.50	0.408
CO_2	2557.40	0.589
CO	0.04	0.000
H_2	0.00	0.000
Sulfur	0.00	0.000
N_2	9.91	0.000
H_2S	3.10	0.000
H_3N	0.00	0.000
COS	0.11	0.000

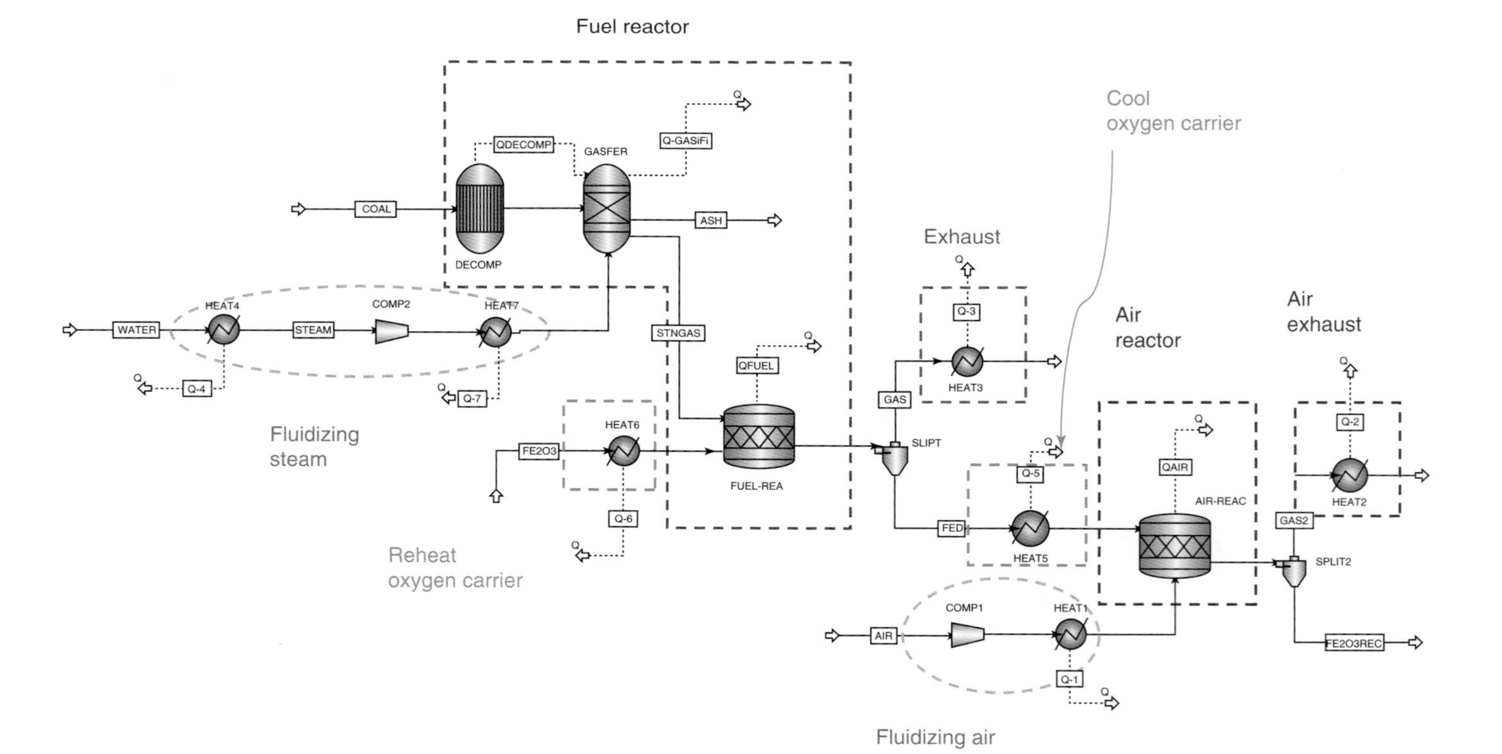

Figure 12.2 Aspen Plus process flow sheet for CLC. Source: Sahir et al. 2014 [5]. Adapted from Elsevier.

of the exhaust from the air reactor, nitrogen made up almost 99% of the stream. The conversion of iron oxides was previously discussed.

This configuration can be compared to that of CLOU where a copper oxide OC on an inert was used. The air reactor was modeled as a stoichiometric reactor, RSTOIC, with a given conversion (Figure 12.3). Conversion was given as 50% Cu_2O to CuO in Sahir et al. [5]. The temperature of the air reactor was 935 °C with a residence time of ~40 seconds.

The fuel reactor, illustrated in Figure 12.3, was similar to that discussed above. First, the solid fuel was decomposed in a RYIELD reactor to its elemental constituents. The elemental constituents were then burned with oxygen, supplied from the OC module, in a RGIBBS reactor where the fuel was combusted entirely. The OC module was, as in the air reactor, an RSTOIC reactor with a given conversion, in this case 54%. Heat exchangers to heat and cool the OC from the fuel reactor were utilized to maintain the fuel reactor exothermic. The residence time was ~30 seconds with a temperature of 950 °C. As discussed above, for the metal oxide stream, the mass flow rate of the air reactor exhaust was equated to that of the inlet to the fuel reactor to simulate the looping of the material.

In the scenario discussed above, the temperature difference between the air and fuel reactor was 15 °C and this resulted in exothermic operation in both reactors. This is a desirable condition, especially when compared to the CLC simulation, which was run where the air reactor needed to be 100 °C higher than the fuel reactor to ensure exothermic conditions in both reactors. The CLOU result is in contrast to conventional CLC, where an iron metal oxide is used, and the fuel reactor is a gasifier, not a combustor. In CLC, the OC carries the heat for the endothermic reactions in the fuel reactor resulting in the need for larger temperature differences between the reactors. The mass of material circulating was approximately 12 700 $kg\,h^{-1}$ where the copper oxide content was about 40% by weight for 100 $kg\,h^{-1}$ of coal, or 127 kg OC/kg coal. For the particular coal studied, this also resulted in approximately 6 $kg\,s^{-1}$ OC per MW_{th}. The loading of OC in the fuel reactor was about 350 kg per MW_{th}.

Table 12.2 shows the exhaust gas concentration from the fuel reactor for this case. As seen in the table, in comparison to Table 12.1 giving the results for CLC, the CLOU exhaust gas has some oxygen in it. In addition, there are combustion products, for example NO, versus products that one would expect from gasification. The air reactor was operated such that the exhaust gas from this unit has an oxygen concentration above the equilibrium value. This is necessary to avoid oxygen uncoupling prior to the fuel reactor.

Sahir et al. [5] also performed a comparative analysis of the capital and operating costs for the two scenarios and a 100 MW_{th} system. In this case, both CLC and CLOU required two fuel reactors. In addition, CLC required two air reactors as compared with CLOU, which only needed one. It is important to note that the residence time in the fuel reactor for CLC was 10 minutes due to the gasification reaction. In terms of capital costs, assuming an attrition of 0.05% per hour for both cases, $6600 per metric ton for the CLOU OC, and $200 per metric ton for the CLC OC, the total capital cost was slightly higher for CLC. This was the result of the need to have a larger compressor for the CLC case and the high residence time in the fuel reactor. In terms of operating cost, the compressor costs were higher, even as compared to the cost of the make-up of the CLOU OC, as

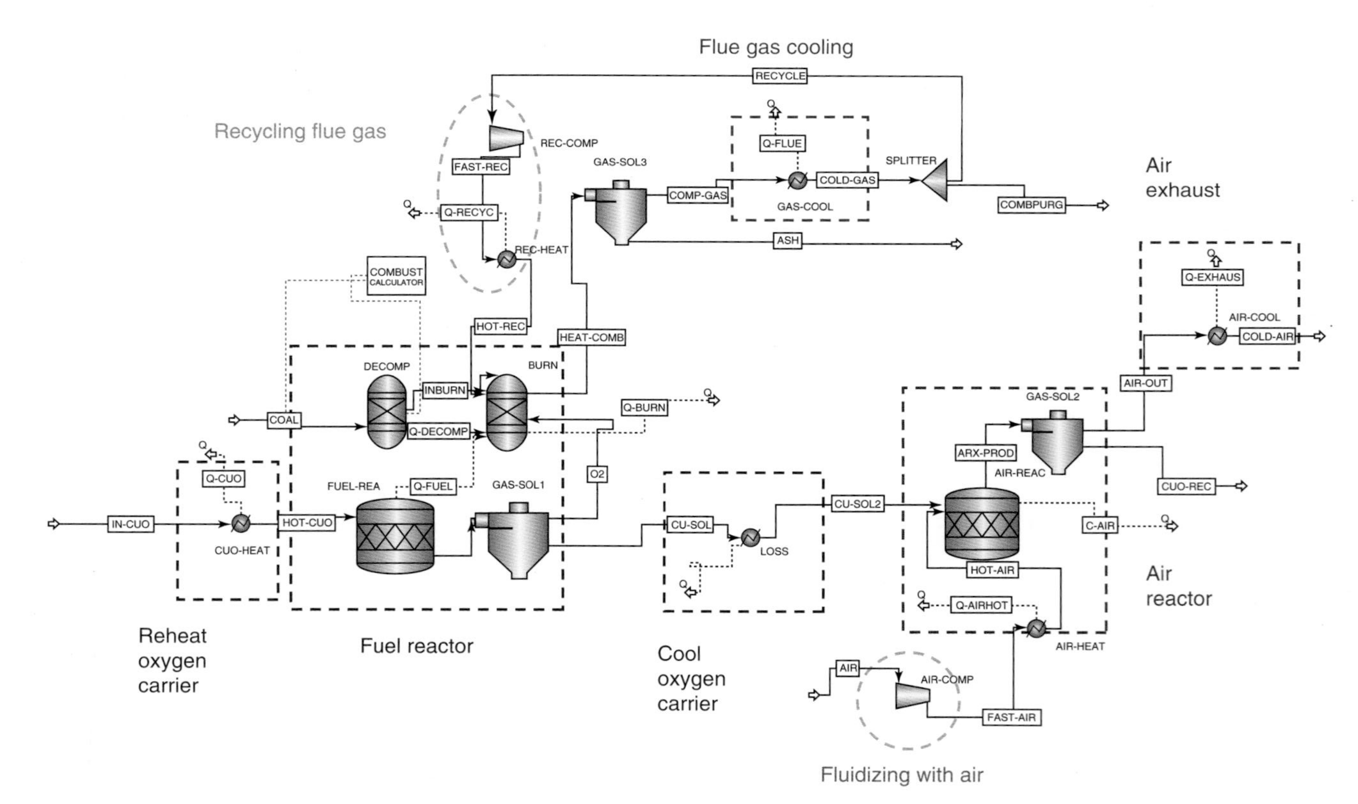

Figure 12.3 Aspen Plus process flow sheet for CLOU. Source: Sahir et al. 2014 [5]. Adapted from Elsevier.

Table 12.2 Flow rate and composition of the exhaust gas from the fuel reactor for the CLOU case.

Constituent	Flow rate (Kg h^{-1})	Mass fraction
O_2	1038.80	0.088
CO_2	8305.00	0.704
Water	2352.34	0.199
NO_2	1.01	0.000
NO	68.31	0.006
SO_2	18.13	0.001
SO_3	1.45	0.000

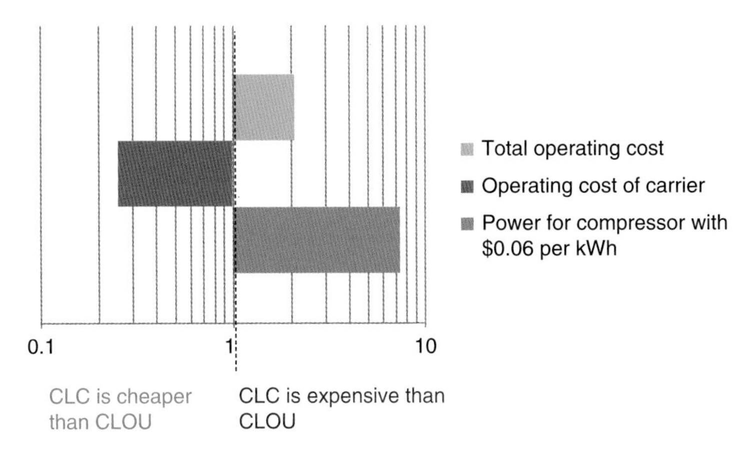

Figure 12.4 Comparative operating costs for CLC versus CLOU. Source: Sahir et al. 2014 [5]. Reproduced with permission of Elsevier.

illustrated in Figure 12.4. Additional operational information is needed to determine the actual scenario, especially the cost of the carriers and the amount of carrier needed.

Zhou et al. [7] performed a similar analysis for CLOU using Aspen Plus. As with their work on CLC [2] Zhou et al. found excellent agreement with experimental data of gas composition and power output. Furthermore, consistent with CLC, they found a nearly linear relationship of coal feed and power output for three types of coal.

12.2.3 Other Approaches to Material and Energy Balance Determinations

In a different approach, Peltola et al. [8] developed a stoichiometric mass, energy, and exergy balance model. Silica-supported CuO OC was used with a medium volatility bituminous coal. Base case temperatures were 850 °C for the air reactor and 950 °C for the fuel reactor. The results showed comparable OC circulation

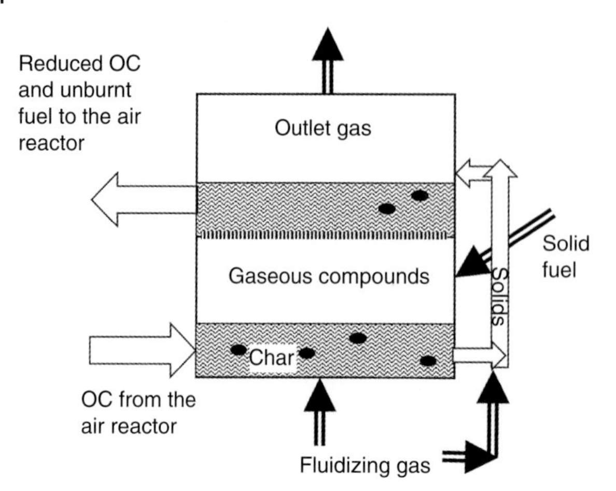

Figure 12.5 Representation of the novel fuel reactor system studied. Source: Coppola et al. 2015 [9]. Reproduced with permission of Elsevier.

rates as discussed above, depending upon the difference between the conversion of OC in the air reactor and the fuel reactor; for 40% CuO on silica, these values ranged from 20.2 to 2.9 $kg\,s^{-1}$ per MW_{th}. These numbers can be compared with the previously reported results of 6 $kg\,s^{-1}$ per MW_{th}. Furthermore, the solids inventory was presented for a range of fuel reactor and air reactor conversions. The resulting solids inventory was a minimum of 400 kg per MW_{th} in Peltola and colleagues' work, comparable to the results from Aspen Plus simulations. The solids inventory was found to be a function of reactor temperature; however, a change to 900 °C for the air reactor and 950 °C for the fuel reactor yielded only slightly higher minimum inventories from the base case of 400 kg per MW_{th}.

Coppola et al. [9] modeled a novel two-stage reactor for CLOU. In this case, the fuel reactor was designed to limit the carryover of carbon char; as such, the bottom of the reactor was designed to convert char while the top reactor handled the volatile matter. A representation of the fuel reactor configuration is shown in Figure 12.5. The OC material was 50% copper oxide on ZrO_2 with bituminous coal as the fuel. The fuel reactor was set at a temperature of 1000 and 950 °C for the air reactor. Results were presented for a range of fuel reactor bed inventories and compared with a single-stage reactor. For comparison with the cases studied above, 400 kg per MW_{th} bed inventory, the conversion of char improved from 50% to 60% from the single-stage fuel reactor to the previously described dual-stage reactor. Notably, methane conversion was 100% in the dual-stage reactor, as compared to about 50% in the single-stage reactor. However, the carbon capture efficiency for the dual-stage reactor was only slightly higher than that for the single stage, approximately 65%. Decreases in the fuel reactor temperature led to decreases in both the char conversion rate and the methane conversion. The decrease in char conversion rate was fairly linear with decrease in fuel reactor temperature. The carbon capture efficiency was approximately the same for fuel reactor temperatures from 900 to 950 °C. Comparisons to a single-stage reactor with a carbon stripper were also made. The carbon capture efficiencies for this system were approximately the same at a bed inventory of 400 kg per MW_{th}. The effect of temperature was minimal for the temperature range discussed above.

12.2.4 Autothermal Operation

An important parameter in CLC is the temperature difference between the fuel and the air reactor, where the goal is to achieve autothermal operation, i.e. both reactors are exothermic and do not require heat addition for operation. Practically, this has been seen in many conventional CLC studies [10, 11] as a limit to stable operation; in particular, conventional CLC utilizes a gasification reaction in the fuel reactor, which requires heat addition. This heat is almost always supplied by the thermal capacity of the OC. In the case of CLOU, the range of heat integration is likely larger due to the fact that the fuel reactor utilizes a combustion reaction. Process modeling is useful for investigating the range of conditions where autothermal operation is possible, given that this range is dependent upon the amount of OC, the fuel studied, amount of preheat of the fluidizing gases in both reactors, and concentration of the metal oxide on the OC.

Sahir et al. [5] investigated autothermal operation where CLC and CLOU were compared as discussed with the details of the model given in the previous paragraphs. For a fuel reactor temperature of 970 °C, an air reactor temperature of 1050 °C was needed to keep the fuel reactor exothermic, a temperature difference of 80 °C. For CLOU, autothermal operation could be achieved with a fuel reactor temperature of 950 °C and an air reactor temperature of 935 °C, a difference of 15 °C, noting that the fuel reactor temperature is now higher, which is useful in terms of air reactor kinetics. The OC oxidation kinetics in the air reactor have been found to be slower as temperature increases [12, 13]. This result is likely due to mass transfer limitations at the higher temperatures.

Peltola et al. [8] also investigated the temperature difference needed between the two reactors. The amount of OC circulating was investigated by utilizing the change in conversion between the two reactors. For an acceptable minimum of 0.2–0.4, the temperature change between the reactors was found to be 50 °C, consistent with the previous results.

Hamilton et al. [14] presented a more detailed comparison. One result is shown in Figure 12.6. The figure shows the autothermal temperature range for a fuel reactor temperature of 950 °C as a function of CuO loading. Additionally, the allothermal range, i.e. the range where some preheating is required, is shown. In this instance, the preheat is either for the fluidizing gases of the fuel reactor, to 450 °C, or the air reactor fluidizing air, to 400 °C. The air reactor temperature is always higher than the fuel reactor, with an increasing autothermal range as the CuO concentration increases, resulting in a decrease of circulating material. For example, for 40% CuO, the air reactor temperature should be 20 °C higher than or nearly the same as the fuel reactor temperature. The only instance where the air reactor temperature could be below the fuel reactor temperature was when the fluidizing gases to the fuel reactor were preheated. Of course, these results depend on numerous operating parameters, making a process model useful as multiple cases can be investigated efficiently.

12.2.5 Using Process Modeling for Steam Production Estimates

While there are several cases where process modeling was used to evaluate the steam production for conventional CLC, Dansie et al. [15] are one of the few

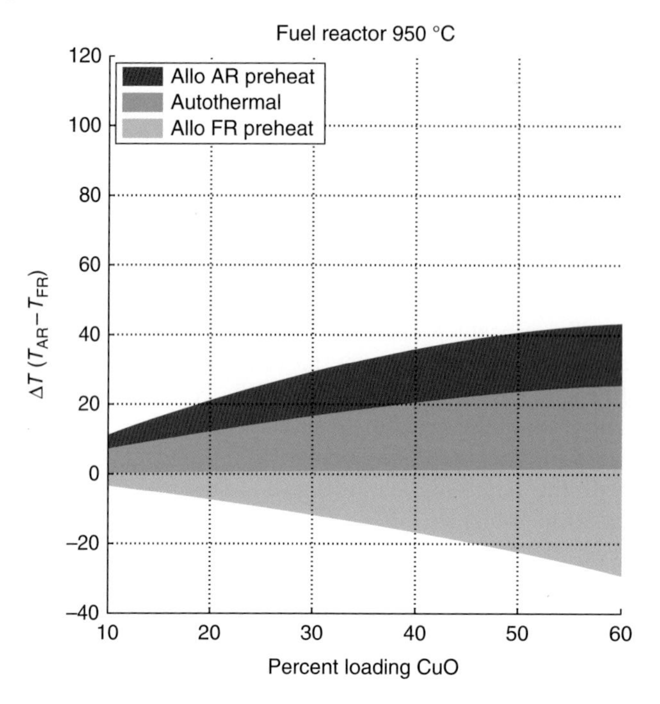

Figure 12.6 Temperature difference between reactors for autothermal operation for CLOU system. Allothermal regions are also shown.

that investigated CLOU. Aspen Plus simulations were the same as previously discussed, but for a 10-MW_{th} scale system. Aspen Energy Analyzer was utilized for the steam production estimates and the assumption was that water entered at 600 psia and 15.6 °C and exited at the same pressure and 307.8 °C, conditions similar to those of other studies of this nature. The largest amount of steam was produced by recovering heat from the air reactor, nearly 40%. The heat recovered from the fuel reactor exhaust was used to produce steam as well, recovering 18% of the heat, for a total of 58% of the heat. The remainder of the heat went to preheat the air fluidizing the air reactor utilizing the air reactor exhaust, 22%; heating the recycle stream to fluidize the fuel reactor utilizing the fuel reactor exhaust, 13%; and a small amount of air preheat from the air reactor, 7%. Clearly, additional studies are needed to understand the range of possible conditions for heat recovery in full-scale systems.

12.2.6 Summary

As discussed, process modeling can help gain insight into many aspects of CLOU system operation and scale-up, namely and primarily, heat integration between the two reactors. The information gained from these simulations allows designers to determine, for a given carrier and fuel, the needed system parameters to ensure autothermal operation. CLOU systems are also complex because of the number of parameters that can change with a direct effect on design, e.g. amount of metal on the OC and kinetics of the carrier. Process modeling is an effective method to "turn the knobs" and determine optimum conditions. Additional operational data, at a larger scale, are needed to determine the amount of material loss; the kinetics and residence times; and, the amount of OC recirculated. Also, additional simulations are needed to determine potential heat recovery systems, with particular interest toward available streams for steam generation.

12.3 Computational Fluid Dynamic Simulations

12.3.1 Background

Process models are a global interpretation of the CLOU system; however, the CLOU system is a complex environment with a range of particles, in terms of size and type, reacting in a fluidized environment. The implementation of simulations that couple the modeling of the kinetics of the OC (see Figure 12.1); the combustion of solid and gaseous fuel constituents; detailed solid-wall contact; and, thermal and fluid transport in highly reactive and high particle concentrations are important to define issues and operating conditions, such as

- the extent of volatiles bypassing the fuel reactor resulting in lower carbon conversion;
- the extent of carryover of solid char resulting in carbon conversion in the air reactor, and lower carbon capture;
- determination of thermal profiles within the reactors to determine local hot/cold spots; and,
- the extent of particle attrition as well as wear of the reactor walls.

There is a wealth of literature on CFD in fluidized environments. A brief synopsis is given here. There are two frames of reference for modeling dense-particle/fluid environments in CFD. In both methods, the fluid is treated as continuous species on an Eulerian grid. The Eulerian–Eulerian (E–E) or two-fluid method, treats the particle phase as a continuous species on an Eulerian grid. In contrast, the Eulerian–Lagrangian (E–L) frame of reference treats the particle phase as parcels on a Lagrangian grid. These parcels can be tracking individual particles or particle clouds, groups of particles that have the same physical quantities of density, size, shape, etc. The E–E frame of reference has a large dependence on the drag model and other particle interaction models [16] and has been used to simulate two- and three-dimensional systems with validated results for CLC. While the E–E frame of reference requires fewer calculations as compared to the E–L frame of reference [17–19], with the advancement of graphical processing units, which can handle large calculations quickly, the utilization of the E–L frame of reference has become feasible. E–L can be divided into discrete element methods (DEMs) to capture the microdynamics of the particles; discrete particle methods (DPMs) where the trajectory of each particle is tracked; and multiphase particle in cell (MP-PIC) methods that allow for a distribution of particle sizes and range of particles taking advantage of the particle clouds previously mentioned. There are a number of commercial and open-source software packages that can be utilized; these are listed in Table 12.3 along with the methodology. As an example, Figure 12.7 illustrates the difference in the parcels for MP-PIC versus coarse-grain DEM in MFIX [20]. As seen in the figure, the single particle in the DEM method is obtained by equating the volume of the tracked parcel to the sum of volumes of individual particles; inter-parcel collisions are considered. In contrast, while MP-PIC can represent a singular particle, it does not consider interparticle or intra-parcel collisions.

Table 12.3 Major commercial and open source particle computational fluid dynamic codes.

Software	School of thought	Method
ANSYS Fluent	E–E	Two fluid
	E–L	DEM
Barracuda-VR	E–L	MP-PIC
MFIX	E–E	Two fluid
	E–L	DEM, MP-PIC

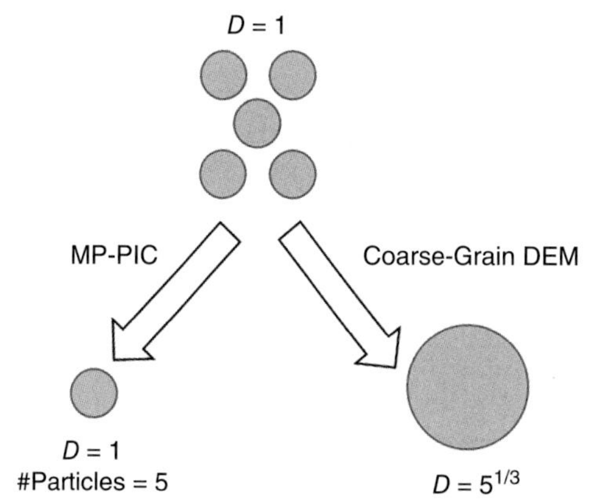

Figure 12.7 The difference between the MP-PIC and coarse-grain DEM approach. In the MP-PIC methodology (left), five spherical particles having diameter of one unit are represented by a single parcel having diameter of one unit and statistical weight equal to five. In the coarse-grain DEM method, the particles are represented by one larger spherical parcel having diameter of $5^{1/3}$ units. Source: Adapted from Garg and Dietiker 2013 [20].

12.3.2 Summary of the Literature

Table 12.4 gives a summary of publications that have used CFD simulations in chemical looping applications. The majority of the articles deal with CLC as opposed to CLOU, and the Eulerian–Eulerian (E–E) frame of reference is more highly utilized. This is likely a result of the more recent advancement of E–L methods, including the introduction of DEM/DPM methods in ANSYS Fluent and MFIX and the introduction of commercial codes such as Barracuda Virtual Reactor (Barracuda VR®), and the improvement of computer hardware.

Jung and Gamwo's work [21] was one of the first published using CFD to simulate a chemical looping system, specifically CLC. The authors used MFIX to introduce a two-fluid method in a 2D space. A nickel-based OC was used to

Table 12.4 Summary of research utilizing CFD to model chemical looping systems.

	Frame of reference	Dimension	Reactor and type of bed	Major conclusion
Jung et al. [21]	E–E	2-D	Fuel reactor, batch bubbling bed	Illustrated the importance of bubble dynamics in fuel conversion for CLC
Kruggel-Emden et al. [22]	E–E	2-D	Air reactor, high velocity riser; fuel reactor, bubbling bed	A time delay to interconnect the reactors for CLC; while there were limitations, could be used as a design basis as the reactors were coupled
Mahalatkar et al. [23]	E–E	2-D	Fuel reactor, batch bubbling bed, coal as fuel	Solid fuel gasification and volatilization models needed improvement as well as drag models
Zhang et al. [24]	E–L	3-D	Fuel reactor, spouted bed, with cyclones and loop seal	Successful validation with cold-flow unit and optimized operation with design modifications
Peng et al. [25]	E–L	3-D	Air reactor and fuel reactor with loop seal	Successful validation with cold-flow unit and solid circulation rate studied
Parker [25]	E–L	3-D	Air reactor and fuel reactor with loop seal	Barracuda VR® simulation on CLC with direct gasification of solid fuel
Hamilton et al. [26]	E–L	3-D	Air reactor and fuel reactor with loop seals, circulating fluid beds	Introduction of CLOU kinetics into Barracuda VR®

oxidize methane gas in a shrinking core model. The fuel reactor was treated as a batch system and there was no circulation or introduction of solids into the reactor. Figure 12.8 shows the solids fraction and the gas compositions, at a residence time of 15 seconds. Solids fraction (a) is shown where red indicates pure gas, a fraction of 1. Gas concentrations of methane (b), carbon dioxide (c), and water (d) ranging from zero (blue) to 0.9, 0.5, and 0.45 respectively are also shown. An investigation of the solids fraction, Figure 12.8a, shows the formation and rise of bubbles, and subsequent bursting of the bubbles within the lower half of the reactor, as indicated by blue, a gas fraction of 0.4, to red, pure gas. The authors point out that this occurred rapidly. The influence of this rapid bubble movement can be seen in Figure 12.8b where methane concentration is shown from 0, blue, to 0.90, red. The exhaust concentration of methane was about 20%, suggesting that the methane did not completely burn out, a result of rapid bubble dynamics, resulting in a conversion of about 50% methane in the system. While the authors did not validate the model with data, they do suggest that this is consistent with the literature data, and suggest a possible reduction of OC particle size to aid in more effective bubble formation. The study shows the importance of the consideration

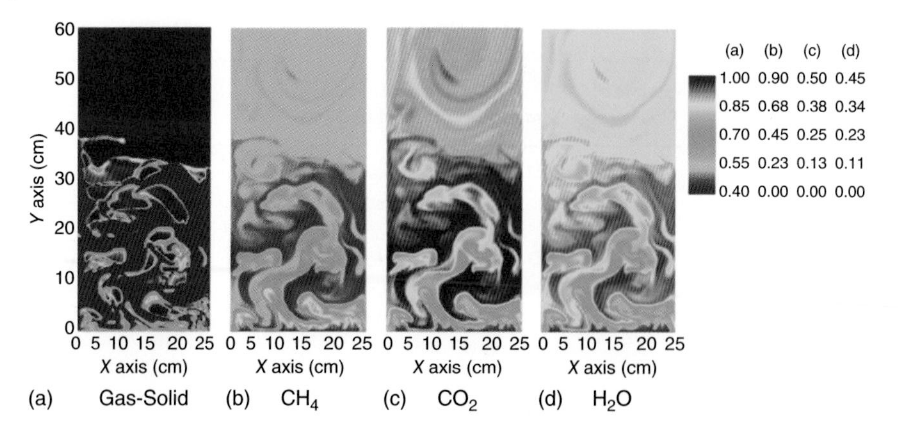

Figure 12.8 Instantaneous concentration profile at a residence time of 15 seconds of a batch fuel reactor utilizing nickel as the oxygen carrier and methane as the fuel. Source: Jung and Gamwo 2008 [21]. Reproduced with permission of Elsevier.

of bubble dynamics, and hence, CFD simulations, in determining the ultimate fuel conversion in the fuel reactor.

Kruggel-Emden et al. [22] were the first to introduce a "connected" chemical looping system, again, in this case CLC. They used ANSYS Fluent 12.1.4, a two-fluid method to model Mn_3O_4 reacting with methane. The simulation included both the fuel reactor, a bubbling fluidized bed, and the air reactor, a high velocity riser, but utilized a time delay between the two reactors to model the loop seals. The simulation showed that at near steady state, the air reactor temperature was 75 °C above the fuel reactor temperature and complete conversion of the methane occurred in the fuel reactor. The previous work of Sahir et al. [5] and Hamilton et al. [14] also showed the importance of this temperature difference in consideration for the system to be at autothermal operation. The authors state that simulations set the stage for many design considerations, but point out that validation of the results is needed.

Mahalatkar et al. [23] simulated a fuel reactor based on the experiments of Leion et al [27]. ANSYS Fluent 6.3 was used and the simulation included the gasification of the solid fuel; devolatilization of the solid fuel; the reduction of the OC, ilmenite, with CO and H_2; and full hydrodynamics. The results showed that a fine and coarse grid had the same numerical uncertainty, thus verifying the simulations. When compared to the experimental results from the study of Leion et al. [27], the simulation rates were close to the experimental rates. Both experimental and simulated results showed a spike of CO and CO_2 at one minute, with similar gas-phase concentrations. Figure 12.9 shows the concentrations of CO and CO_2 for the simulations and experiments at a height of 3 m with the reactor at 1223 K and 50% steam. The concentrations of CO_2 were higher in the simulations as compared to the data, possibly as a result of CO converting to CO_2 more rapidly in the simulations. Author suggestions for improvements included better kinetics for the char gasification and devolatilization, as well as drag law changes, consistent with the previously mentioned limitation that E–E frameworks are dependent on drag models. In addition, the simulation did not account

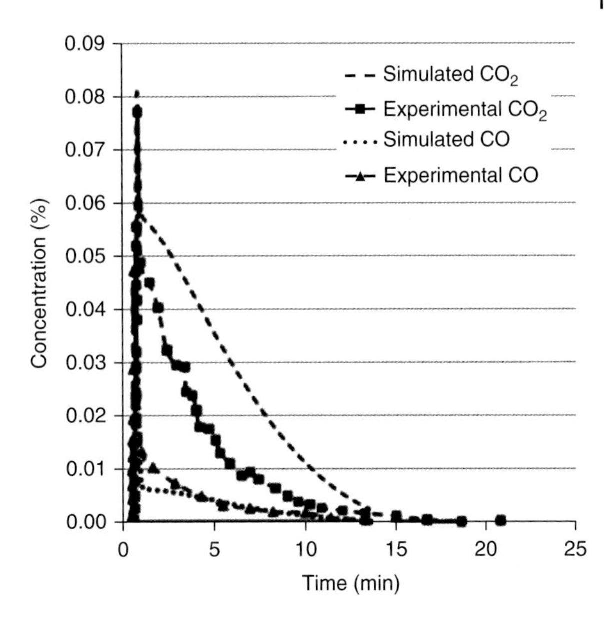

Figure 12.9 Gas phase concentration profile of CO_2 and CO comparing simulations and experimental work. Source: Mahalatkar et al. 2011 [23]. Reproduced with permission of Elsevier.

for changes in coal particle density and diameter, another shortcoming in the E–E methods. This study shows the importance of validation of the models, especially when they will be used to scale up results.

After 2013, several E–L based simulations were published related to chemical looping. These ranged from studying particle segregation [28] to irregularly shaped particles [29]. The introduction of the Lagrangian parcels to model the solid phase allowed for these studies to be performed with a decreased number of empirical models. Zhang et al. [24] utilized the E–L framework to simulate a spouted fluidized bed, which was used as the fuel reactor in a CLC simulation. A cyclone and loop seal completed the system to allow for recirculation of the particles. Spouted beds have intense particle–particle and particle–wall interactions. ANSYS Fluent 14.5, with DEM and the Syamlal-O'Brian drag model, was employed and simulations were compared to cold-flow unit experimental data. The simulation yielded good results and comparisons. However, recirculation back to the fuel reactor was limited; therefore, the same inputs were used with a different geometry to study the static pressure, streamlines, and solid concentrations with success. The study proved that the development of a validated CFD/DEM model, with "high fidelity," could be used to effectively optimize the design of a CLC system with coal. In reality, it is often difficult to perform experiments with a variety of design considerations. A detailed, validated, CFD model is an effective and efficient way of optimizing design for operation.

Peng et al. [25] also used ANSYS Fluent 14.5 and DEM-based simulations to simulate a 10-kW system, with validation using a cold-flow unit. Both the simulations and the experiments showed a build-up of material on the horizontal transport from the air reactor to the air reactor cyclone. Solid circulation rates were found to be proportional to fuel reactor flow rate. For the air reactor, flow increases were found to be related to solid circulation by a logarithmic law. Additionally, Figure 12.10 shows that residence time in the riser of the air

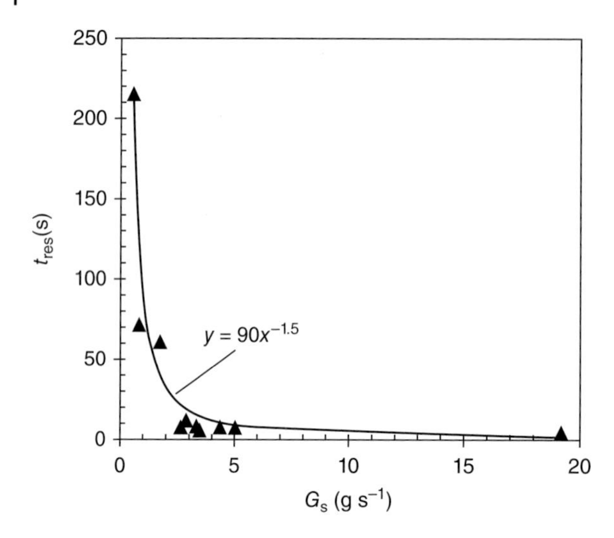

Figure 12.10 Residence time as a function of solid recirculation rate [25]. Data versus simulation predictions are shown.

reactor decreased dramatically as the solid recirculation rate increased. This result demonstrated the complex nature of a CLC system. The authors note that the higher solid recirculation rate and shorter residence time may reduce the conversion of the OC, which in turn requires more material in the system and, perhaps a larger reactor; however, the higher solid recirculation rate improves the heat transfer between the two reactors. As noted, a CLC or CLOU system is a complex interaction of many parameters, as in this case, the optimum solids recirculation rate where the residence time is long enough for conversion and heat transfer is enhanced. CFD simulations coupled with heat transfer and reactions can be used to "turn the knobs" toward the ideal design and operation.

As shown in Table 12.4, Barracuda VR® is a commercial software package that utilizes E–L framework and MP-PIC. Parker [30] used Barracuda VR® to run a 3D simulation of an iron-based CLC system in the Department of Energy, National Energy Technology Laboratory's 50-kW system. The simulations reached steady state for pressures and hydrodynamics, but the temperatures were still in a transient regime. The transient temperatures did not show an effect on the oxidation or reduction rates. Parker's simulations were not validated, but this work represents one of the first studies of a CLOU system with two integrated reactors.

To explore the hydrodynamics of two different CLOU configurations, Hamilton et al. [31] utilized Barracuda VR® simulations. The two systems were a dual, bubbling bed system, which was an existing 10-kW laboratory system and a cold-flow unit that was a scaled representation of the University of Utah's 200-kW Process Development Unit, two circulating fluidized beds. Experimental data were used to validate the simulations. For the dual circulating fluidized bed system, the simulations showed higher global circulation rates, the result of the implementation of the Wen–Yu drag model. Other results showed similarities between the simulations and the experiments as trends and magnitudes of the results were close. For the dual bubbling beds, the global circulation rate data did not agree with the simulations. A potential reason for this disagreement is the fact that

Barracuda-VR was developed for large-scale units and the smaller reactor size of the dual bubbling bed system's wall effects were perhaps underpredicted.

A CLOU system burning a solid fuel is complicated, not only by the changing particle diameter of the fuel, representing a range of particle sizes, but also by the fact that the reduction and oxidation kinetics are not easily incorporated into software packages. For example, Clayton et al. [12] and Sahir et al. [13] both found that the kinetics were limited by the driving force for the equilibrium concentration of oxygen in the fuel reactor and, in some cases, mass transfer. Equilibrium concentration is a function of vapor pressure, which is logarithmic. This creates a highly nonlinear kinetic relationship, something that cannot be incorporated into Barracuda-VR. To enable the use of these kinetics, Hamilton et al. [26] linearized the equations and incorporated them into Barracuda-VR. The system was the dual, bubbling bed system discussed above. The OC was 40% copper oxide on SiO_2 with a particle size distribution around 200 μm. The coal size particles had a size distribution with an average of approximately 150 μm. Figure 12.11 illustrates the results for the simulation at 40 seconds (see [26] for 60 seconds results). As seen in this figure, OC particles were predominately in the bed in both reactors. The bed concentration peaked at 60% particles (see Figure 12.9a), a value representative of the batch reactor in Jung and Gamwo's work [21]. Coal particles, shown in Figure 12.9b, were not present in the air reactor, confirmed by the CO_2 emissions (Figure 12.9d), where the concentration of CO_2 was zero in the air reactor. The average carbon capture efficiency was determined to be 94%. These results have yet to be validated with experimental data, but show the ability to predict system operation with complicated kinetics as well as two different particle size distributions.

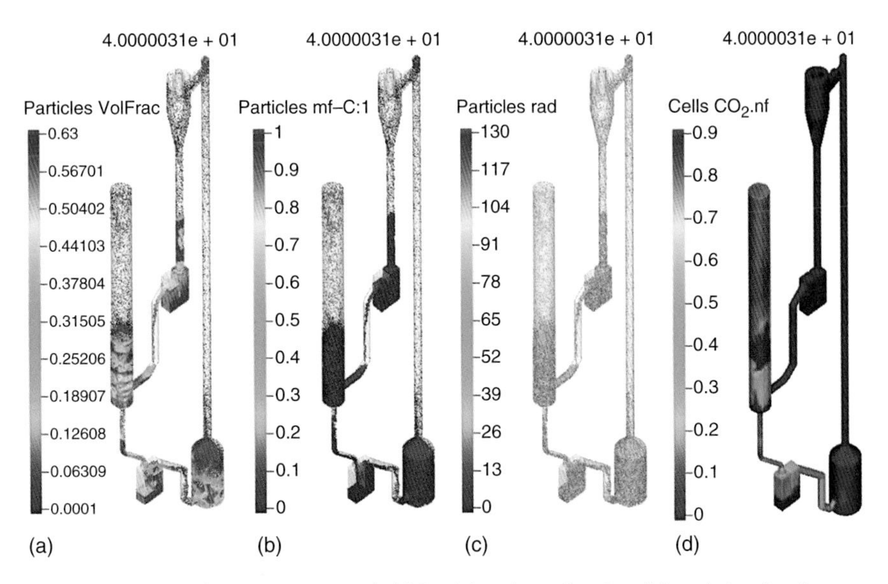

Figure 12.11 Conditions at 40 seconds (a) Particle volume fraction, (b) coal char fraction, (c) particle radius, (d) CO_2 mole fraction. Fuel reactor is on the left.

12.3.3 Conclusions

This section has discussed the present state of CFD modeling for CLC systems, with a particular focus on the differences between E–E and E–L frames of reference. E–E methods are not as computationally extensive; they are dependent on drag models and do not have the capability of tracking particles to the extent of E–L methods, which may be a limitation. With the implementation of MP-PIC in commercial codes, the use of these codes for a complicated system such as CLOU, especially in solid fuel combustion, is useful, as the ability to input a particle size distribution that can change with time for multiple types of particles is important.

The models have shown their ability to aid in and optimize system designs, especially with regard to supplying high-fidelity information. For example, the simulations can demonstrate operational or design considerations that limit the carbon conversion of the fuel and potential carryover into the air reactors. However, the simulations are still in need of experimental validation for both the hydrodynamics and reacting flows, especially at scales greater than 10 kW.

References

1 Mattisson, T., Lyngfelt, A., and Leion, H. (2009). Chemical-looping with oxygen uncoupling for combustion of solid fuels. *Int. J. Greenhouse Gas Control* 3: 11–19.

2 Zhou, L., Zhang, Z., and Agarwal, R.K. (2014). Simulation and validation of chemical-looping combustion using ASPEN plus. *Int. J. Energy Environ.* 5 (1): 53–58.

3 Meng, W.X., Banerjee, S., Zhang, X., and Agarwal, R.K. (2015). Process simulation of multi-stage chemical-looping combustion using Aspen plus. *Energy* 90: 1869–1877.

4 Porrazzo, R., White, G., and Ocone, R. (2014). Aspen plus simulations of fluidised beds for chemical looping combustion. *Fuel* 136: 46–56.

5 Sahir, A.H., Dansie, J.K., Cadore, A.L., and Lighty, J.S. (2014). A comparative process study of chemical-looping combustion (CLC) and chemical-looping with oxygen uncoupling (CLOU) for solid fuels. *Int. J. Greenhouse Gas Control* 22: 237–243.

6 Eyring, E.M., Konya, G., Lighty, J.S. et al. (2011). Chemical looping with copper oxide as carrier and coal as fuel. *Oil Gas Sci. Technol.* 66 (2): 209–221.

7 Zhou, L., Zhang, Z., Chivetta, C., and Agarwal, R. (2013). Process simulation and validation of chemical-looping with oxygen uncoupling (CLOU) process using Cu-based oxygen carrier. *Energy Fuels* 27 (11): 6906–6912.

8 Peltola, P., Tynjälä, T., Ritvanen, J., and Hyppänen, T. (2014). Mass, energy, and exergy balance analysis of chemical looping with oxygen uncoupling (CLOU) process. *Energy Convers. Manage.* 87: 483–494.

9 Coppola, A., Solimene, R., Bareschino, P., and Salatino, P. (2015). Mathematical modeling of a two-stage fuel reactor for chemical looping combustion with oxygen uncoupling of solid fuels. *Appl. Energy* 157: 449–461.

10 Ohlemüller, P., Busch, J.-P., Reitz, M. et al. (2016). Chemical-looping combustion of hard coal: autothermal operation of a 1 MW_{th} pilot plant. *J. Energy Resour. Technol.* 138 (4): 42203.
11 Linderholm, C. and Schmitz, M. (2016). Chemical-looping combustion of solid fuels in a 100kW dual circulating fluidized bed system using iron ore as oxygen carrier. *J. Environ. Chem. Eng.* 4 (1).
12 Clayton, C.K., Sohn, H.Y., and Whitty, K.J. (2014). Oxidation kinetics of Cu_2O in oxygen carriers for chemical looping with oxygen uncoupling. *Ind. Eng. Chem. Res.* 53 (8): 2976–2986.
13 Sahir, A.H., Sohn, H.Y., Leion, H., and Lighty, J.S. (2012). Rate analysis of chemical-looping with oxygen uncoupling (CLOU) for solid fuels. *Energy Fuels* 26 (7): 4395–4404.
14 Hamilton, M.A., and Lighty, J.S. (2016). Determination of autothermal and allothermal state for 10 MW CLC and CLOU with Powder River Basin coal as fuel. Poster presented at the 4th International Conference on Chemical Looping.
15 Dansie, J.K., Sahir, A.H., Hamilton, M.A., and Lighty, J.S. (2015). An investigation of steam production in chemical-looping combustion (CLC) and chemical-looping with oxygen uncoupling (CLOU) for solid fuels. *Chem. Eng. Res. Des.* 94: 12–17.
16 Hannes, J.P. (1996). *Mathematical Modelling of Circulating Fluidized Bed Combustion*. Aachen: City-Print Verlag.
17 Wang, S., Gao, J., Lu, H. et al. (2012). Simulation of flow behavior of particles by cluster structure-dependent drag coefficient model for chemical looping combustion process: air reactor modeling. *Fuel Process. Technol.* 104: 219–233.
18 Guan, Y., Chang, J., Zhang, K. et al. (2014). Three-dimensional CFD simulation of hydrodynamics in an interconnected fluidized bed for chemical looping combustion. *Powder Technol.* 268: 316–328.
19 Wang, S., Lu, H., Zhao, F., and Liu, G. (2014). CFD studies of dual circulating fluidized bed reactors for chemical looping combustion processes. *Chem. Eng. J.* 236: 121–130.
20 Garg, R., and Dietiker, J.F. (2013). Documentation of open-source MFIX–PIC software for gas-solids flows, From URL https://mfix.netl.doe.gov/documentation/mfix_pic_doc.pdf.
21 Jung, J. and Gamwo, I.K. (2008). Multiphase CFD-based models for chemical looping combustion process: fuel reactor modeling. *Powder Technol.* 183 (3): 401–409.
22 Kruggel-Emden, H., Rickelt, S., Stepanek, F., and Munjiza, A. (2010). Development and testing of an interconnected multiphase CFD-model for chemical looping combustion. *Chem. Eng. Sci.* 65 (16): 4732–4745.
23 Mahalatkar, K., Kuhlman, J., Huckaby, E.D., and O'Brien, T. (2011). CFD simulation of a chemical-looping fuel reactor utilizing solid fuel. *Chem. Eng. Sci.* 66 (16): 3617–3627.
24 Zhang, Z., Zhou, L., and Agarwal, R. (2014). Transient simulations of spouted fluidized bed for coal-direct chemical looping combustion. *Energy Fuels* 28 (2): 1548–1560.

25 Peng, Z., Doroodchi, E., Alghamdi, Y.A. et al. (2015). CFD–DEM simulation of solid circulation rate in the cold flow model of chemical looping systems. *Chem. Eng. Res. Des.* 95: 262–280.
26 Hamilton, M.A., Whitty, K.J., and Lighty, J.S. (2016). Incorporating oxygen uncoupling kinetics into computational fluid dynamic simulations of a chemical looping system. *Energy Technol.* 4 (10): 1237–1246.
27 Leion, H., Lyngfelt, A., Johansson, M. et al. (2008). The use of ilmenite as an oxygen carrier in chemical-looping combustion. *Chem. Eng. Res. Des.* 86 (9): 1017–1026.
28 Peng, Z., Doroodchi, E., Alghamdi, Y., and Moghtaderi, B. (2013). Mixing and segregation of solid mixtures in bubbling fluidized beds under conditions pertinent to the fuel reactor of a chemical looping system. *Powder Technol.* 235: 823–837.
29 Tabib, M.V., Johansen, S.T., and Amini, S. (2013). A 3D CFD-DEM methodology for simulating industrial scale packed bed chemical looping combustion reactors. *Ind. Eng. Chem. Res.* 52 (34): 12041–12058.
30 Parker, J.M. (2014). CFD model for the simulation of chemical looping combustion. *Powder Technol.* 265: 47–53.
31 Hamilton, M.A., Whitty, K.J., and Lighty, J.S. (2016). Numerical simulation comparison of two reactor configurations for chemical looping combustion and chemical looping with oxygen uncoupling. *J. Energy Resour. Technol.* 138 (4): 42213.

Section 4

Other Chemical Looping Processes

13

Calcium Looping Carbon Capture Process

Yiang-Chen Chou, Wan-Hsia Liu, and Heng-Wen Hsu

Industrial Technology Research Institute, Green Energy and Environment Research Laboratories, 195, Sec. 4, Chung Hsing Road, Chutung, Hsinchu 31040, Taiwan

13.1 Introduction

The world needs ever-increasing energy supplies to sustain economic growth and development. Meanwhile, excessive emissions of carbon dioxide (CO_2) are released from fossil fuel combustion for today's energy use. The increasing carbon dioxide emission level in the atmosphere absorbs the heat radiated off the surface of the earth, possibly causing global warming gradually, and threatening our climate. Since fossil energy continues to be needed in the world to meet energy demands in the near future, increasing concerns on the CO_2 level in the atmosphere and energy demand around the world have stimulated vital development of carbon mitigation technology for facing global climate change issues. A cyclic carbonation–calcination process using calcium-based sorbent (calcium looping process) has been regarded as one of the promising carbon mitigation technologies for large-scale CO_2 capture due to the characteristics of nontoxic sorbent, high CO_2 capture capacity, high-quality heat recovery, and low sorbent cost. This process is conducted by two individual reactions in the carbonator and calciner. The carbonator and calciner play the roles of sorbent carbonation (CO_2 capture) and sorbent calcination (regeneration), respectively.

A successful operation of the calcium looping process strongly depends on the reaction characteristics in the reactors, e.g. reaction kinetics, thermodynamic equilibrium behavior, heat transport, and sorbent reactivation, as well as the system configuration, e.g. reactor type, heating type, combustion type, unit integration, solid transportation device, and gas–solid contact scheme. This chapter describes the concept of calcium looping carbon capture process and its thermodynamic and kinetic properties primarily. The current status of this technology around the world is introduced. Finally, strategies for further improving the process performance are revealed.

13.1.1 Fundamental Principles of Calcium Looping Process

The calcium looping process is a carbon capture technology using calcium-based sorbents cyclically transported between a carbonator and a calciner for in situ

Handbook of Chemical Looping Technology, First Edition. Edited by Ronald W. Breault.

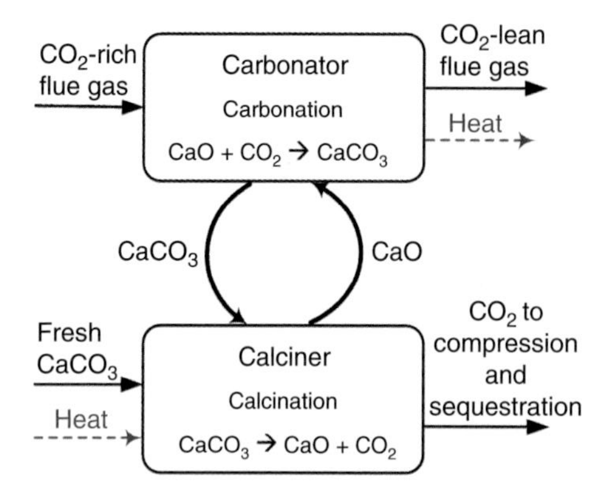

Figure 13.1 Schematic diagram of calcium looping process.

removal of CO_2 from the flue gas exhausted from a combustion or industrial process, as shown in Figure 13.1. The fundamental concept of this carbon capture process is the employment of carbonation reaction of calcium oxide (CaO) with gaseous CO_2 at high temperature in the carbonator, yielding the reaction product of calcium carbonate ($CaCO_3$). The carbonated sorbent ($CaCO_3$) is transferred to the calciner for sorbent regeneration. The concept of this technique was first applied in the sorption-enhanced hydrogen production [1]. The author utilized CaO + MgO and $CaCO_3$ + MgO to enhance the performance of coal gasification with steam. Han and Harrison [2] applied this concept in a CO_2 acceptor gasification process consisting of an interconnected fluidized bed coal gasifier and a combustor. The carbonation reaction in this process produced a H_2-rich stream during gasification of lignite coal. However, this process was not considered as a CO_2 capture technology until CO_2 was credited by some as being responsible for global warming and climate change issues. Shimizu et al. [3] initiated application of the carbonation–calcination process into CO_2 mitigation from flue gas. The heat demand of calcium-based sorbent regeneration was provided by coal combustion with pure oxygen; therefore, CO_2 concentration from the calciner was higher than 95%. Thereafter, numerous researchers have gradually dedicated their studies to develop the cyclic carbonation–calcination technology in capturing CO_2 from flue gas.

CaO derived from natural limestone is considered as a viable sorbent candidate because of its nontoxic, cost-effective, and high CO_2 adsorption capacity characteristics. The reaction between CaO and CO_2 occurs immediately during carbonation, and the carbonation reaction of CaO could be divided into two stages depending on the reaction rate of CaO carbonation. At the beginning of the carbonation reaction, the reaction of CO_2 with CaO is carried out under a fast chemically reaction-limited condition (kinetic-controlled reaction stage). Most CaO existing on the surface of the calcium-based sorbents is carbonated to form $CaCO_3$ in this stage. In the second stage, the carbonation rate is decreased severely because the carbonation reaction is gradually limited by the mass transfer of CO_2 diffused through the carbonation layer on the sorbent surface (diffusion-controlled reaction stage). Thereafter, the CO_2 molecules are no longer diffused through the carbonation layer for the carbonation of CaO inside the calcium-based sorbent because severe diffusion resistance is caused by the formation of a dense carbonation layer over a critical value. For any commercial applications, only the fast chemically reaction-limited region could be considered in order to design a compact reactor for CO_2 capture.

One of technical barriers for applying calcium looping process into a real industrial process is sorbent decay with carbonation–calcination cycles. Limestone-derived CaO loses around 15–20% of CO_2 capture capacity for each cycle because of the loss of sorbent surface area and the inaccessibility of internal pore volume caused by sorbent sintering during calcination [4–6]. Most calcium-based sorbents retained a carbonation conversion of approximately 20% after 20 continuous carbonation–calcination cycles, as shown in Figure 13.2 [5]. Such a sorbent decay requires a replenishment of fresh limestone to maintain an adequate level of CO_2 carrying capacity of sorbent in the carbonator. The purge of surplus inactive sorbent is necessary to handle the solid circulation through the whole system harmoniously. Moreover, the decay in conversion is likely only dependent on the number of cycles rather than the reaction conditions and reaction times [7]. The conversion of lime ($x_{c,N}$) as a function of cycle number (N) was proposed as

$$x_{c,N} = f^{N+1} + b \tag{13.1}$$

The constants $f = 0.782$ and $b = 0.174$ were determined by regressing experimental data for a multicycle carbonation–calcination experiment. A design equation to calculate the maximum carbon capture efficiency (E_{CO_2}) by cyclic carbonation/calcination process containing a continuous sorbent purge and a make-up of fresh sorbent was developed:

$$E_{CO_2} = \frac{1 + \left(F_0/F_R\right)}{\left(F_0/F_R\right) + \left(F_{CO_2}/F_R\right)} \left[\frac{f\left(F_0/F_R\right)}{\left(F_0/F_R\right) + 1 - f} + b \right] \tag{13.2}$$

where F_0 (mol s^{-1}) and F_{CO_2} (mol s^{-1}) are the flow of fresh feed sorbent and CO_2, respectively, entering the system; F_R (mol s^{-1}) is the total amount of sorbent required to react with the CO_2 in the system. The increases of the ratios of

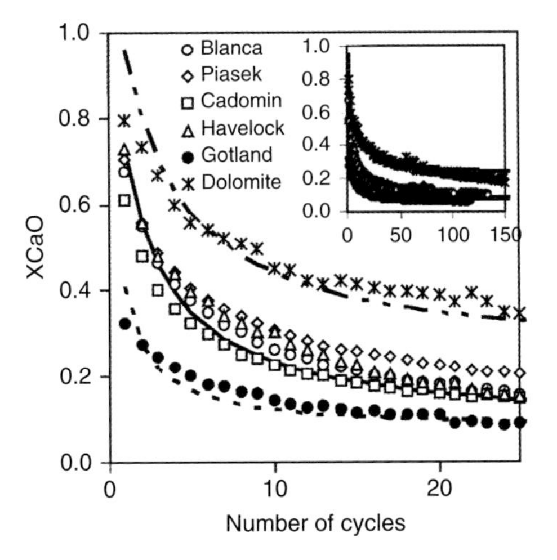

Figure 13.2 Comparisons of carbonation conversion over multiple carbonation–calcination cycles for experiments conducted using different types of limestones.

F_R/F_{CO_2} and F_0/F_R enhance the maximum carbon capture efficiency. However, more energy for sorbent regeneration in the calciner is also required in order to heat up the solids coming from the carbonator and the vessel containing fresh limestone.

The characteristics of calcium looping system include the design of the carbonator, calciner, elutriation, and their interconnections, to capture CO_2 through high-temperature limestone-derived sorbents. The reactivity of the sorbent depends on a variety of factors, including the type of combustion (direct or indirect), mode of reactor, type of limestone, reaction temperature, and gas atmosphere within the reactors. The use of reactors with homogeneous flow patterns of solid–gas two-phase flows, e.g. a fluidized bed reactor or an entrained bed reactor, is the most developed solution to carry out the carbonation and calcination reactions. Numerous industrial applications have demonstrated the significant advantage of utilizing fluidized bed reactors in a solid–gas reaction. The majority of research in developing the calcium looping process, therefore, utilizes circulating fluidized beds (CFBs) as carbonator or both as carbonator and calciner. The reactor dimensions for achieving an appropriate degree of carbonation and calcination conversions are designed on the basis of both reaction chemistries and hydrodynamic phenomena inside the reactor. The stoichiometry and reaction rate of carbonation and calcination reactions and the maximum allowable gas velocity through the vessel govern the diameter of a fluidized bed reactor. The gas–solid flow pattern, carbonation and calcination rates, and freeboard height for suppressing solid entrainment determine the height of the reactor [8].

The major energy consumption of a calcium looping system comes from heat demand in the calciner. The heat demand corresponds to the endothermic reaction responsible for sorbent regeneration and fresh limestone calcination, and the sensible heat responsible for heating the sorbent transported from the carbonator. The supply and handling of the large flow of heat in the calcination system is a challenging engineering task. The solution driving the calcination reaction is commonly a calciner integrated with an indirect- or direct-fired combustor. The heat supplied form the indirect-fired combustor does not directly contact the sorbent in the calciner, allowing the generation of a pure CO_2 stream after sorbent regeneration in this configuration. However, application of the indirect firing calcination in a calcium looping system is rare because of lower heat utilization efficiency. In contrast to the indirect-fired combustion, the heat is directly brought to bear on sorbents in a direct-fired calcination system. The direct-fired hydrogen or oxy-fuel combustions result in a highly pure CO_2 stream that can be separated from the combustion stream after steam condensation.

13.1.2 Thermodynamics and Reaction Equilibrium of CaO and $CaCO_3$

In a calcium looping process, flue gas exhausted from an industrial process is conveyed to the carbonator and contacts with calcium oxide to carry out the carbonation reaction, producing calcium carbonate and a CO_2-lean stream. Undesired reactions, e.g. hydration, sulfidation, and chlorination, may take place when the flue gas contains H_2O, H_2S, COS, SO_2, and HCl. The carbonated sorbent is then regenerated in the calciner, while highly pure CO_2 is produced under direct-fired oxy-fuel or indirect-fired calcination. The carbonation and calcination reactions are thermodynamically spontaneous and reversible. The levels of these reactions are determined on the basis of the equilibrium curve of thermodynamic analysis for the partial pressure of CO_2 as a function of temperature. As the thermodynamic data shown in Figure 13.3, the curve represents the equilibrium between CaO and $CaCO_3$ relative to the CO_2 partial pressure shown in the left axis. If the CO_2 partial pressure contained in a flue gas is higher than the equilibrium CO_2 partial pressure corresponding to an operating temperature, the condition of carbonation of CaO would occur. The calcination of CaO is relatively altered when the CO_2 partial pressure is lower than the equilibrium CO_2 partial pressure. For example, the carbonation reaction takes place at temperatures of 650 °C with CO_2 partial pressures higher than 0.01 atm, and the calcination reaction is observed at temperatures higher than 890 °C even with the CO_2 partial pressure approaching 1 atm. Moreover, the thermodynamic equilibrium curves for hydration, sulfidation, and chlorination of CaO also predict their reaction temperature zones under specific H_2O, H_2S, COS, SO_2, and HCl concentrations. Therefore, controlling the temperature inside the carbonator is possible to avoid these undesired reactions when the flue gas contains H_2O, H_2S, COS, SO_2, and HCl.

The calcium looping process uses the reversible reaction between ($CaO + CO_2$) and $CaCO_3$ to achieve carbon capture and sorbent regeneration. Exothermic heat is liberated from the carbonation reaction. The liberated heat can be used to generate steam or/and electricity to reduce the energy penalty of the CO_2 capture system [9, 10]. In contrast, additional energy is required to drive the endothermic calcination reaction. The most developed solution for the heat demand in the sorbent regeneration is oxy-fuel combustion although an extra energy penalty is associated with the air separation unit (ASU) for oxygen production. The niche of applying the oxy-fuel combustion is to avoid the dilution of the resultant gas with nitrogen in air; thus, a highly pure CO_2 stream is subsequently produced from the calcination process. However, one of the critical challenges in developing and scaling up a carbon capture process is the high energy requirements for sorbent or solvent regeneration. The energy

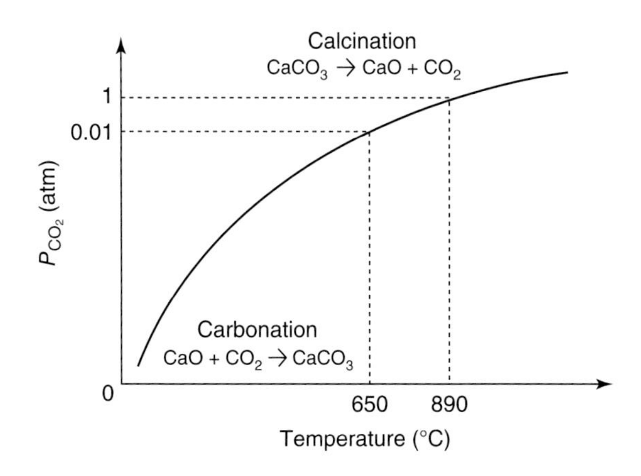

Figure 13.3 The thermodynamic curve for the carbonation of CaO.

requirements may lead to a significant and variable increase in the levelized cost of electricity when the carbon capture process is associated with a power plant. By comparison with the amine absorption process, the calcium looping process provides a possible alternative that is compatible with the heat recovery facility because of high-temperature streams produced from this system. The sources of recoverable energy from a calcium looping process include the exothermic carbonation reaction (>650 °C), the CO_2-lean stream from the carbonator (>650 °C), the purge of surplus inactive sorbent from the carbonator (>650 °C), and the concentrated CO_2 stream from the calciner (>900 °C). A large quantity of sensible energy is recovered through cooling the high-temperature streams to temperatures that depend on the restraints of the equipment valves, e.g. gas purification or CO_2 compression unit. The recovered heat can be used to retrofit a power plant or to drive a steam cycle for increasing the gross electricity and diminishing the energy penalty. Every 10% decrease in thermal extraction efficiency may increase the energy penalty between 3% and 5% for indirect-fired, natural gas-fired, and coal-fired calciner. Moreover, the energy penalty of the calcium looping process is significantly associated with its carbon capture efficiency because the higher flow rates of circulating sorbent and fresh limestone keep a higher level of carbon capture capacity as well as increase the heat requirement for sorbent regeneration [10]. The energy required for compressing the huge volume of captured CO_2 has also a significant impact on the overall energy penalty [11]. Therefore, enhancing the thermal extraction efficiency of this system and reducing the CO_2 compression energy diminish the energy penalty of the carbon capture process and shorten its commercial progress.

13.2 Current Status of Calcium Looping Process

The calcium looping process using limestone derived CaO has emerged recently as a potentially economically advantageous technology to achieve sustainable CO_2 capture efficiencies. Certain aspects of this process in laboratory-scale experiments have been conducted to investigate the thermodynamic, kinetic, reaction mechanism, sorbent properties, and technical feasibility of this process. The calcium looping technology is progressing quickly from laboratory-scale tests toward bench-scale and pilot-scale demonstrations. At the bench-scale and pilot-scale facilities, all industrial constraints and engineering tasks have to be taken into account, e.g. impurities in flue gas and raw sorbents, sorbent attrition phenomena, high-temperature solid transportation, heating type, heat integration, equipment reliability, and operation over long periods. The developed kilowatt-scale and megawatt-scale plants of calcium looping technology are summarized in Table 13.1.

13.2.1 Kilowatt-Scale Calcium Looping Facility

The kilowatt-scale calcium looping systems include the 3 kW_{th} unit at Industrial Technology Research Institute (ITRI), the 10 and 200 kW_{th} units at IFK (University of Stuttgart), the 20 kW_{th} unit at Technical University of Darmstadt,

Table 13.1 The facilities of calcium looping technology around the world.

Organization	Capacity	Type (Carbonator/calciner/interlinking)	References
Kilowatt-scale facility			
ITRI (Taiwan)	3 kW_{th}	Bubbling fluidized bed/rotary kiln/pneumatic transport	[12, 13]
IFK, University of Stuggart (Germany)	10 kW_{th}	Bubbling fluidized bed/riser/cone valve-integrated double exit loop seal	[14]
	200 kW_{th}	Turbulent fluidized bed/circulating fluidized bed/loop seal with L-valve	[15–17]
INCAR-CSIC (Spain)	30 kW_{th}	Circulating fluidized bed/circulating fluidized bed/loop Seals	[18, 19]
	300 kW_{th}	Circulating fluidized bed/circulating fluidized bed/loop seals	[20, 21]
CANMET Energy (Canada)	75 kW_{th}	Circulating fluidized bed/two stage fluidized bed/pneumatic transport with L-valve	[22]
Ohio State University (USA)	120 kW_{th}	Entrained bed/rotary kiln/volumetric hopper	[23]
Megawatt-scale facility			
Technical University of Darmstadt (Germany)	1 MW_{th}	Circulating fluidized bed/circulating fluidized bed/loop seal with screw conveyor	[24–26]
INCAR-CSIC (Spain)	1.7 MW_{th}	Circulating fluidized bed/circulating fluidized bed/loop seals	[27–31]
ITRI (Taiwan)	1.9 MW_{th}	Bubbling fluidized bed/rotary kiln/pneumatic transport	[12, 13]

the 30 and 300 kW_{th} units at Instituto Naclonal del Carbon Consejo Superior de Investigaciones Cientificas (INCAR-CSIC), the 75 kW_{th} unit at CANMET Energy, and the 120 kW_{th} unit at Ohio State University [12–23]. The system designed by ITRI consists of a bubbling fluidized bed (BFB) carbonator and a rotary kiln calciner connected with a pneumatic pipe transport system (as shown in Figure 13.4) [12]. A direct-liquefied petroleum gas-fired combustor with an integrated flue gas recirculation system provides the heat for sorbent regeneration in the calciner. The configuration of this calcination system offers more uniform temperature distribution through the calciner and thus enhances the heat utilization efficiency of this system. Moreover, CO_2 capture levels higher than 85% were achieved in a 100-h continuous operation. The rotary kiln, as commonly used in cement process for clinker production, is also used as the calciner in the 120 kW_{th} facility developed by The Ohio State University (OSU) (as shown in Figure 13.5). The OSU system is an outgrowth of two other processes developed at OSU: the Ohio State Carbonation Ash Reactivation (OSCAR) process and the Calcium-based Reaction Separation for CO_2 ($CaRS-CO_2$) process. Therefore, sulfur, trace heavy metal (arsenic, selenium, and mercury), SO_2, and CO_2 can

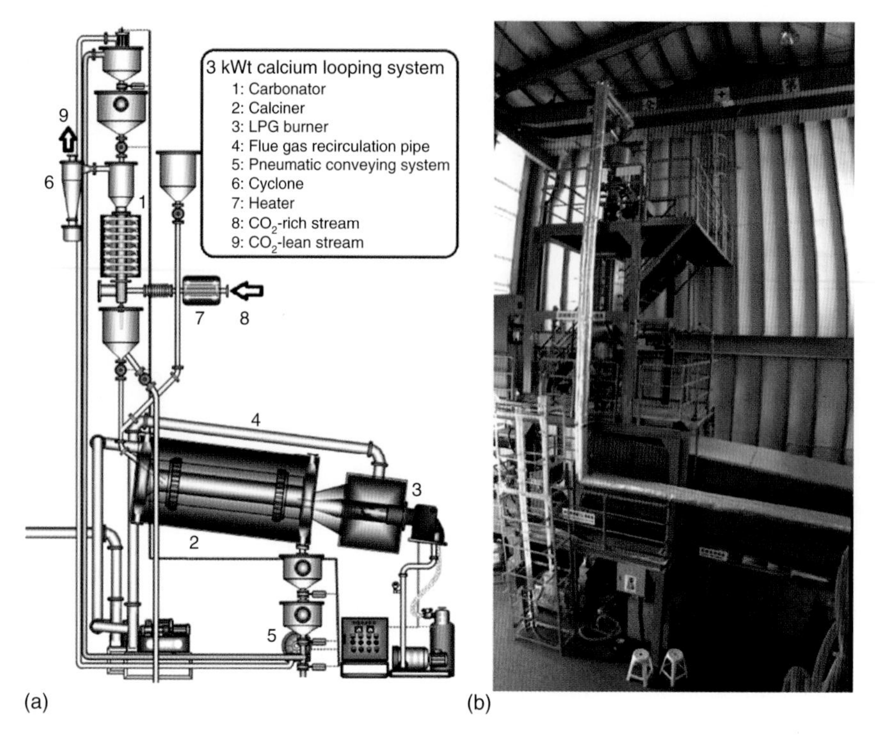

Figure 13.4 (a) Schematic and (b) view of 3 kW_{th} bench-scale system at ITRI.

Figure 13.5 Photo of 120 kW –calcination reaction (CCR) process facility for CO_2 and SO_2 capture at Ohio State University initiated in 2005 with coal-fired stoker, rotary calciner, and entrained bed carbonator shown and with the same facility used also for Carbonation–Calcium–Hydration Cyclic Process operation. Source: With permission from Professor L.-S. Fan at Ohio State University.

be simultaneously captured by calcium-based sorbents. An additional hydration process integrated in the OSU carbon capture system demonstrated that the sorbent activity could be maintained over multiple cycles. Fly ash in the carbonator did not affect the carbonation reaction nor did sulfation reaction in the entrained bed carbonator. Over 90% CO_2 and near 100% SO_2 could be captured on a once-through basis at a Ca:C molar ratio of 1.3 [23].

Dual fluidized bed (DFB) calcium looping system has been developed by numerous research groups [14–22]. The Institute of Combustion and Power Plant Technology (IFK) at the University of Stuttgart has developed a 10 kW_{th} system that consists of a riser calciner and a BFB carbonator [14]. The circulating rate of calcium-based sorbents between the beds is controlled by a cone valve-integrated loop seal. The carbonator temperature window of 600–660 °C was suitable to achieve CO_2 capture efficiencies higher than 90% and to drive a state-of-the-art steam cycle. CO_2 capture efficiency was enhanced by increasing Ca:C molar ratio in the carbonator for a given CO_2 molar flow. Ca:C molar ratios higher than 14 led to the maximum CO_2 capture efficiency. Inspired by these results [14], a 200 kW_{th} calcium looping facility has been constructed for the investigation of long-term performance of this process under real combustion conditions (as shown in Figure 13.6) [15–17]. This system includes two operational configurations: (i) CFB carbonator and CFB calciner; (ii) turbulent fluidized bed carbonator and CFB calciner. Two cone valves and the combination of an L-valve and a loop seal are respectively used for controlling the sorbent circulating rate in the systems. In the turbulent fluidized bed carbonator, the heat from the exothermic reaction is removed using a heat exchanger; the heat released in the CFB carbonator is removed via a water cooled heat exchanger in the dense bed region and a bayonet cooler in the lean bed region. The heat demand in the CFB calciner is provided from an oxy-fuel-fired combustor with flue gas recycle. CO_2 capture efficiencies higher than 90% were achieved consistently over a broad range of process conditions during multiple hours of operation. This system has passed more than 600 hours for operating in CO_2 capture and more than 300 hours for stable oxy-fuel calcination with flue gas recycle in the calciner [16, 17].

A 30 kW_{th} system consisting of DFB reactors has been developed by the INCAR-CSIC [18, 19]. The calciner and the carbonator are connected to high-efficiency primary cyclones, respectively. Sorbents collected from the cyclones are transported to the opposite reactor through BFB loop seals. Electric ovens surround the first 2.5 m of the risers and also the loop seals. The calcium carbonate formed in the carbonator was regenerated in the calciner integrating with a direct-fired combustor under typical conditions (800–900 °C). The insufficient separation efficiency of the cyclones and the intense sorbent attrition in the first calcinations led to unstable operation of the system because the solid looping rates and the solids inventory could not be kept constant for a long period of time. The operating stability of this system was improved through the extension of the riser heights and reconstruction of the high-efficiency cyclones. Moreover, another 300 kW_{th} calcium looping facility following the design of 30 kW_{th} system has been constructed in La Robla, property of Gas Natural Fenosa, to capture CO_2 "in situ" during the combustion of biomass in a fluidized

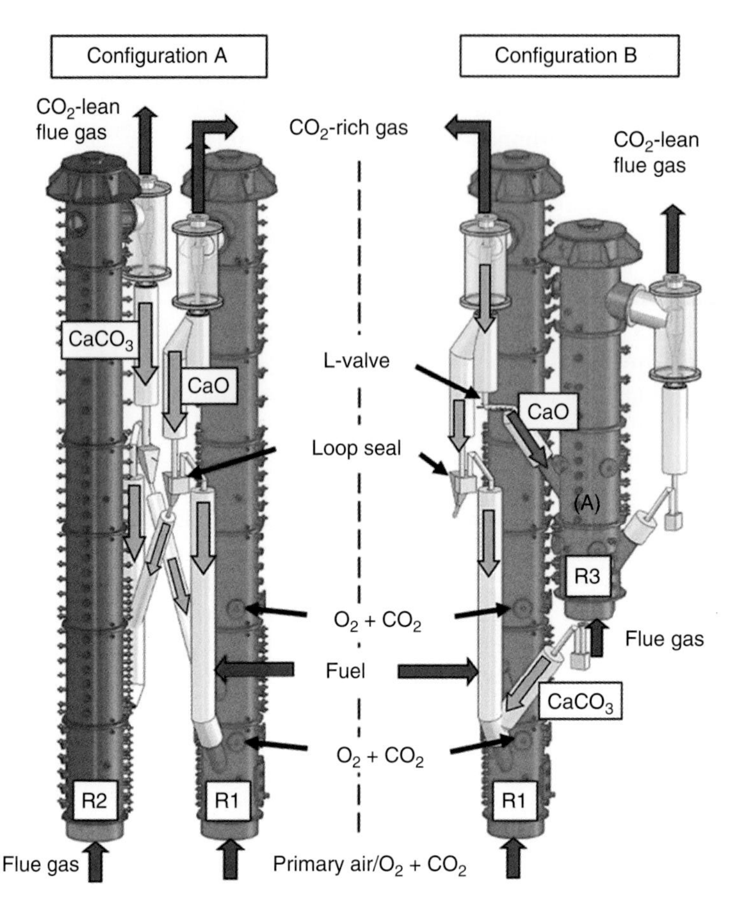

Figure 13.6 200 kW_{th} calcium looping system at University of Stuttgart. Source: Dieter et al. 2014 [17]. Reproduced with permission of Elsevier.

bed (as shown in Figure 13.7) [20, 21]. Continuous and stable operation was achieved in the facility for 360 hours. The stable CO_2 capture efficiency (65–85%) demonstrated the technical and operational feasibility of the carbon capture concept presented in their work.

Another DFB calcium looping system in the scale of 75 kW_{th} has been constructed by CANMET Energy (as shown in Figure 13.8) [22]. The bottom of the reactors is surrounded with electric heat (15 kW) that provides supplemental heating during warm-up operation. A solids transport system assembled at the bottom of the calciner is used to transport regenerated sorbent to the carbonator. The sorbent flow is controlled by a solenoid valve activated by a computer and is lifted by conveying air to a cyclone equipped at a height of 6 m. The solid sorbent collected from the cyclone is then introduced into the carbonation bed through an L-valve. A 45° "T" line assembled at the bottom of the carbonator allows the solids to return back to the calciner or back to the bottom stage combustor for SO_2 capture (if necessary, depending on the fuel used). This facility has been operated for more than 50 hours. CO_2 capture levels higher than 95%

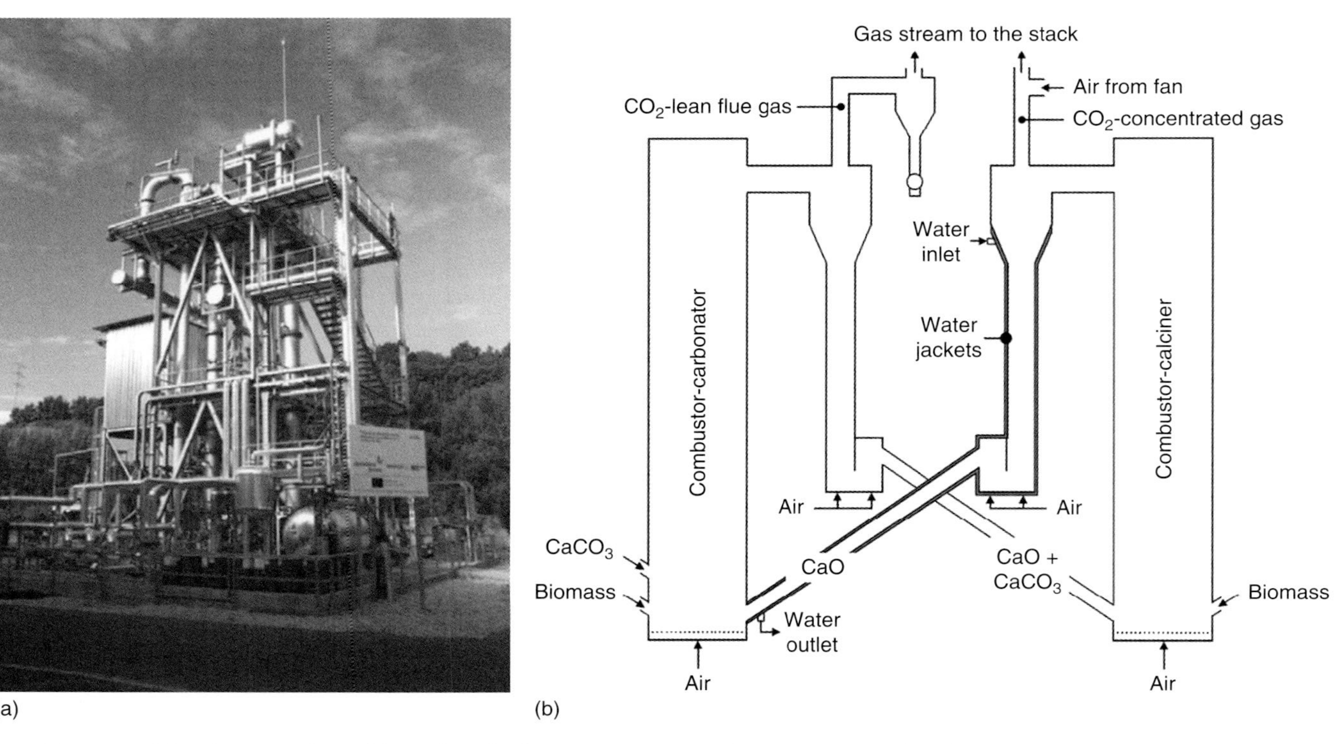

Figure 13.7 (a) View and (b) schematic of La Robla 300 kW_{th} facility. Source: Diego et al. 2016 [21]. Reproduced with permission of Elsevier.

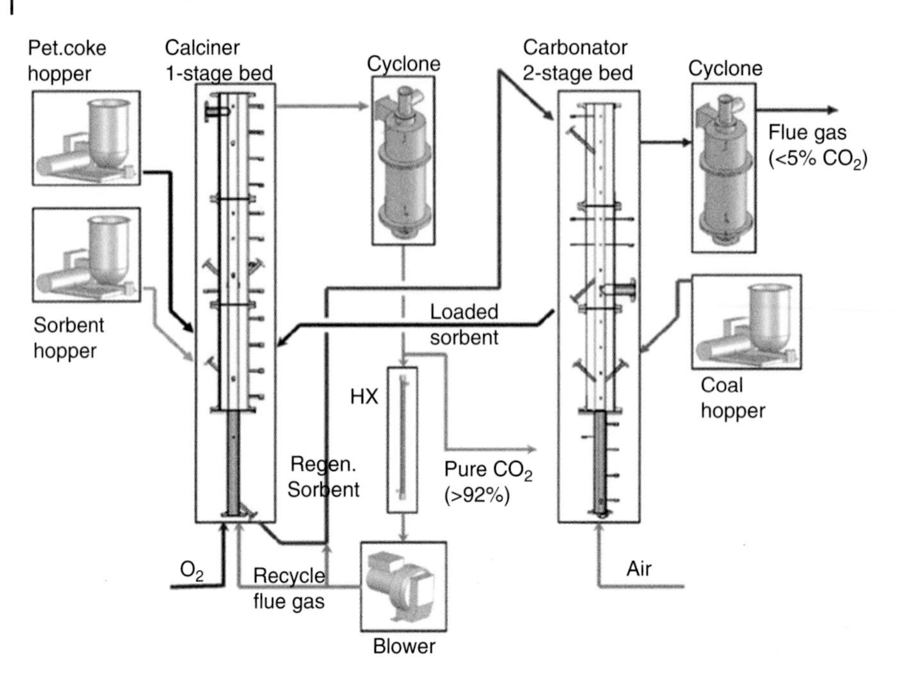

Figure 13.8 75 kW_{th} calcium looping system developed by CANMET Energy. Source: Lu et al. 2008 [22]. Reproduced with permission of Elsevier.

were achieved within the temperature window of 580–600 °C for the first several cycles, which decreased to a lower level (>72%) after more than 25 cycles due to sorbent deactivation. CO_2 capture efficiency decreased with increase in superficial gas velocity in the carbonator. Moreover, oxy-fuel combustion of biomass and coal with flue gas recycle was appropriate for sorbent regeneration in the calciner. The CO_2 concentration from the calciner off-gas of ~85 vol.% was achieved, and even higher levels are expected as the unit design is improved to minimize reactor leakage.

13.2.2 Megawatt-Scale Calcium Looping Plant

Encouraging results from the kilowatt-scale experiments provided a strong motivation to demonstrate the calcium looping process under realistic conditions at a megawatt pilot plant. Several projects that address their studies on developing pilot-scale processes have been undertaken in recent years. The pilot-scale systems include the 1 MW_{th} facility at Technical University (TU) of Darmstadt (Germany), the 1.7 MW_{th} facility at La Pereda (Spain), and the 1.9 MW_{th} facility at Heping (Taiwan). The 1 MW_{th} TU Darmstadt pilot plant consisting of two interconnected CFB reactors (as shown in Figure 13.9) has been operated for more than 1500 hours in fluidized bed mode and more than 400 hours in CO_2 capture mode [24–26]. A synthetic flue gas, which is heated up to 350 °C using auxiliary electric heaters, is used as a fluidizing medium in the carbonator to perform experiments. An adjustable, internal heat-removing system is assembled in the carbonator to remove the heat generated from exothermic carbonation reaction

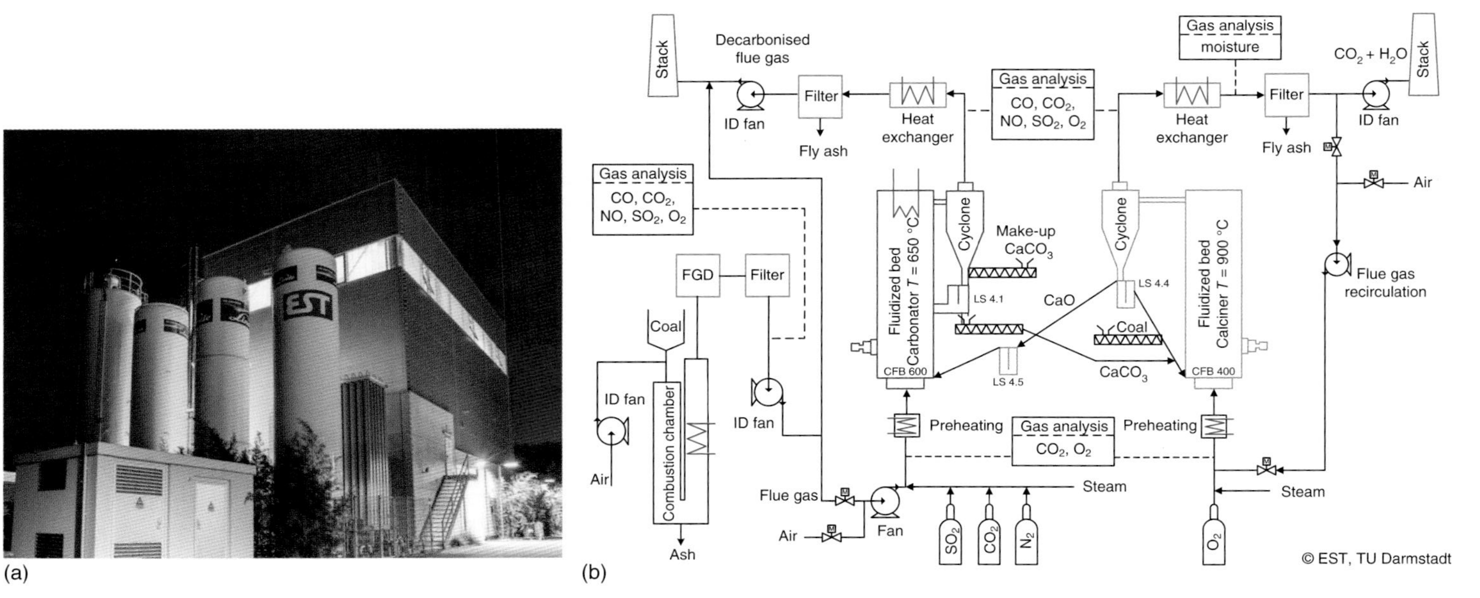

Figure 13.9 (a) View and (b) schematic of 1 MW_{th} TU Darmstadt pilot plant. Source: With permission from Dr Martin Helbig at Technische Universität Darmstadt.

and to control the desired temperature in the carbonator. The heat demand for the endothermic calcination reaction in the calciner is provided by either propane or coal combustion. The whole system is refractory lined to minimize heat lost. Make-up limestone is fed into the carbonator by a gravimetric feed system. The solid circulating rate between the loop seals of these two reactors is controlled by screw conveyors. In the case of propane-fired mode, the CO_2 capture efficiency was higher than 80% until the reduction of propane vaporization capacity caused by ambient restriction (the temperature in Darmstadt was about −15 °C) because the limited propane vaporization capacity could not provide sufficient heat for complete sorbent regeneration. A CO_2 capture efficiency of 83–91% could be obtained when this system was operated in coal-fired mode. Once the temperature in the carbonator was decreased from 650 to 610 °C, the CO_2 capture efficiency dropped to about 60% [24]. The long-term effects of fuel, sorbent, flue gas composition, reactor design, and operating conditions on the carbon capture performance have been scheduled in their further experiments [26]. Moreover, a 20 MW_{th} pilot plant is expected to be designed to bring the process even closer to industrial application under the European Union 7th Framework Programme (EU-FP7) project SCARLET.

Another project (CaOling) founded by the EU-FP7 has been started in December 2009 as a three-year project to promote the progress toward the industrial application of Ca-looping technology before 2020. Consejo Superior de Investigaciones Cientificas (CSIC), Empresa Nacional de Electricidad Sociedad Anónima (ENDESA), Foster Wheeler, and Hulleras del Norte Sociedad Anónima (HUNOSA) preliminary reached an agreement in 2009, and this consortium was substantially reinforced by leading R&D supports from IFK of the University of Stuttgart in Germany, Lappeenranta University in Finland, Imperial College in the UK, and the University of Ottawa and CANMET Energy in Canada. A 1.7 MW_{th} plant has been subsequently designed to process approximately 1% of the flue gas produced from a 50 MW_e La Pereda CFB power plant in Asturias, Spain (as shown in Figure 13.10) [27–31]. The core of the facility consists of two CFB reactors, 15 m in height with the diameters of 0.75 m in the calciner and 0.65 m in the carbonator. The typical operating temperatures for the carbonator and calciner are 600–715 °C and 820–950 °C, respectively. The heat demand in the calciner comes from coal-fired air combustion or from coal-fired oxy-fuel combustion. The carbonator is assembled with removable cooling bayonet tubes to control the carbonator temperature. Typical gas velocities inside these two reactors are designed in the range of 2–6 $m\,s^{-1}$. Each reactor is equipped with a high efficiency cyclone (cut size 5 μm) and a loop seal that can collect and deliver the collected solids to the same reactor (internal circulation) or to the opposite reactor [27]. Sánchez-Biezma et al. [28] reported that a total of 800 hours of operation with coal combustion in the calciner has been achieved, including approximately 160 hours in carbon capture experiments. The CO_2 capture efficiency was close to the maximum allowed by the equilibrium at 660 °C, resulting in approximately 90% carbon capture efficiency as well as SO_2 capture efficiencies higher than 95%. The CO_2 concentration detected at the exit of the calciner equipped with an oxy-fuel combustor was around 85%. Moreover, the operation of this pilot plant in combustion mode has been extended for more

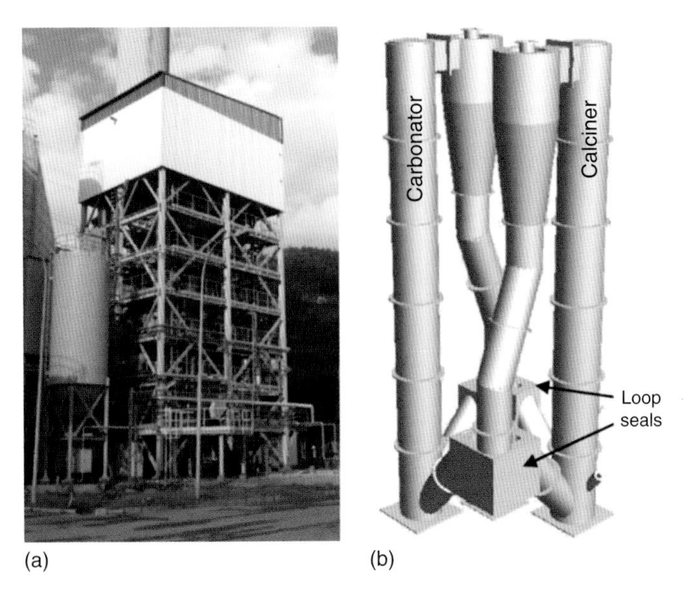

Figure 13.10 (a) View and (b) schematic of La Pereda 1.7 MW_{th} facility. Source: Diego et al. 2016 [31]. Reproduced with permission of Elsevier.

than 1800 hours, including 380 hours in CO_2 capture mode with a CO_2 capture efficiency of 40–95%. Another 170 hours operation has also been achieved in CO_2 capture mode with stable operation under oxy-fuel combustion of coal [29]. The CO_2 capture efficiencies were found to be lower than 40% at the beginning of the experiment due to the high content of non-calcined sorbent in the solid inventory when the calcination temperatures were lower than 920 °C. As the CaO content in the solid inventory increased, the carbon capture efficiency rose to a value close to that limited by thermodynamic equilibrium. The increase of calcination temperature, however, tended to cause a decrease in carbon capture level because the higher temperature sorbent arriving to the carbonator induced the increase in equilibrium CO_2 partial pressure in flue gas. The removable cooling bayonet tubes could be used to stabilize the temperature toward the target carbonation temperature. Moreover, the replenishment of fresh limestone and the extraction of solids were necessary to maintain the sorbent activity in the inventory of solids. Moreover, a novel reactivation process based on the recarbonation principle has been integrated into the 1.7 MW_{th} system to enhance the CO_2 carrying capacity of sorbent [31]. An existing loop seal was retrofitted to the recarbonator for further carbonating the solid stream coming from the carbonator. More than 100 hours of operation have been accumulated in the recarbonation experiments. The introduction of the recarbonation step led to an improvement in the CO_2 carrying capacity of the sorbent. The recarbonation reaction was under the preferred operating conditions of the solid residence time in the range of one to three minutes, temperatures inside this reactor between 750 and 800 °C, and CO_2 inlet concentration of 50–70 vol%. The CO_2 carrying capacity of sorbent increased by up to 10% under reasonable and scalable recarbonation conditions.

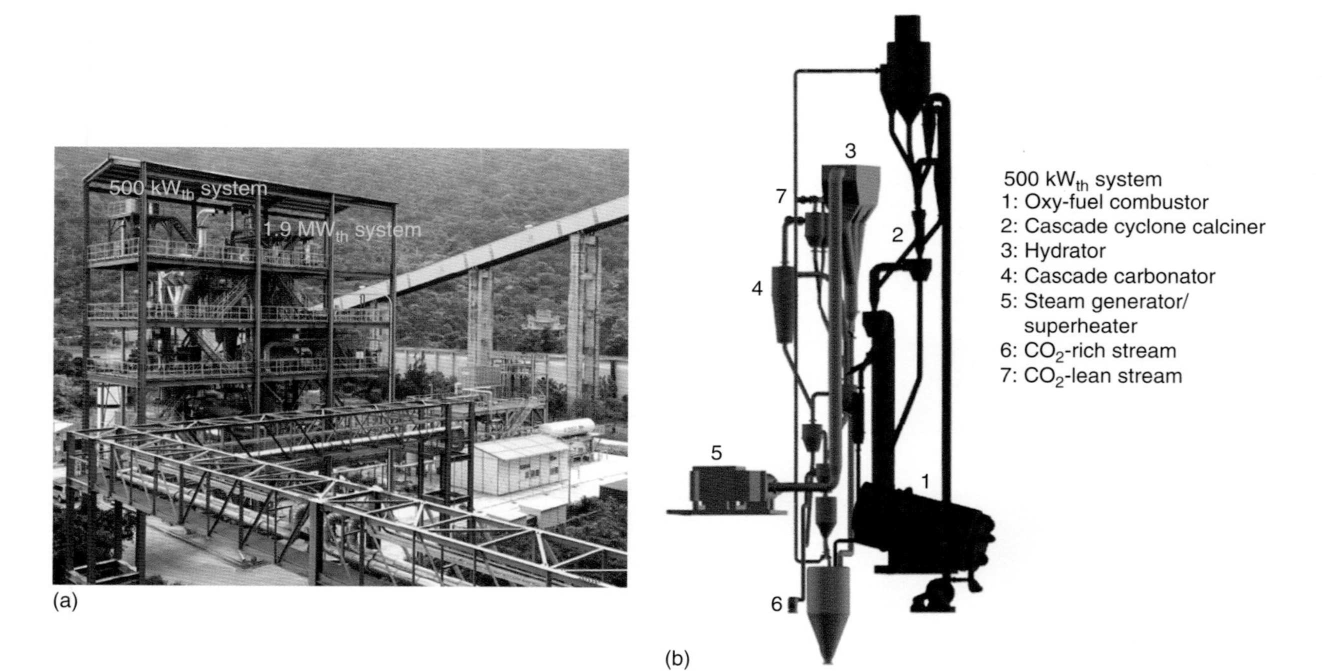

Figure 13.11 (a) View of 1.9 MW$_{th}$ and 500 kW$_{th}$ facilities; (b) schematic of 500 kW$_{th}$ cascade cyclone system.

The calcium looping process is noted to have potential for integration into the cement manufacturing industry because the inactive sorbent from the system can be used as the raw material for the cement plant, and the high-grade "waste heat" extracted from the carbon capture system can be utilized to preheat coal, air, limestone, and other materials used in the cement process. Therefore, ITRI, in cooperation with the Taiwan Cement Company (TCC), has erected a 1.9 MW_{th} calcium looping pilot plant at TCC's cement plant in Heping, Taiwan (as shown in Figure 13.11a). The pilot plant was designed to capture 1 tonne of CO_2 per hour from the cement plant flue gas containing 20–25% CO_2. The configuration of this system is based on experience with their 3 kW_{th} unit. The BFB carbonator has a diameter of 3.3 m and a height of 4.2 m, and a double-layered perforated plate is assembled in the carbonator. The rotary kiln calciner inclines at an angle 5° to the horizontal and has a length of 5 m and a diameter of 0.9 m. A diesel-fired oxy-combustor, in 1–4 MMBtu/h capacity, equipped with a flue gas circulation system provides the heat for sorbent regeneration and to produce a near pure stream of CO_2. The regenerated sorbent from the calciner is transported to the carbonator through a pneumatic conveying system. Until now, this system has passed more than 1000 hours for unit operations and another 700 hours for continuous experiment in CO_2 capture [13]. Approximately 50% CO_2 capture efficiency was obtained for experiments conducted using the sorbent with the calcination efficiency of 30–40%. Enhancement of the sorbent calcination efficiency led to carbon capture levels higher than 85%. In recent years, ITRI has focused attention on improving the performance of this technology by a combined cascade cyclone and steam hydration process. The reactors in the configuration of cascade cyclone contribute greatly to promote the solid–gas contact pattern corresponding to improving heat utilization in the calciner and to enhance reaction conversion in the carbonator. Steam hydration is used to improve sorbent reactivity in the carbonation reaction. A hydration-integrated cascade cyclone calcium looping system in the scale of 500 kW_{th} has been erected by the side of the original 1.9 MW_{th} facility (as shown in Figure 13.11). The trial operations of the 500 kW_{th} facility have been accomplished. Experimental results indicated that higher heat utilization efficiency was achieved in the calciner, and the temperature gradient through the calciner was less than 200 °C under appropriate operating conditions. The calcination efficiency higher than 80% could be realized for experiments reasonably controlling sorbent transportation and energy supply conditions. Steam hydration significantly promoted the sorbent activity in carbon capture. 91.37% CO_2 could be captured in the cascade cyclone carbonator. This project is focused on the long-term continuous operation of the 500 kW_{th} system in 2018. The operating experience will be applied to the design of a 10 MW_{th} system in the near future.

13.3 Strategies for Enhancing Sorbent Recyclability and Activity

Calcium looping processes are attractive in carbon capture applications because they can achieve in situ combustion or post-combustion CO_2 removal with inherent high-quality heat generation. A number of these processes at the pilot scale have been built to demonstrate their feasibility to be suitably applied in large-scale facilities. One of the key challenges for operating a calcium looping process is the decrease of carbon capture capacity with carbonation–calcination cycles because limestone-derived sorbent reveals a rapid decay in its CO_2 uptake with cycle number. The lower sorbent capacity and reactivity lead to higher solid circulation rates as well as a larger reaction system, resulting in higher operating and capital costs. Therefore, the ability of the sorbent to maintain stable reactivity in cyclic reactions is one of critical criteria for improving the process performance and shortening its commercial progress.

The decay of calcium-based sorbents in their recyclability and reactivity is mainly attributed to the deteriorating effects coming from the variation of sorbent macrostructural and microstructural properties, e.g. surface area, pore volume, particle size, and other physical factors, at high calcination temperatures [32]. Abanades and Alvarez [33] explained the sorbent decay in terms of a fractional loss of microporosity along with a cumulative gain in mesoporosity. Macropores are generated instead of micropores, and the macropores on the surface of CaO are filled by the carbonated product formed during each cycle of carbonation. The residual CaO conversion as a function of cycle number could be expressed by a semiempirical equation [5]. In a subsequent study [34], a textural transformation mechanism and another empirical parameter denoted as the sintering exponent were proposed to explain their experimental adsorption data. The qualitative mechanism representing the formation of an interconnected calcium oxide network (skeleton) over repeat carbonation–calcination cycles is shown in Figure 13.12. In this mechanism, the first calcination of calcium carbonate produces the highly dispersed reactive calcium oxide, followed by incomplete recarbonation in the next step. The grains of calcium oxide asymptotically grow, agglomerate, and ultimately form a rigid interconnected calcium oxide skeleton in the following cycles. The final texture of the calcium oxide skeleton prevents further sintering of the adsorbent and retains a residual

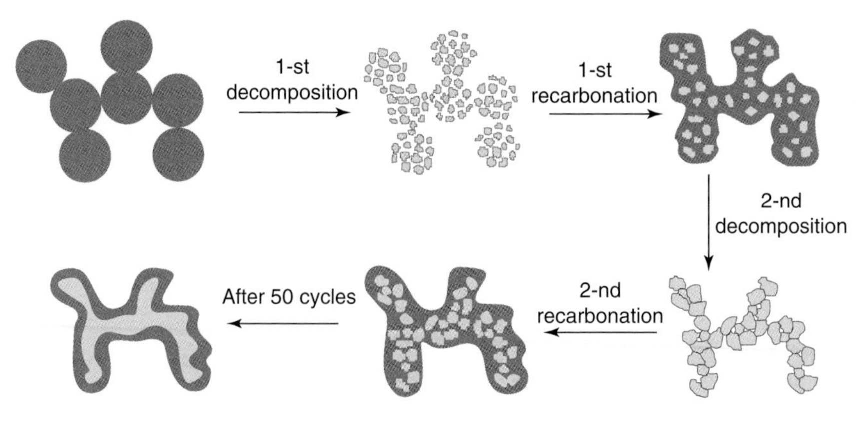

Figure 13.12 Scheme of textural transformation of the CaO sorbent in calcination–carbonation cycles. Source: Lysikov et al. 2007 [34]. Reproduced with permission of American Chemical Society.

CaO conversion after a large number of cycles. In conclusion, improving initial sorbent properties and rejuvenating deactivated sorbents during the cyclic carbonation–calcination steps are equally important. Several investigations have been extensively made to enhance the recyclability and reactivity of the calcium-based sorbent. The approaches of these studies can be categorized into three aspects: (i) synthesis of CaO sorbent from inorganic or organometallic precursors, (ii) incorporation of dopant or inert stabilizer with calcium-based sorbents, and (iii) sorbent reactivation through additional processing.

13.3.1 Synthesis of CaO Sorbent from Inorganic or Organometallic Precursors

The decay of the CO_2 adsorption capacity of CaO derived from natural limestone mostly comes from the deteriorating effects of sorbent sintering at high temperatures. A strategy to enhance the stability of calcium oxide adsorbents derived from an inorganic precursor, natural limestone, over multiple calcination–carbonation cycles by reducing the particle size of the adsorbents down to the nanometer scale was proposed by Barker [35]. The calcium oxide (ca. 10 nm in diameter) showed excellent recyclability (approximately keep 93% carbonation conversion) after several carbonation–calcination cycles because the carbonation reaction completes within the fast kinetic-controlled regime for porous CaO with critical particle size less than 44 nm, and the slow diffusion-controlled regime does not exist. However, the pure CaO derived from nano-sized particles may be susceptive to sintering because their higher surface area leads to numerous contact points to facilitate mass transfer [36]. Another strategy to enhance the sorbent stability is the synthesis of mesoporous calcium carbonate derived from another inorganic precursor, $Ca(OH)_2$, in the presence of N40V dispersant, a sodium salt of a carboxylic acid, by a wet precipitation process in a slurry bubble column [37]. CO_2 dissolved in water provides carbonate ions (CO_3^{2-}) reacted with Ca^{2+} ions to form $CaCO_3$. The formed $CaCO_3$ subsequently precipitates out because the solubility of $CaCO_3$ in water is much lower than that of $Ca(OH)_2$. The specific ratio of N40V to $Ca(OH)_2$ allows the surface charge of zero on the incipiently formed calcium carbonate particles, leading to a predominantly mesoporous structure precipitated calcium carbonate (PCC). The PCC allows the heterogeneous reaction to occur on a larger CaO surface area, and the carbonate layer is not able to plug all of the pore mouths, ultimately resulting in a carbonation conversion higher than 85% as well as superior recyclability up to three carbonation–calcination cycles.

A number of studies explored the relation between the characteristics of calcium-based sorbents synthesized from organometallic precursors and their CO_2 uptake performance [38–41]. Lu et al. [38] identified the preliminary screening of precursors, e.g. $Ca(NO_3)_2{\cdot}4H_2O$, CaO, $Ca(OH)_2$, $CaCO_3$, and $Ca(CH_3COO)_2{\cdot}H_2O$, for their application in the calcium looping process. The highest uptake characteristics for CO_2 could be observed from a "fluffy" structure sorbent ($CaAc_2$–CaO) derived from $Ca(CH_3COO)_2{\cdot}H_2O$ as shown in Figure 13.13. The CO_2 uptake capacity remained stable within the first 10 carbonation–calcination cycles and began to decrease slowly in the

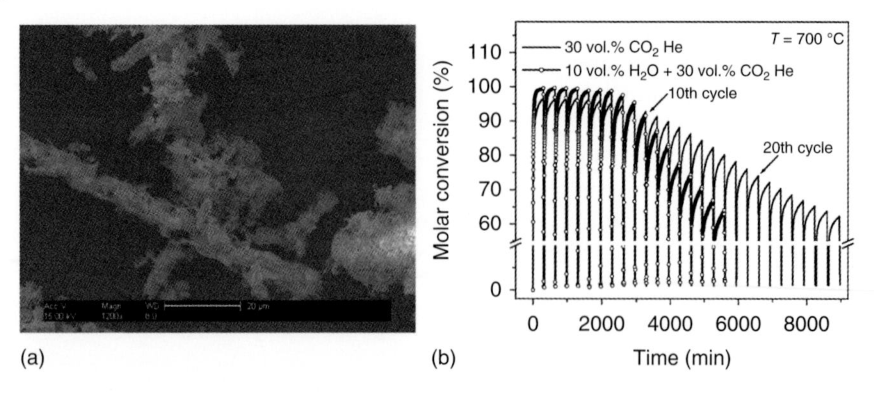

Figure 13.13 Sorbent (a) scanning electron microscope (SEM) image of $CaAc_2$–CaO; (b) CO_2 uptake capacity of $CaAc_2$–CaO. Source: Lu et al. 2006 [38]. Reproduced with permission of American Chemical Society.

subsequent cycles. The $CaAc_2$–CaO still maintained a relatively high carbonation conversion (about 62%) after 27 cycles. The CO_2 uptake capacity of a series of calcium-based sorbents synthesized from various organometallic precursors, including $Ca(C_2H_5COO)_2$, $Ca(CH_3COO)_2$, $Ca(CH_3COCHCOCH_3)_2$, $Ca(COO)_2$, and $Ca(C_7H_{15}COO)_2$, by a simple calcination technique was examined in a following study [39]. CaO sorbents derived from $Ca(C_2H_5COO)_2$ and $Ca(CH_3COO)_2$ exhibited higher CO_2 uptake capacity among the five organometallic precursors-originated sorbents in their work because of the high surface area and large pore volume arised from their mesoporous structure. On the contrary, the poor performance of CaO sorbents derived from $Ca(COO)_2$ and $Ca(C_7H_{15}COO)_2$ is due to their microporous morphology. Moreover, sorbents in nanosize synthesized from $Ca(NO_3)_2{\cdot}4H_2O$ and $C_6H_8O_7{\cdot}H_2O$ by sol–gel method was reported by Luo et al. [41]. Well-dispersed uniform particles (200 nm) within the fluffy and porous structural sorbents exhibited much higher reactivity and recyclability for cyclic reactions compared to the sorbents derived from commercial micro- and nano-sized calcium carbonate. A high carbonation rate of sorbent (60% conversion within 20 s) as well as a better sintering-resistant property could be achieved during cyclic high-temperature reactions. All the abovementioned studies indicate the importance of surface area, pore volume, and particle size from sorbent morphology in relation to CO_2 capture capacity. The physical properties of CaO derived from inorganic and organometallic precursors are summarized in Table 13.2.

13.3.2 Incorporation of Dopant or Inert Stabilizer with Calcium-Based Sorbents

An alternative strategy to promote reaction activity during the carbonation step or to retard sorbent sintering during the high-temperature regeneration step is the use of dopant or inert stabilizer incorporated in calcium oxide precursors. Reddy and Smirniotis [42] used alkali metals as dopants to synthesize a number of calcium-based sorbents for CO_2 capture. The alkali metals-doped CaO sorbents

Table 13.2 Physical properties of CaO derived from various precursors.

Precursor	BET surface area ($m^2\ g^{-1}$)	Pore volume ($cm^3\ g^{-1}$)	References
Inorganic precursor			
$CaCO_3$	12.8	0.027	[37]
$CaCO_3$	17.8	0.078	[37]
$Ca(OH)_2$	33.3	0.1	[37]
CaO	9.84	0.05	[40]
$CaCO_3$	11.62	0.16	[40]
$Ca(OH)_2$	7.67	0.09	[40]
$CaCO_3$	6.84	0.03	[40]
CaO	4.2	0.02	[38]
$CaCO_3$	5.3	0.08	[38]
$Ca(OH)_2$	13.9	0.15	[38]
$CaCO_3$	12.4	n/a[a]	[41]
$CaCO_3$	17.0	n/a[a]	[41]
Organometallic precursors			
$Ca(CH_3COO)_2{\cdot}H_2O$	20.2	0.23	[38]
$Ca(C_2H_5COO)_2$	15	0.18	[39]
$Ca(CH_3COO)_2$	20	0.22	[39]
$Ca(CH_3COCHCOCH_3)_2$	12	0.09	[39]
$Ca(COO)_2$	5.9	0.02	[39]
$Ca(C_7H_{15}COO)_2$	9.3	0.015	[39]
$Ca(CH_3COO)_2{\cdot}xH_2O$	12.16	0.14	[40]
$Ca[CH_3CH(OH)COO]_2{\cdot}xH_2O$	12.39	0.22	[40]
$CaC_{12}H_{22}O_{14}{\cdot}H_2O$	16.96	0.27	[40]
$Ca_3(C_6H_5O_7)_2{\cdot}4H_2O$	11.42	0.06	[40]
$Ca(HCO_2)_2$	10.10	0.08	[40]
$Ca(NO_3)_2{\cdot}4H_2O$	12.6	n/a[a]	[41]

a) Not available.

showed high CO_2 adsorption capacities and rapid adsorption/desorption characteristics due to the increase in basic sites on the CaO surface, which allows for the selective chemisorption of CO_2. Figure 13.14 shows the sequence of the performance of alkali metals as dopants on CaO as Cs > Rb > K > Na > Li, corresponding to the relationship between the adsorption characteristics and the increase in the electropositivity, or equivalently atomic radii, of the alkali metals. Cs-doped CaO sorbent containing 20 wt% Cs exhibited the highest CO_2 uptake capacity due to its zero affinity for N_2 and O_2 and very low affinity for water. For the Cs containing higher than 20 wt%, the sorbent performance was not improved further because of the formation of paracrystalline Cs_2O on the

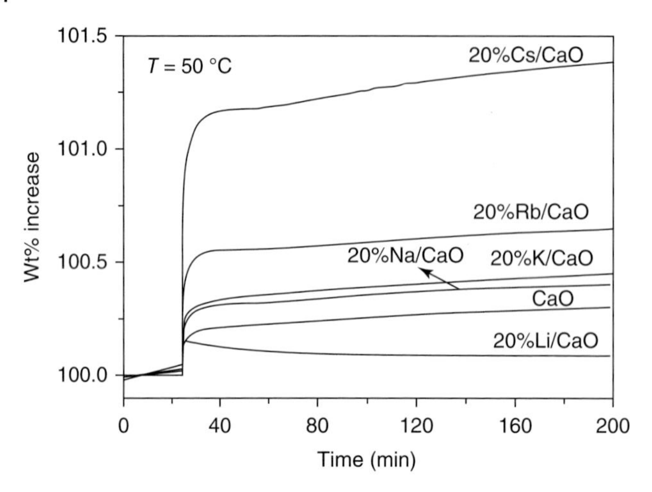

Figure 13.14 CO_2 adsorption capacity of CaO doped with alkali metals. Source: Reddy and Smirniotis [42]. Reproduced with permission of American Chemical Society.

surface of CaO. Moreover, prospective refractory dopants with relatively high Tammann temperatures are also expected to stabilize the CO_2 uptake capacity of CaO in the high-temperature cyclic reactions. The synthesis of calcium-based sorbents with a wide range of refractory dopants, e.g. Si, Ti, Cr, Co, Zr, and Ce, by flame spray pyrolysis (FSP) was explored to address the rapid performance deterioration issue and to delay the sintering-agglomeration phenomena [43]. Among all of the doped CaO sorbents, Zr-doped CaO exhibited the best CO_2 capture performance and thermal stability under identical conditions of operation. The carbonation conversions of the Zr–Ca containing 30 mol% Zr did not decay (kept at 64%) even after 100 cycles. The high uptake for CO_2 is attributed to the nanosize of the particles, the high surface area, and the large pore volume of the sorbents, and the great stable performance is ascribed to the refractory nature of the dopant.

Recently, the incorporation of inert stabilizers, e.g. Al_2O_3, MgO, ZrO_2, TiO_2, SiO_2, and Y_2O_3, with the calcium-based sorbent has also been investigated to improve the recyclability of sorbents derived from synthetic precursors. The inert materials dispersed among the sorbent particles during synthesis are able to inhibit sorbent sintering during cyclic calcination–carbonation reactions due to their property of high Tammann temperature. Calcium-based sorbents incorporated with aluminum-based stabilizer are the most studied metal-stabilized sorbents. Various synthesis methods have been conducted to achieve the aim of increasing surface area and pore volume and obtaining high dispersion of inert stabilizers within the sorbent particles [44–61], including wet mixing, limestone acidification by citric acid followed by two-step calcination, solid-state reaction, ultrasonic spray pyrolysis (USP), combination of precipitation and hydration, co-precipitation, citrate preparation, sol–gel, citrate-assisted sol–gel technique followed by two-step calcination, single nozzle FSP, and precipitation.

In one of the earliest works, Li et al. [45] synthesized the aluminum-stabilized calcium-based sorbents ($Ca_{12}Al_{14}O_{33}$/CaO) by the wet mixing method. A

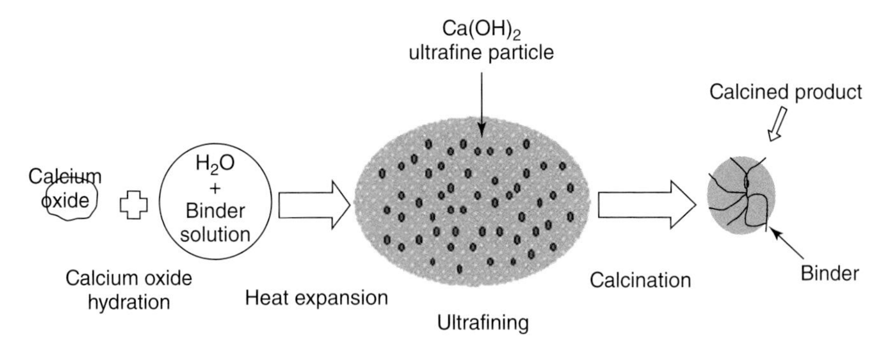

Figure 13.15 Mechanism of $Ca_{12}Al_{14}O_{33}$/CaO preparation. Source: Li et al. 2005 [45]. Reproduced with permission of American Chemical Society.

solution of 2-propanol containing aluminum nitrate nonahydrate ($Al(NO_3)_3 \cdot 9H_2O$) and powdered calcium oxide were well mixed and then dried and calcined at 500 °C. The possible mechanism of sorbent synthesis is presented in Figure 13.15, including the chemical reaction (CaO hydration) → thermal expansion → calcination reaction → material structure reestablishment (formation of binder among CaO particles) → formation of high surface area. The release of NO_2 and H_2O came from the decomposition of 2-propanol and nitric acid, and the dehydration of $Ca(OH)_2$ during synthesis yielded an ultrafine and porous powder. The Al_2O_3 then reacted with CaO to form a binder ($Ca_{12}Al_{14}O_{33}$, mayenite) within the sorbent thermodynamically, leading to a stable structure among CaO micrograins. The CO_2 uptake capacity of the $Ca_{12}Al_{14}O_{33}$/CaO containing 25 wt% $Ca_{12}Al_{14}O_{33}$ attained 0.41 g CO_2 per gram of sorbent at the first calcination–carbonation cycle and then slightly increased over the first five cycles due to an increase in the pore volume inside particles caused by structural changes. The adsorption capacity of the $Ca_{12}Al_{14}O_{33}$/CaO remained at 0.45 g CO_2 per gram of sorbent at the end of the 13th cycle. The superior characteristics in CO_2 capture capacity is attributed to the formation of ultrafine CaO particles stabilized by an inert $Ca_{12}Al_{14}O_{33}$ support. However, excess of inert $Ca_{12}Al_{14}O_{33}$ reacted with CaO to form $Ca_3Al_2O_6$. The formation of $Ca_3Al_2O_6$ decreased the ratio of CaO to $Ca_{12}Al_{14}O_{33}$ in sorbents and then led to the decline of cyclic CO_2 capture capacity [46]. Figure 13.16 shows the reaction scheme of $Ca_3Al_2O_6$ formation from $Ca_{12}Al_{14}O_{33}$ and the formation mechanism of aluminum-stabilized calcium-based sorbent under various calcination temperatures. Ca^{2+} diffusing through mayenite layers reacts with $Ca_{12}Al_{14}O_{33}$ to form $Ca_3Al_2O_6$ at temperatures higher than 1100 °C. Severe sorbent sintering among grains of pure $Ca_3Al_2O_6$ may occur at temperatures higher than 1200 °C.

The transition between $Ca_{12}Al_{14}O_{33}$ and $Ca_3Al_2O_6$ depends on the calcium and aluminum precursors as well as synthesis method, which can limit Ca^{2+} diffusion into the stabilizer structure for further reaction [56]. Broda et al. [56] investigated the CO_2 uptake capacity of aluminum-stabilized CaO synthesized by using different calcium precursors, such as $Ca(OH)_2$, $Ca(NO_3)_2 \cdot 4H_2O$, $Ca(CH_3COO)_2 \cdot H_2O$, and $Ca(C_5H_7O_2)$, as shown in Figure 13.17. The decrease in the ratio of Ca^{2+} to Al^{3+} enhanced the surface area and pore volume of the synthetic sorbent,

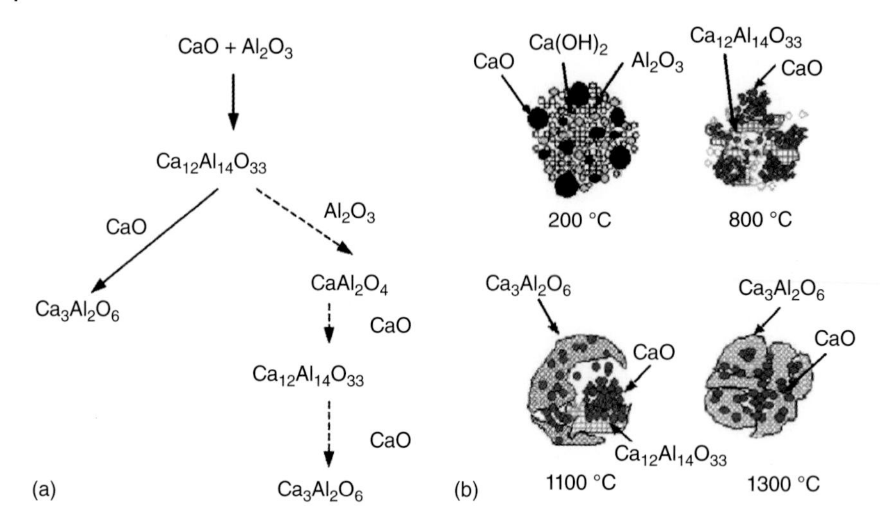

Figure 13.16 (a) The reaction scheme of the formation of $Ca_3Al_2O_6$ [62]; (b) the formation mechanism of the aluminum-stabilized calcium-based sorbents under various calcination temperatures. Source: Li et al. 2006 [46]. Reproduced with permission of American Chemical Society.

followed by the decrease of overall CO_2 uptake of the sorbent due to the increase of mayenite ($Ca_{12}Al_{14}O_{33}$) quantity. The sorbents with nanostructural morphology associated with a high surface area and pore volume were obtained from the strongly basic and weakly acidic precursors, e.g. ($Ca(OH)_2$, $Ca(CH_3COO)_2{\cdot}H_2O$, and $Ca(C_5H_7O_2)$), yielding excellent sorbent recyclability. A CO_2 uptake capacity of 0.51 g of CO_2 per gram of sorbent could be retained after 30 cycles. In the following study [57], the aluminum-stabilized CaO derived from the calcium precursor of $Ca(OH)_2$ was applied for carbon capture in a fluidized bed reactor. The aluminum-stabilized CaO sorbent was 60% higher than that of a reference limestone. The superior cyclic CO_2 uptake of the sorbent is because of their nanostructural morphology and the homogeneously dispersed aluminum-based stabilizer (mostly in the form of the mixed oxide mayenite). The high Tammann temperature stabilizer stabilizes the nanostructured morphology over multiple reaction cycles.

The CO_2 uptake performance of a dolomite-derived CaO was reported to be higher than that of a limestone-derived CaO over multiple cycles due to the presence of $MgCO_3$ in the dolomite [62, 63]. $MgCO_3$ in dolomite provides additional pore volume during its decomposition to MgO. The formation of MgO is capable of stabilizing the pore structure of sorbent and then reducing the rate of sorbent sintering during multiple calcination–carbonation cycles because of its high Tammann temperature (1276 °C). Therefore, MgO is considered as another promising stabilizer to incorporate with calcium-based sorbents. Li et al. [64] synthesized magnesia-stabilized CaO sorbents by co-precipitation, dry physical mixing, wet physical mixing, and solution mixing methods. MgO/CaO sorbents synthesized by the dry and wet physical mixing of the calcium-based precursor ($Ca(CH_3COO)_2$) with MgO particles followed by high-temperature

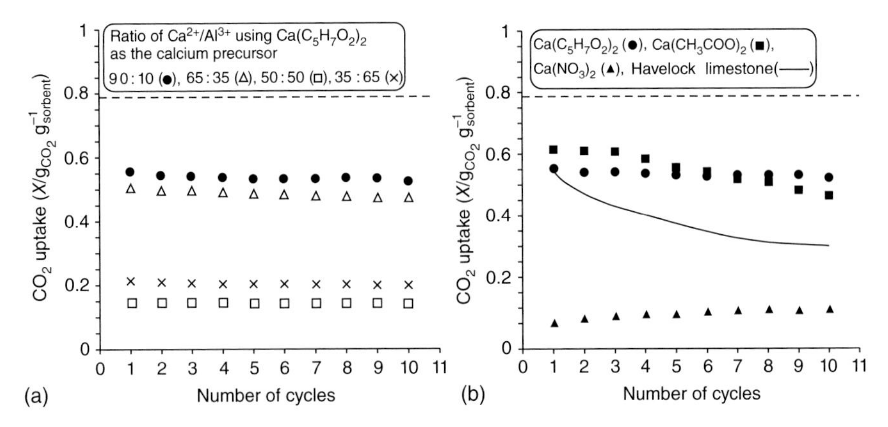

Figure 13.17 CO_2 uptake capacity of sorbents synthesized using various Ca^{2+}/Al^{3+} ratios and precursors. Source: Broda et al. 2012 [56]. Reproduced with permission of John Wiley and Sons.

calcination exhibited the most durable performance in CO_2 capture. The performance of synthetically prepared MgO/CaO was higher than that of CaO derived from dolomite. The CO_2 capture capacity of the MgO/CaO containing 26 wt% MgO attained 53 wt% after 50 calcination–carbonation cycles, which is twofold higher than pure CaO obtained from the same source under the same test conditions. Moreover, other materials such as ZrO_2, TiO_2, SiO_2, $KMnO_4$, and Y_2O_3 have been also considered as inert stabilizer to promote the cyclic reactivity of CaO. Most of these studies indicated that the inert stabilizers are capable of maintaining sorbent structure under cyclic calcination–carbonation reactions. However, the content of stabilizer in synthetic sorbent remains as the main challenge of this approach because the overall CO_2 uptake capacity of sorbents may suffer from the quantity of inert stabilizer contained in the sorbent. The cost of synthetic sorbent is also higher than that of CaO derived from natural limestone. The impacts from using a metal-stabilized sorbent have to be taken into consideration to assess the economic feasibility of the carbon capture process.

13.3.3 Sorbent Reactivation Through Additional Processing

Instead of sorbent synthesis, there are several approaches in process modification to improve the CO_2 uptake capacity of natural limestone in multiple calcination–carbonation cycles, including steam carbonation/calcination, hydration, and recarbonation. Effects of steam on the carbonation and/or calcination reactions have been investigated because 5–20% steam is typically contained in the flue gas produced from combustion. Most researchers have reported that adding steam during carbonation or calcination produces a notably positive effect on the sorbent recyclability. Although steam pretreatment may not improve CaO physical properties such as specific surface area and pore structure, the hydroxyl groups generated from steam dissociation on the CaO surface and bicarbonates produced through interaction of CO_2 with surface hydroxyl groups are considered as important intermediates in steam catalysis for CaO carbonation [65, 66]. Arias et al. [67] focused their study on the effect of steam over a very short reaction time where the carbonation reaction occurred under a kinetic-controlled regime. Similar values of carbonation rate constant were calculated for experiments conducted with and without the presence of steam (20 vol%) in the reacting gas. However, the presence of steam in the carbonation reaction enhanced CO_2 diffusion through the carbonation layer. Thus, steam addition in carbonation reaction retards the transition from the kinetic-controlled reaction stage to the diffusion-controlled reaction stage, i.e. when the product layer is thicker [6]. Broda et al. [68] further assessed the effect of steam on the cyclic CO_2 capture capacity of pelletized $Ca_{12}Al_{14}O_{33}/CaO$ sorbent. The addition of steam promoted CO_2 uptake of the pellet sorbent from 0.20 g of CO_2 per gram of sorbent to 0.31 g of CO_2 per gram of sorbent after 10 cycles of repeated calcination and carbonation reactions due to the enhancement of solid-state diffusion through the $CaCO_3$ product layer.

A number of studies have also explored the effect of steam injection during calcination on the reactivity of CaO for carbon capture. The presence of steam in sorbent calcination has been reported to change the sorbent morphology, probably causing a shift from smaller to larger pores and resulting in a structure that increases CO_2 uptake capacity [69]. Although Champagne et al. [69] indicated that the total surface area of sorbents was substantially reduced by sintering during calcination with steam concentrations exceeding 40%, another study [70] indicated that the steam content of 50% in their bench-scale rotary bed calciner aided in reducing the extent of sorbent sintering and resulted in the production of a more reactive sorbent with a CO_2 capture capacity of 45 wt%. A potential explanation to describe the effects of steam addition during carbonation or/and calcination on the morphology of CaO sorbent was proposed as shown in Figure 13.18 [71]. The open pore structure of CaO sorbent is susceptible to pore blockage and plugging during calcination and carbonation reactions. The presence of steam during carbonation mitigates the severe pore mouth plugging and accelerates the solid-state diffusion-controlled reaction, leading to a higher carbonation conversion. Thus, sorbents with greater pore volume could be obtained in the following calcination. The presence of steam during calcination causes slightly larger pores (~50 nm) in CaO due to steam-enhanced sintering and enables the pore structure to be maintained in a relatively stable state for carbonation. These larger pores are less susceptible to pore blockage and thus allow for higher carbonation conversion. A synergistic effect of a shift to larger pores and the lower diffusion resistance of carbonate layer on the sorbent reactivity may be observed when steam is present for both calcination and carbonation.

The reactivation of spent sorbent using a hydration process, forming $Ca(OH)_2$ from CaO, is another promising approach to improve the sorbent recyclability and reactivity in multicyclic calcination–carbonation operations. The steam partial pressure operating in a normal carbonator and calciner cannot induce the hydration reaction because the hydration reaction must be conducted at a steam partial pressure exceeding 1 atm at temperatures greater than 510 °C [70]. An intermediate hydrator between carbonator and calciner is required. However, in some precombustion applications, e.g. hydrogen production by

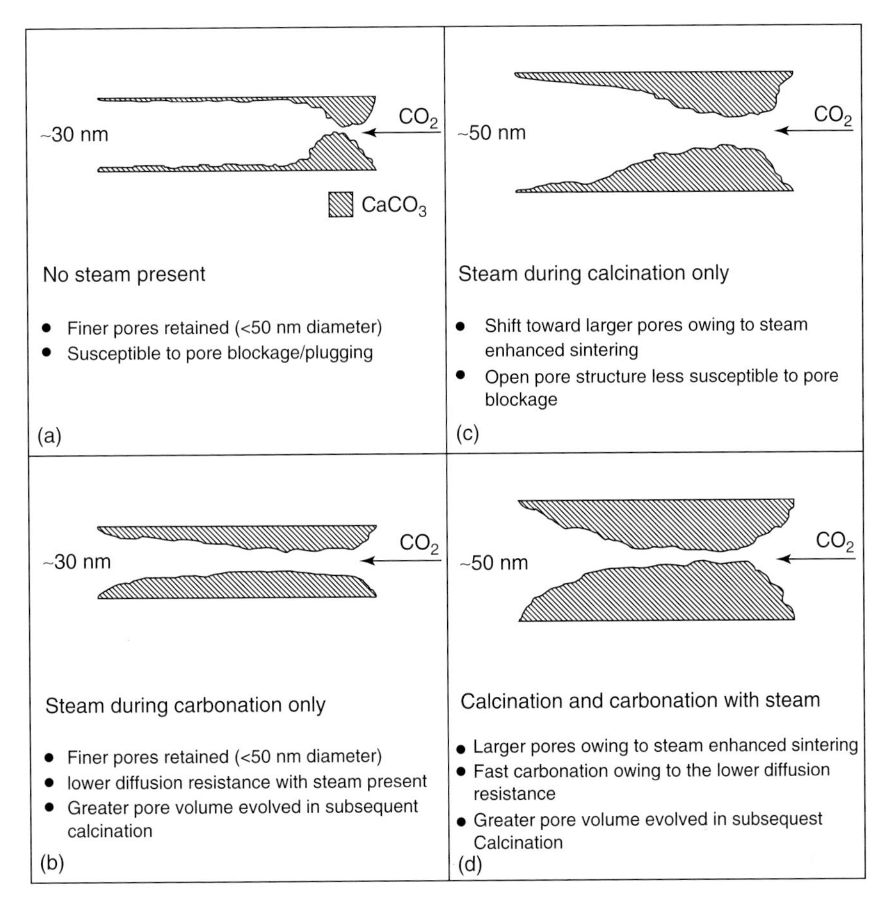

Figure 13.18 Schematic representation of the sorbent morphology under various reaction environments: (a) no steam present; (b) steam present during carbonation only; (c) steam during calcination only; and (d) steam present for carbonation and calcination. Source: Donat et al. 2012 [71]. Reproduced with permission of American Chemical Society.

reaction-integrated novel gasification (HyPr-RING) process [72], a pressurized carbonator containing steam partial pressures higher than 1 atm will be such that hydration can occur in situ. In the HyPr-RING process, the sorbent is more durable in repetitive CO_2 sorption with the intermediate hydration step even under eutectic conditions. The study reported by Manovic and Anthony [73] is one of the earliest works to investigate the possibility of reactivation of spent calcium-based sorbent using steam hydration (200 °C) in a pressurized reactor. The hydrated spent sorbent showed even better characteristics in reactivity and recyclability for carbon capture than the natural sorbent. This is because the specific volume of the $Ca(OH)_2$ produced by hydration is 1.98 times greater than that of CaO. $Ca(OH)_2$ fractures are formed during the crystalline expansion due to its low tensile strength and weak crack resistance, resulting in sorbent reactivation [74]. Moreover, the carbonation of hydrated calcium-based sorbent, $Ca(OH)_2$, could be categorized into direct carbonation and indirect carbonation [75]. Direct carbonation refers to the direct reaction of CO_2 with $Ca(OH)_2$ to produce $CaCO_3$ and H_2O. On the contrary, indirection carbonation of $Ca(OH)_2$ refers to a reaction of CO_2 with CaO derived from dehydration of $Ca(OH)_2$ under specific operating conditions. Much of the initial work on sorbent reactivation using hydration for calcium looping process was performed using indirect carbonation. However, the conversion from direct carbonation of $Ca(OH)_2$ was higher than that from indirect carbonation of $Ca(OH)_2$. This is because a thin and dynamic water film is formed on the sorbent surface at the dehydration temperature. Carbonate ions and H^+ ions formed from CO_2 dissolution in the liquid film are injected into the crystal lattice effectively, forming a precipitate ($CaCO_3$) rapidly. The heat released from carbonation drives more water formation from $Ca(OH)_2$ dehydration, resulting in the continuous absorption of CO_2 in the liquid film to form $CaCO_3$ [76]. Other work reported that the direct carbonation of $Ca(OH)_2$ exhibited higher conversion at temperatures in the range of 200–650 °C as compared to that of CaO; however, the carbonation conversion of $Ca(OH)_2$ would be contrarily lower than that of CaO when $Ca(OH)_2$ was dehydrated prior to carbonation [77]. Therefore, two operating strategies were proposed for operating the hydration-integrated calcium looping process: (i) operating the $Ca(OH)_2$ carbonation at temperatures lower than dehydration temperature; (ii) conducting the direct carbonation of $Ca(OH)_2$ immediately when $Ca(OH)_2$ is heated higher than the dehydration temperature.

Figure 13.19 shows the process flow diagram of a calcium looping process integrated into a 500 MW_e subcritical coal-fired power plant with a target efficiency of 35.8% high heating value (HHV). The process analysis from ASPEN Plus simulations in this case indicated that the energy penalty of this integrated system could be reduced from 20.7% to 16.1% when the heat released from the highly exothermic hydration reaction was extracted for electricity production [11]. Cost assessments for a capture process integrated by industrial symbiosis of the power plant, cement plant, and calcium looping process revealed that although extra costs were required for the hydrator and piping as well as steam consumption in a hydration-integrated system, the overall equipment cost and operating cost would be significantly reduced because the enhancement of carbonation conversion induced by hydration significantly reduced the sizes of the whole carbon capture system [78]. This design indicated that the CO_2 capture cost was possible lower than €25 per ton of CO_2. In conclusion, by introducing a hydration step between calcination and carbonation it is possible to reduce the calcium to carbon molar ratio, solids circulation, energy consumption, and reactor size, all of which improve the process economics and ability to integrate into an industrial process. However, the size of sorbent in the range of 200–300 μm was possibly reduced to <20 μm after the first hydration [79]. The attrition of hydrated particles poses a challenge for operating the sorbents in the solid circulating system. Therefore, the improvement of solid handling technique and the mitigation of sorbent attrition are still the essential steps for integrating hydration into a real calcium looping system.

Another intermediate step, recarbonation, has been suggested in an attempt to maintain the reactivity of limestone needed to sustain sufficient CO_2 capture

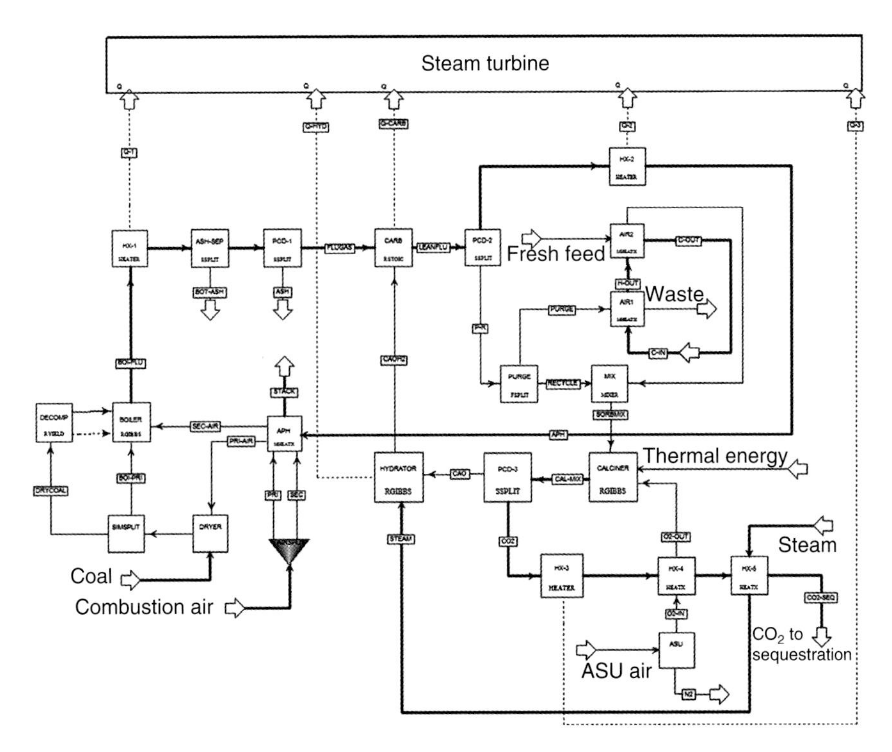

Figure 13.19 Process flow diagram of a calcium looping process integrated into a 500 MW_e coal-fired power plant. Source: Wang et al. 2013 [11]. Reproduced with permission of Elsevier.

efficiency in a calcium looping process. One of the earliest studies demonstrated that if the carbonation of CaO completed within its slow diffusion-controlled regime, the subsequent calcination produced a more porous and higher surface CaO sorbent because the CO_2 flowed from the core of the sorbent during calcination [32]. Inspired by this observation, Sun et al. [80] conducted cyclic carbon capture by a limestone-derived sorbent for more than 1000 cycles and revealed that approximately treble carbonation conversion could be observed after 1000 calcination–carbonation cycles when the carbonation duration was extended from 3.5 to 9 minutes. Grasa et al. [81] proposed a calcium looping system incorporating an extended recarbonator for carbon capture as shown in Figure 13.20. The partially carbonated sorbent delivered from the carbonator is placed in a recarbonator to contact with high temperature stream containing concentrated CO_2 (generated from the calciner). An additional conversion of the sorbents under the diffusion-controlled reaction stage mitigates the sorbent decay in the CO_2 capture capacity during cyclic calcination–carbonation cycles. A short (100–200 s) recarbonation stage on partially carbonated particles of CaO allowed for the stabilization of CO_2 carrying capacities at 15–20 mol%. However, in practice, sorbent reactivation through recarbonation reaction may lead to a greater portion of the carbonation time corresponding to the prolonged carbonation stage. Therefore, understanding the extent of sorbent reactivation correlating with its recarbonation conversion is essential to optimize

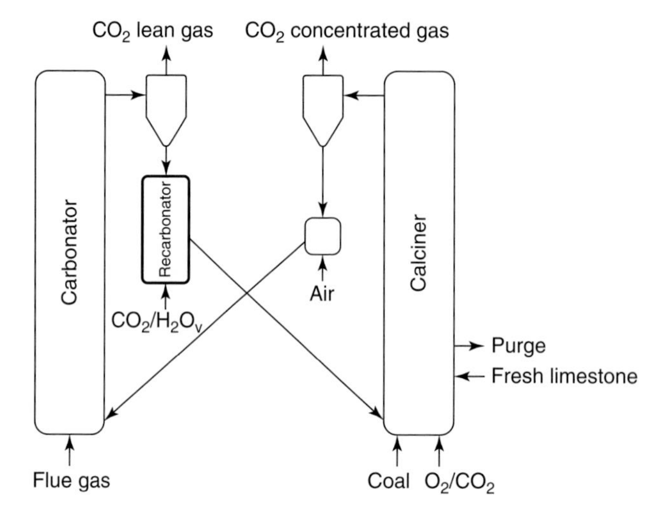

Figure 13.20 Schematic diagram of a recarbonation-integrated calcium looping process. Source: Grasa et al. 2014 [81]. Reproduced with permission of American Chemical Society.

the residence time in a carbonator and operate this carbon capture system in a commercially economical consideration.

References

1 Squires, A.M. (1967). Cyclic use of calcined dolomite to desulfurize fuels undergoing gasification. *Adv. Chem. Ser.* 69: 205–229.

2 Han, C. and Harrison, D.P. (1994). Simultaneous shift reaction and carbon dioxide separation for the direct production of hydrogen. *Chem. Eng. Sci.* 49 (24): 5875–5883.

3 Shimizu, T., Hirama, T., Hosoda, H. et al. (1999). A twin fluid-bed reactor for removal of CO_2 from combustion processes. *Chem. Eng. Res. Des.* 77 (1): 62–68.

4 Abanades, J.C., Anthony, E.J., Wang, J., and Oakey, J.E. (2005). Fluidized bed combustion systems integrating CO_2 capture with CaO. *Environ. Sci. Technol.* 39: 2861–2866.

5 Grasa, G.S. and Abanades, J.C. (2006). CO_2 capture capacity of CaO in long series of carbonation/calcination cycles. *Ind. Eng. Chem. Res.* 45 (26): 8846–8851.

6 Manovic, V. and Anthony, E.J. (2010). Carbonation of CaO-based sorbents enhanced by steam addition. *Ind. Eng. Chem. Res.* 49: 9105–9110.

7 Abanades, J.C. (2002). The maximum capture efficiency of CO_2 using a carbonation/calcination cycle of $CaO/CaCO_3$. *Chem. Eng. J.* 90: 303–306.

8 Gupta, C.K. and Sathiyamoorthy, D. (1999). *Fluid Bed Technology in Material Processing*. Boca Raton, FL: CRC Press.

9 Romeo, L.M., Catalina, D., Lisbona, P. et al. (2011). Reduction of greenhouse gas emissions by integration of cement plants, power plants, and CO_2 capture systems. *Greenhouse Gases Sci. Technol.* 1 (1): 72–82.

10 Rodriguez, N., Alonso, M., Grasa, G., and Abanades, J.C. (2008). Heat requirements in a calciner of $CaCO_3$ integrated in a CO_2 capture system using CaO. *Chem. Eng. J.* 138: 148–154.

11 Wang, W., Ramkumar, S., and Fan, L.S. (2013). Energy penalty of CO_2 capture for the carbonation–calcination reaction (CCR) process: parametric effects and comparisons with alternative processes. *Fuel* 104: 561–574.

12 Chang, M.H., Huang, C.M., Liu, W.H. et al. (2013). Design and experimental investigation of calcium looping process for 3 kW_{th} and 1.9 MW_{th} facilities. *Chem. Eng. Technol.* 36 (9): 1525–1532.

13 Chang, M.H., Chen, W.C., Huang, C.M. et al. (2014). Design and experimental testing of a 1.9 MW_{th} calcium looping pilot plant. *Energy Procedia* 63: 2100–2108.

14 Charitos, A., Hawthorne, C., Bidwe, A.R. et al. (2010). Parametric investigation of the calcium looping process for CO_2 capture in a 10 kW_{th} dual fluidizedbed. *Int. J. Greenhouse Gas Control* 4: 776–784.

15 Hawthornea, C., Dietera, H., Bidwea, A. et al. (2011). CO_2 capture with CaO in a 200 kW_{th} dual fluidized bed pilot plant. *Energy Procedia* 4: 441–448.

16 Dieter, H., Hawthorne, C., Zieba, M., and Scheffknechta, G. (2013). Process in calcium looping post combustion CO_2 capture successful pilot scale demonstration. *Energy Procedia* 37: 48–56.

17 Dieter, H., Beirow, M., Schweitzer, D. et al. (2014). Efficiency and flexibility potential of calcium looping CO_2 capture. *Energy Procedia* 63: 2129–2137.

18 Abanades, J.C., Alonso, M., Rodriguez, N. et al. (2009). Capture CO_2 from combustion flue gases with a carbonation calcination loop. Experimental results and process development. *Energy Procedia* 1: 1147–1154.

19 Alonso, M., Rodriguez, N., Gonzalez, B. et al. (2010). Carbon dioxide capture from combustion flue gases with a calcium oxide chemical loop. Experimental results and process development. *Int. J. Greenhouse Gas Control* 4: 167–173.

20 Alonso, M., Diego, M.E., Pérez, C. et al. (2014). Biomass combustion with in situ CO_2 capture by CaO in a 300 kW_{th} circulating fluidized bed facility. *Int. J. Greenhouse Gas Control* 29: 142–152.

21 Diego, M.E. and Alonso, M. (2016). Operational feasibility of biomass combustion with in situ CO_2 capture by CaO during 360 h in a 300 kW_{th} calcium looping facility. *Fuel* 181: 325–329.

22 Lu, D.Y., Hughes, R.W., and Anthony, E.J. (2008). Ca-base sorbent looping combustion for CO_2 capture in pilot plant-scale dual fluidized beds. *Flue Proc. Technol.* 89: 1386–1395.

23 Wang, W., Ramkumar, S., Li, S. et al. (2010). Sub-pilot demonstration of the carbonation calcination reaction (CCR) process: high-temperature CO_2 and sulfur capture from coal-fired power plants. *Ind. Eng. Chem. Res.* 49: 5094–5101.

24 Kremer, J., Galloy, A., Dtrohle, J., and Epple, B. (2013). Continuous CO_2 capture in a 1-MW_{th} carbonate looping pilot plant. *Chem. Eng. Technol.* 33 (9): 1518–1524.

25 Ströhle, J., Junk, M., Kremer, J. et al. (2014). Carbonate looping experiments in a 1 MW_{th} pilot plant and model validation. *Fuel* 127: 13–22.

26 Hilz, J., Helbing, M., Stroh, A., et al. (2015). 1MW$_{th}$ pilot testing and scale-up of the carbonate looping process in the SCARLET project. 3rd Post Combustion Capture Conference (PCCC3) (8–11 September 2015).

27 Sánchez-Biezma, A., Ballesteros, J.C., Diaz, L. et al. (2011). Postcombustion CO_2 capture with CaO status of the technology and next steps towards large scale demonstration. *Energy Procedia* 4: 852–859.

28 Sánchez-Biezma, A., Ballesteros, J.C., Diaz, L. et al. (2013). Testing postcombustion CO_2 capture with CaO in a 1.7 MW$_t$ pilot facility. *Energy Procedia* 37: 1–8.

29 Arias, B., Diego, M.E., Abanades, J.C. et al. (2013). Demonstration of steady state CO_2 capture in a 1.7 MW$_{th}$ calcium looping pilot. *Int. J. Greenhouse Gas Control* 18: 237–245.

30 Abanades, J.C., Arias, B., Lyngfelt, A. et al. (2015). Emerging CO_2 capture systems. *Int. J. Greenhouse Gas Control* 40: 126–166.

31 Diego, M.E., Arias, B., Mendz, A. et al. (2016). Experimental testing of a sorbent reactivation process in La Pereda 1.7 MW$_{th}$ calcium looping pilot plant. *Int. J. Greenhouse Gas Control* 50: 14–22.

32 Barker, J. (1973). Reversibility of the reaction of calcium carbonate to give calcium oxide and carbon dioxide. *Appl. Chem. Biotechnol.* 23 (10): 733–742.

33 Abanades, J.C. and Alvarez, D. (2003). Conversion limits in the reaction of CO_2 with lim. *Energy Fuels* 17 (2): 308–315.

34 Lysikov, A.I., Salanov, A.N., and Okynev, A.G. (2007). Change of CO_2 carrying capacity of CaO in isothermal recarbonation–decomposition cycles. *Ind. Eng. Chem. Res.* 46 (13): 4633–4638.

35 Barker, R. (1974). The reactivity of calcium oxide towards carbon dioxide and its use for energy storage. *J. Appl. Chem. Biotechnol.* 24: 221–227.

36 Florin, N.H. and Harris, A.T. (2009). Reactivity of CaO derived from nano-sized $CaCO_3$ particles through multiple CO_2 capture-and-release cycles. *Chem. Eng. Sci.* 64 (2): 187–191.

37 Gupta, H. and Fan, L.S. (2002). Carbonation–calcination cycle using high reactivity calcium oxide for carbon dioxide separation from flue gas. *Ind. Eng. Chem. Res.* 41: 4035–4042.

38 Lu, H., Reddy, E.P., and Smirniotis, P.G. (2006). Calcium oxide based sorbents for capture of carbon dioxide at high temperatures. *Ind. Eng. Chem. Res.* 45 (11): 3944–3949.

39 Lu, H., Khan, A., and Smirniotis, P.G. (2008). Relationship between structural properties and CO_2 capture performance of CaO-based sorbents obtained from different organometallic precursors. *Ind. Eng. Chem. Res.* 47: 6216–6220.

40 Liu, W., Low, N.W., Feng, B. et al. (2010). Calcium precursors for the production of CaO sorbents for multicycle CO_2 capture. *Environ. Sci. Technol.* 44: 841–847.

41 Luo, C., Zheng, Y., Zheng, C. et al. (2013). Manufacture of calcium-based sorbents for high temperature cyclic CO_2 capture via a sol–gel process. *Int. J. Greenhouse Gas Control* 12: 193–199.

42 Reddy, E.P. and Smirniotis, P.G. (2004). High-temperature sorbents for CO_2 made of alkali metals doped on CaO supports. *J. Phy. Chem. B* 108: 7794–7800.

43 Lu, H., Khan, A., Pratsinis, S.E., and Smirniotis, P.G. (2009). Flame-made durable doped-CaO nanosorbents for CO_2 capture. *Energy Fuels* 23 (2): 1093–1100.
44 Yancheshmeh, M.S., Radfarnia, H.R., and Iliuta, M.C. (2016). High temperature CO_2 sorbents and their application for hydrogen production by sorption enhanced steam reforming process. *Chem. Eng. J.* 283: 420–444.
45 Li, Z.S., Cai, N.S., Huang, Y.Y., and Han, H.J. (2005). Synthesis, experimental studies, and analysis of a new calcium-based carbon dioxide absorbent. *Energy Fuels* 19: 1447–1452.
46 Li, Z.S., Cai, N.S., and Huang, Y.Y. (2006). Effect of preparation temperature on cyclic CO_2 capture and multiple carbonation calcination cycles for a new Ca-based CO_2 sorbent. *Ind. Eng. Chem. Res.* 45: 1911–1917.
47 Pacciani, R., Muller, C.R., Davidson, J.F. et al. (2008). Synthetic Ca-based solid sorbents suitable for capturing CO_2 in a fluidized bed. *Can. J. Chem. Eng.* 86: 356–366.
48 Dennis, J.S. and Pacciani, R. (2009). The rate and extent of uptake of CO_2 by a synthetic, CaO-containing sorbent. *Chem. Eng. Sci.* 64: 2147–2157.
49 Zhang, X.Y., Li, Z.G., Peng, Y. et al. (2014). Investigation on a novel $CaOAY_2O_3$ sorbent for efficient CO_2 mitigation. *Chem. Eng. J.* 243: 297–304.
50 Radfarnia, H.R. and Iliuta, M.C. (2013). Metal oxide-stabilized calcium oxide CO_2 sorbent for multicycle operation. *Chem. Eng. J.* 232: 280–289.
51 Stendardo, S., Andersen, L.K., and Herce, C. (2013). Self-activation and effect of regeneration conditions in CO_2-carbonate looping with $CaOACa_{12}Al_{14}O_{33}$ sorbent. *Chem. Eng. J.* 220: 383–394.
52 Sayyah, M., Ito, B.R., Rostam-Abadi, M. et al. (2013). CaO-based sorbents for CO_2 capture prepared by ultrasonic spray pyrolysis. *RSC Adv.* 3: 19872–19875.
53 Kim, J.N., Ko, C.H., and Yi, K.B. (2013). Sorption enhanced hydrogen production using one-body $CaOACa_{12}Al_{14}O_{33}ANi$ composite as catalytic absorbent. *Int. J. Hydrogen Energy* 38: 6072–6078.
54 Kierzkowska, A.M., Poulikakos, L.V., Broda, M., and Muller, C.R. (2013). Synthesis of calcium-based, AlO_3-stabilized sorbents for CO_2 capture using a co-precipitation technique. *Int. J. Greenhouse Gas Control* 15: 48–54.
55 Zhang, M.M., Peng, Y.X., Sun, Y.Z. et al. (2013). Preparation of $CaOAAl_2O_3$ sorbent and CO_2 capture performance at high temperature. *Fuel* 111: 636–642.
56 Broda, M., Kierzkowska, A.M., and Müller, C.R. (2012). Application of the sol–gel technique to develop synthetic calcium-based sorbents with excellent carbon dioxide capture characteristics. *ChemSusChem* 5 (2): 411–418.
57 Broda, M., Kierzkowska, A.M., and Müller, C.R. (2012). The influence of the calcination and carbonation conditions on the CO_2 uptake of synthetic Ca-based sorbents. *Environ. Sci. Technol.* 46 (19): 10849–10856.
58 Angeli, S.D., Martavaltzi, C.S., and Lemonidou, A.A. (2014). Development of a novel-synthesized Ca-based CO_2 sorbent for multicycle operation: parametric study of sorption. *Fuel* 127: 62–69.

59 Radfarnia, H.R. and Sayari, A. (2015). A highly efficient CaO-based CO_2 sorbent prepared by a citrate-assisted sol–gel technique. *Chem. Eng. J.* 262: 913–920.
60 Koirala, R., Reddy, G.K., and Smirniotis, P.G. (2012). Single nozzle flame-made highly durable metal doped Ca-based sorbents for CO_2 capture at high temperature. *Energy Fuels* 26: 3103–3109.
61 Florin, N.H., Blamey, J., and Fennell, P.S. (2010). Synthetic CaO-based sorbent for CO_2 capture from large-point sources. *Energy Fuels* 24: 4598–4604.
62 Silaban, A., Narcida, M., and Harrison, D.P. (1996). Characteristics of the reversible reaction between CO_2 and calcined dolomite. *Chem. Eng. Commun.* 146 (1): 149–162.
63 Chen, Z., Song, H.S., Portillo, M. et al. (2009). Long-term calcination/carbonation cycling and thermal pretreatment for CO_2 capture by limestone and dolomite. *Energy Fuels* 23: 1437–1444.
64 Li, L., King, D.L., Nie, Z., and Howard, C. (2009). Magnesia-stabilized calcium oxide absorbents with improved durability for high temperature CO_2 capture. *Ind. Eng. Chem. Res.* 48 (23): 10604–10613.
65 Yang, S.J. and Xiao, Y.H. (2008). Steam catalysis in CaO carbonation under low steam partial pressure. *Ind. Eng. Chem. Res.* 47 (12): 4043–4048.
66 Borgwardt, R.H. (1989). Calcium oxide sintering in atmospheres containing water and carbon dioxide. *Ind. Eng. Chem. Res.* 28: 493–500.
67 Arias, B., Grasa, G., Abanades, J.C. et al. (2012). The effect of steam on the fast carbonation reaction rates of CaO. *Ind. Eng. Chem. Res.* 51 (5): 2478–2482.
68 Broda, M., Manovic, V., Anthony, E.J., and Müller, C.R. (2014). Effect of pelletization and addition of steam on the cyclic performance of carbon-templated, CaO-based CO_2 sorbents. *Environ. Sci. Technol.* 48 (9): 5322–5328.
69 Champagne, S., Lu, D.Y., Macchi, A. et al. (2013). Influence of steam injection during calcination on the reactivity of CaO-based sorbent for carbon capture. *Ind. Eng. Chem. Res.* 52: 2241–2246.
70 Ramkumar, S. and Fan, L.S. (2010). Thermodynamic and experimental analyses of the three-stage calcium looping process. *Ind. Eng. Chem. Res.* 49: 7563–7573.
71 Donat, F., Florin, N.H., Anthony, E.J., and Fennell, P.S. (2012). Influence of high-temperature steam on the reactivity of CaO sorbent for CO_2 capture. *Environ. Sci. Technol.* 46: 1262–1269.
72 Lin, S., Harada, M., Suzuki, Y., and Hatano, H. (2005). Process analysis for hydrogen production by reaction integrated novel gasification (HyPr-RING). *Energy Convers. Manage.* 46: 869–880.
73 Manovic, V. and Anthony, E.J. (2007). Steam reactivation of spent CaO-based sorbent for multiple CO_2 capture cycles. *Environ. Sci. Technol.* 41: 1420–1425.
74 Glasson, D.R. (1958). Reactivity of lime and related oxides. II. Sorption of water. Vapor on calcium oxide. *J. Appl. Chem.* 8: 798–803.
75 Blamey, J., Dennis, Y., Paul, L. et al. (2011). Reactivation of CaO-based sorbents for CO_2 capture: mechanism for the carbonation of $Ca(OH)_2$. *Ind. Eng. Chem. Res.* 50 (17): 10329–10334.

76 Materic, V.S. and Smedley, I. (2011). High temperature carbonation of $Ca(OH)_2$. *Ind. Eng. Chem. Res.* 50 (10): 5927–5932.

77 Chou, Y.C., Cheng, J.Y., Chang, M.H., et al. (2015). Device and method for capturing carbon dioxide. Taiwan Patent, TW 1,499,449.

78 Chou, Y.C., Chang, M.H., Chen, W.C., et al. (2012). Improvement strategy and cost assessment for CO_2 capture by calcium looping process. 2012 Taiwan Symposium on Carbon Dioxide Capture, Storage and Utilization (25–27 November 2012). Taipei, Taiwan.

79 Phalak, N., Deshpande, N., and Fan, L.S. (2012). Investigation of high-temperature steam hydration of naturally derived calcium oxide for improved carbon dioxide capture capacity over multiple cycles. *Energy Fuels* 26: 3903–3909.

80 Sun, P., Lim, C.J., and Grace, J.R. (2008). Cyclic CO_2 capture by limestone-derived sorbent during prolonged calcination carbonation cycling. *AIChE J.* 54 (6): 1668–1677.

81 Grasa, G., Martínez, I., Diego, M.E., and Abanades, J.C. (2014). Determination of CaO carbonation kinetics under recarbonation conditions. *Energy Fuels* 28: 4033–4042.

14

Chemical Looping of Low-Cost MgO-Based Sorbents for CO_2 Capture in IGCC

Hamid Arastoopour and Javad Abbasian

Department of Chemical and Biological Engineering, Wanger Institute for Sustainable Energy Research (WISER), Illinois Institute of Technology, Chicago, IL, USA

14.1 Introduction

Biomass and fossil fuels including coal continue to play a significant role in the total energy picture. Advanced power generation technologies such as integrated gasification-combined cycle (IGCC) are among the leading contenders for power generation conversion in the twenty-first century because such processes offer significantly higher efficiencies and superior environmental performance compared to coal and biomass combustion processes. It is envisioned that these advanced systems can competitively produce low-cost electricity at efficiencies higher than 60% while achieving "near-zero discharge" energy plants if the environmental concerns associated with these processes, including climate change, can be effectively eliminated at competitive costs [1].

Near-term applications of CO_2 capture from pre-combustion systems will likely involve physical or chemical absorption processes. However, these commercially available processes (e.g. SELEXOL) operate at low temperatures, imparting a severe energy penalty on the system and, consequently, their use could significantly increase the costs of electricity production. Therefore, development of high temperature regenerative processes based on solid sorbents offers an attractive alternative option for carbon capture at competitive costs.

Illinois Institute of Technology (IIT) has developed a regenerative high-temperature CO_2 capture process that is capable of removing more than 98% of CO_2 from a simulated water–gas shift (WGS) mixture at IGCC conditions using highly reactive and mechanically strong MgO-based sorbents and a circulating fluidized bed (CFB) loop. Furthermore, the sorbent exhibited some WGS catalytic activity at 300 °C, increasing hydrogen concentration from 37% (inlet) to about 70% in the reactor exit [2]. The results of theoretical modeling of the sorbent/catalyst performance in a packed bed indicated that hydrogen concentration in the simulated WGS mixture can achieve more than 95%

Handbook of Chemical Looping Technology, First Edition. Edited by Ronald W. Breault.

conversion. The cyclic chemical reactions for CO_2 capture involving magnesium oxide are

$$MgO + CO_2 \rightarrow MgCO_3 \quad (CO_2 \text{ absorption reaction}) \tag{14.1}$$

$$MgCO_3 \rightarrow MgO + CO_2 \quad (\text{regeneration reaction}) \tag{14.2}$$

$$CO + H_2O \leftrightarrow CO_2 + H_2 \quad (\text{water–gas-shift reaction}) \tag{14.3}$$

The regenerability and long-term durability of the sorbent have been demonstrated over 25 consecutive absorption/regeneration cycles, indicating that the sorbent is suitable for long-term applications. It was also shown that only the outer layer (40–50 μm thick) of the sorbent (particle diameter about 500 μm) reacted with CO_2 [3]. Therefore, it is expected that, by reducing the particle size to about 100–200 μm, the CO_2 absorption capacity of the sorbent can be significantly improved. The results of theoretical modeling of the sorbent/catalyst performance in a packed bed indicated that the hydrogen concentration in the simulated WGS mixture can achieve more than 95%.

CFB loop-based reactors have the potential to be among the most important devices in the chemical and energy industries. CFB reactors are currently used in fluid catalytic cracking (FCC) applications, with more than seven decades of history and more than 400 units in operation worldwide today [4]. Furthermore, gasification of coal and biomass, synthesis of olefin from methanol, and chemical looping are among the relatively new applications of CFB reactors [5]. Thus, the CFB reactor ensures a continuous carbon dioxide removal process in a relatively compact unit using solid particles, which makes it an excellent candidate for chemical looping of MgO-based sorbents for CO_2 capture and regeneration. The basic configuration of a CFB reactor consists of a riser where the particles are transported by the gas flow, a cyclone to separate gas and solid at the top of the riser, a standpipe (down-comer) to return the separated solid to the riser inlet, and a flow-controlling device (e.g. L-valve) to control the solid flow. In processes that include a regenerable sorbent or catalyst, a second fluidized bed reactor can be added between the down-comer and the L-valve to serve as a regenerator reactor. Furthermore, computational fluid dynamics (CFD) provides an excellent approach to the reactor design in a systematic and economically feasible way. To use CFD to perform simulations of the CO_2 capture regenerative process, a model based on the multiphase flow dynamics governing equations taking into account the absorption/regeneration and the WGS kinetics is needed. Therefore, in this work a CFD approach was used to describe CO_2 sorption and regeneration in CFB loop reactors using an MgO-based sorbent.

The key experimentally verified parameters needed for the CFD modeling of the absorption/regeneration reactor system including the absorption/regeneration and the WGS reaction rates and their dependence on the operating conditions (i.e. temperature, pressure, gas composition, catalyst/sorbent ratio, etc.) were also developed.

This chapter describes a chemical looping of MgO-based sorbents for CO_2 capture in IGCC, which uses a process similar to the Nation Energy Technology Laboratory (NETL) carbon capture unit (C2U) experimental setup [6]. Gas containing CO_2 enters the bottom of the riser sorption reactor and mixes with fresh sorbent. The MgO-based sorbent particles mix with the coal gases and absorb CO_2 through chemical reaction. The CO_2-laden particles flow up the riser, turn, and flow into the cyclone. CO_2-lean gas is separated from particles in the cyclone and exits the system, and the CO_2-laden particles pass through a loop-seal and enter the regenerator where CO_2 is released from the sorbent particles by heating the spent sorbent with high-temperature steam. The CO_2 lean gas exits the carbon capture unit (C2U) system and the regenerated sorbent particles continue through the loop to the next loop-seal. The fresh sorbent particles pass through the loop-seal to the riser and the process continues. To maintain gas-particle flow in a CO_2 capture loop, gases need to be injected around the system to keep particles fluidized. Figure 14.1 shows the schematic diagram of our CFB loop.

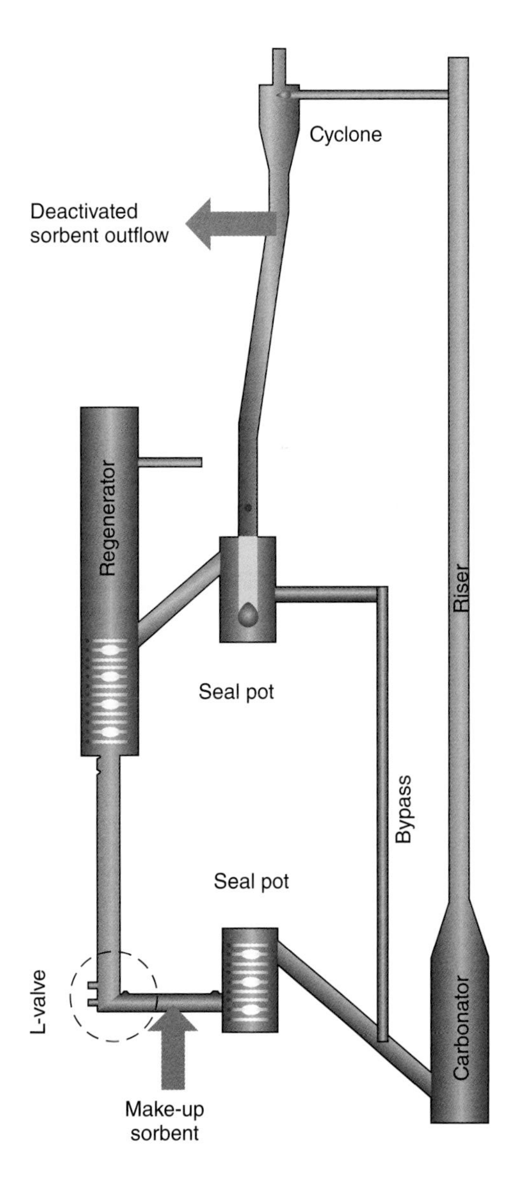

Figure 14.1 Schematic diagram of a circulating fluidized bed (CFB) loop.

14.2 MgO-Based Sorbent

The low-cost MgO-based sorbents were prepared by crushing and screening dolomite into the size range of 150–180 μm. The dolomite particles were dried at 120 °C and partially calcined at 550 °C to decompose the $MgCO_3$ while maintaining $CaCO_3$ in the sorbent. This method of calcination increases the porosity and surface area of the dolomite while maintaining its high mechanical strength. The partially calcined dolomite was impregnated with a solution of potassium carbonate to enhance the reactivity of the sorbent toward CO_2. The impregnated sorbents were dried and re-calcined in the same temperature range to stabilize the sorbent properties for carbonation/regeneration reactions at high temperatures. The phase components and elemental composition of the sorbents were determined using X-ray diffraction (XRD), indicating that the main components are $CaCO_3$, MgO, $K_2Ca(CO_3)_2$, and $K_2Ca_2(CO_3)_3$, while $CaMg(CO_3)_2$ exists in a negligible amount.

The reactivity of the sorbent toward CO_2 was determined in a dispersed bed reactor. A schematic diagram of the dispersed bed reactor unit is presented in Figure 14.2. The unit consists of three main subsections: the gas feeding section, the reactor section, and the gas effluent section. For the tests involving N_2/CO_2 mixtures, the feed gases are supplied by the pressurized cylinders flow through pre-calibrated mass flow controllers (MFCs) to accurately achieve the desired mixture composition before entering the reactor. Stainless steel tubing with 0.318 cm outside diameter (OD) is used to deliver the gas mixture to the reactor.

A highly sensitive backpressure regulator was installed downstream of the reactor to maintain the total pressure of the reactor at a predetermined setting. A manual bubble flow meter was placed at the very end of the unit to measure the changes in the flow rate of the reactor effluent.

The reactor is a custom-made 316 stainless steel tubular reactor with 2.54 cm OD × 1.905 cm inside diameter (ID) and 75 cm length. A porous frit with 2 μm openings made of Hastelloy®-276 alloys was repaired 30 cm above the bottom of the reactor. Approximately 45 cm of the reactor was externally heated with a single-zone electric tubular furnace. Two thermocouples were placed inside the bed of sorbents and below the porous frit and were connected to the data acquisition system (Figure 14.2). The thermocouple inside the bed, with 0.102 cm OD, is placed inside a 0.318 cm thermo-well and is moved along the bed during each run to measure the temperature of the bed at different axial positions.

To evaluate the sorbent in a well-dispersed differential reactor, about 1 g of the sorbent was distributed in 20 cm^3 of quartz beads (diameter = 850–1200 μm) along the reactor. To enhance a uniform radial distribution of the gas composition, as well as the gas flow rate (plug flow), 7 cm of quartz beads were placed on the top of the sorbent bed. The test is initiated by setting the backpressure regulator at the desired value and feeding nitrogen to the reactor until the reactor reaches the desired pressure and becomes stabilized. The bed temperature is gradually increased to reach the desired settings. To determine the bed temperature during the CO_2 adsorption, the bed temperature was monitored and measured at seven locations along the sorbent bed. After reaching the desired temperature and pressure inside the reactor, the test begins by switching the gas

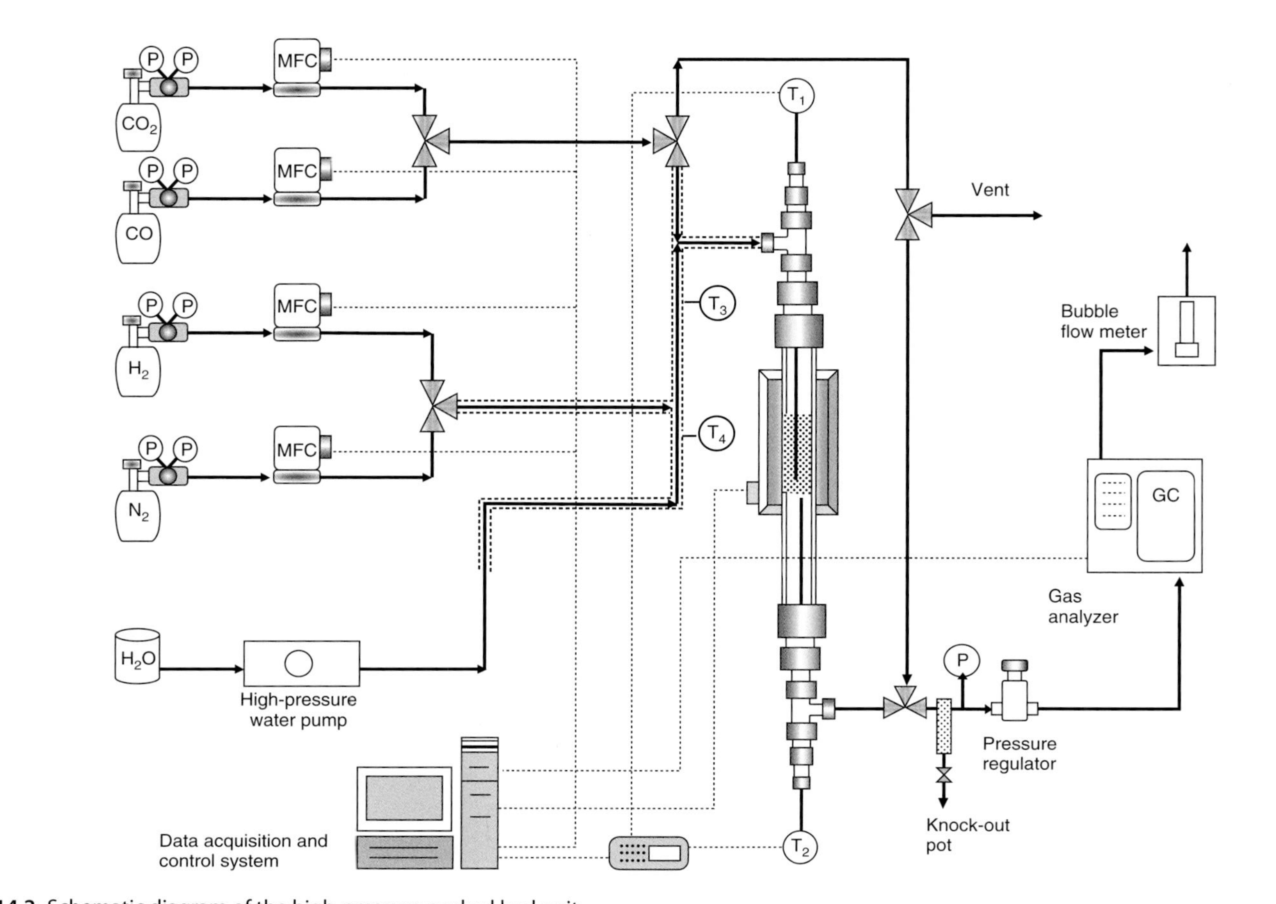

Figure 14.2 Schematic diagram of the high-pressure packed bed unit.

to a gas mixture containing CO_2 for a predetermined period. Following the test, the reactor is rapidly cooled and the reacted sample is removed from the reactor and analyzed by XRD.

The effect of various preparation parameters on the reactivity of the sorbent was studied to identify the best sorbent formulation. The parameters studied included drying, calcination, and re-calcination temperatures, heat-up rates, and solution molality during impregnation and the resulting potassium to magnesium ratio.

Among all of the preparation parameters studied, the potassium/magnesium (K/Mg) ratio was identified as the key variable affecting the CO_2 capacity of the sorbent. The effect of the potassium/magnesium (K/Mg) ratio on the CO_2 capacity of all sorbents after 40 minutes residence time in the dispersed bed reactor is shown in Figure 14.3. The result clearly indicates that the optimum K/Mg ratio is about 0.15, which has generally been achieved when the sorbent was impregnated using a 1 M solution of potassium carbonate. The results also indicate that the optimum drying temperature during sorbent preparation is in the range of 70–100 °C and that the CO_2 capacity of the sorbent decreases as the re-calcination temperature is increased from 500 to 550 °C, suggesting possible sintering of the sorbent at the higher temperature. The preparation parameters for the best sorbent formulation (i.e. HD52-P2) are presented in Table 14.1.

The effect of the reaction temperature on the carbonation reaction rate involving HD52-P2 sorbent was investigated in a series of tests in which the sorbent was exposed to pure CO_2 and 20 atm at different temperatures and sorbent residence times. The results are presented in Figure 14.4, indicating that as long as the absorption temperature is below 450 °C, the reactivity of the sorbent improves with increasing temperature (i.e. 340–450 °C). The lower conversion in the sorbent reactivity at 490 °C is due to the significantly higher equilibrium CO_2 partial pressure in carbonation reaction at higher temperatures. The higher equilibrium CO_2 partial pressure results in lower concentration driving force, which is the difference between the CO_2 partial pressure in the gas mixture and the CO_2

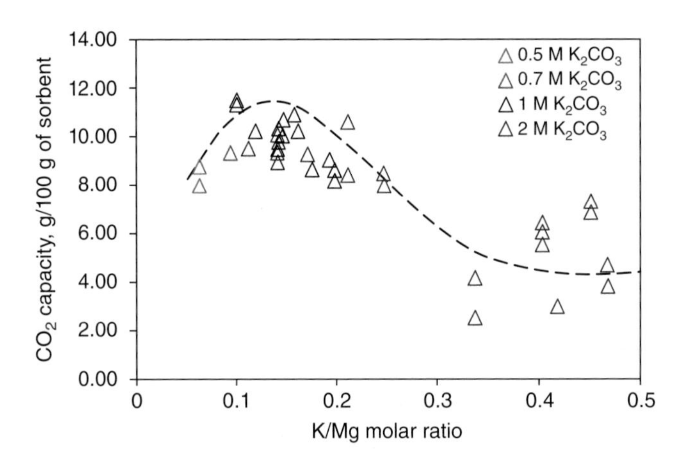

Figure 14.3 Effect of K/Mg ratio on the capacity of the sorbent.

Table 14.1 Preparation parameters for HD52-P2 sorbent.

Preparation parameters	HD52-P2
Sorbent particle diameter (μm)	150–180
Calcination temperature (°C)	520
Calcination temperature ramp (°C min^{-1})	1
Duration of calcination (h)	8
Concentration of potassium carbonate in the impregnation solution (mol l^{-1} or M)	1
Duration of impregnation (h)	20
Drying temperature (°C, post-impregnation)	90
Humidity during drying (%)	Ambient
Duration of drying (h)	24
Re-calcination temperature (°C, post-drying)	470
Calcination temperature ramp (°C min^{-1})	1
Duration of re-calcination (h)	4

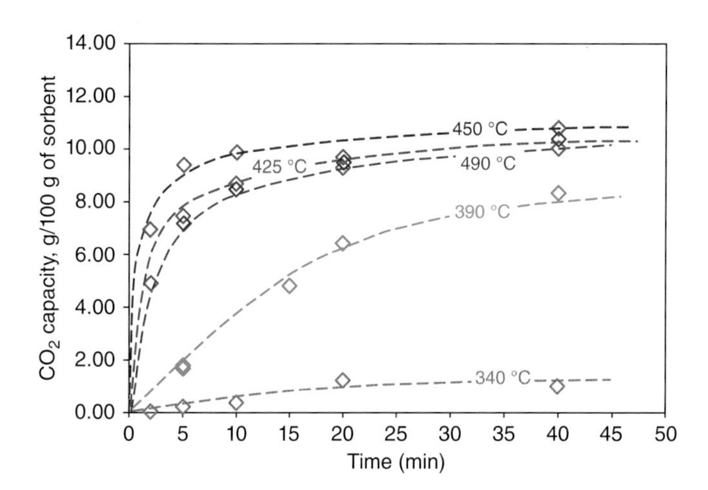

Figure 14.4 Effect of temperature on the absorption reactivity of the HD52-P2 sorbent.

equilibrium pressure at the reaction operating temperature (PCO_2 – Pe), leading to a lower reaction rate at higher temperatures.

The reversible chemical reaction between the porous MgO-based sorbent and CO_2 can be expressed as follows:

$$MgO\ (s) + CO_2(g) \leftrightarrow MgCO_3(s) \quad (14.4)$$

The intrinsic rate of reaction can generally be described by the following equation:

$$r_{MgO} = -[1/(4\pi r_i^{\,2})] \cdot [(dN_{MgO})/dt] = k_s \cdot (C_i - C_e)^n \quad (14.5)$$

The dependence of the intrinsic reaction rate constant k_s on temperature is expressed in terms of the Arrhenius equation:

$$k_s = k_{s0} \cdot e(-E_a/RT) \tag{14.6}$$

To determine the order of reaction (n), the intrinsic reaction rate constant (k_s), and the activation energy (E_a), the initial slopes of the conversion curves (i.e. $(dX/dt)_{t=0}$) were obtained from the dispersed bed experimental data and the order of reaction was obtained by the slope of $\text{Ln}(dX/dt)_{t=0}$ versus $\ln(C_b - C_e)$. The results indicate that the order of the carbonation reaction with respect to CO_2 concentration is one. Similar gas/solid reactions involving alkali/alkaline material with reactant gases such as CO_2 and SO_2 generally have been reported as first order in the literature. The activation energy of the reaction was obtained by the Arrhenius plot of the intrinsic rate to be 134 kJ mol^{-1} for the temperature range of 350–420 °C.

The effect of steam on the CO_2 reactivities of the sorbent is presented in Figure 14.5. The results indicate that the presence of steam significantly enhances the sorbent reactivity and capacity. A similar trend was observed at other temperatures below 420 °C, and the strongest effect was observed at 360 °C. The effect of steam in increasing the porosity and changing the pore size distribution has been reported for dolomite and limestone by others [7–9]. Wolff et al. [10] found that the adsorption of H_2O on the surface of Pt/MgO catalyst used for the WGS reaction is very strong, and a transient $Mg(OH)_2$ component is created as the catalyst is exposed to the H_2O environment. Liu and Shih [11], who studied calcium sulfation reactions in the presence of humid simulated flue gases, postulated that H_2O enhances the carbonation reaction by forming a region of dense water vapor around the reacting MgO particles where CO_2 reacts to form carbonate ions and H^+ ions, and the released Mg^{2+} ions can react with the carbonate ions to form $MgCO_3$.

To determine the effect of temperature on the rate of the regeneration reaction, the carbonated sorbent was heated in the CO_2/N_2 (50/50 mol%) environment up

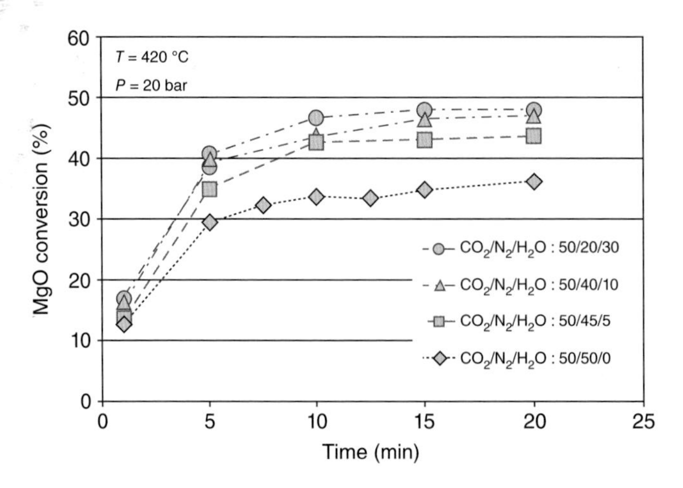

Figure 14.5 Effect of steam on sorbent reactivity and capacity.

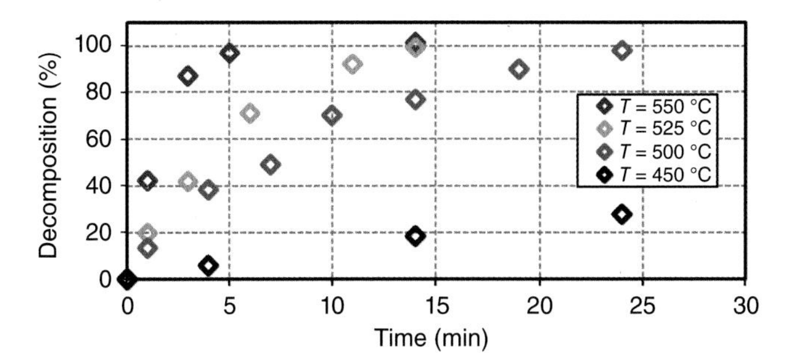

Figure 14.6 Effect of temperature on regeneration reaction.

to the regeneration temperature. The reactor pressure was maintained at 20 atm through the entire regeneration test. The regeneration step was started by switching the gas mixture from a 50/50 mixture of CO_2/N_2 to pure nitrogen gas. Following each test, the sorbent was removed from the reactor and analyzed by XRD. The results are shown in Figure 14.6, indicating that, as expected, the reactivity of the regeneration reaction improves with increasing temperature (i.e. between 450 and 550 °C).

14.3 Reaction Model for Carbon Capture and Regeneration

To assess the sorbent carbonation in the regenerative MgO-based process, the variable diffusivity shrinking core model (VDM) [12] was selected to capture the behavior of the sorbent at different conditions to predict the long-term performance of the sorbent in various reactor configurations. A schematic diagram of the model is illustrated in Figure 14.7. In this model, the reaction occurs at the interface between the shrinking unreacted core and the product layer. As the reaction proceeds, the difference in the larger molar volumes of the solid product and the smaller molar volume of the reactant leads to decreasing pore volume of the product layer, decreasing the diffusion rate of the gaseous reactant through

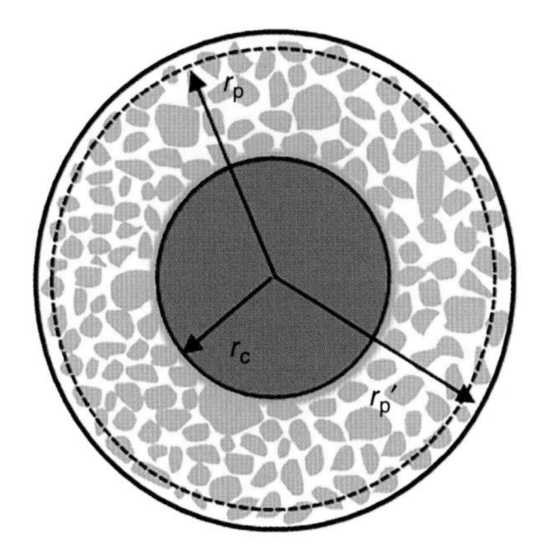

Figure 14.7 Schematic diagram of expanded particle.

the sorbent particle. The rate of carbonation reaction at the surface of the shrinking core can be described as

$$r_{MgO} = -\frac{1}{4\pi r^2}\frac{dN_{MgO}}{dt} = k_s(C_b - C_e) \tag{14.7}$$

The governing differential equations describing the transport of CO_2 through the porous particle and its reaction at the surface of the individual particles, along with the relevant initial and boundary conditions, are given below:

$$D_e\left[\frac{1}{r^2}\frac{\partial}{\partial r}\left(r^2\frac{\partial C}{\partial r}\right)\right] = 0 \tag{14.8}$$

$$C = C_R \quad \text{at } r = R \tag{14.9}$$

$$D_e\frac{\partial C}{\partial r} = k_s(C_i - C_e) \text{ at } r = r_i \tag{14.10}$$

$$\frac{dr_i}{dt} = \frac{-k_s}{C_{MgO}}\left[\frac{C_b - C_e}{1 + \frac{k_s}{D_e}r_i\left(1 - \frac{r_i}{r'_p}\right)}\right] \tag{14.11}$$

$$r'_p = \sqrt[3]{r_i^3 + Z_v(r_p^3 - r_i^3)} \tag{14.12}$$

$$Z_v = \frac{MW_p/\rho_p(1-\varepsilon)}{MW_r/\rho_r(1-\varepsilon_s)} \tag{14.13}$$

$$\frac{dX}{dt} = -\frac{\frac{3}{r_p}\frac{k_s}{N^{\circ}_{MgO}}(C_b - C_e)(1-X)^{\frac{2}{3}}}{1 + \frac{k_s}{D_e}r_p(1-X)^{\frac{1}{3}}\left(1 - \sqrt[3]{\frac{1-X}{1-X+XZ}}\right)} \tag{14.14}$$

In the VDM, the extent of the non-catalytic gas–solid reaction is dictated by the intrinsic rate of reaction as well as the rate of diffusion of the reactant gas through the solid particle. The reactant gas (i.e. CO_2) has to diffuse through the particle product layer before reaching the reaction interface. The diffusion rate through this layer changes during the carbonation reaction because of the changes in the product layer properties, resulting in changes in the porosity. Therefore, product layer diffusivity can be a significant function of sorbent physical properties affected by changes in the overall conversion and/or the duration of exposure to the reaction environment. In this study, we assumed that the diffusivity through the product layer is an exponential function of the conversion of the particle.

$$D_e = D_{e_0}\exp(-\alpha \cdot X^2) \tag{14.15}$$

The VDM was used to fit the experimental data obtained in the dispersed bed reactor at different operating conditions. The two adjustable parameters of the VDM (D_{eo} and α) were estimated by determination of the best fit to the experimental results through minimization of the least squares of the errors. The results are presented in Figures 14.8 and 14.9 and indicate that the model can provide an excellent fit to the experimental data.

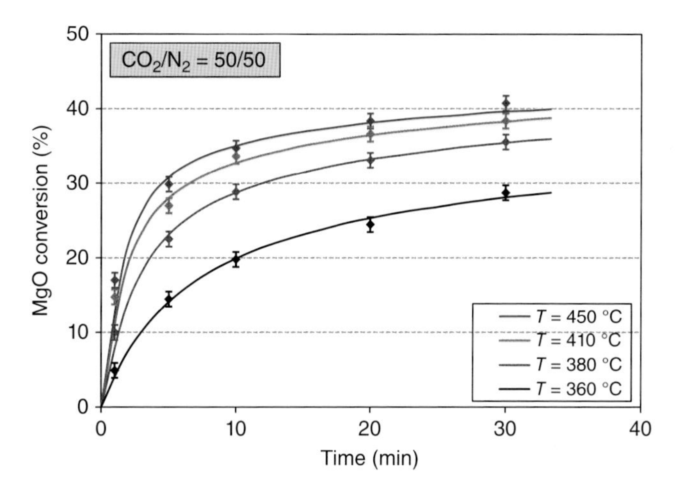

Figure 14.8 VDM fit to carbonation reaction data (dry carbonation).

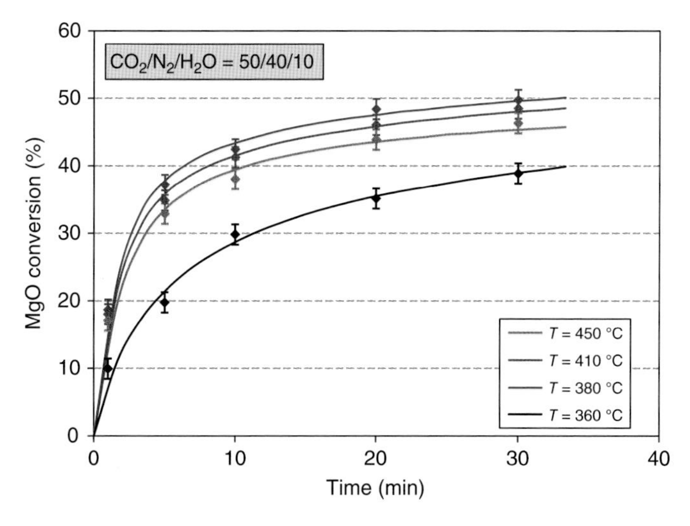

Figure 14.9 VDM fit to carbonation reaction data (10 mol% steam).

To assess the sorbent regeneration in the regenerative MgO-based process, the shrinking core model (SCM) by Yagi and Kunii [13] was selected to capture the behavior of the sorbent at different conditions. A schematic diagram of the model is illustrated in Figure 14.10. The reaction occurs at a sharp interface, which divides the reacted outer layer (product layer) from the un-reacted core of the solid.

The rate of carbonation reaction at the surface of the shrinking core can be described as

$$r_{\mathrm{MgCO_3}} = k_{\mathrm{d}}\left(1 - \frac{C_{\mathrm{i}}}{C_{\mathrm{e}}}\right) \tag{14.16}$$

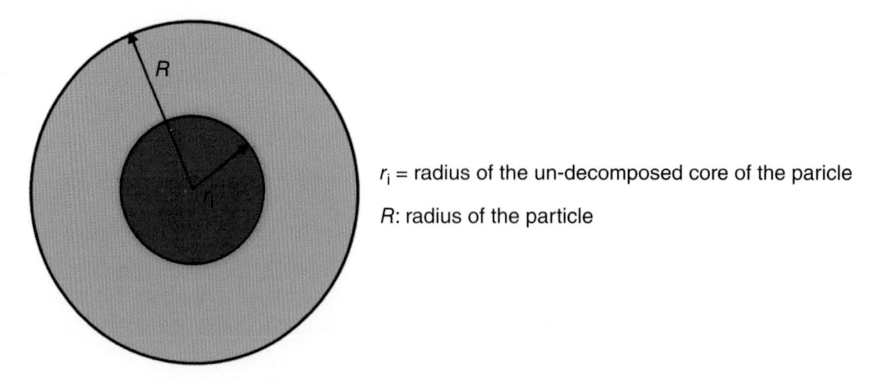

Figure 14.10 Schematic diagram of regenerated particle.

The governing differential equations describing the transport of CO_2 through the porous particle and its reaction at the surface of the individual particles, along with the relevant initial and boundary conditions, are given below:

$$D_e\left[\frac{1}{r^2}\frac{\partial}{\partial r}\left(r^2\frac{\partial C}{\partial r}\right)\right] = 0 \tag{14.17}$$

$$C = 0 \quad \text{at } r = R \tag{14.18}$$

$$D_e\frac{\partial C}{\partial r} = k_d\left(1 - \frac{C_i}{C_e}\right) \quad \text{at } r = r_i \tag{14.19}$$

$$\frac{\partial X}{\partial t} = \frac{3r_i^2 k_d}{R^3 N^\circ_{MgCO_3}}\left(1 - \frac{C_i}{C_e}\right) \tag{14.20}$$

$$\frac{\partial X}{\partial t} = \frac{\frac{3}{R}\frac{k_d}{N^\circ_{MgCO_3}}\sqrt[3]{(1-X)^2}}{1 + \frac{k_d}{D_e C_e}R\sqrt[3]{1-X}\left(1 - \sqrt[3]{1-X}\right)} \tag{14.21}$$

To model the sorbent regeneration behavior, the two adjustable parameters of the shrinking core model (D_e and k_d) were estimated by determination of the

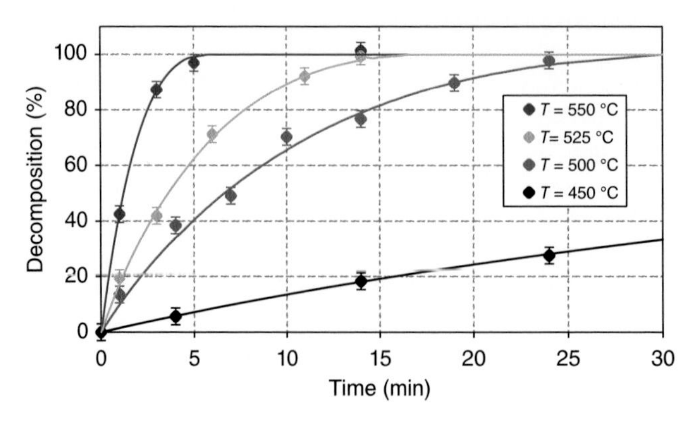

Figure 14.11 Shrinking core model fit to regeneration reaction data.

best fit to the experimental data through minimization of the least squares of the errors. The results are presented in Figure 14.11 and indicate that the model can provide an excellent fit to the experimental data.

14.4 CFD Simulations of the Regenerative Carbon Dioxide Capture Process

During the last three decades, extraordinary advances in the CFD approach to fluid-particle flow systems [14–16] have significantly impacted our approach in the design and scale-up of processes based on fluid-particle including CFB loop for sorption and regeneration. CFD models for fluid-particle flow systems are based on the continuum theory and conservation laws for mass, momentum, and energy. In order to close the conservation equations for the momentum, one needs to calculate the stress tensors for solid phase. The most accepted approach is based on the kinetic theory of granular flow [17, 18].

The kinetic theory approach, which is based on the oscillation of the particles, uses a granular temperature equation to determine the turbulent kinetic energy of the particles, assumes a distribution function for instantaneous particle velocity, and defines a constitutive equation based on particle collision, interaction, and fluctuation. In fact, the kinetic theory approach for granular flow allows the determination of, for example, particle phase stress, pressure, and viscosity in place of the empirical equations.

However, in a more concentrated fluid-particle flow system, not only should the flow be characterized by a two- or three-dimensional flow equation, but also the formation of large structures such as clusters should be included in modeling of transport phenomena of such flow systems. To solve these microscopic two-fluid model equations, very small grid sizes of less than 10 particle diameter are needed. For most processes of practical interest, such fine spatial grids and small time steps require significant computational time. Thus, the effect of the large-scale structures using coarse grids should be accounted for by using approaches such as filtering equations or energy minimization multi-scale, EMMS [19–22]. In this study, the kinetic theory approach along with EMMS for drag reduction, $k - \varepsilon$ model for turbulence, Sinclair and Mallo [23] expression for granular energy exchange with turbulent gas flow, expressions of Laux [24] and Lun et al. [25] for frictional and bulk viscosities, and Johnson and Jackson [26] for boundary conditions were used. For more details about constitutive relations, see Arastoopour et al. [27].

Table 14.2 summarizes the governing equations used in this work.

14.4.1 Two-dimensional Simulation of the Regenerator and the Carbonator in the CFB Loop

The objective is to numerically simulate the CO_2 capture process using CFB reactors at elevated temperature and pressure by including both regeneration reaction and carbonation reaction. However, it is computationally expensive to conduct the simulations in a three-dimensional domain because of inclusion of two

Table 14.2 Two-fluid model governing equations based on the kinetic theory approach.

Conservation of mass

Gas phase

$$\frac{\partial}{\partial t}(\varepsilon_g \rho_g) + \nabla \cdot (\varepsilon_g \rho_g \boldsymbol{v}_g) = \dot{m}_g \tag{14.22}$$

Solid phase

$$\frac{\partial}{\partial t}(\varepsilon_s \rho_s) + \nabla \cdot (\varepsilon_s \rho_s \boldsymbol{v}_s) = \dot{m}_s \tag{14.23}$$

$$\varepsilon_g + \varepsilon_s = 1 \tag{14.24}$$

Conservation of momentum

Gas phase

$$\frac{\partial}{\partial t}(\varepsilon_g \rho_s \boldsymbol{v}_g) + \nabla \cdot (\varepsilon_g \rho_g \boldsymbol{v}_g \boldsymbol{v}_g) = \varepsilon_g \nabla P + \nabla \cdot \tau_g + \varepsilon_g \rho_g g - \beta_{gs}(\boldsymbol{v}_g - \boldsymbol{v}_s) \tag{14.25}$$

Solid phase

$$\frac{\partial}{\partial t}(\varepsilon_s \rho_s \boldsymbol{v}_s) + \nabla \cdot (\varepsilon_s \rho_s \boldsymbol{v}_s \boldsymbol{v}_s) = \varepsilon_s \nabla P + \nabla \cdot \tau_s - \nabla p_s + \varepsilon_s \rho_s g - \beta_{gs}(\boldsymbol{v}_g - \boldsymbol{v}_s) \tag{14.26}$$

Conservation of species

Gas phase

$$\frac{\partial}{\partial t}(\varepsilon_g \rho_g y_j) + \nabla \cdot (\varepsilon_g \rho_g \boldsymbol{v}_g y_j) = R_j \tag{14.27}$$

Solid phase

$$\frac{\partial}{\partial t}(\varepsilon_s \rho_s y_i) + \nabla . (\varepsilon_s \rho_s \boldsymbol{v}_s y_i) = R_j \tag{14.28}$$

Conservation of solid phase fluctuating energy

$$\frac{3}{2}\left[\frac{\partial(\rho_s \varepsilon_s \theta)}{\partial t} + \nabla \cdot (\rho_s \varepsilon_s \theta v_s)\right] = (-P_s \mathbf{I} + \tau_s) : \nabla v_s + \nabla \cdot (k_s \nabla \theta) - \gamma + \phi_{gs} \tag{14.29}$$

heterogeneous reactions as mentioned above. Therefore, to obtain similar results as three-dimensional simulation using two-dimensional simulations and significantly reduced computational time, two-dimensional simulations with a higher specularity coefficient were used. In order to further reduce the computational time, the carbonation reaction was assumed to take place only in fluidized bed No. 1 (carbonator) at a temperature of 380 °C, while the regeneration reaction was assumed to take place only in fluidized bed No. 2 (regenerator) at a temperature of 500 °C. The system pressure was set to 50 atm. Figure 14.12 shows a schematic diagram of a CFB loop under study. Initially the L-valve, standpipe, and fluidized bed No. 2 were filled with particles at 45% voidage, while all other parts were filled with particles at 50% voidage (see Figure 14.12). The particles used in this study were 185 μm in diameter and of 2500 kg m^{-3} density. In addition, the initial sorbents in the regenerator and loop-seal 1 were assumed to be 65% carbonated, while all other parts of the system were filled with the fresh sorbent (see Table 14.3). The inlet gas in the regenerator was pure steam and in

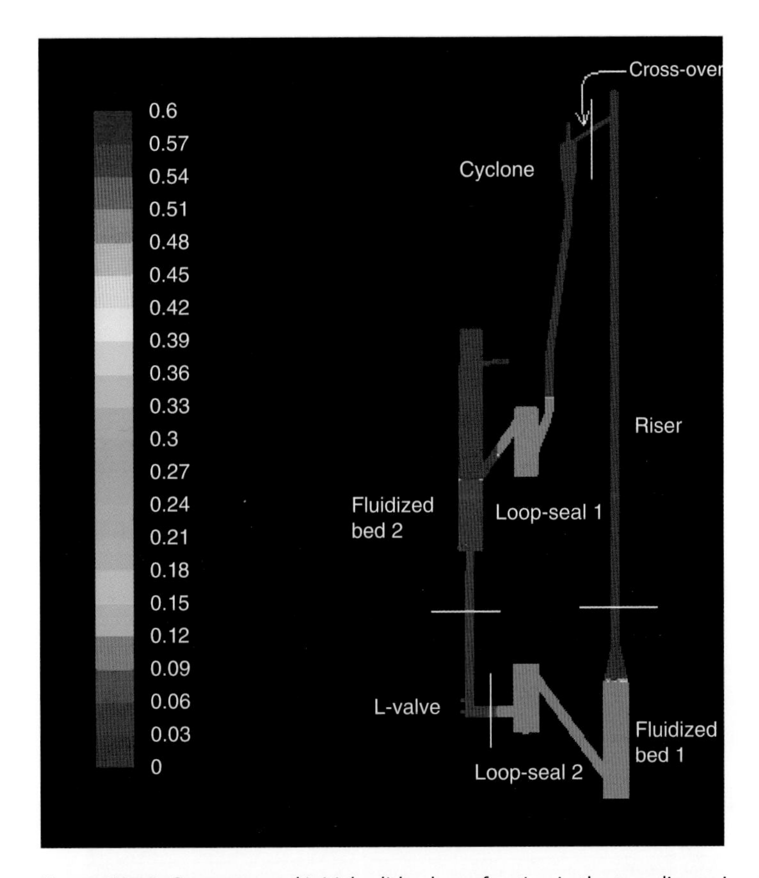

Figure 14.12 Geometry and initial solid volume fraction in the two dimensional circulating fluidized bed (CFB) loop.

Table 14.3 Fresh MgO-based sorbent analysis.

Analysis	wt%
$MgCO_3$	0.00
MgO	25.20
$CaCO_3$	49.70
$CaMg(CO_3)_2$	13.00
$K_2Ca_2(CO_3)_3$	0.60
$K_2Ca(CO_3)_2$	11.50
Total	100.00

the carbonator inlet, the CO_2 mole fraction was 14.26%, and the rest was steam. Table 14.4 shows the boundary conditions used in the simulations.

Figure 14.13 shows the contours of the CO_2 mole fraction at different times. The mole fraction of CO_2 is almost identical after 10 seconds, suggesting that steady-state operation of the CFB loop has been reached.

Table 14.4 Boundary conditions used for the two-dimensional simulation of the CFB with reactions.

Fluidized bed No. 1 inlet gas velocity (m s^{-1})	0.267
Fluidized bed No. 2 inlet gas velocity (m s^{-1})	0.100
Loop-seal 1 inlet gas velocity (m s^{-1})	0.030
Loop-seal 2 inlet gas velocity (m s^{-1})	0.030
L-valve aeration inlet gas velocity (m s^{-1})	0.004
Pressure (atm)	50
Gauge pressure at the cyclone outlet (atm)	0
Gauge pressure at the regenerator outlet (atm)	0
Temperature in the regenerator (°C)	500
Temperature in the carbonator (°C)	380

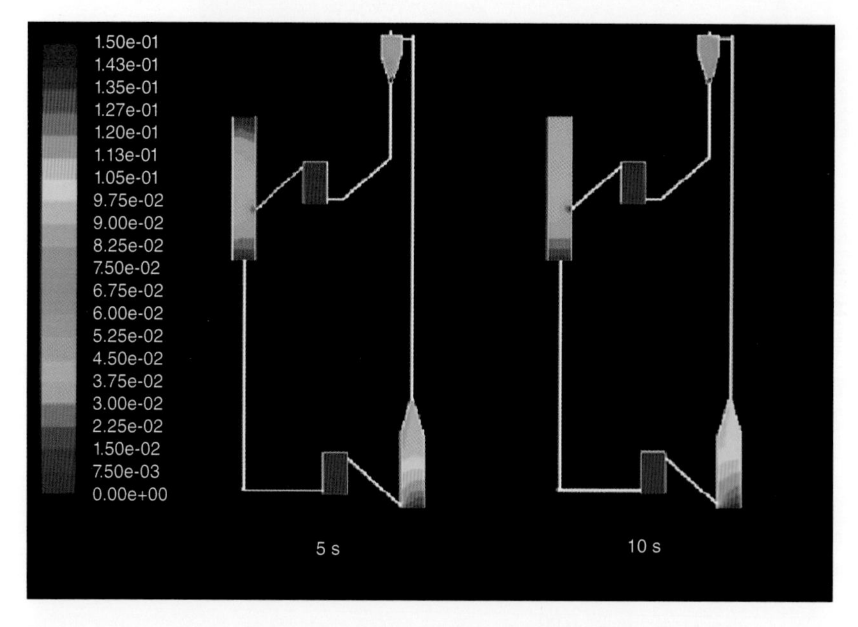

Figure 14.13 Contours of CO_2 mole fraction in the system as a function of time.

14.4.2 Three-dimensional Simulation of Carbon Capture and Regeneration in the Absorber and Regenerator Reactors

Figure 14.14 shows the contours of the instantaneous solid volume fraction and CO_2 mole fraction at $t = 20$ seconds for a solid circulation rate of 220 g s^{-1} in the CO_2 sorption reactor. The solid volume fraction contours show a very dense and well-mixed solid phase in the carbonator and a dilute region in the riser. Reaction rate contours also show that most of the CO_2 capture takes place in the carbonator with very little reaction in the riser.

The effect of pressure on the regeneration of MgO-based sorbent in the presence of steam is presented in Figure 14.15. The results shown in Figure 14.15 indicate that as the system pressure decreases, the mole fraction of CO_2 increases,

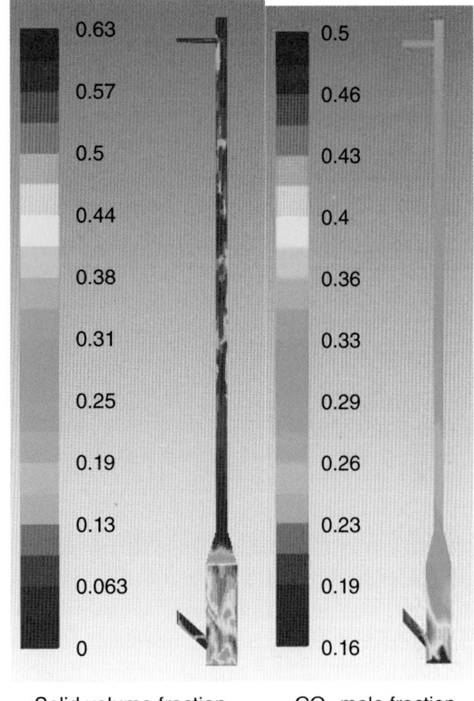

Figure 14.14 Contours of instantaneous solid volume fraction and CO_2 mole fraction in the CO_2 absorber (fluidized bed No. 1) at $t = 20$ seconds, for a solid circulation rate of $220\,g\,s^{-1}$. (Source: This figure was originally published in *Powder Technology* 286, 2015 and has been reused with permission.)

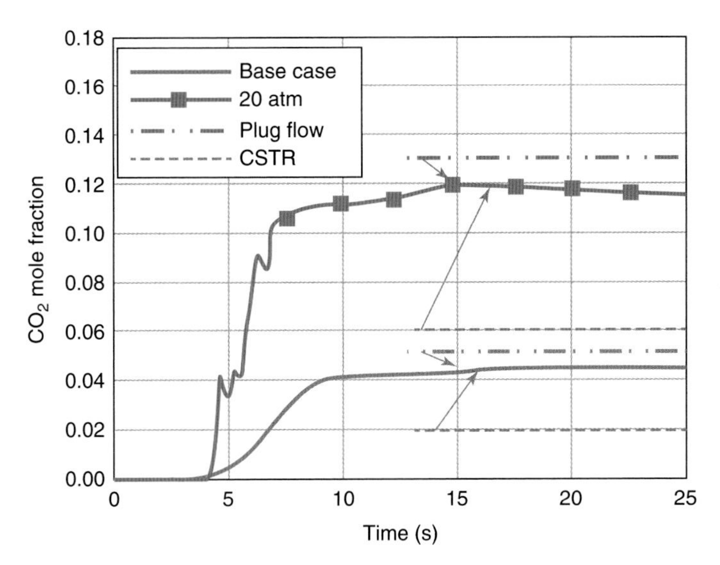

Figure 14.15 Comparison of generated CO_2 mole fraction as a function of time at two different pressures of 50 and 20 atm. (Source: This figure was originally published in *Powder Technology*, 2017 and has been reused with permission.)

which is partially due to the dilution at lower pressure (with regard to CO_2 concentration) and partially due to lower mass flow rate of the gas at lower pressure. The dashed lines in Figure 14.15 show the limiting cases for plug flow reactor (top) and CSTR (bottom) at steady-state conditions. Figure 14.15 shows that the fluidized bed behavior operating at elevated pressures is close to that of the plug flow reactor, which should be attributed to low levels of mixing in the reactor.

14.5 Preliminary Economic Assessment

A preliminary base case design was developed for a regenerative MgO-based pre-combustion carbon capture process that can be incorporated in a 500 MW IGCC power plant. The coal used for this study is a typical Illinois #6 bituminous coal. The composition of the coal is presented in Table 14.5 along with the high heating value (HHV) of the coal, which was calculated using the DuLong correlation. The overall efficiency of the IGCC plant was assumed to be 40%. The oxygen-blown gasifier was assumed to operate at 1000 °C and 50 atm using a steam-to-carbon ratio of one. The oxygen requirement and the syngas composition after high-temperature desulfurization (with regenerable mixed metal oxide sorbent) was determined by the FlexFuel Gasifier Simulation Model, FFGSM [28], developed by the coauthor at the process design and gas processing laboratory (PDGPL) at IIT. The operating condition of the gasifier and the composition of the syngas (calculated by FFGSM) are presented in Table 14.6. The key properties of the regenerable MgO-based sorbent used in this process are presented in Table 14.7.

In the regenerative MgO-based process envisioned, the sorbent is used in a CFB system, consisting of a number of carbonators and regenerators, cyclones, L-valves, seal pots, and down-comers. The rates of reaction for carbonation and regeneration are based on the reaction models described above. The "cleaned" syngas enters the carbonator where the solid sorbent reacts with the gas and captures a significant fraction of the CO_2 in the syngas and the CO_2-lean stream leaves the reactor through a riser. The solid sorbents in the carbonator are carbonated and then accelerated and pneumatically transported through the riser and

Table 14.5 Properties of Illinois #6 Coal.

Ultimate analyzes	wt% (moisture free)
C	70.2
H	4.8
N	0.9
S	3.1
O	9.9
A	11.1
Total	100
Moisture content (%)	5
HHV (btu lb^{-1})	12 545

Table 14.6 Operating condition of the gasifier and syngas composition.

Temperature (°C)	1000
Pressure (atm)	50
Steam-to-carbon ratio	1
Oxygen-to-carbon ratio	0.306
Coal feed rate ($lbm\ h^{-1}$)	3.58 e5
Steam feed rate ($lbmol\ h^{-1}$)	20.9 e3
Oxidant feed rate ($lbmol\ h^{-1}$)	6.4 e3
Product gas ($lbmolh^{-1}$)	4.24 e4
CO content (mol%)	25.09
H_2O content (mol%)	26.55
CO_2 content (mol%)	14.26
H_2 content (mol%)	27.16
N_2 content (mol%)	Balance

Table 14.7 Key sorbent properties.

Particle diameter (μm)	185
Density ($kg\ m^{-3}$)	2500
Minimum fluidization velocity ($m\ s^{-1}$)	0.03
Terminal velocity ($m\ s^{-1}$)	1.42

recycled in a cyclone. A fraction of the carbonated sorbent leaving the cyclone is fed to the regenerator and the rest (bypass) is fed to the carbonator to capture CO_2 in the following cycle. The ratio of the flow rate of solids bypassing the regenerator to the flow rate of the sorbent going through the regenerator is defined as the bypass ratio and is one of the most important parameters of this process. Figure 14.1 shows the schematic of the system for a single carbonator and a single regenerator. A sorbent make-up stream is introduced into the system to make up for deactivation of the sorbent as it goes through multiple cycles, and a deactivated sorbent flow is removed from the system to ensure the steady-state operation of the system. Table 14.8 shows the operating condition of the carbonators and risers used in the base case design calculations.

The fraction of CO_2 removal determines the change in the sorbent conversion as it passes through the carbonator. The solid circulating rate can be changed over a wide range depending on the application and is determined separately with the design of an appropriate L-valve. The temperature of the carbonator is set to result in the optimum rate of reaction in the carbonator. The flow rate of the sorbent make-up will specify the average number of cycles that the solid sorbent will be going through before it leaves the system, which has a strong effect on the reaction rate, and therefore the carbonator size. Obviously, higher make-up flow rate will result in lower cycle number, higher rate of reaction, and lower reactor size, but will increase the sorbent cost. The gas velocity in the carbonator should be higher than the minimum fluidization velocity of sorbents to ensure a smooth

Table 14.8 Operating and design variables in the carbonator and riser.

Number of carbonators and risers	10
Temperature (°C)	380
Pressure (atm)	50
Carbonator inlet gas velocity ($m\,s^{-1}$)	0.04
Riser gas velocity ($m\,s^{-1}$)	20
Solid circulating rate in each riser ($kg\,s^{-1}$)	20
Bypass ratio	0.15
Desired CO_2 removal (%)	90
Make-up flow rate ($kg\,s^{-1}$)	5

Table 14.9 Carbonator and riser sizing and design conditions.

Carbonator diameter (m)	4.3
Carbonator height (m)	20.7
Riser diameter (m)	0.19
Riser height (m)	3
Riser solid hold up (%)	0.05
Sorbent cycle number	33
Sorbent inlet conversion	10
Sorbent outlet conversion	65
CO_2 inlet concentration ($kmol\,m^{-3}$)	0.13
CO_2 outlet concentration ($kmol\,m^{-3}$)	0.02

flow of solid in the carbonator. The gas velocity in the riser is also selected to be higher than the terminal velocity of the solids to effectively transport the solids up to the cyclone. The riser gas velocity has a direct impact on the solid hold-up in the riser. A proper design will have a solid hold-up of less than 5% in the riser. The height of the riser, on the other hand, should be long enough to be able to accelerate the solids to the specified velocity. Table 14.9 shows the sizing and design calculations of each carbonator and riser.

Steam is used as a regeneration medium, mainly because the presence of steam improves the rate of the regeneration reaction (as discussed above), cost of the gas, and the ease of separation of CO_2 from steam (in a condenser). Also, the steam temperature in the inlet of the regenerator should be higher than the regenerator temperature in order to compensate partially for the energy needed to warm the solids to the regenerator temperature and provide the energy required for the endothermic regeneration reaction. The operating conditions of the regenerator are presented in Table 14.10.

The flow rate of steam was calculated based on the energy needs in the system and will be discussed below. Because the rate of regeneration is higher than

Table 14.10 Regenerator operating and design conditions.

Number of carbonators and risers	2
Temperature (°C)	500
Pressure (atm)	50
Regenerator inlet gas velocity ($m\,s^{-1}$)	0.13
Steam flow rate in each regenerator ($kg\,s^{-1}$)	10.07
Solid circulating rate in each regenerator ($kg\,s^{-1}$)	82.5
Extent of reaction (%)	100
Solid loss percentage in the cyclone (%)	0

Table 14.11 Regenerator sizing and design conditions.

Regenerator diameter (m)	5
Regenerator height (m)	15
Sorbent inlet conversion	65
Sorbent outlet conversion	0
CO_2 inlet concentration ($kmol\,m^{-3}$)	0
CO_2 outlet concentration ($kmol\,m^{-3}$)	0.12

the rate of carbonation, the number of regenerators is considerably less than the number of carbonators. The regeneration temperature was assumed to be 500 °C because of the significantly lower rate of sorbent deterioration (compared to the sorbents regenerated at 550 °C) as discussed above. In the base case design, the rate of solid loss in the cyclone was assumed to be negligible. It should be noted that, as long as the rate of solid loss is lower than the fresh sorbent make-up rate, any loss incurred in the cyclone can be offset by a reduction in the rate of spent sorbent removal from the system. The extent of sorbent regeneration is the major parameter in specifying the regenerator size. Table 14.11 shows the sizing and design calculations of each regenerator.

To produce the steam required in the regenerator, water is used and heated through an extensive heat integration network inside the system as well as utilization of external fuel (e.g. natural gas). The heat integration network was developed in Aspen/HYSYS® and is shown in Figure 14.16. A make-up water stream is used and some water is also purged to prevent CO_2 accumulation inside the system. Table 14.12 shows the important parameters of the heat integration network. The liquid CO_2 produced is at 50 atm pressure and can easily be pumped and sequestered or stored.

In the preliminary economic analyses performed in this task, the cost associated with the process consisted of operating costs and capital investment costs. The major components of the operating costs included sorbent cost, labor cost, and fuel cost, while the major components of the capital costs included those associated with the carbonators and regenerators and the heat integration network. Table 14.13 shows cost analyses for three different scenarios. The cost of

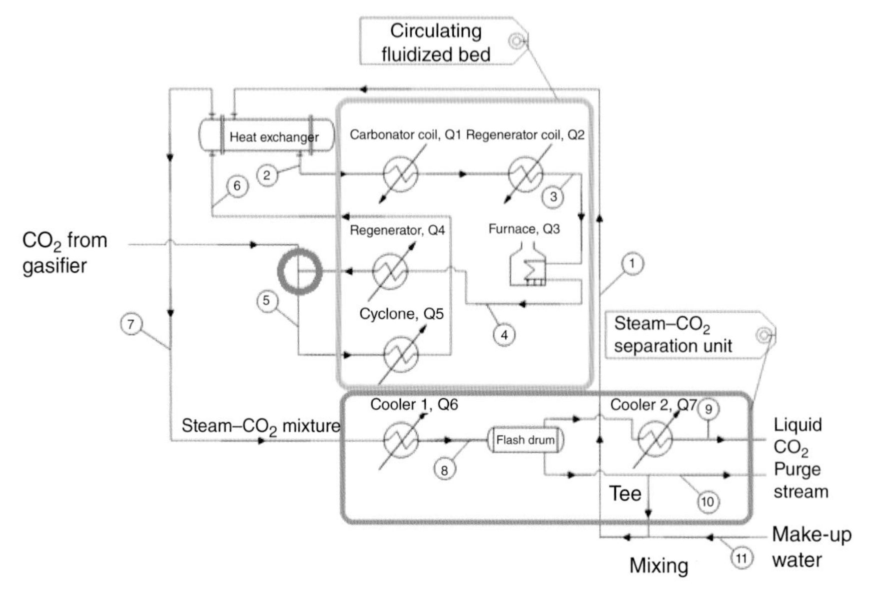

Figure 14.16 Heat integration scheme in the process.

Table 14.12 Heat integration parameters.

Fuel energy ($kJ\ s^{-1}$)	7.68e4
Steam temperature in the inlet of regenerator (°C)	592
Regenerator outflow temperature (°C)	500
Make-up water flow rate ($kg\ s^{-1}$)	0.3
Make-up water temperature (°C)	30
Purge flow rate ($kg\ s^{-1}$)	0.1
Purge temperature (°C)	95
Liquid CO_2 flow rate ($kg\ s^{-1}$)	19.6
Liquid CO_2 temperature (°C)	15

Table 14.13 Cost basis.

	Scenarios		
Cost basis	**Optimistic**	**Average**	**Pessimistic**
Sorbent cost (\$/ton)	60	80	100
Spent sorbent disposal cost (\$/ton)	−15[a)]	10	25
Cost of electricity ($\$\ kW^{-1}$)	0.1	0.1	0.1
Cost of natural gas (\$/MMBTU)	5	5	5
Useful plant life (yr)	50	50	50

a) Saleable byproduct.

Table 14.14 Major cost components of the process.

Costs	Scenarios		
	Optimistic	Average	Pessimistic
Sorbent cost (\$million yr^{-1})	6.42	12.8	17.8
Labor cost (\$million yr^{-1})	4	4	4
Fuel cost (\$million yr^{-1})	10.4	10.4	10.4
Operating life of the system (yr)	50	50	50
Total bare-module cost of the equipment (\$million)	68.1	68.1	68.1
Total capitalized cost (\$million yr^{-1})	5.91	5.91	5.91
CO_2 capture cost (\$ ton^{-1} CO_2)	31	38	44
Increase in the cost of electricity (%)	9.65	11.97	13.77

carbon capture (liquid CO_2 at 50 atm) was determined under three scenarios presented in Table 14.14 with the calculated costs of the major cost components of the process.

Acknowledgment

Thanks to Emad Abbasi, Emad Ghadirian, Armin Hassanzadeh, and Shahin-Zarghami for their contributions to this chapter as part of their PhD research.

References

1 U.S. Department of Energy (2009). Financial Assistance Funding Opportunity Announcement for Support of Advanced Coal Research at U.S. Colleges and universities. FOA Number: DE-FOA-0000146.

2 Hassanzadeh, A. and Abbasian, J. (2009). A regenerative process for pre-combustion CO_2 capture and hydrogen production in IGCC. Proceedings of the 26th Annual International Pittsburgh Coal Conference.

3 Hassanzadeh, A. and Abbasian, J. (2010). Regenerable MgO-based sorbents for high-temperature CO_2 removal from syngas: sorbent development, evaluation, and reaction modeling. *Fuels* 89 (6): 1287–1297.

4 Chen, Y.M. (2011). Evolution of FCC: past, present, and future-and the challenges of operating a high-temperature CFB system. In: *Proceeding of the 10th International Conference on Circulating Fluidized Beds and Fluidization Technology - CFB-10* (ed. T.M. Knowlton). New York: ECI.

5 Sundaresan, S. (2011). Reflections on mathematical models and simulation of gas-particle flows. In: *Proceeding of the 10th International Conference on Circulating Fluidized Beds and Fluidization Technology - CFB-10* (ed. T.M. Knowlton). New York: ECI.

6 Shadle, L., Spenik, J., Monazam, E. et al. (2010). Approach to developing the C2U: a bench-scale carbon capture test unit using supported amine sorbents. NETL CO_2 Capture Technology Meeting, Pittsburgh, PA (13–17 September 2010).
7 Dobner, S., Sterns, L., Gralf, R.A., and Squires, A.M. (1977). Cyclic calcination and recarbonation of calcined dolomite. *Ind. Eng. Chem.* 16 (4): 479–486.
8 Hughes, R.W., Lu, D., Anthony, E.J., and Wu, Y. (2004). Improved long-term conversion of limestone-derived sorbents for in-situ capture of CO_2 in a fluidized bed combustor. *Ind. Eng. Chem. Res.* 43: 5529–5539.
9 Laursen, K., Duo, W., Grace, J.R., and Lim, C.J. (2004). Cyclic steam reactivation of spent limestone. *Ind. Eng. Chem. Res.* 43: 5715–5720.
10 Wolff, E.H.P., Gerritsen, A.W., and Verheijen, P.J.T. (1993). Attrition of an alumina-based sorbent for regenerative sulfur capture from flue gas in a fixed bed. *Powder Technol.* 76: 47.
11 Liu, C. and Shih, S. (2009). Kinetics of the reaction of iron blast furnace slag/hydrated lime sorbents with SO_2 at low temperatures: effects of the presence of CO_2, O_2, and NO. *Ind. Eng. Chem. Res.* 48: 8335–8340.
12 Abbasi, E., Hassanzadeh, A., and Abbasian, J. (2013). Regenerable MgO-based sorbent for high-temperature CO_2 removal from syngas: 2. Two-zone variable diffusivity shrinking core model with expanding product layer. *Fuel* 105: 128–134.
13 Yagi, S. and Kunii, D. (1955). *5th International Symposium on Combustion*, vol. 231. New York: Reinhold.
14 Arastoopour, H. and Gidaspow, D. (1979). Vertical pneumatic conveying using four hydrodynamic models. *Ind. Eng. Chem. Fundam.* 18 (2): 123–130.
15 Arastoopour, H., Lin, S.C., and Weil, S.A. (1982). Analysis of vertical pneumatic conveying of solids using multiphase flow models. *AIChE J.* 28 (3): 467–473.
16 Abbasi, E., Abbasian, J., and Arastoopour, H. (2015). CFD–PBE numerical simulation of CO_2 capture using MgO-based sorbent. *Powder Technol.* 286: 616–628.
17 Arastoopour, H. (2001). Numerical simulation and experimental analysis of gas/solid flow systems: 1999 Fluor-Daniel plenary lecture. *Powder Technol.* 119 (2): 59–67.
18 Gidaspow, D. (1994). *Multiphase Flow and Fluidization: Continuum and Kinetic Theory Descriptions.* San Diego, CA: Academic Press.
19 Benyahia, S. (2012). Analysis of model parameters affecting the pressure profile in a circulating fluidized bed. *AIChE J.* 58 (2): 427–439.
20 Milioli, C.C., Milioli, F.E., Holloway, W. et al. (2013). Filtered two-fluid models of fluidized gas-particle flows: new constitutive relations. *AIChE J.* 59 (9): 3265–3275.
21 Abbasi, E. and Arastoopour, H. (2011). CFD simulation of CO_2 sorption in a circulating fluidized bed using deactivation kinetic model. In: *Proceeding of the 10th International Conference on Circulating Fluidized Beds and Fluidization Technology - CFB-10* (ed. T.M. Knowlton), 736–743. New York: ECI.

22 Ghadirian, E. and Arastoopour, H. (2016). CFD simulation of a fluidized bed using the EMMS approach for the gas–solid drag force. *Powder Technol.* 288: 35–44.

23 Sinclair, J.L. and Mallo, T. (1998). Describing particle-turbulence interaction in a two-fluid modelling framework. Proceedings of ASME Fluids Engineering Division Summer Meeting (FEDSM'98), Washington, DC (21–25 June 1998), pp. 7–14.

24 Laux, H. (1998). *Modeling of Dilute and Dense Dispersed Fluid-Particle Flow.* Trondheim: Norwegian University of Science and Technology.

25 Lun, C.K.K., Savage, S.B., Jeffrey, D.J., and Chepurniy, N. (1984). Kinetic theories for granular flow: inelastic particles in Couette flow and slightly inelastic particles in a general flowfield. *J. Fluid Mech.* 140: 223–256.

26 Johnson, P.C. and Jackson, R. (1987). Frictional-collisional constitutive relations for granular materials, with application to plane shearing. *J. Fluid Mech.* 176: 67–93.

27 Arastoopour, H., Gidaspow, D., and Abbasi, E. (2017). *Computational Transport Phenomena of Fluid-Particle Systems,* Mechanical Engineering Series. Springer.

28 FFGSM. http://mypages.iit.edu/~abbasian/ffgsm.html (accessed 21 March 2018).

Index

Handbook of Chemical Looping Technology, First Edition. Edited by Ronald W. Breault.

d

n

o

p

q

r

t